Lecture Notes in Computer Science 14828

The series Lecture Notes in Computer Science (LNCS), including its subseries Lecture Notes in Artificial Intelligence (LNAI) and Lecture Notes in Bioinformatics (LNBI), has established itself as a medium for the publication of new developments in computer science and information technology research, teaching, and education.

LNCS enjoys close cooperation with the computer science R & D community, the series counts many renowned academics among its volume editors and paper authors, and collaborates with prestigious societies. Its mission is to serve this international community by providing an invaluable service, mainly focused on the publication of conference and workshop proceedings and postproceedings. LNCS commenced publication in 1973.

Federico Maggi · Manuel Egele · Mathias Payer ·
Michele Carminati
Editors

Detection of Intrusions and Malware, and Vulnerability Assessment

21st International Conference, DIMVA 2024
Lausanne, Switzerland, July 17–19, 2024
Proceedings

Editors
Federico Maggi
AWS
San Diego, CA, USA

Manuel Egele
Boston University
Boston, MA, USA

Mathias Payer
EPFL
Lausanne, Switzerland

Michele Carminati
Politecnico di Milano
Milan, Italy

ISSN 0302-9743 ISSN 1611-3349 (electronic)
Lecture Notes in Computer Science
ISBN 978-3-031-64170-1 ISBN 978-3-031-64171-8 (eBook)
https://doi.org/10.1007/978-3-031-64171-8

This Springer imprint is published by the registered company Springer Nature Switzerland AG
The registered company address is: Gewerbestrasse 11, 6330 Cham, Switzerland

Preface

On behalf of the Program Committee, we are delighted to present the proceedings of the 21st Conference on Detection of Intrusions and Malware & Vulnerability Assessment, DIMVA 2024. Over the past two decades, DIMVA has become a well-established security conference, consistently attracting high-quality research and fostering collaboration between academia, industry, and government. DIMVA is organized by the Special Interest Group – Security, Intrusion Detection, and Response (SIDAR) - of the German Informatics Society (GI).

This year, we received 110 valid submissions and accepted 28 papers (22 full papers and 6 short papers), resulting in a competitive acceptance rate of 25.45%. All papers received three double-blind reviews, and each PC member was assigned up to 8 papers to review. For the second year, DIMVA introduced multiple submission deadlines, with a balanced acceptance rate across both cycles: 7 out of 41 papers were accepted in the first cycle, and 21 out of 76 papers were accepted in the second cycle. The number of submissions increased with respect to the previous year, and in response, we implemented a revision model which closely follows the concept of the previous "accept with shepherding" decision but allows for more significant changes in the revised papers.

We would like to express our deep gratitude to the Program Committee members for their countless hours spent reviewing submissions for both cycles, engaging in online and in-person discussions, and shepherding numerous papers. Online alone, PC members and external reviewers exchanged more than 500 comments. We also would like to thank the Organizing Committee for their dedication and hard work, laying the foundation for DIMVA's successful 21st edition. We appreciate the support from our sponsors and host institutions.

Furthermore, this year's DIMVA conference introduced a special focus on the application of Artificial Intelligence (AI) and Machine Learning (ML) to security. Recognizing the growing importance of AI and ML in addressing contemporary security challenges, DIMVA aimed to explore the latest advancements, methodologies, and applications within this field. Through a dedicated session, presentations, and discussions, participants had the opportunity to delve into the intersection of AI/ML and security, fostering a deeper understanding of their potential implications for mitigating cyber threats.

Finally, we would like to thank the authors who submitted their work and presented their contributions, as well as the attendees who participated in the conference. Your enthusiasm and commitment have made DIMVA the thriving and enriching event it is today. We look forward to your future contributions.

DIMVA was held on campus at EPFL in Lausanne, Switzerland. In addition to authors presenting their papers, we had two exciting keynote sessions, and introduced a poster session where attendees discussed their up and coming work. The first keynote by Andrea Continella from University of Twente focused on firmware security in the era of smart devices. The second keynote by Veelasha Moonsamy from Ruhr University

Bochum focused on wireless physical-layer sensing. Together, this program allowed the attendees to discuss, engage, and socialize in a relaxed environment close to Lac Léman.

May 2024

Federico Maggi
Manuel Egele
Mathias Payer

Organization

General Chair

Mathias Payer	EPFL, Switzerland

Program Committee Chairs

Federico Maggi	Amazon Web Services, USA
Manuel Egele	Boston University, USA

Publication Chair

Michele Carminati	Politecnico di Milano, Italy

Poster Chair

Flavio Toffalini	EPFL, Switzerland

Conference Coordinator

Natascha Fontana	EPFL, Switzerland

Steering Committee Chairs

Ulrich Flegel	Infineon Technologies, Germany
Michael Meier	University of Bonn and Fraunhofer FKIE, Germany

Steering Committee

Magnus Almgren	Chalmers University of Technology, Sweden
Sébastien Bardin	CEA, France

Leyla Bilge	Gen Digital, France
Gregory Blanc	Télécom SudParis, France
Herbert Bos	Vrije Universiteit Amsterdam, Netherlands
Danilo M. Bruschi	Università degli Studi di Milano, Italy
Roland Büschkes	RWE AG, Germany
Juan Caballero	IMDEA Software Institute, Spain
Lorenzo Cavallaro	King's College London, UK
Hervé Debar	Télécom SudParis, France
Sven Dietrich	City University of New York, USA
Mathias Fischer	Universität Hamburg, Germany
Giorgio Giacinto	University of Cagliari, Italy
Cristiano Giuffrida	Vrije Universiteit Amsterdam, Netherlands
Daniel Gruss	TU Graz, Austria
Bernhard Haemmerli	Acris GmbH and HSLU Lucerne, Switzerland
Thorsten Holz	CISPA Helmholtz Center for Information Security, Germany
Marko Jahnke	CSIRT, German Federal Authority, Germany
Klaus Julisch	Deloitte, Switzerland
Christian Kreibich	ICSI, USA
Christopher Kruegel	UC Santa Barbara, USA
Pavel Laskov	University of Liechtenstein, Liechtenstein
Federico Maggi	Amazon Web Services, USA
Clémentine Maurice	CNRS, IRISA, France
Nuno Neves	University of Lisboa, Portugal
Roberto Perdisci	University of Georgia and Georgia Institute of Technology, USA
Michalis Polychronakis	Stony Brook University, USA
Konrad Rieck	TU Braunschweig, Germany
Jean-Pierre Seifert	Technical University of Berlin, Germany
Robin Sommer	Corelight, USA
Urko Zurutuza	Mondragon University, Spain

Program Committee

Daniele Antonioli	EURECOM, France
Sébastien Bardin	CEA, LIST, Université Paris-Saclay, France
Kevin Borgolte	Ruhr University Bochum, Germany
Marcel Busch	EPFL, Switzerland
Yinzhi Cao	Johns Hopkins University, USA
Michele Carminati	Politecnico di Milano, Italy
Daniele Cono D'Elia	Sapienza University of Rome, Italy

Roland Yap	National University of Singapore, Singapore
Stefano Zanero	Politecnico di Milano, Italy

Additional Reviewer

Alexander Bulekov	AWS, USA

Contents

Vulnerability Detection and Defense

Exceptional Interprocedural Control Flow Graphs for x86-64 Binaries

Joshua Bockenek[1(✉)], Freek Verbeek[1,2], and Binoy Ravindran[1]

[1] Virginia Tech, Blacksburg, USA
jabocken@vt.edu

[2] Open University of The Netherlands, Heerlen, The Netherlands

Abstract. Standard control flow graphs (CFGs) extracted from binaries by state-of-the-art disassembly/decompilation tools do not include information about exception-related control flow. However, such information is useful when statically analyzing programs that utilize structured exceptions. To fill that gap, we propose the concept of *Exceptional Interprocedural Control Flow Graphs (EICFGs).* These graphs extend traditional CFGs by adding edges for stack unwinding, frame cleanup, and try/catch behavior caused by thrown exceptions. We provide an approach for generating EICFGs from x86-64 binaries featuring C++ exceptions. The approach is based on symbolically executing an abstract semantics that includes binary-level exception-related function calls. We validated our abstract semantics by generating concrete test cases that were then evaluated using real binaries. We applied an implementation of our approach to 341 off-the-shelf x86-64 binaries compiled from C++ as well as C and Fortran source code. From those binaries, we identified 2574 unique throws and successfully resolved the exceptional control flow for every one of them. We show that resolving throws leads to increased instruction reachability and uncovers edges not found by state-of-the-art tools such as Ghidra.

Keywords: C++ exceptions · Binary Analysis · Control Flow Graphs

1 Introduction

Control flow graphs (CFGs) are key components in both compilation and analysis of software. They provide information on the runtime execution of a program, i.e., which control flow paths are dynamically possible. Typically, they are constructed during compilation using the source code as ground truth. In the field of binary analysis, however, they must be *reconstructed* from a binary. This is a challenge, since control flow transfers may be *indirect*, i.e., their jump target may be resolved at runtime. This makes generating CFG an undecidable problem, as the exact values of certain state parts can only be determined at runtime.

This problem is exacerbated when dealing with *exceptional control flow*, induced mainly by the C++ `throw` and `catch` statements. The target of a throw, i.e., to which instruction address the control flow is transferred after execution

F. Maggi et al. (Eds.): DIMVA 2024, LNCS 14828, pp. 3–22, 2024.
https://doi.org/10.1007/978-3-031-64171-8_1

of a throw statement, is decided dynamically at runtime. It is based on, among other things, the current call/return stack, the current caught exception stack, and low-level information pertaining to rethrows and catch statements. Moreover, it inherently requires *interprocedural* analysis, as the function call history is relevant for assessing the throws' targets. Interprocedural analysis is challenging, as one cannot simply isolate a function and do analysis, but must consider the binary as a whole.

Existing state-of-the-art disassemblers/decompilers, such as IDA Pro, Ghidra, and Binary Ninja, do not provide sufficient information on exceptional control flow. Typically, they are able to extract exception information statically available in binaries (e.g., landing pad locations). They can even provide interprocedural control flow graphs. However, they do not perform the interprocedural static analysis required for reconstructing *exceptional* control flow. That is, they cannot trace the path from an exception throw site to the landing pad instructions that exception goes to in the process of unwinding. CFGs under such analyses will have *throw sites as terminating locations* with no outgoing edges.

This paper provides a tool for static, interprocedural, automated C++-exception-aware control flow analysis on the binary level. That tool produces Exceptional Interprocedural Control Flow Graphs (EICFGs), which document the direct exception-handling-related components of the state of the program, such as exception objects currently allocated, number of uncaught exceptions, and which exceptions are currently in a caught state (see Sects. 2 and 3).

Effectively, EICFGs expose control flow edges not found by other work. The motivation behind exposing such edges is that it leads to more accurate reachability analysis: instructions that were deemed unreachable may now be considered reachable, or instructions that were considered reachable only through a single path may now be considered reachable through other paths as well. In general, we argue that improved binary-level reachability analysis is useful for security analysis, verification, and patching. Some more concrete examples of use-cases are:

- A disassembler/decompiler leverages reachability information to decide which parts of the binary are lifted. The information stored in EICFGs enable a decompiler to lift paths that were not found by existing tools.
- Binary-level verification, be it theorem proving, symbolic execution, or model-checking, typically aims to be overapproximative. This requires insight into *all* possible execution paths. An EICFG can aid such verification efforts.
- ROP-programming aims at finding executable sequences (gadgets) that end with an instruction that modifies control-flow. More insight into the possible execution paths of a binary increases may lead to a larger attack surface.

The tool itself targets stripped binaries using C++ exception handling that were compiled for the x86-64 instruction set architecture (ISA) and System V application binary interface (ABI). We assume that external calls do not produce exceptions themselves. We also do not model `setjmp`/`longjmp`.

A desired characteristic is that the produced EICFGs are *overapproximative*: every concrete path, i.e., every path possible during dynamic execution, is included in the EICFG. We informally – through test case generation – argue that the abstract semantics that are symbolically executed overapproximate the concrete semantics [2]. However, it may be the case that an indirection is unresolved, in which case not all paths are explored. In those cases, the CFGs are clearly annotated. We thus argue that EICFGs are overapproximative *modulo unresolved indirections*. To strengthen this claim, we validated our abstract transition rules for exception handling against the concrete implementations of the corresponding library functions (Sect. 4). This was done by generating abstract states, concretization to a concrete CPU state, running the function under validation in an instrumented binary, observing the CPU state, and verifying that the concrete transition is correctly contained in the abstract transition.

We applied the tool to 341 off-the-shelf x86-64 binaries compiled from C++, C, and Fortran source code (Sect. 5). The implemented tool was able to identify and trace 2574 unique throws. Dealing with exceptional control flow can increase average instruction reachability by 14 instructions per unique throw, with 188 average unwind edges in the possible unwinding paths from each throw. Those edges are ones tools such as Ghidra do not produce.

All code, scripts, and tested binaries are publicly available at https://doi.org/10.5281/zenodo.11081942.

2 EICFGs

A normal CFG provides users with information on control flow transfers induced by *jumps* and *calls*. Given a specific jump or call instruction, one can look up in the CFG the set of successor instructions and a state predicate on which that successor selection is based. Such state predicates are typically represented by expressions over flags (for example, `CF` and `ZF` for the carry and zero flag) or a jump table calculation.

A summarized reproduction of the interprocedural control flow graph generated by Ghidra for an example program can be seen in Fig. 2, generated from a binary compiled from the source code in Fig. 1. In this example, function `main` can indirectly call both functions `foo` and `bar`. Each of these two can throw an exception, and if that happens, `main` can either rethrow that exception (the red box) or normally return (the green box).

While Ghidra can identify catch and cleanup landing pads (boxes LP in Fig. 2), it cannot show that the throws will unwind there. IDA Pro and Binary Ninja produce similar results; they can identify landing pad locations intraprocedurally, but they do not trace exceptional control flow interprocedurally. We have verified that this holds for more complicated programs as well (some of the programs used in Sect. 5).

In contrast, we produce an EICFG which augments the normal CFG with edges related to exceptional control flow (the dashed edges in Fig. 2). These edges actually form a series of edges that lead from a `throw` to its landing pad, with labels containing predicates over, e.g., the current caught exception stack. Moreover, while not claiming it as a contribution in itself, we have taken care to resolve indirections whenever possible. The edge from `main` to `foo` and `bar` are due to indirect calls. Neither Ghidra, IDA-PRO or Binary Ninja generated these edges for this example.

```
int (*const FOOBAR[])(int) = {foo, bar};

int main(int argc, char* argv[]) {
  try {
    if (argc < 2)
      return FOOBAR[argc](argc);
  }
  catch (const std::exception&) {
    if (argc < 0)
      throw;
    else
      return 0;
  }
}
```

Fig. 1. Part of the C++ source code that was compiled to binary

Instead of simple labels, EICFGs use *exceptional state predicates* as labels. Informally, the information on which exceptional control flow is based is:

1. the exception object, e.g., type info and rethrown state;
2. the current set of return addresses on the stack;
3. the current uncaught exception count;
4. the current caught exception stack; and
5. a static address (landing pad) table for the unwinding process.

Exception type info is used to determine which catch blocks, if any, are applicable to the exception being thrown. Rethrown status is necessary when determining behavior when dealing with the binary equivalent of an argumentless `throw`. The return address stack is necessary to provide context for unwinding. The uncaught exception count keeps track of how many thrown exceptions are currently uncaught. This is useful for diagnostic information. Next, the caught exception stack provides information to set up implicit rethrowing. Finally, the landing pad table (LPT) maps from potential unwind spots in a binary to locations where unwinding can exit.

Definition 1. *An EICFG is a directed graph with instruction addresses as vertices and edges labelled with exceptional state predicates. There is an edge between instruction addresses a_0 and a_1 with label P if the instruction at address a_0 leads to instruction address a_1 for any state that satisfies predicate P.*

Some of the additional information provided by EICFG is illustrated in Fig. 3, which models the process of throwing an exception from the same example program as Fig. 2. The representation in the figure indicates the process of unwinding from one throw site in the control flow graph to a try-catch block or cleanup

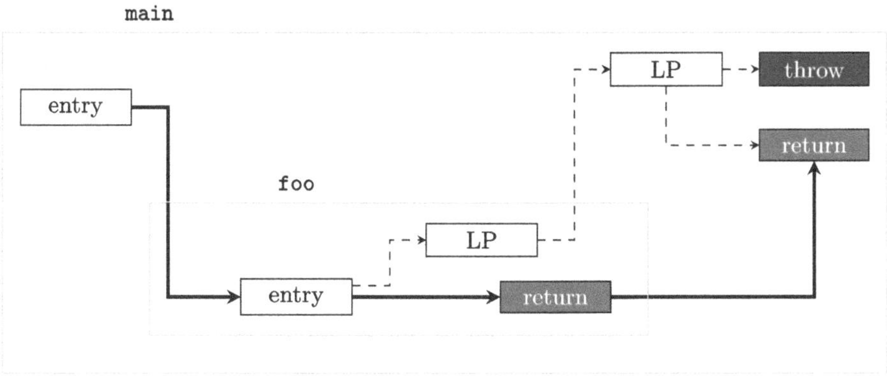

Fig. 2. Summarized view of a subpart of the control flow of Fig. 1. Each box is itself a control flow graph over various instructions. The thick black edges are also generated by existing works, the dashed edges are not. (Color figure online)

landing pad. This path was triggered by the snippet of assembly shown in Fig. 4, which allocates (`0x125b`), initializes (`0x126d`), and throws (`0x1286`) an exception. The landing pad table of the binary, LPT, directs the unwinding process: when stack unwinding reaches address i and $j \in \mathsf{LPT}(i)$, control flow branches to address j. Otherwise, another frame is popped off the stack. This will be elaborated on in Sect. 3.2. For this example, we get unwinding from address (`0x1286`) to `0x137e` to `0x138b`.

Due to overapproximation, there are two possible paths from that point. One is the path for an exception object that is not of the caught type, some checks of which occur via the assembly instructions `0x138f` and `0x1393` of Listing 1.1. This path results in unwinding being resumed (`0x1398`) and leads to a bad termination case at instruction `0x116e`. The other path continues from node `0x139d`. It eventually leads to `0x116e` from `0x13dd` (intervening nodes elided). This is a good termination case as reaching that halting instruction occurred outside unwinding.

Resolving of indirections is shown in Fig. 5. Whenever an indirection occurs, we detect whether the state is reached only via a conditional jump that positively bounds an index. If so, we concretize by all possible values for that index. In the given example, we detect values being read from a table with an index j strictly bounded by 2. This results in a computable indirection, and resolve an indirect call to either `foo` or `bar`.

3 Technical Formulation

The derivation of EICFGs from a binary requires four additional things: 1) a static landing pad table, 2) an abstract state model, 3) an abstract transition relation, and 4) a symbolic execution engine to apply the rules making up our abstract transition relation. We describe those here,using the following types:

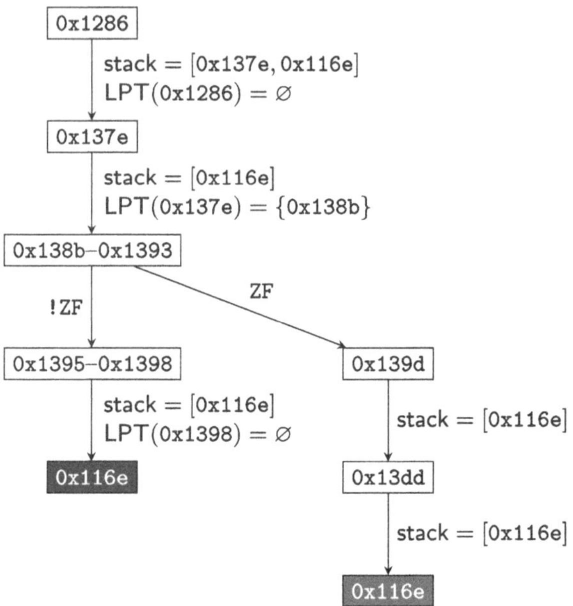

Fig. 3. EICFG fragment.

$\mathbb{B}$, $\mathbb{N}$, $\mathbb{Z}$, denote Boolean, natural and integral numbers respectively. $\mathbb{V}$ denotes *symbolic expressions*, which may be $\top$, indicating any or an unknown or undefined value. $\mathbb{P}$ denotes *immediate 64-bit addresses.* $\mathbb{R}$ denotes *registers.* $\mathbb{W}$ denotes *exception IDs*, and finally we have an enumeration $\mathbb{T} := \{\top, \texttt{Good}, \texttt{Bad}\}$ for the program termination type.

3.1 Landing Pad Table

This information describes how unwinding should proceed given the unwinding reaching specific locations in a binary. It is extracted from the Common Information Entries (CIEs), Frame Description Entries (FDEs), and language-specific data areas (LSDAs) of the binary under test and assumed to be correct. In our current formulation, an entry in the catch table is merely a pointer to the corresponding landing pad for this entry. While type pointers and exception specifications exist within the LSDAs as well, we do not currently utilize that information.

Thrown exceptions in C++ can be caught by catch blocks. Individual stack frames may require cleanup during the process of unwinding as well. The addresses of those catch blocks and cleanup routines are called landing pads. To accomplish reaching those addresses during unwinding, we require a *landing pad table*.

```
125b: call 10d0 <__cxa_allocate_exception>
1260: mov  rbx,rax
1263: lea  rsi,[rip+0xd9b] # 2005
126a: mov  rdi,rbx
126d: call 10e0 #std::domain_error init
1272: mov  rax,QWORD PTR [rip+0x2d4f]
1279: mov  rdx,rax
127c: lea  rsi,[rip+0x2abd] # 3d40
1283: mov  rdi,rbx
1286: call 1120 <__cxa_throw>
...
129c: call 10f0 <__cxa_free_exception>
```

Fig. 4. Assembly corresponding to Fig. 3.

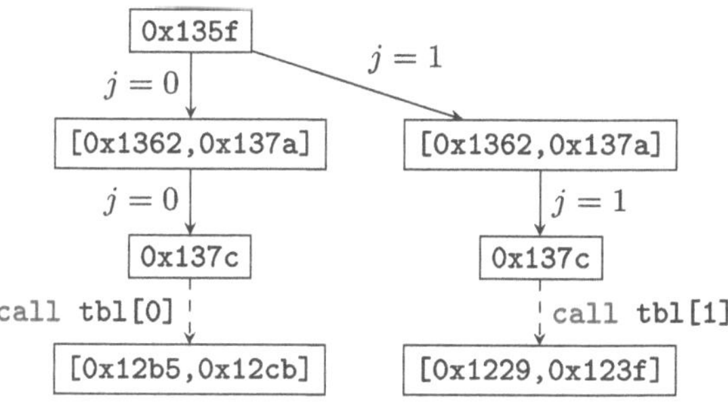

Fig. 5. Identifying indirection.

Definition 2. *A landing pad table* LPT *is a static map from instruction addresses to set of possible landing pads. Formally,* LPT*s have type* $\mathbb{P} \rightarrow 2^{\mathbb{P}}$.

Currently, we overapproximate and do not include exception type when dete mining landing pads.

Example 1. One of our running example landing pad entries is LPT(0x137 {0x138b}. Thus, when an unwinding routine reaches instruction 0x137e, routine will jump to 0x138b.

Exception Objects. Exceptions of type $\mathbb{E}$ are records with named fi

Definition 3.

$$\mathbb{E} := \begin{cases} \mathsf{rethrown} & : \mathbb{B} \\ \mathsf{handlerCount} & : \mathbb{Z} \end{cases}$$

This record has two fields. The Boolean field rethrown *indicate status of the exception. Natural field* handlerCount *stores the catch block handlers for the exception.*

Listing 1.1. Example throw landing pad.

```
1387:    mov      eax, ebx
1389:    jmp      13d7
138f:    cmp      rdx, 0x1 # check for matching exception
1393:    je       139d
1395:    mov      rdi, rax
1398:    call     _Unwind_Resume
         # exception not handled by this catch block
139d:    mov      rdi, rax
13a0:    call     __cxa_begin_catch # start of catch block
13a5:    mov      QWORD PTR [rbp-0x18], rax
13a9:    cmp      DWORD PTR [rbp-0x24], 0x0
13ad:    jns      13b4
13af:    call     __cxa_rethrow # exception rethrow
13b4:    mov      ebx, 0x0
13b9:    call     __cxa_end_catch # end of catch block
13be:    jmp      1387
13c0:    endbr64
...
13d7:    # normally return
```

3.2 Abstract State

Our exception-containing abstract states, type Σ, are records as well, with the llowing fields:

the register map. This field has type $\mathbb{R} \rightarrow \mathbb{V}$. As previously stated, and writing registers smaller than 64 bits (e.g. `ebp` versus `rbp`) requires ing and shifting the underlying 64-bit register's value. This behavior ed into our symbolic execution engine. Larger registers (e.g. `xmm` and r registers) exist as operands in our instruction representation but l for state updates or reads.

eturn address stack. Maintaining the current list of return cessary in order to perform stack unwinding and handle thrown s also helpful in detecting recursion. We make use of standard tiofunctions. For the initial state, we have an empty stack. A `call` ss of the next instruction to the stack, a `ret` pops. Some func- bc_start_main are given special treatment.

on mapping. This field has type $\mathbb{W} \rightarrow \mathbb{E}$. When an excep- eives an ID based on its creation location and is stored in the key.

ination state. This field has type $\mathbb{T}$. It defaults to program or function has not terminated yet. When a tes, it is set to either `Good` or `Bad`, indicating either normal or abnormal termination, respectively. We treat cases where an exception propagates to the top of the stack without being caught to be such "bad" cases.

Auxiliary Exception Variables. Σ also contains a count of the number of currently-uncaught exceptions (uncaught: $\mathbb{N}$) and a stack of the currently-caught exception IDs (caught: $[\mathbb{W}]$). These fields are manipulated and used during entry to and exit from catch blocks as well as when rethrowing exceptions. This handling comes into play when dealing with nested catch blocks, exceptions (re)thrown within such blocks, etc.

Definition 4. *The type of abstract states, notation Σ, is a record with fields:*

$$\begin{array}{llllll} \mathsf{rmap} & : \mathbb{R} \rightarrow \mathbb{V} & \mathsf{stack} & : [\mathbb{P}] & \mathsf{emap} & : \mathbb{W} \rightarrow \mathbb{E} \\ \mathsf{terminated} & : \mathbb{T} & \mathsf{uncaught} & : \mathbb{N} & \mathsf{caught} & : [\mathbb{W}] \end{array}$$

To ease register references, for some state σ and named register r, the notation $\sigma.\mathtt{r}$ is shorthand for $\sigma.\mathsf{rmap}(\mathtt{r})$, e.g. $\sigma.\mathtt{rdi} \equiv \sigma.\mathsf{rmap}(\mathtt{rdi})$.

3.3 Abstract Transition Rules

We define an abstract transition relation over abstract states. For various regular instructions we define semantics (e.g., `mov` and `add`). There is no need to provide semantics for the entire instruction set, as we can overapproximate values with $\top$ for irrelevant instructions. Conditional jumps are modeled non-deterministically. Unknown external functions are overapproximated by trashing the state according to a calling convention. Recursion is treated as follows: assume a call to a function inside the binary for some state σ. Then, if the return address to be pushed on the stack is already in σ.stack, we instead treat that call as an unmodeled external call and continue execution past it.

Figure 6 provides a set of rules modeling exception-related ABI calls. The following abbreviations are utilized in those rules:

$$\begin{aligned} \text{handler}(id, \sigma) &\equiv \sigma.\mathsf{emap}(id).\mathsf{handlerCount} \\ \text{rethrown}(id, \sigma) &\equiv \sigma.\mathsf{emap}(id).\mathsf{rethrown} \\ \text{pushStack}(fr, \sigma, \sigma') &\equiv \sigma'.\mathsf{stack} = \text{push}(fr, \sigma.\mathsf{stack}) \\ \text{popStack}(\sigma, \sigma') &\equiv \sigma'.\mathsf{stack} = \text{pop}(\sigma.\mathsf{stack}) \\ \text{pushCaught}(c, \sigma, \sigma') &\equiv \sigma'.\mathsf{caught} = \text{push}(c, \sigma.\mathsf{caught}) \\ \text{popCaught}(\sigma, \sigma') &\equiv \sigma'.\mathsf{caught} = \text{pop}(\sigma.\mathsf{caught}) \end{aligned}$$

Notation for increment/etc.:

$$\begin{aligned} \text{handler}(id, \sigma')\text{++} &\equiv \text{handler}(id, \sigma') = \text{handler}(id, \sigma) + 1 \\ \text{handler}(id, \sigma')\oplus &= \text{handler}(id, \sigma') = |\text{handler}(id, \sigma)| + 1 \end{aligned}$$

Notation X−− follows the same concept, but the alternate for decrementing is $\text{handler}(id, \sigma')\ominus \equiv \text{handler}(id, \sigma') = -\,\text{sign}(\text{handler}(id, \sigma)) * (|\text{handler}(id, \sigma)| - 1)$. As a special case, $0\ominus = 0$.

Non-unwinding Rules. Figure 6a shows the rule for the special starting function that initiates the runtime used for standard features of C and C++, `__libc_start_main`. For this rule, we require the post-state's current instruction pointer be restricted to whatever was previously stored in `rdi`, `rcx`, or `r8`. We also require the stored return address to be on the top of a stack frame.

Next, Fig. 6b illustrates the rule for function `__cxa_allocate_exception`. Our modeling assumes a system where virtual memory allocations always succeed (and the runtime terminates programs when they use up too much memory). It results in an exception object added to the exception map with the post-state σ''s instruction pointer as its ID. The object starts in a non-rethrown state and with no handlers. The ID is also set as the return value of the function in σ'`.rax`.

Example 2. After instruction `0x125b` of Fig. 4, we have: σ'`.rax` $=$ `0x1260`, σ'.emap(0x1260).handlerCount $= 0$, and $\neg\sigma'$.emap(0x1260).rethrown.

The rule in Fig. 6c is for the memory-related function `__cxa_free_exception`. This rule ensures the absence of an exception in the exception map based on the given ID. At our level of abstraction, the function `_Unwind_DeleteException` exhibits the same semantics and is thus elided.

Example 3. Consider instruction `0x129c` of Fig. 4. The result of this instruction is σ'.emap $= \varnothing$.

The rules in Figs. 6d and 6e define `__cxa_begin_catch` behavior for different cases. For an exception not already caught, the associated rule pushes it onto the caught-exception stack. The rule for already-caught exceptions does not do this. However, both rules increment that exception's handler count and decrement the state's count of uncaught exceptions. Though not listed, there is also a rule for an ID not currently in the exception map. That rule operates the same as our (elided) rule for unmodeled external calls. This allows for safe overapproximation.

Example 4. Consider instruction `0x13a0` of Listing 1.1. Assuming the existence of a valid exception object with ID id that was just thrown, the post-state σ' will satisfy: handler$(id, \sigma') = 1$, σ'.caught $= [id]$, and σ'.uncaught $= 0$.

The rules listed in Figs. 6f and 6g then define some `__cxa_end_catch` behavior. The first rule applies when an exception ID is available on top of the caught stack, there are no more handlers for the corresponding exception object, and it is being rethrown. In this case, it is popped off the caught stack and no longer treated as being rethrown. The second rule applies when an exception is available, has no more handlers, and is not being rethrown. In that case, it is popped off the caught stack and removed from the exception map. Not shown is the rule for an exception that still has handlers remaining. In that case, its handler count is decremented but no other changes are made. Additionally, the case for an empty σ.caught again operates as an unmodeled external call for the sake of overapproximation.

$$\frac{\sigma'.\texttt{rip} \in \sigma.\{\texttt{rdi},\texttt{rcx},\texttt{r8}\} \quad \text{pushStack}(\sigma.\texttt{rip}+5,\sigma,\sigma')}{\sigma \xrightarrow{\mathsf{A}} \sigma'}$$

(a) `__libc_start_main`

$$\frac{id = \sigma'.\texttt{rip} \quad \sigma'.\texttt{rax} = id \quad \sigma'.\mathsf{emap}(id) = e \quad \neg e.\mathsf{rethrown}}{\sigma \xrightarrow{\mathsf{A}} \sigma'}$$

(b) `__cxa_allocate_exception`

$$\frac{\sigma'.\mathsf{emap}(\sigma.\texttt{rdi}) = \top}{\sigma \xrightarrow{\mathsf{A}} \sigma'}$$

(c) `__cxa_free_exception`

$$\frac{id = \sigma.\texttt{rdi} \quad id \notin \sigma.\mathsf{caught} \quad \text{handler}(id,\sigma')\oplus \quad \text{pushCaught}(id,\sigma,\sigma') \quad \sigma'.\mathsf{uncaught}{-}{-}}{\sigma \xrightarrow{\mathsf{A}} \sigma'}$$

(d) `__cxa_begin_catch` (not already caught)

$$\frac{id = \sigma.\texttt{rdi} \quad id \in \sigma.\mathsf{caught} \quad \text{handler}(id,\sigma')\oplus \quad \sigma'.\mathsf{uncaught}{-}{-}}{\sigma \xrightarrow{\mathsf{A}} \sigma'}$$

(e) `__cxa_begin_catch` (already caught)

$$\frac{id = \text{peek}(\sigma.\mathsf{caught}) \quad \text{handler}(id,\sigma) = 1 \quad \text{rethrown}(id,\sigma) \quad \text{handler}(id,\sigma') = 1 \quad \neg\,\text{rethrown}(id,\sigma') \quad \text{popCaught}(\sigma,\sigma')}{\sigma \xrightarrow{\mathsf{A}} \sigma'}$$

(f) `__cxa_end_catch` (have caughts, last handler, rethrown)

$$\frac{id = \text{peek}(\sigma.\mathsf{caught}) \quad \text{handler}(id,\sigma) = 1 \quad \neg\,\text{rethrown}(id,\sigma) \quad \text{handler}(id,\sigma') = 1 \quad \sigma'.\mathsf{emap}(id) = \top \quad \text{popCaught}(\sigma,\sigma')}{\sigma \xrightarrow{\mathsf{A}} \sigma'}$$

(g) `__cxa_end_catch` (have caughts, last handler, not rethrown)

$$\frac{\sigma.\mathsf{caught} = [] \quad \sigma'.\mathsf{terminated} = \mathsf{Bad}}{\sigma \xrightarrow{\mathsf{A}} \sigma'}$$

(h) `__cxa_rethrow` alt

$$\frac{\sigma \xRightarrow{\mathsf{U}^+} \sigma' \quad \sigma'.\texttt{rax} = \sigma.\texttt{rdi}}{\sigma \xrightarrow{\mathsf{A}} \sigma'}$$

(i) `_Unwind_Resume` (LP(s))

$$\frac{\sigma \xRightarrow{\mathsf{U}^-} \sigma' \quad \sigma'.\mathsf{terminated} = \mathsf{Bad}}{\sigma \xrightarrow{\mathsf{A}} \sigma'}$$

(j) `_Unwind_Resume` (no LPs)

$$\frac{\sigma \xRightarrow{\mathsf{U}^+} \sigma' \quad \sigma'.\mathsf{uncaught}{+}{+} \quad \sigma'.\texttt{rax} = \sigma.\texttt{rdi}}{\sigma \xrightarrow{\mathsf{A}} \sigma'}$$

(k) `__cxa_throw` (LP(s))

$$\frac{\sigma \xRightarrow{\mathsf{U}^-} \sigma' \quad \sigma'.\mathsf{terminated} = \mathsf{Bad}}{\sigma \xrightarrow{\mathsf{A}} \sigma'}$$

(l) `__cxa_throw` (no LPs)

$$\frac{\sigma \xRightarrow{\mathsf{U}^+} \sigma' \quad id = \text{peek}(\sigma.\mathsf{caught}) \quad \sigma'.\mathsf{uncaught}{+}{+} \quad \text{handler}(id,\sigma')\ominus \quad \text{rethrown}(id,\sigma') \quad \sigma'.\texttt{rax} = id}{\sigma \xrightarrow{\mathsf{A}} \sigma'}$$

(m) `__cxa_rethrow` (caught+LP(s))

Fig. 6. Abstract transition rules (unchanged state parts mostly elided).

Example 5. Consider instruction `0x13b9` of Listing 1.1. Assume the statements in Example 4 hold for the pre-state. Then, the post-state σ' for that instruction will satisfy σ'.emap(id) = $\top$ and σ'.caught = [].

Unwinding Rules. In case of stack unwinding, the stack is recursively popped until one of two conditions occurs: a landing path is found or not. We respectively use $\overset{\mathsf{U}^+}{\Longrightarrow}$ and $\overset{\mathsf{U}^-}{\Longrightarrow}$ to indicate the compound stack unwinding transition from a state until one of those conditions is met.

Example 6. Assume σ.stack = [`0x116e`]. Then, for $\sigma \overset{\mathsf{U}}{\longrightarrow} \sigma'$ to hold, we must have σ'.stack = [] and σ'.`rip` = `0x116e`.

Figures 6i and 6j show the simplest unwinding function rules, those for `_Unwind_Resume`. The main addition to the general unwinding transition is that, when landing pads are found, the original function argument (σ.`rdi`) is preserved in the result state's return register (σ'.`rax`). This models the concrete handling for carrying through exceptions during unwinding.

Example 7. Consider instruction `0x1398` of Listing 1.1. As previously described in Sect. 2, this instruction is intended to continue unwinding for exceptions that do not satisfy the source code's catch type specification. Assuming no more applicable landing pad table entries, the only valid post-states for the transition here match σ'.stack = [] and σ'.terminated = Bad.

The rules for the initiating function `__cxa_throw`, shown in Figs. 6k and 6l, expand on those for `_Unwind_Resume`. They add the condition that the post-state's uncaught exception count is incremented. As before, given our level of abstraction the function `_Unwind_RaiseException` is semantically equivalent to the base throw function and thus shares its rules.

Example 8. Consider instruction `0x1286` of Fig. 4. We previously stepped through the process of throwing using this instruction in Sect. 2, so we merely state the results here. As this is the first throw at this time, we have σ'.uncaught = 1. Additionally, the unwinding process stops for σ'.`rip` $\in$ LPT(`0x137e`) = {`0x138b`}, giving us σ'.`rip` = `0x138b`.

The rules for `__cxa_rethrow` in Figs. 6h and 6m add a twist by utilizing the current caught-exception stack. When an exception object ID is available on the top of the caught stack, unwinding proceeds as usual. Furthermore, the corresponding exception object is marked as being rethrown and its ID is stored in `rax` for later usage. By contrast, when no caught exception objects are available, `__cxa_rethrow` must lead to an abnormal termination for strict modeling. However, that second rule can be relaxed for additional overapproximation by using the `_Unwind_Resume` rules instead.

Example 9. Consider instruction `0x13af` of Listing 1.1. Assume: σ.caught = [id], σ.emap(id) = e, and LPT(`0x13af`) = {`0x13c0`}. Then we end up with: σ'.`rip` = `0x13c0` and σ'.emap(id).rethrown.

Additional rules exist for the process of *forced unwinding*, or manual stack unwinding. Those are summarized here, starting with `_Unwind_ForcedUnwind`. That function functions similarly to `__cxa_throw` (Figs. 6k and 6l). However, instead of stopping based on landing pad table information, it executes the function stored in σ.`rsi` in each frame and uses the result to determine when to stop. `_Unwind_DeleteException` functions like `cxa_free_exception` (Fig. 6c) at the end of that process. The helper func `_Unwind_GetIP` stores the current frame's instruction pointer in σ'.`rax`. Finally, the other helper function `_Unwind_GetRegionStart` stores the current procedure fragment's starting address in σ'.`rax`.

3.4 Symbolic Execution

We perform symbolic execution by application of the rules making up our abstract transition relation. The symbolic execution engine was developped in Haskell, specifically tailored for this purpose. For some initial abstract state σ_0, σ_0.`rip` is either manually provided or obtained from the binary's Executable and Linking Format (ELF) info. Then we iteratively fetch the instruction at that address, increment `rip` appropriately, and apply the applicable abstract transition rule to obtain successor states. If the transition rule results in multiple possible continuing states, we apply the symbolic execution step to each successor state. If no non-terminating states result, this path of execution ends.

To prevent infinite loops and alleviate some of the state space issues that can occur with such non-deterministic evaluation, we provide a join operation. This join operation is focused on exceptional state. From Σ it preserves emap, uncaught, and caught. To maintain contextual awareness, it also preserves `rip`, stack, and terminated. As an implementation detail, it also includes the temporary indices used by our jump table heuristic to ensure proper separation. All other state parts are combined, with priority given to the first equivalent state produced. For a more aggressive join, emap, uncaught, and caught can be excluded from the preserved state parts. The abstract transition rules are also simplified to support this exclusion.

3.5 Argument for Overapproximation

We consider a formal definition of the concrete transition rules our abstract ones overapproximate outside the scope of this paper. This is because our abstraction focuses on the domain of exceptional control flow in terms of its ABI-level definition. By contrast, concrete rules require a concrete implementation. Instead, we provide an informal argument for why our abstract transition rules are overapproximative.

First, for normal (non-exception-related) assembly instructions, our abstract transition rules default to assigning $\top$ to destination operands, overapproximating their effect. Only those instructions whose arguments affect exception- and stack-related behavior as well as global memory operations receive full modeling. They include `mov` and its relatives, `push`+`pop` and related instructions, and basic arithmetic/bitwise instructions. If we did not model those instructions, we could lose too much information concerning exceptional or even regular control flow.

Second, for exception-related function calls, the semantics in Fig. 6 purposefully omit information from the abstract state. An example is the type of the exception being allocated. The abstract step function, then, considers *all* possible next states for *any* exception type.

With respect to indirections, we deploy a tactic for resolving them that, *if* successful, *then* it is overapproximative. Our tactic aims to find a program point (i.e., an instruction address) from which a variant of under-constrained and overapproximative symbolic execution [8] is executed from that program point up to the indirection. If the resulting symbolic state provides a finite bound to the set of jump targets of the indirection, the indirection is marked as resolved. Otherwise, it is deemed unresolvable.

In the case of unresolvable indirection, we do not apply additional heuristics or guesses. Instead, we stop further exploration at the indirection, if a jump, and clearly annotate the output accordingly. Unresolved indirect calls are treated as unmodeled external calls, but the same principle applies. We thus informally argue that the produced EICFG is overapproximative *modulo* unresolved indirections. If the EICFG is not annotated with any unresolved indirections, it is an overapproximation.

4 Validation

To increase trustworthiness, we validated some of our abstract transition rules against the corresponding real-world implementations. Specifically, we generated abstract states σ, and validated that:

$$\sigma \xrightarrow{\mathsf{A}} \sigma' \wedge \gamma(\sigma) \xrightarrow{\mathsf{C}} s' \implies \alpha(s') = \sigma' \tag{1}$$

Here α and γ denote abstraction and concretization functions (from the field of abstract interpretation [3]), and $\xrightarrow{\mathsf{C}}$ denotes concrete execution.

Abstract states σ are obtained through via *test case generation* [1]. For each rule under validation, we generated 10 000 arbitrary initial abstract states (σ) and then applied the rule to obtain the corresponding abstract post states (σ'). Then, using a test harness implemented as a combination of Python and the GNU Project debugger (GDB) scripts, we ran constructed real-world binaries featuring the desired concrete functions. The usage of GDB allowed easy interceding at specific points in the binaries in order to set up the initial state and extract the state after the step. Function γ operates before the concrete library function is executed, setting the state parts in Table 1 to their test case values. Function α operates after the concrete library function is executed, extracting the listed state parts from the concrete program state. The test harness then verifies that the abstracted state parts match the expected ones generated previously, satisfying Eq. (1). Table 1 shows our validation status.

Table 1. Validated state parts.

Rule	rip	in/out regs	handlerCount	uncaught	handlerSwitchValue	caught
`__cxa_throw`	✓	✓	✓	✓		
`__cxa_begin_catch`	✓	✓	✓	✓	✓	
`__cxa_end_catch`	✓	N/A	✓	✓	✓	Partial
`__cxa_rethrow`	✓	✓	✓	✓		Partial
`_Unwind_Resume`		✓	✓	✓	✓	

Our constructed test programs are designed to be minimal but still call the specific library functions we provided abstract transition rules for. To easily read and write memory using GDB, we provided dummy versions of certain structs. Specifically, the hidden library structs for individual exception objects as well as global exception information. To utilize those structs within GDB, the programs must be built in debug mode.

4.1 Concretization and Abstraction

Next, the abstract start and end states are generated by a small wrapper around our abstract transition rules. We used components of the property testing library QuickCheck [1] to instrument the start state generation. The initial starting addresses are defined by the binary versions of the above-mentioned test programs. The end state generation is performed by applying a single step of our methodology to those test programs using those start states. The start states as well as the end states for each step are then exported for use by the test harness.

The concretization and abstraction functions γ and α are part of that test harness. As in abstract interpretation, they interface between the generated abstract states and the concrete memory layouts mentioned above. Both functions operate via GDB breakpoints that are set depending on the rule under test. Function γ operates before the concrete library function is executed, setting the state parts in Table 1 to their test case values. Function α operates after the concrete library function is executed, extracting the listed state parts from the concrete program state after the library function is executed. The test harness then verifies that the abstracted end state parts match the expected ones generated previously, satisfying Eq. (1).

4.2 Observations and Challenges

During the process of this validation, we uncovered several implementation quirks that were not obvious. For example, the field handlerCount is actually a signed integer. This means that, when generating initial states, a negative value may be produced. As it turns out, the concrete implementation of `__cxa_begin_catch` takes the absolute value of negative handler counts supplied to it before incrementing that value. `__cxa_rethrow` performs a similar, but stranger, transformation. It decreases the magnitude by one, then inverts the sign; if the magnitude is 0, handlerCount is unchanged. The implementation of our abstract transition rules was updated to reflect those unearthed quirks.

We also did not cover those cases where an exception does not get caught and results in program termination due to stack unwinding. This is because we were unable to easily check the desired state parts in such cases. Similarly, we could not validate the `rip` modification of the function `_Unwind_Resume`. When running in GDB with our test program constructed to utilize `_Unwind_Resume` , handler switch value manipulation is required to trigger that path. That in itself is not necessarily an issue, but when that path is taken, control flow is ultimately redirected to the landing pad for the catch block that leads to the `_Unwind_Resume` rather than an appropriate parent landing pad. This prevents us from validating the target landing pad (or lack thereof) for that function (i.e. σ'`.rip`). However, we were still able to validate the other exceptional state components manipulated by that function.

5 Experimental Results

Here we present the results of generating EICFGs for 341 real-world programs and libraries. These programs and libraries have a variety of sizes and use cases and were sourced from places like GitHub and the Advanced Package Tool (APT) repositories. 49 of these programs utilize C++ exception handling (several despite the lack of exception handling tables) while all have been compiled for the x86-64 ISA and the System V ABI. The EICFG generation for each binary was executed on a server with four Intel® Xeon® E7-8890v4 CPUs (for a total of 96 cores) running at 2.20 GHz with 252 GiB of RAM. The server's OS was Ubuntu 18.04.5 LTS. Execution timeout was set to eight hours.

Ideally we would directly compare the number of edges found by our tool with the number of edges found by related work. That, however, does not give insight into what is gained by generating EICFGs. Existing tools, such as Ghidra, IDA Pro, and Binary Ninja produce fewer edges related to interprocedural exceptional control flow. However, they may produce more edges as they apply heuristics, pattern recognition, and best-guesses to resolve indirections.

We therefore directly compare the number of novel edges that are found using our approach versus an approach that does not consider exceptional control flow. This is a *baseline comparison*: a summary of the results comparing EICFG generation to the results obtained by running our tool with exception handling *disabled* (see Table 2). That version treats throw-related functions as terminating functions, while catch-related functions were treated as no-ops.

30 of the binaries we analyzed are not included in the table as the tool ran out of memory or timed out due to state space explosion, in one case just in the baseline version. We identified 2574 unique throws and traced the exceptional control flow of each one. Based on our analysis, we were able to identify 760 of them as uncaught; the remaining 1814 all had a potential catch block in their unwinding path. On average, dealing with exceptional control flow can increase coverage of reachable instructions by 14 instructions per unique throw, with each throw averaging 188 edges in the potential unwinding paths from that throw. Those edges are ones tools such as Ghidra do not produce.

Table 2. Case study results.

Groups	Absolute Numbers							Baseline Comparison
	Binaries	Instructions	Unwind Edges	Unique Throws	Caught Throws	Time (s)	Mem (GB)	Instruction Differential
NASA	13/14	1742K	1410	167	136	5056	36	1027
Xen	82/90	181K	0	0	0	112	1	0
Magick	15/17	172K	14	14	2	73	1	15
Cups	163/164	317K	3938	33	0	4049	2	0
Other	18/23	763K	63 260	830	526	5521	20	11 766
caf	5/6	632K	7952	466	254	918	12	4251
art	1/4	57K	216	31	30	151	10	1885
audio	5/7	100K	49 199	523	380	639	8	8509
drives	3/3	21K	21 683	69	61	273	6	674
games	3/7	35K	4081	260	254	55	1	3759
science	1/1	23K	3567	94	92	43	3	3009
tasking	1/1	23K	6	4	3	359	17	119
torrent	1/5	18K	328 948	83	76	4306	243	1195
	311/341	4089K	484274	2574	1814	21447	30	36 200

The Xen binaries exhibited no change, as none contained any exceptional control flow. We have decided to cover some non-trivial binaries without exceptional behavior as well, to validate that our approach never decreases the number of edges found. Thus, other than the time it takes to run EICFG generation, there is no cost.

6 Related Work

Bottom-Up Approaches: Decompilers and Disassemblers. We provide a summary of works in Table 3. The table describes whether or not the works do intra- or interprocedural exception analysis and if so, if it is static or dynamic. Static approaches, such as the one presented in this paper, typically aim for overapproximation. They utilize abstraction or other methods to model paths symbolically. In contrast, dynamic approaches are inherently underapproximative. This is because they rely on concrete runtime behavior and evaluating all possible concrete paths is infeasible.

While Binary Ninja, IDA Pro, and Ghidra all support some form of intraprocedural exception handling analysis, they can only perform interprocedural exception handling analysis dynamically via debugging. For example, Ghidra provides default, platform-dependent analyses that extract try-catch block information and other landing pad information for exception handling. However, it does not provide unwinding control flow in the generated block/call graphs, only cross-references for the LPT. This means that it can identify landing pads and analyze the code following them, but it cannot identify the exceptions that will

Table 3. Bottom-up Exceptional Analysis Comparisons

Program	Intraprocedural†	Interprocedural‡	URL
This work	**Statically**	**Statically**	
Binary Ninja	Statically	Dynamically	https://binary.ninja
IDA Pro	Statically	Dynamically	https://hex-rays.com/ida-pro/
Ghidra	Statically	Dynamically	https://ghidra-sre.org
McSema	Statically	No	https://github.com/lifting-bits/mcsema
RetDec	Unknown	No	https://github.com/avast/retdec

† Can it identify the landing pads in a function?
‡ Can it trace from (re)throw to landing pad?

reach them. Its built-in debugger that can perform unwinding is also ultimately just an interface to external, dynamic debugging tools such as GDB or LLDB, which perform dynamic analysis and instrumentation. Traces are also supported, but those require the program to have been run previously.

McSema is an executable lifter: It lifts machine code to LLVM bitcode. It provides intraprocedural exception analysis by generating the LLVM representation for exception landing pads. However, it does not perform interprocedural analysis as it merely lifts to LLVM bitcode, rather than tracing the execution of exceptions from throw site to landing pad. RetDec functions similarly to McSema as a decompiler-to-LLVM, though it also supports C output. However, we could not find information about its ability, or lack thereof, to deal with exception handling. As with McSema, it does not perform interprocedural exception analysis.

Top-Down Approaches. By contrast, there are tools that analyze exceptions from the source-code side [6,7,9]. This prevents the analysis of legacy code without source but allows for better static analysis during development, or even formal proofs of correctness of the exceptional semantics.

Hutton and Write [4] provided basic formal semantics for source-level C++-like exceptions. They accompanied this with a compiler for a small language with exceptions and a proof of correctness of that compilation. This is different from our approach as we do not aim to verify the correctness of exceptional behavior due to our lack of ground truth (source code or some program specification). Additionally, Prabhu et al. [7] provide source-level generation of interprocedural exception control-flow graphs (IECFGs) for C++ exceptions. These IECFGs are much like our EICFGs, but extracted from source code instead. Unlike our work, however, IECFGs are used to *eliminate* exceptions when compiling to a binary to make static analysis easier. They cannot be used to analyze already-compiled programs with exceptions. Jiang et al. [5] used the additional edges exposed by IECFGs to design test cases for path testing and branch testing that take exceptional behavior into account.

Zhang et al. provide semantics of exceptional control flow of C++ source code [9], and show how symbolic execution of their semantics can be used to discover exception handling bugs in real-life open-source software such as a JSON parser and an SQL server. They argue that the number of edges in a call graph increases by 22% when exception handling is taken into account, and note that the density of exception handling bugs is relatively high. This emphasizes the importance of properly dealing with exceptional control flow.

7 Conclusion

Many C++ programs exhibit exceptional control flow that standard CFG extraction tools in disassemblers and decompilers do not identify. To deal with that issue, we have provided EICFGs and a tool for generating them. EICFGs extend standard CFGs extracted from binaries by including nodes and edges for exceptional control flow. Our abstract transition relation for exceptional control flow has been informally – through testing – shown to overapproximate the concrete semantics, modulo unresolved indirections. Furthermore, we have applied our EICFG generator to 341 real-world programs and libraries. We identified 2574 unique throws and were able to trace each one's exceptional control flow: 1814 were potentially caught while 760 had no identified potential for being caught. On average, dealing with exceptional control flow can increase coverage of reachable instructions by 14 instructions per unique throw, with each throw averaging 188 edges in the potential unwinding paths from that throw. Those edges are ones tools such as Ghidra do not produce.

One of the main drawbacks of our work is a propensity for state space explosion. While we were able to target programs with over 400 000 instructions, our EICFGs generally do not scale far beyond that. Even for smaller programs, we experienced timeouts and out-of-memory cases when a significant number of control flow nodes and edges were generated. Methods of reducing the state space while maintaining interprocedural exceptional analysis would provide for increased scalability and the ability to target even larger programs. For example, modeling of exception type info and integrating it into the LPT determinations would allow pruning of dead branches, reducing the tool's overapproximation without introducing unsoundness.

Our approach is specific to C++. The main challenge in applying a similar analysis to binaries compiled from different languages with exception handling, is to define abstract transition rules for their binary-level implementations (see Fig. 6). We argue that these semantics must be thoroughly tested, as small implementation details may have large impact on the soundness of the approach.

There are several potential client analyses for our work. These include (semi-) formal and verified decompilation, binary-level data-flow analysis tools, and binary-level security analysis tools that require knowledge about the reachability of instructions. These use cases align with the use cases of mature reverse engineering tools such as Ghidra, and we envision that this work will be integrated in such a tool suite.

Acknowledgment. We would like to thank the anonymous reviewers and our shepherd Andrea Mambretti for their insightful comments and suggestions, which helped to greatly improve the paper. This work is supported by the Defense Advanced Research Projects Agency (DARPA) and Naval Information Warfare Center Pacific (NIWC Pacific) under Contract No. N66001-21-C-4028, by DARPA under Agreement No. HR00112090028, the US Office of Naval Research (ONR) under grants N00014-17-1-2297 and N00014-18-1-2665, NAVSEA/NEEC under grant N00174-16-C-0018, and US Naval Surface Warfare Center Dahlgren Division under grant N00174-20-1-0009.

References

1. Claessen, K., Hughes, J.: QuickCheck: a lightweight tool for random testing of Haskell programs. In: Proceedings of the fifth ACM SIGPLAN International Conference on Functional Programming, pp. 268–279 (2000)
2. Cousot, P.: Abstract interpretation. ACM Comput. Surv. **28**(2), 324–328 (1996). https://doi.org/10.1145/234528.234740
3. Cousot, P., Cousot, R.: Abstract interpretation: a unified lattice model for static analysis of programs by construction or approximation of fixpoints. In: Principles of Programming Languages, pp. 238–252 (1977). https://doi.org/10.1145/512950.512973
4. Hutton, G., Wright, J.: Compiling exceptions correctly. In: Kozen, D. (ed.) MPC 2004. LNCS, vol. 3125, pp. 211–227. Springer, Heidelberg (2004). https://doi.org/10.1007/978-3-540-27764-4_12
5. Jiang, S., Jiang, Y.: An analysis approach for testing exception handling programs. ACM SIGPLAN Not. **42**(4), 3–8 (2007)
6. Kechagia, M., Devroey, X., Panichella, A., Gousios, G., van Deursen, A.: Effective and efficient API misuse detection via exception propagation and search-based testing. In: Software Testing and Analysis, pp. 192–203 (2019). https://doi.org/10.1145/3293882.3330552
7. Prabhu, P., Maeda, N., Balakrishnan, G., Ivančić, F., Gupta, A.: Interprocedural exception analysis for C++. In: Mezini, M. (ed.) ECOOP 2011. LNCS, vol. 6813, pp. 583–608. Springer, Heidelberg (2011). https://doi.org/10.1007/978-3-642-22655-7_27
8. Ramos, D.A., Engler, D.: Under-constrained symbolic execution: correctness checking for real code. In: 24th USENIX Security Symposium (USENIX Security 2015), pp. 49–64 (2015)
9. Zhang, H., Luo, J., Hu, M., Yan, J., Zhang, J., Qiu, Z.: Detecting exception handling bugs in C++ programs. In: 2023 IEEE/ACM 45th International Conference on Software Engineering (ICSE), pp. 1084–1095. IEEE (2023)

S2malloc: Statistically Secure Allocator for Use-After-Free Protection and More

Ruizhe Wang, Meng Xu(✉), and N. Asokan

University of Waterloo, Waterloo, Canada
{ruizhe.wang,meng.xu.cs}@uwaterloo.ca, asokan@acm.org

Abstract. Attacks on heap memory are ever-increasing. Existing entropy-based secure memory allocators can provide statistical defenses against various heap-based exploits, including use-after-free (UAF). However, although UAF mitigation is in scope, such allocators are not tailored to *detect* failed UAF attempts. Consequently, an attacker can beat entropy-based protection by repeating the same attack, possibly in combination with heap spraying, to improve their chance of success further.

We introduce S2MALLOC, aiming to enhance UAF-attempt detection without compromising other security guarantees or introducing significant overhead. S2MALLOC consists of three new constructs in the secure allocator design space: **free block canaries (FBC)** to detect UAF attempts, **random in-block offset (RIO)** to stop the attacker from accurately overwriting the victim object, and **random bag layout (RBL)** to impede attackers from estimating the block size based on its address. We show that S2MALLOC can detect UAF attempts with a reasonable probability and is practical, with only a 2.8% CPU overhead on PARSEC and an 11.5% CPU overhead on SPEC.

Keywords: Secure Memory Allocator · Use-After-Free

1 Introduction

Heap-related vulnerabilities are severe and common threats that can be leveraged to launch attacks resulting in arbitrary code execution or information leakage. Heap overflow, double and invalid free, and use-after-free (UAF) are among the most common types of these vulnerabilities. According to the Common Vulnerabilities and Exposures (CVEs) report of 2022, they were ranked 19th, 11th, and 7th in terms of bug prevalence [25].

Secure memory allocators play a crucial role in defending against heap exploitations. They can be broadly categorized into two types: entropy-based generic allocators, which use randomness to enhance security, and UAF-mitigating allocators, which specifically target UAF vulnerabilities.

UAF occurs when a previously freed memory block is used to store data. It receives special attention in secure allocator design due to its prevalence

F. Maggi et al. (Eds.): DIMVA 2024, LNCS 14828, pp. 23–43, 2024.
https://doi.org/10.1007/978-3-031-64171-8_2

and the powerful exploitation primitives (e.g., arbitrary read/write) it enables. Chromium has reported that more than a third of its security issues are related to UAF, which is more prevalent than other types of memory errors combined [36]. Prior works have demonstrated that effective UAF mitigation is not only possible but can be achieved in several ways, such as tracking and nullifying pointers on de-allocation (e.g., MarkUs [2]), forward-only allocation (e.g., FFmalloc [41]), and context-based allocation (e.g., Cling [3]). However, these approaches also have drawbacks, such as substantial CPU overheads (e.g., about 40% for MarkUs), downgraded protection in edge cases (e.g., same-context UAF is still possible with Cling), or new attack vectors exposed due to complexity in implementation (e.g., FFmalloc) albeit preventable.

Entropy-based heap allocation is another way to mitigate UAF exploits. Allocators in this theme share one limitation: probabilistic protection. However, the relaxed security requirement also enables them to protect most, if not all, common heap vulnerabilities with simpler and binary-compatible designs. Specifically on UAF mitigation, entropy-based memory allocators typically use delayed free-lists [20,30] to prevent the same memory block from being *immediately* reallocated after being freed. Attackers now face a moving target even when they obtain a dangling pointer, as they have less confidence in knowing when this pointer becomes valid again and/or which object it might point to.

While achieving relatively low CPU overhead, especially compared with UAF-mitigating allocators (details in Sect. 2), existing entropy-based allocators still face the challenge of entropy loss, further compromising the probabilistic protection. To illustrate how entropy loss can occur: 1) an attacker could control multiple dangling pointers via heap spraying, increasing the chance of corrupting a target object through one of the dangling pointers; 2) an attacker could infer block size or type based on pointer addresses; 3) an attacker could retry the same attack repetitively if a failed attempt only causes silent failure (more details in Sect. 3).

To overcome these challenges, we propose S2malloc, an allocator that reduces entropy loss by detecting UAF *attempts* with low CPU and memory overhead on par with state-of-the-art entropy-based memory allocators. S2malloc achieves its promises by combining several new realizations of existing concepts: **randomized in-slot offset (RIO)**, **free-block canary (FBC)**, and **random bag layout (RBL)**. RIO mitigates UAF attacks by allocating blocks with random offsets, obstructing the attacker from locating the target field in a data structure. FBC puts cryptographically secure canaries in free blocks to detect illegal writes, turning a failed UAF exploitation attempt into a clear signal. RBL organizes blocks of the same size range using sub-bags. Only blocks within the same memory page are guaranteed to be in the same sub-bag.

Summary. We claim the following contributions:

- We analyze current entropy-based allocators in real-world UAF attack scenarios and show their protection is weaker than claimed (Sect. 3).
- We present S2malloc, a drop-in solution that addresses the above weaknesses while protecting against other commonly observed heap memory vul-

nerabilities. S2MALLOC does not require special hardware, program recompilation, or elevated privileges and works on x86 and AARCH (Sect. 4).
- Through various real-world CVEs and benchmarks, we show that S2MALLOC can successfully detect all attacks (Sect. 5) while incurring 2.8% CPU overhead and 27% memory overhead on the PARSEC benchmark and 11.5% CPU overhead and 37% memory overhead on the SPEC benchmark (Sect. 6).

Both the software artifact and an extended version of the paper are available. In the extended version, we 1) discuss the implementation details more thoroughly, 2) document finer-grained evaluation results, and 3) provide a more formal analysis on the security of S2MALLOC.

2 Background and Related Work

2.1 Entropy-Based Allocators

Entropy-based heap allocators strive to protect against almost all heap-based exploits mentioned above by minimizing the attacker's success rate. Ideally, the success rate should be low enough to deter attackers from trying to attack the system. However, as demonstrated in Guarder [30], practical implementations face limitations in memory and computation resources.

BIBOP. State-of-the-art entropy-based allocators typically build on the *Big Bag of Pages* (BIBOP) [16] mechanism with various security enhancements. BIBOP-style allocators allocate blocks of the same size class together using one or more continuous memory pages and pre-emptively allocate sub-pools for each size class. Each sub-pool is divided into several slots pre-emptively, and each allocated object takes one slot. A slot is not further divided or merged so that memory fragmentation can occur only within a slot. For a long-running program, blocks from newly mapped pages continue to be added to the pools until the program reaches a steady state, after which most allocations and deallocations involve existing blocks only. No performance degradation will happen as the program executes. A bitmap monitors the status of each slot and can be used to defend against double or invalid frees. Bag metadata is stored separately to avoid metadata-based attacks [29], and UAF only occurs within the same size class.

Extended Secure Features. Prior works [20,27,30] have introduced a diverse set of security enhancements over the basic BIBOP-style allocation, including:

- *Guard page*: A guard page is a single-page virtual memory block not mapped to physical memory. Therefore, any attempt to dereference an address in a guard page triggers a segmentation fault. Guard pages can be placed after each bag or randomly within bags to protect against out-of-bound accesses.
- *Heap canary*: This is a small object set to a secure value and placed at the end of each slot. Heap overflow can be detected if the canary value is modified.
- *Random allocation*: This guarantees that slots within each bag are not allocated sequentially (i.e., linear allocations). Instead, each allocated slot is randomly chosen from at least r free slots (where r is the entropy).

2.2 UAF-Mitigating Allocators

UAF-mitigating allocators specialize in UAF-protection only and can be broadly categorized into the following types:

A *Pointer validation on dereference* [11,39,42]: i.e., checks whether a pointer is valid and safe for a read/write operation upon dereference.
B *Pointer invalidation on free* [8,12,13,28,41]: i.e., whenever `free` is called on a pointer, it revokes its capability to access the associated virtual address.
C *Memory sweeping* [2,10,18]: i.e., checks all pointers stored in memory and either actively removes all dangling pointers or reuses a freed memory chunk with the assurance that no dangling pointer to this freed chunk exists.
D *Context-based allocation* [3,9,38]: i.e., permits the re-allocation of freed memory chunks only to objects allocated in the same "context".

Not all UAF-mitigating allocators are drop-in replacements of the system memory allocator, as some of them require recompilation of the protected programs [3,38,40], special hardware [12], or kernel modifications [13].

Security Guarantee. UAF-mitigating allocators in categories A, B, and C can eliminate all UAF exploits (hence complete mitigation), although some of them might incur large overheads or require customized kernel or hardware. Context-based allocators (category D), first proposed in [9] as a heap-safety property, typically run faster but offer incomplete protection. For example, Cling [3] defines the allocation context based on the two innermost return addresses on the call stack when `malloc` is invoked. Suppose two memory allocations occur in the same context; in this case, the object allocated the second time may occupy the same memory chunk allocated the first time (if the first object is freed). TypeAfterType [38] defines the allocation context based on object types (e.g., the type `int` passed in `malloc(sizeof(int))`, and objects may be reallocated on freed memory chunks originating from the same allocation context. In summary, UAF is still possible in context-based allocators.

However, unlike entropy-based allocators (including S2MALLOC), which also mitigate UAF imperfectly, the protection from context-based allocators is deterministic. Specifically, if the dangling pointer and the target object are allocated in different contexts, there is zero chance of success in corrupting the target object, regardless of the attackers' strategies. On the other hand, context-based allocators provide no protection if the dangling pointer and the target object fall in the same context. Entropy-based allocators provide probabilistic protection in both cases. We will discuss the implications via CVEs in Sect. 5.3.

Porting for UAF-Write Defense Only. Allocators that instrument runtime checks for pointer validity (i.e., category A) can opt to trade protections of UAF-read attacks for performance by checking the pointer validity on write accesses only. In fact, as reported in [31], only 5% to 25% of memory accesses are write accesses in the SPEC 2017 benchmark, indicating a potential reduction in overhead for category-A allocators. On the other hand, for allocators in categories B, C, and D, splitting UAF-read and UAF-write protection is inherently hard as their designs do not differentiate between read and write accesses.

2.3 Widely Deployed Secure Allocators

Secure allocators have been adopted in industrial codebases as well. Scudo [22] is a solution from LLVM [21] that balances security and performance and supports random allocation and delayed free-list (the latter is disabled by default), a subset of security features provided by entropy-based allocators [20,30]. Hardened_malloc [15] is the default allocator of GrapheneOS [14], a privacy and security-focused Android-based OS. Hardened_malloc incorporates all security features discussed in Sect. 2.1 and focuses on UAF-write protections. Similar to other entropy-based allocators, Hardened_malloc suffers from reduced randomness. It also incurs increased overheads for larger blocks, which limits its use scenarios.

3 Motivation

While effective in defending against various heap exploits, entropy-based allocators are not ideally suited to protect against UAF attacks. Specifically, easy-to-identify blind spots exist that drastically reduce the efficacy of defending UAF using random allocation and increase attackers' confidence in launching a successful attack without trying to predict the next allocation slot.

3.1 Adversary Model

We assume that the attacker can analyze the source code and binary executable to determine the implementation details of the target program, including vulnerabilities and other relevant information, such as the size and layout of critical data structures. We also assume that the attacker can identify when a victim object is allocated or de-allocated and is aware that the target runs with S2MALLOC.

We assume that the underlying OS kernel and hardware are trusted, and an attacker cannot utilize a data leakage channel, such as `/proc/$pid/maps`, to discover the location of the heap allocator's metadata. The attacker cannot compromise the random number generator, nor can they take control of the heap allocator. Exploiting bugs of the allocator itself is out of scope. These assumptions are similar to that of other entropy-based allocators [20,30].

We allow attackers to use any existing heap feng-shui [32] technique (e.g., heap spray) to prepare or manipulate heap layout to facilitate UAF exploits. Attackers can retry an exploit if previous attempts fail silently. These assumptions make our adversary model more robust than those assumed in entropy-based allocators and on par with the adversary models in UAF-mitigating allocators [2,41].

3.2 Challenge 1: Entropy Loss

Entropy-based allocators thwart UAF by avoiding instant memory reuse, as that makes where a new object is allocated more predictable. However, if 1)

```
1  mrb_io_initialize_copy(mrb_state *mrb, mrb_value copy) {
2    mrb_value orig;
3    struct mrb_io *fptr_copy, *fptr_orig;
4    fptr_copy = (struct mrb_io *)DATA_PTR(copy);
5    if (fptr_copy != NULL) {
6      mrb_free(mrb, fptr_copy);
7    }
8    fptr_copy = (struct mrb_io *)mrb_io_alloc(mrb);
9    fptr_orig = io_get_open_fptr(mrb, orig);
10   DATA_PTR(copy) = fptr_copy;
11 }
```

Fig. 1. Example UAF attack based on mRuby issue 4001 [34].

the attacker can continue to retry the attack when the previous trial fails, or 2) heap memory can be spoofed with either dangling pointers or victim objects, it is guaranteed that the attack would eventually succeed without the victim's notice.

Figure 1 is an example, abstracted from mRuby issue 4001 [34], a UAF vulnerability in the Ruby compiler. The function `mrb_io_initialize_copy` is called when opening a file. It first frees the existing data pointer of the `copy` object (`DATA_PTR(copy)`) (line 7) and allocates a new object to it (line 9 and 11). If an invalid argument is given, calling `io_get_open_fptr` will throw an exception (line 10), causing an early return, which makes `DATA_PTR(copy)` a dangling pointer. The attacker can then close the file in Ruby, which will set the first word of the pointed memory to `INF`. If a string object takes this memory, its length will be overwritten to `INF`, enabling arbitrary memory reads and writes.

In the above scenario, random (i.e., non-sequential) allocation or delayed freelists available in existing entropy-based allocators [20,27,30] merely increase the attack difficulty: as long as the attacker can wait, the memory chunk referred to by the dangling pointer will eventually be reallocated, allowing the UAF exploit to proceed after some delay. Furthermore, the entropy diminishes if an attacker can repeat the same attack *without penalty* (e.g., when a failed attempt does not crash the target program or trigger attention). Similarly, if attackers can spray the heap with either dangling pointers of victim objects, the probability of success increases significantly.

This UAF exploit strategy motivates us to design an allocator that 1) actively searches for UAF attempts and raises signals if evidence is found and 2) stops the attacker from locating critical information in memory even if the attacker manages to obtain a dangling pointer. In this example, we can stop the attacker from locating the string length deterministically even if the attacker manages to obtain a dangling pointer for a string object initially pointed to by `DATA_PTR(copy)`. In addition, any attempts to write to unallocated memory will be detected with a high probability. If an attacker attempts to spray the heap with arbitrary write to increase success rates, we can raise a signal on or even before the UAF happens.

3.3 Challenge 2: Information Leak

Existing entropy-based allocators [20,30] create a memory pool for each block size range, resulting in the leakage of block size via their address, possibly revealing the victim program's internal state to the attacker. For example, each BIBOP bag is assigned an 8 GB virtual memory pool in SlimGuard [20]. Any objects belonging to this bag will be allocated from this pool. As a result, obtaining a known-sized block will be sufficient to infer the size of any blocks sharing the upper 32-bit address. Further, block addresses in Guarder [30]are size-aligned, allowing the attacker to infer the block size based on its address. For example, a block with an address value ending with `0x10` is of size 1 to 16 bytes, and a block with an address value ending with `0x300` is highly likely of size 129 to 256 bytes.

We mitigate this threat by dividing bags into sub-bags, limiting the size leakage only if the attacker-controlled block resides in the same sub-bag as the victim block. Furthermore, we assign random guard pages within sub-bags to make their boundaries unpredictable.

4 Design and Implementation

4.1 Architectural Overview

At its core, S2MALLOC adopts BIBOP to manage memory blocks. Figure 2 shows an overview of S2MALLOC. S2MALLOC maintains per-thread metadata (Ⓐ) stored in a memory chunk requested directly from the kernel. Huge blocks are obtained or released from the OS directly (Ⓑ) and stored using a linked list. Small objects are maintained using bags, claiming memory indirectly from a segregated memory pool (Ⓒ). Each **regular bag** maintains the metadata of blocks of a size range, including the number of free slots and a list of **sub-bags**. We take **sub-bag** as the basic unit of a group of slots (Sect. 4.3).

The data field points to a memory chunk requested from the memory pool to store the objects allocated to this sub-bag. The bitmap indicates whether the current slot is taken or not. If the current slot is taken, the offset table cell stores an offset indicating where the data stored in the bag starts (Sect. 4.2). Otherwise, the slot is free, and the offset table stores the location of the FBC (Sect. 4.5).

When a `free()` call is received by S2MALLOC (step ① → ②), S2MALLOC checks whether: (1) the bitmap indicates the current block is taken; (2) the offset stored in the offset table matches with the freed pointer address; and (3) the canary value is not modified. If all checks pass, the block will be freed: the bitmap cell will be set to free, and an FBC will be put at a random location in the current block. The offset table will be updated to store the location of the FBC.

When a `malloc()` call is received (step ② → ③), S2MALLOC randomly selects one free block of the corresponding size and checks the FBC of current and nearby free blocks. A random offset will be generated, indicating where the data starts within the block, and the offset table will be updated accordingly. The heap

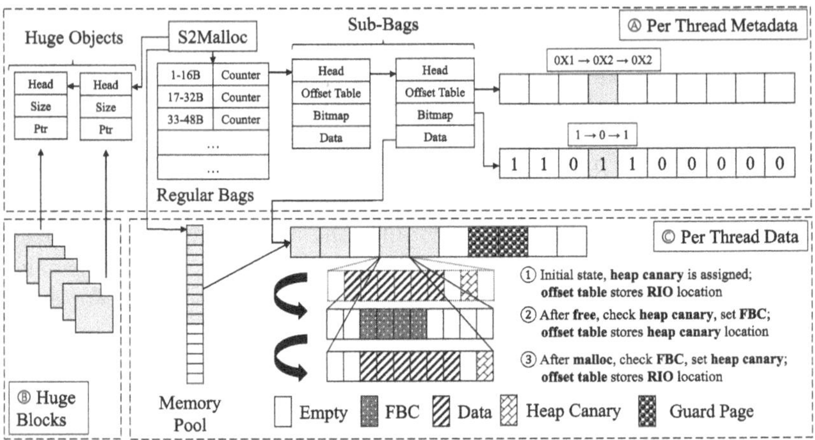

Fig. 2. Overview of S2MALLOC with an example of free and malloc. Ⓐ, Ⓑ, and Ⓒ show three S2MALLOC segments, stored in segregated memory. ①, ②, and ③ show how an allocated bag slot is freed and then allocated.

canary will be set after the last data byte of the current block, and the bitmap will be updated. This example assumes that each block stores at most 7 bytes of data. The heap canary is initially set to the 9th byte as the offset is one and then set to the 10th after reallocation as the offset is changed to two.

4.2 Randomized In-Slot Offset (RIO)

In S2MALLOC, an object is stored with a random offset p, and the first p bytes of the assigned slot will be left empty. After the slot is freed and allocated to another object, the offset p will be re-computed. Thus, the relative offset between these two objects cannot be accurately predicted, and attackers cannot reuse a dangling pointer to read or write the newly allocated object with predictable alignment. This design aligns with the BIBOP concept and does not cause fragmentation within large memory chunks.

We define k to be the RIO entropy. For each bag with blocks of b bytes, $e = b/k$ bytes are not used to take data, and each block can take at most $b - e$ bytes of data. We refer to these extra bytes as entropy bytes.

Suppose this block is `malloc`-ed with an object of s bytes ($s < b - e$). Before this block is allocated, $p \in (0, b - s)$ is computed to decide the starting byte of the data object. p is 16-byte aligned following the minimum default alignment of GNU C implementation [37]. It is compatible with special data structures, such as atomic structs which must be aligned with registers and cache lines. The offset of each block will be stored separately in an offset table.

4.3 Random Bag Layout (RBL)

As with existing secure allocators, S2MALLOC employs BIBOP to manage small-size blocks. Blocks larger than 64KB are mapped and unmapped directly from the OS and are managed using a linked list. Smaller blocks are further classified as small, medium, and large blocks to decrease the number of size classes. Small bags contain blocks smaller or equal to 1 KB, and a bag is created every 16 bytes (16 bytes granularity). Medium bags contain blocks within the range of 1 KB and 8 KB with a granularity of 512 bytes; large bags contain blocks within the range of 8KB and 64 KB with a granularity of 4 KB.

S2MALLOC does not link block sizes to block addresses. Instead of allocating a dedicated virtual memory pool for each bag (as shown in prior works [20,30]), we create a single virtual memory address pool for all bags. We further divide each bag into sub-bags, each containing 256 slots. Each bag creates new sub-bags upon need, and a newly created sub-bag will request corresponding memory from the pool. We use a bump pointer to track the available memory in the pool and sequentially allocate pool memory to sub-bags.

S2MALLOC randomly inserts guard pages within sub-bags to thwart overflow, spraying, and random pointer access. This makes memory pool allocation even less predictable. S2MALLOC does not leak the block size unless both a known and a victim block reside on the same memory page. Adjacent blocks may not be within the same sub-bag, and RIO guarantees block start addresses cannot be deterministically predicted. On the contrary, two blocks will likely be in the same size range in SlimGuard if their address difference is smaller than 8 GB.

Table 1. Overview of existing secure memory allocators and S2MALLOC to illustrate how S2MALLOC fills the gap (MD: metadata, ◐: One-shot attack only, ●: Repeated/spray attacks). Overheads are measured on PARSEC [5], detailed in Sect. 6. Overheads of MarkUs and FFmalloc (numbers marked with $*$) are reported in [41] instead of measured by us.

Secure Allocators	Guard Pages	Rand. Alloc.	Segre. MD	Heap Canary	Ptr Inval.	UAF Miti-gate	UAF Detect	Overheads	
								Memory	Runtime
DieHarder	Y	Y	Y	N	N	◐	N	21.3%	2.1%
Guarde	Y	Y	Y	Y	N	◐	N	58.1%	2.4%
SlimGuard	Y	N	N	Y	N	◐	N	22.5%	4.4%
S2MALLOC	Y	Y	Y	Y	N	●	High prob.	26.8%	2.8%
MarkUs	N	N	N	N	Y	●	N	13.0%*	42.9%*
FFmalloc	N	N	N	N	Y	●	Y	50.5%*	33.1%*

4.4 Hardening Heap Canaries

A canary is a small data block placed after the allocated memory to detect overflow. It was initially introduced in StackGuard [6] to protect the stack and has been adopted to protect the heap [26]. The canary is set to a specific value when a memory slot is allocated. This value is checked when the slot is freed, and memory overflow is detected if the canary value changes.

However, in previous entropy-based allocators, this value is set to be either globally identical [30] or bound with slots [20] so that a knowledgeable attacker can trivially break. We follow the previous secure canary designs to use the secure MAC of the memory address as the canary [19,24]. Specifically, we take the CMAC-AES-128 encrypted block address as the canary implemented using AES-NI [17] (on ×86) or Neon [4] (on AARCH) to keep the canary confidential and compact. Even if the attacker learns a canary value, they can only use it to break the current object or any further objects allocated to this slot with the same RIO. In S2MALLOC, we put a ι-byte canary immediately after the last data-storage byte in the allocated slot (i.e., the $(p + b - e)$th byte), and the canary will be checked upon `free`.

4.5 Free Block Canaries (FBC)

Existing entropy-based allocators defend against UAF-write attacks by statistically avoiding allocating a victim object in a block pointed to by a dangling pointer. Although a failed attack attempt only modifies a free block without causing any harm, it is not detectable either, and given that the same attack can be retried, the attacker will eventually succeed.

To detect such attempts, we put a canary of length c, also computed using CMAC-AES-128, in each free block. Its value is checked before the block is allocated and will be reset after it is freed. We also check the FBC of d nearby blocks to improve the detection rate. FBC guarantees that accessing a freed block is not risk-free. We detail its protection rate in Sect. 5. In Fig. 2, we further illustrate how the two kinds of canaries (FBC and regular heap canary) are set and cleared when an allocated block is freed and then allocated again.

Initially, in S2MALLOC, we create the memory pool using the `mmap` system call with the `ANON` flag, and all allocated memories are initialized to zero in the Linux environment. We take this advantage and use the zeros as the initial FBC to avoid additional CPU and memory usage for unallocated blocks. We also use the whole slot as the canary with improved security. As encrypted canaries may be more costly than zeroing out memory, S2MALLOC chooses to zero out small blocks and checks the whole block before it is allocated to a new object, bringing both security and computation benefits.

4.6 Summary and Comparison

Table 1 summarizes S2MALLOC and state-of-the-art secure allocators along the two defense lines closely related to S2MALLOC (as discussed in Sect. 2.1 and Sect. 2.2).

S2MALLOC is an entropy-based allocator. It has nearly identical heap exploitation protection features *except* for UAF protection compared with existing works [20,27,30]. S2MALLOC provides a much more robust security assurance in the presence of UAF vulnerabilities. In particular, S2MALLOC addresses the two entropy-loss cases (discussed in Sect. 3.2 and Sect. 3.3) with RIO (Sect. 4.2) and RBL (Sect. 4.3), respectively, and hence, providing much higher effectiveness on UAF mitigation. In addition, S2MALLOC is designed to monitor the integrity of the heap actively and watch for UAF attempts, including heap spraying practices that aim to prepare the heap data and layout for UAF exploits. S2MALLOC achieves this through a synergy of regular heap canaries (Sect. 4.4) and FBC (Sect. 4.5). Consequently, S2MALLOC provides a robust security guarantee and cannot be easily evaded.

Table 1 also shows a sheer contrast between entropy-based allocators and allocators specialized in complete UAF-mitigation. Notably, although providing a complete mitigation guarantee toward UAF, allocators in this line [2,41] might impair program efficiency significantly and are less suitable for deployment in time-sensitive use cases. In contrast, as will be presented in Sect. 6, S2MALLOC incurs a considerably lower overhead, which is typical for entropy-based allocators, making S2MALLOC practical and deployable on production systems if the end-user can tolerate a marginal chance of protection failure (less than 10% in the default setting of S2MALLOC, discussed in Sect. 5).

5 Security Evaluation

5.1 Formal Analysis

To mathematically model how S2MALLOC provides defense against UAF, we make the following assumptions for the attacker and target program (which are consistent with our adversary model in Sect. 3.1):

① The goal of the attacker is to modify the *victim field*, a sensitive field (e.g., a function pointer or an `is_admin` flag) in a specific type of object, a.k.a., *a victim object*, via memory writes over a dangling pointer (i.e., UAF-writes).
② The attacker can obtain a dangling pointer through a bug in the program at any point of time during execution.
③ The program repetitively allocates and frees the type of objects targeted by the attacker (i.e., victim objects) during its execution. However, we do not assume that each victim object is freed before the next victim is allocated.
④ The attacker can either indirectly monitor or directly control the allocations of victim objects, i.e., the attacker knows when a victim object is allocated, but does not know the address of the allocation.
⑤ Any memory writes through the dangling pointer is conducted after the victim object is allocated.
⑥ If the intended sensitive field of a victim object is overridden, the attack succeeds; otherwise, the program continues, allowing the attacker to repeat the exploitation effort unless detected by S2MALLOC (condition ⑦).

⑦ S2MALLOC checks FBCs on each heap allocation and detects the attack if any FBC is modified.

To simplify the illustration, we assume that the above execution logic is the only code logic that involves heap management. Below, we introduce two simple and yet realistic strategies a reasonable attacker might consider:

S1: Repeat UAF-Writes Through the Same Dangling Pointer. Suppose an attacker is confident that the memory block pointed to by the dangling pointer is not taken by a long-living irrelevant object. In that case, the attacker may prefer to keep reusing that dangling pointer for future attack attempts and hope that a victim object is allocated in that block.

S2: Repeat UAF-Writes with Freshly Obtained Dangling Pointers. Suppose an attacker worries that irrelevant objects can be long-lasting (i.e., holding a block that may be allocated to victim objects). In that case, the attacker may prefer to use a fresh dangling pointer per each attempt. However, we note that this is only an attack strategy, and we assume that there is no heap allocation other than victim objects.

For each scenario above, we discuss the rates of whether the attacker adopts the heap spray technique separately, introducing four scenarios in total. We refer readers to our full paper for an extended discussion.

5.2 Illustrate the Protection Rates

We take the mRuby issue 4001 (Fig. 1) as an example and show how its protection rate is computed. The size of object `mrb_io` is 16 bytes, and the attacker's goal in each run is to overwrite 4 bytes of it. With the default settings ($r = 256$, $s = 32$ to store a 16-byte object), the attack success rate of each trial is approximately 0.002, and the probability of overwriting FBC in a free block is approximately 0.16. In Table 2, we show how the rates change as the number of attack rounds goes up. The attack is 64% likely to be detected if the attacker adopts the first attack strategy and 95% likely to be detected using the second strategy after running it 500 times.

Table 2. Protection and attack success rates of attack rounds in mRuby issue 4001 using the two strategies.

round	Attack Strategy 1						Attack Strategy 2					
	1	5	10	50	100	500	1	5	10	50	100	500
$p_{protection}$	1.4%	4.1%	7.4%	28%	43%	64%	0.8%	12%	37%	95%	95%	95%
p_{attack}	1.2%	2.6%	4.4%	15%	24%	35%	1.2%	2.6%	4.0%	5.5%	5.5%	5.5%

Table 3. Summary of how different memory allocators defend against eight exploitation techniques on seven vulnerabilities. Vanilla BIBOP allocator and Scudo [22] are vulnerable to all attacks and behave similarly to Guarder [30] (DP: dangling pointer, ○: no defense, ◐: detect at the end of execution, ◑: defense via zero-out, ●: detect via FBC change, ◉ : non-deterministic leak (RIO), ■: thwart the exploitation ability).

Vulnerability	Attack pattern	[30]	[27]	[20]	S2malloc	[38]	[3]
CVE-2015-6831	DP → Write	◐	○	○	●	○	○
CVE-2015-6834	DP → Write	◐	○	○	●	○	○
CVE-2015-6835	DP → Write	◐	○	○	●	○	○
CVE-2015-6835	DP → Write → sleep	○	○	○	●	○	○
CVE-2020-24346	DP → Write	○	○	○	●	○	○
Python-91153	DP → Write	○	○	○	●	■	○
mruby-4001	DP → Write	○	○	○	●	■	■
CVE-2022-22620	DP → Read	○	◑	○	◉	○	○

5.3 Defending Against Real-World CVEs

In this section, we compare how S2MALLOC, Guarder, DieHarder, SlimGuard, Cling, and TypeAfterType perform in defending against seven UAF vulnerabilities. We select these vulnerabilities based on the following criteria:

- The vulnerabilities target the Linux platform and can be mitigated in the user space (i.e., not a Linux kernel bug);
- The vulnerabilities can be deterministically triggered (i.e., not racy);
- Public exploits for these vulnerabilities are available and the exploit breaks the program information integrity (i.e., not just causing DoS).

Out of the seven vulnerabilities, six are UAF-write bugs and one (CVE-2022-22620) is a UAF-read only bug. We also found seven exploits against these vulnerabilities (two exploits for CVE-2015-6835 with different attack patterns). All CVEs except Python-91153 and mruby-4001 can cause arbitrary code execution (ACE) if properly exploited. However, a powerful attack (e.g., ACE by overriding a function pointer) can succeed when the target object is precisely allocated to a freed memory chunk that is still referred to by a dangling pointer[1]. S2MALLOC can mitigate attacks by reducing the chance that a target object is

[1] An attacker may attack blindly, e.g., overriding a code pointer through the dangling pointer regardless of whether it points to a target object or not. This will have three consequences: 1) corrupting FBC, which may cause the attack to be detected upon future `mallocs`, 2) overwriting a wrong field due to RIO which may cause the program to enter a weird state (e.g., crash), 3) a successful attack. If the program can be recovered from a weird state automatically (e.g., crash resilience), the attacker can retry the same attack and eventually case 1 or case 3 will occur. However, without the probabilistic detection on UAF attempts by S2MALLOC, only case 3 will occur.

referred to by a dangling pointer. Evaluation results of the eight exploits are in Table 3.

Entropy-Based Allocators. S2MALLOC can thwart all UAF-write attacks evaluated, Guarder can detect three exploits by recognizing double-free attempts, but DieHarder and SlimGuard fail to thwart these exploits. For the UAF-read attack, S2MALLOC uses RIO to stop the attacker from reusing the memory chunk with accurate object alignment, causing the data read by the dangling pointer to be not sensible. DieHarder zeros out the memory after it is freed, which is effective if the attacker tries to over-read a freed block.

Context-Based Allocators. While we expect context-based allocators to demonstrate strong and stable protection, some of the exploits, unfortunately, hit on certain blind spots in Cling and TypeAfterType by accident.

In the case of Cling, if both the dangling pointer and the target object (i.e., the object an attacker hope to corrupt) are allocated through the same multi-layer function call sequence, they are considered to fall under the same allocation context, causing the target object to be possibly accessed by the dangling pointer. We will illustrate this limitation through the CVE-2015-6835 case study presented later. In fact, all examined CVEs, except mruby-4001, hit this limitation of Cling. Cling mitigates mruby-4001 by limiting the attacker to target objects of type `mrb_io`, which prevents the attacker from creating a powerful attack primitive.

TypeAfterType can unpack `malloc` wrappers with an arbitrary number of layers until it finds a `sizeof(T)` in the function argument, and an ID `i` is given to each allocation site of `T`. The tuple (`i`,`T`) makes the allocation context, and all memory allocations through this call sequence will be allocated from a memory pool dedicated to this context. However, if the dangling pointer and the target object share the same context in an exploit, UAF can still occur. We will illustrate this limitation through the CVE-2015-6835 case study. TypeAfterType mitigates Python-91153 by limiting the target object to be a reallocated `string`, and mruby-4001 by limiting the target object to be an `mrb_io`.

Case Study: CVE-2015-6835. This CVE is a UAF bug in the PHP session deserializer, which reconstructs a session from a serialized string. (see Fig. 7 in the extended version for simplified code snippets to illustrate this CVE and its exploit in details). An attacker can exploit this vulnerability to control a dangling pointer to a freed `zval` object. This is possible as the return value (a `zval` pointer) of `php_var_unserialize` is freed in its caller without noticing that the same pointer might also be stored in a global variable `SESSION_VARS`.

The `zval` type, unfortunately, is a reference-counting wrapper over nearly all other objects involved in the PHP engine (see Fig. 4 in the extended version for the type definition of `zval`). Therefore, an attacker might corrupt any `zval` object that may be reallocated to the freed slot. They can simply uses the `echo(..)` function to dump a newly allocated `zval` in the freed memory.

1) Protection from entropy-based allocators. S2MALLOC checks FBC on every `malloc()`. In this exploit, when the attacker tries to use the dangling pointer in `zend_echo_handler`, its `refcount` field is increased, causing the FBC to be modified. This enables S2MALLOC to detect the UAF attempt when the corrupted slot or a nearby slot is about to be reallocated. If the `refcount` change does not corrupt FBC (simulated by disabling the FBC check) and this corrupted block is reallocated, S2MALLOC can still stop the exploit as RIO causes misalignment between the dangling pointer and new object, causing the `type` field to have value `UNKNOWN` and prevents echo printing.

Guarder and DieHarder try to mitigate this attack by random allocation: hoping the new object will not be referred to by a dangling pointer. However, our experiment shows that Guarder fails if the attacker re-runs the attack multiple times or spray the heap with victim objects. SlimGuard fails to provide protections as it always allocates the most recently freed objects to the program. It does not implement the claimed random allocation feature and does not have any other security features that could detect UAF. DieHarder zeros out the memory chunk that stops information leakage of the freed `zval`, but it cannot prevent an attacker to corrupt the newly allocated `zval` over the freed chunk.

2) Protection from context-based allocators. In this exploit, both the dangling pointer and the target object (i.e., the object the attacker wish to dump information via `zend_echo_handler`) are allocated by the the same multi-layer `malloc` wrapper: `php_var_unserialize`→`emalloc`→`malloc`. This is critical to understand why Cling and TypeAfterType fail to mitigate this exploit.

For Cling, this `malloc` wrapper implies that the allocation of many `zval` objects will be sharing the same context (measured by the two innermost return addresses on the call stack). This leaves the dangling pointer plenty of candidate objects to refer to after several rounds of deserialization in PHP. TypeAfterType can inline `malloc` wrappers but the inlining stops at `php_var_unserialize` because it sees the `sizeof(zval)` argument in `emalloc` and hence, will allocate all `zval` objects originating from this `malloc` wrapper from the same pool. Unfortunately, the dangling pointer is also allocated this way, enabling UAF among the dangling pointer to other `zval` objects as well.

Summary. The combination of random allocation and delayed free-list provided by previous entropy-based allocators focus on one-time attacks only, Hence, repeating the same attack is a simple yet effective solution to undermine their protection. Context-based allocators, on the other hand, might fail to detect UAF among objects allocated of the same context. While these results highlight the effectiveness of S2MALLOC, we note that information leakage through corrupted pointers might diminish this guarantee, which is discussed in Sect. 7.

6 Performance Evaluation

To evaluate its performance, we align with previous works [2,13,20,30] that run S2MALLOC on macro, micro, and real-world programs, covering computation-heavy, memory-heavy, real-world simulating, and dedicated memory allocator benchmarks. We also analyze how different parameter values influence it.

Experiment Setup. Experiments are performed on both x86 and AARCH servers for macro benchmarks. Others are only on the x86 server. The x86 server is configured with 160-core 2.40 GHz Intel CPUs (x86 architecture) with 1 TB system memory. We set up the AARCH server on Amazon Web Service (AWS) using the `im4gn.4xlarge` machine with 16 vCPU cores and 64 GB memory. Benchmarks are measured in the Docker environment with Debian 11 on both machines, and they are used exclusively for this evaluation task. We note that benchmarking within a Docker container only introduce negligible overheads [23]. We measure the overheads using the GNU `time` binary and setting the `LD_PRELOAD` environment variable to substitute the system default allocator.

We obtain SlimGuard, Guarder, and DieHarder from their GitHub repository. To provide a fair result, we reduce the allocation entropy bit of Guarder to eight (same as the default value of SlimGuard and S2MALLOC). We also disable DieHarder from zeroing out freed blocks (slightly accelerates DieHarder).

S2MALLOC is measured with the settings of checking two nearby blocks ($d = 2$), 10% random guard page, and taking 1/4 of the block size as random offset entropy ($e = 0.25b$). We zero out blocks smaller than a memory page (4096 bytes) and take the whole block as FBC. For blocks larger than a memory page, we set an 8-byte FBC ($c = 8$) in the corresponding blocks. We set the heap canary length to be one byte ($\iota = 1$) following the design of SlimGuard and Guarder. All reported times and memory usage are normalized using the baseline (glibc) output. We use geometric averages to compute average overheads and report the means of five runs. We note that the standard deviations are minimal.

6.1 Macro Benchmarks

PARSEC. We use the PARSEC3 [5] benchmark. We exclude three network tests and one test (x264) that fails to be compiled in the baseline scenario, and only report the results of the remaining 12 benchmarks. Additionally, we exclude "raytrace" from execution for the AARCH server as it cannot be compiled.

SPEC CPU2017. We use SPEC CPU2017 [33] version 1.1.9. We report the results of 12 C/C++-only tests.

Results. On the AARCH machine, we exclude Guarder from analysis, noticing that Guarder relies on AES-NI [17], , which is not available on AARCH machines.

For the SPEC benchmark, on the x86 machine, S2MALLOC introduces 11.5% run-time overhead, smaller than two allocators - SlimGuard and DieHarder- and the least memory overhead at 37.4%. On the AARCH machine, S2MALLOC introduces a similar 15.5% run-time overhead, larger than the SlimGuard since SlimGuard fails to run tests with frequent heap memory management operations. Running the PARSEC benchmark gives similar results, with smaller memory and run-time overheads (Table 4).

Table 4. Normalized overheads on SPEC and PARSEC benchmarks.

	Run-time Overhead				Memory Overhead			
	x86		AARCH		x86		AARCH	
	SPEC	PARSEC	SPEC	PARSEC	SPEC	PARSEC	SPEC	PARSEC
S2malloc	**12%**	**2.8%**	**16%**	**1.8%**	**37%**	**27%**	**38%**	**28%**
SlimGuard	17%	4.4%	7.7%	2.6%	57%	23%	57%	24%
DieHarder	31%	2.1%	–	2.5%	59%	21%	–	21%
Guarder	3.5%	2.4%	–	–	56%	58%	–	–

Table 5. Normalized run-time and memory overheads of running mimalloc-benchmark.

	Run-time Overhead	Memory Overhead
S2MALLOC	189%	343%
SlimGuard	298%	250%
DieHarder	229%	92%
Guarder	56%	980%

Table 6. Normalized overhead changes compared with the default settings.

	Δ Memory Overhead	Δ Run-time Overhead
0 Nearby	−0.05%	−0.42%
4 Nearby	−0.13%	+0.4%
4B Random N	−12.35%	−0.33%
12B Random N	+68.23%	+0.73%
50% Entropy	+9.57%	+0.49%

6.2 Micro Benchmarks

To further understand the overheads, we investigate its performance using mimalloc-bench [7], composed of real-world and calibrated programs that frequently allocate and free heap memory. Results are shown in Table 5.

All secure memory allocators incur larger overheads compared to running real-world (see Sect. 6.3) or general-purpose benchmarks as mimalloc-bench tests operate the heap memory in a biased frequent way, and some tests (e.g., "leanN" generates the largest run-time overhead with S2MALLOC) count the CPU ticks instead of seconds of finishing each call.

We take one test, `glibc-simple`, originally from glibc-bench [1], to further investigate the two most common heap object management functions – `malloc()` and `free()`. The test times the execution of allocating and freeing a large number of blocks of a given size. We modify the benchmark and monitor the execution

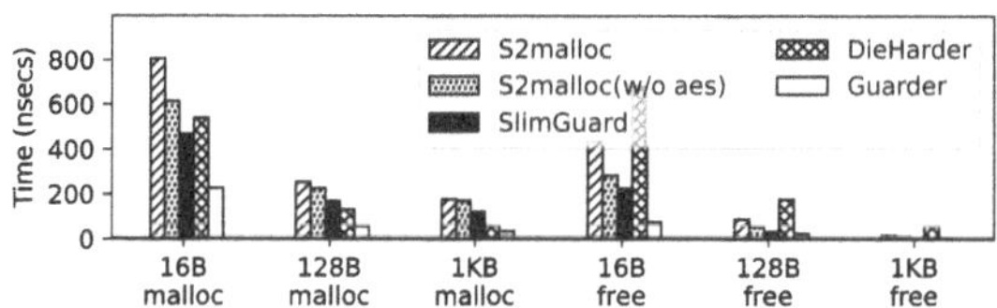

Fig. 3. Execution times of glibc-simple.

Table 7. Real-world programs' memory usage (KB) and throughput (requests/second).

	Nginx			Lighttpd			Redis	
	Throughput	Memory	p50	Throughput	Memory	p50	Throughput	Memory
S2MALLOC	9705	7867	56	11050	9425	44	218460	44689
Guarder	9496	13724	52	11146	11088	44	221245	46545
SlimGuard	6159	4935	81	11153	6428	44	219986	47398
DieHarder	8769	7396	57	11128	13626	44	221734	52419
Glibc	9564	3400	52	10742	5069	44	218156	50909

time of calling `malloc()` and `free()` separately. To investigate the time consumption of different sizes, we vary the block size `S` to be 16B, 128B, and 1KB and change the number of allocated blocks `N` correspondingly so that the total allocation size (`N * S`) is always 1000 MB following the benchmark design of SlimGuard [20]. Results are presented in Fig. 3, and is the average of 100 runs.

Generally, S2MALLOC takes more time to execute `malloc()` than other compared allocators and takes less time to execute `free()` than DieHarder but longer time than Guarder and SlimGuard. However, a significant overhead comes with our cryptographically secure canary implementation, which all allocators should adopt. Although using hardware acceleration, the canary value is computed in each `malloc()` and `free()` call, introducing a nonnegligible computation tax. After disabling this feature and using a fixed value as the canary, following Guarder's implementation, the execution time of both calls is close to other allocators, and the increased overhead is expected as S2MALLOC introduces extra security guarantees. For example, in 16B `malloc` and `free` calls, S2MALLOC is 31% and 26% slower than SlimGuard respectively.

6.3 Performance on Real-World Programs

To evaluate the performance of S2MALLOC in real-world environments, we run Nginx (1.18.0), Lighttpd (1.4.71), and Redis (7.2.1) on the x86 machine. We use ApacheBench (ab) [35] 2.3 to test the throughput and delays using the Nginx default root page of 613 bytes as the requested page to minimize the I/O times. We send 500 concurrent requests, which reaches the maximum capacity of our server. On Redis, we use the same settings as mimalloc-bench [7]. Results are in Table 7. Applying S2MALLOC on these programs results in minimal throughput influence (even better throughput on Nginx and Redis). Running S2MALLOC delays the request response time on Nginx but not Lighttpd. Applying all compared allocators increases memory consumption.

6.4 Influence with Different Parameters

In addition to the default settings, we also measure how different parameters, namely, nearby checking range, random offset entropy, and RIO entropy, influence the run-time and memory overheads. We take 0 and 4 blocks for the nearby checking range compared to the default setting: 2. We take 4 bytes and 12 bytes for the random allocation entropy compared to the default 8 bytes. For random offset entropy, we reserve 50% of the block size for RIO compared to the default 25%. Table 6 shows how different parameters influence the overheads.

Changing the nearby checking range does not introduce observable differences in the memory overhead. The introduced delta is likely due to server fluctuations. Using a larger nearby checking range introduces a greater run-time overhead, as S2MALLOC needs to compute and check more canary values. A larger random allocation entropy or RIO introduces larger memory and run-time overheads.

7 Discussion

Limited UAF-Read Protection. An attacker can always read through a dangling pointer (unless the page is unmapped) and none of the entropy-based allocators, including S2MALLOC, can prevent this. S2MALLOC mitigates UAF-read exploits based on the assumption that the attacker cannot distinguish memory content stored in target data fields from the content stored in other data fields. However, in the scenario where the target object contains a field that the attacker could craft, e.g., a marker value such as `0xdeadbeef`, the target field can be located by probing the marker via UAF read(s). Mitigation of UAF-read exploits can be achieved completely with allocators in categories A, B, C, and partially with allocators in category D (see Sect. 2.2).

Increasing Canary Checking Frequency. As the core of detecting invalid UAF writes, S2MALLOC checks the nearby canaries when a block is freed. While this design could virtually cover all memory slots for a program with frequent heap operations, UAF attack attempts cannot be detected if there is no `malloc` call that triggers S2MALLOC to check FBC. Executing FBC checks in other memory operations (e.g., `free`) is one option to increase checking frequency. Another option is to integrate FBC checks into program logic. Both options require a careful balance between checking frequency (i.e., security) and overhead.

8 Conclusion

While statistically effective against all common heap exploitation techniques, state-of-the-art entropy-based heap allocators are not tailored to *actively detect* unsuccessful exploitation attempts. As a result, to beat a randomization-based moving-target scheme, an attacker can simply launch the same attack repeatedly, potentially with heap spraying, until success.

In this paper, we present S2MALLOC to fill the gap of exploitation attempt detection without compromising security and performance. In particular, we introduce three novel primitives to the design space of heap allocators: random in-block offset (RIO), free block canaries (FBC), and random bag layout (RBL). Combined with conventional BIBOP-style random allocation and heap canaries, S2MALLOC can maintain at least the same level of protection against other heap exploitations (e.g., overflows) and yet still achieves 69% and 96% protection rate in two attack scenarios, respectively, against UAF exploitation attempts targeting a 64 bytes object, while only incurs marginal performance overhead, making S2MALLOC practical to even production systems.

References

1. Standalone Glibc-Benchtests. https://github.com/xCuri0/glibc-benchtests
2. Ainsworth, S., Jones, T.M.: MarkUs: drop-in use-after-free prevention for low-level languages. In: Proceedings of the 41st IEEE Symposium on Security and Privacy (Oakland), San Francisco, CA (2020)
3. Akritidis, P.: Cling: a memory allocator to mitigate dangling pointers. In: 19th USENIX Security Symposium, pp. 177–192 (2010)
4. ARM Developer: Neon. https://developer.arm.com/Architectures/Neon
5. Bienia, C., Kumar, S., Singh, J.P., Li, K.: The PARSEC benchmark suite: characterization and architectural implications. Technical report (2008)
6. Cowan, C., et al.: StackGuard: automatic adaptive detection and prevention of buffer-overflow attacks. In: Proceedings of the 7th Conference on USENIX Security Symposium (1998)
7. daanx: mimalloc-bench—Suite for benchmarking malloc implementations
8. Dang, T.H.Y., Maniatis, P., Wagner, D.: Oscar: a practical page-permissions-based scheme for thwarting dangling pointers. In: Proceedings of the 26th USENIX Conference on Security Symposium (2017)
9. Dhurjati, D., Kowshik, S., Adve, V., Lattner, C.: Memory safety without runtime checks or garbage collection. In: Proceedings of the 2003 ACM SIGPLAN Conference on Language, Compiler, and Tool for Embedded Systems (2003)
10. Erdős, M., Ainsworth, S., Jones, T.M.: MineSweeper: a "clean sweep" for drop-in use-after-free prevention. In: Proceedings of the 27th ACM International Conference on Architectural Support for Programming Languages and Operating Systems (2022)
11. Farkhani, R.M., Ahmadi, M., Lu, L.: PTAuth: temporal memory safety via robust points-to authentication. CoRR (2020)
12. Filardo, N., et al.: Cornucopia: temporal safety for CHERI heaps. In: Proceedings of the 41st IEEE Symposium on Security and Privacy. IEEE Computer Society (2020)
13. Gorter, F., Koning, K., Bos, H., Giuffrida, C.: DangZero: efficient use-after-free detection via direct page table access. In: Proceedings of the 29th ACM Conference on Computer and Communications Security (CCS) (2022)
14. GrapheneOS: GrapheneOS: the private and secure OS. https://grapheneos.org/
15. GrapheneOS: Hardened_malloc
16. Hanson, D.R.: A portable storage management system for the ICON programming language. Softw.: Pract. Exp. **10**(6), 489–500 (1980)
17. Intel Corporation: Intel Advanced Encryption Standard Instructions (AES-NI)
18. Lee, B., et al.: Preventing use-after-free with dangling pointers nullification. In: Proceedings of the 2015 Annual Network and Distributed System Security Symposium (NDSS) (2015)
19. Liljestrand, H., Gauhar, Z., Nyman, T., Ekberg, J.E., Asokan, N.: Protecting the stack with paced canaries. In: Proceedings of the 4th Workshop on System Software for Trusted Execution (2019). https://doi.org/10.1145/3342559.3365336

20. Liu, B., Olivier, P., Ravindran, B.: SlimGuard: a secure and memory-efficient heap allocator. In: Proceedings of the 20th International Middleware Conference (2019)
21. LLVM Project: Clang: a C language family frontend for LLVM
22. LLVM Project: Scudo Hardened Allocator
23. Martin, J.P., Kandasamy, A., Chandrasekaran, K.: Exploring the support for high performance applications in the container runtime environment. Hum.-Cent. Comput. Inf. Sci. **8**(1) (2018)
24. Mashtizadeh, A.J., Bittau, A., Boneh, D., Mazières, D.: CCFI: cryptographically enforced control flow integrity. In: CCS 2015, New York, NY, USA (2015)
25. National Vulnerability Database: 2022 CWE Top 25 Most Dangerous Software Weaknesses. https://cwe.mitre.org/top25/archive/2022/2022_cwe_top25.html
26. Nikiforakis, N., Piessens, F., Joosen, W.: HeapSentry: kernel-assisted protection against heap overflows. In: Detection of Intrusions and Malware & Vulnerability Assessment (2013)
27. Novark, G., Berger, E.D.: DieHarder: securing the heap. In: Proceedings of the 17th ACM Conference on Computer and Communications Security (CCS) (2010)
28. Qiang, W., Li, W., Jin, H., Surbiryala, J.: Mpchecker: use-after-free vulnerabilities protection based on multi-level pointers. IEEE Access **7**, 45961–45977 (2019)
29. Shellphish: shellphish/how2heap: A repository for learning various heap exploitation techniques. https://github.com/shellphish/how2heap
30. Silvestro, S., Liu, H., Liu, T., Lin, Z., Liu, T.: Guarder: a tunable secure allocator. In: Proceedings of the 27th USENIX Security Symposium (Security) (2018)
31. Singh, S., Awasthi, M.: Memory centric characterization and analysis of spec CPU2017 suite. In: Proceedings of the 2019 ACM/SPEC International Conference on Performance Engineering, ICPE 2019. ACM (2019)
32. Sotirov, A.: Heap Feng Shui in Javascript. Black Hat Europe (2007)
33. Standard Performance Evaluation Corporation: SPEC CPU® 2017
34. Teuchert, D., Aschermann, C. and Frassetto, T., Abera T.: Use after free in File#initilialize_copy (2018). https://github.com/mruby/mruby/issues/4001
35. The Apache Software Foundation: Apache HTTP Server Documentation: ab - Apache HTTP Server Benchmarking Tool (2023)
36. The Chromium Projects: Memory safety. https://www.chromium.org/Home/chromium-security/memory-safety/
37. The GNU Project: Aligned Memory Blocks (The GNU C Library). https://www.gnu.org/software/libc/manual/html_node/Aligned-Memory-Blocks.html
38. Van Der Kouwe, E., Kroes, T., Ouwehand, C., Bos, H., Giuffrida, C.: Type-after-type: practical and complete type-safe memory reuse. In: Proceedings of the 34th Annual Computer Security Applications Conference (ACSAC) (2018)
39. Van Der Kouwe, E., Nigade, V., Giuffrida, C.: DangSan: scalable use-after-free detection. In: Proceedings of the Twelfth European Conference on Computer Systems, pp. 405-419. Association for Computing Machinery (2017)
40. Wang, R., Xu, M., Asokan, N.: SeMalloc: semantics-informed memory allocator (2024)
41. Wickman, B., et al.: Preventing use-after-free attacks with fast forward allocation. In: Proceedings of the 30th USENIX Security Symposium (Security) (2021)
42. Younan, Y.: FreeSentry: protecting against use-after-free vulnerabilities due to dangling pointers. In: Network and Distributed System Security Symposium (2015)

Acoustic Side-Channel Attacks on a Computer Mouse

Mauro Conti[1,2], Marin Duroyon[2], Gabriele Orazi[1,4(✉)], and Gene Tsudik[3]

[1] University of Padua, Padua, Italy
mauro.conti@unipd.it, gabriele.orazi@phd.unipd.it
[2] Delft University of Technology, Delft, The Netherlands
[3] University of California, Irvine, USA
gts@ics.uci.edu
[4] FDM Business Services, Milan, Italy

Abstract. Acoustic Side-Channel Attacks (ASCAs) extract sensitive information by using audio emitted from a computing devices and their peripherals. Attacks targeting keyboards are popular and have been explored in the literature. However, similar attacks targeting other human-interface peripherals, such as computer mice, are under-explored. To this end, this paper considers security leakage via acoustic signals emanating from normal mouse usage.

We first confirm feasibility of such attacks by showing a proof-of-concept attack that classifies four mouse movements with 97% accuracy in a controlled environment. We then evolve the attack towards discerning twelve unique mouse movements using a smartphone to record the experiment. Using Machine Learning (ML) techniques, the model is trained on an experiment with six participants to be generalizable and discern among twelve movements with 94% accuracy. In addition, we experiment with an attack that detects a user action of closing a full-screen window on a laptop. Achieving an accuracy of 91%, this experiment highlights exploiting audio leakage from computer mouse movements in a realistic scenario.

Keywords: Cybersecurity · Human Interface Devices · Computer Mouse · Acoustic Signals · Acoustic Side-channel · Input Devices · Audio Leakage · Machine Learning · Mouse Movement

1 Introduction

In general, Side-Channel Attacks (SCAs) exploit unintended leaks, such as patterns of power consumption [12,14], electromagnetic emissions [19], visual information (e.g., from screen content or flickering), or audio (e.g., noise from speakers, CPU or disk) from computers or their peripherals to extract sensitive information. Of those, ASCAs typically focus on Human Interface Devices (HIDs) – mechanical devices that produce sound when used by humans. A notable example keyboard acoustic emanations. Characteristic clicking sounds of a keyboard,

F. Maggi et al. (Eds.): DIMVA 2024, LNCS 14828, pp. 44–63, 2024.
https://doi.org/10.1007/978-3-031-64171-8_3

generally considered innocuous, can be analyzed to infer user keystrokes. Such attacks occur in the presence of an attacker-placed (or attacker-compromised) microphone near the target device. While considerable research has been conducted on keyboard-based attacks [3], ASCAs on mouse movements remain relatively unexplored.

Since computer mice are a very popular means of interaction with laptop and desktop computers, this work seems to understand potential security vulnerabilities associated with ASCAs. Our objective is to evaluate the correlation between acoustic signals and mouse movements thus assessing the feasibility of mouse-based SCAs. We begin formulating three research questions:

RQ1. *Can mouse sounds leak its motion path and/or activity?*
RQ2. *If so, with what accuracy?*
RQ3. *In which real-world scenarios do mouse-based ASCAs pose a security risk?*

This paper is organized as follows: Sect. 2 explores related work. Next, Sect. 3 details our research methodology for the experiments. Then, Sect. 4 describes the core experiments, followed by Sect. 5, which discusses security implications of the ASCA. Section 7 concludes the paper and outlines some directions for future works.

2 Related Work

SCAs exploit unintended information leakage from computing devices or their peripherals. For instance, a pattern power consumption of a USB-powered device can be used to profile its activity [18]. SCA can even be used to infer the structure of a neural network [4]. This section overviews ASCAs – the category of SCA that subsumes our work.

Passive ASCAs operate by interpreting audio emanations. The present work falls into this category since the equipment we use (microphone) captures audio without any interactions with the target computer, its peripherals or the user.

Gupta *et al.* [11] use audio Time Difference of Arrival (TDoA) at two different points (using two microphones on the same device) to infer the location of the sound source. Similar studies [15,22] explore various methods to recover keystrokes from acoustic emanations. Balagani *et al.* [2] present a novel ASCA on ATM PIN entry, called the PinDrop attack. It involves two steps: (1) an acoustic profile is created for each key on the target PIN pad, and (2) the attacker records audio emitted by each pressed key during PIN entry and compares these recordings to the acoustic profiles to identify the keys pressed. The resulting classifier is tested on a dataset comprising ten samples for all 26 English alphabet keys, achieving 90.61% accuracy.

Cecconello *et al.* [3] present an attack on popular Voice-over-IP (VoIP) software (Skype) that captures and transmits all acoustic emanations, including keyboard sounds. This attack can lead to leakage of sensitive information, such as passwords, that a victim user might type during a Skype call. Given some knowledge of the victim's typing style and the keyboard model, an attacker can achieve a top-5 accuracy of 91.7% in learning a random key pressed by the victim. We note that [3] is relevant to our work since the keyboard is typically used in tandem with the mouse and our attack scenario is comparable to that in [3].

PoKeMon [9] is a new keystroke monitoring method for smartphones using Mel-Frequency Cepstral Coefficient (MFCC) [21]. Nandakumar *et al.* [16] propose an innovative finger tracking technique by transforming a smartphone into an active sonar system. The system transmits inaudible sounds in the 18–20 kilohertz (kHz) range and tracks the echoes of the finger with its microphones. The system achieves 2-D finger tracking accuracy of 8 millimeters (mm) at 169 frames/sec with a smartphone prototype.

Cheng *et al.* [6] use device speakers to emit inaudible acoustic signals, which are then reflected off the user's fingers and recorded by the smartphone microphones. Similarly, Orthogonal Frequency-Division Multiplexing (OFDM) [8] sounds emitted from device's speakers, while the microphones on the same device are then used to record the echos of these sounds. Also, [6] demonstrates that the number of unlock patterns an attacker must try until a successful authentication can be reduced by up to 70% using this ASCA.

The two results most relevant to our work are Synesthesia [10] and Behavicker [5]. In the former, Genkin *et al.* examine how ASCAs can leak screen content via audio emanations, discovering that LCD screens emit content-dependent audio signals which can be captured by nearby microphones. Experiments conducted using both built-in and webcam microphones reveal acoustic leakage from audio recordings and video-conferencing, e.g., Google Hangouts. The authors determine that the power consumption of the monitor's circuits changes depending on the screen content, causing internal components to vibrate and emit sounds. The authors iteratively simulate the acoustic leakage during hundreds of key presses on the on-screen keyboard, which lead to a 100% accuracy rate in detecting key presses. In their attempt at text extraction they achieve an accuracy rate between 88% to 98% for most individual characters. When extracting words from the screen, the correct word is identified in the top five most probable words in 72% of the cases.

The goal of Behavicker (Chen *et al.* [5]) is to determine user activities from keyboard and mouse clicking and scrolling events. Sounds produced by keyboard typing or mouse clicking reveal information about users' behavior. Actions, such as browsing a news website or playing a video game, produce distinct keyboard and mouse usage patterns, each with its own unique acoustic footprint. Behavicker identifies six basic interaction events with a 88.3% accuracy and differentiates between seven computer-usage activities with a 82.7% accuracy. It utilizes two functional modules: Acoustic-based Interaction Event Recognition and Computer-Usage Recognition. The first uses signal processing and ML to recognize interaction events, while the second employs hierarchical classifiers via time-series analysis to distinguish between types of activities.

Research Gap: Our review of related work reveals an appreciable research gap with respect to the analysis of audio leakage from mouse activities. However, mouse usage remains widespread and, similar to keyboards, mice produce a litany of sounds. Therefore, security of mouse movements deserves and needs to be studied. This is the key motivation for our work.

3 Methodology

Our work involves multiple phases based on research questions posed in Sect. 1. The initial phase is an investigation of mouse-based acoustic leakage. The second phase is the refinement of our predictive models. The third and final phase validates these models in real-world scenarios by recording and analyzing audio from mouse activity in more complex and realistic settings, such as an external mouse used on a laptop.

We do not assume that all mouse activities reflect typical user behavior. Moreover, since this research is exploratory, the datasets of samples consisted of one participant. Since collecting these datasets take several days and do not always provide fruitful results, we conducted initial tests with a single sample size. Thus, while we initially ignored plausibility of mouse movements and sample sizes, we considered this later on in experiments exploring real-world security risks.

We emphasize that all experiments aim to infer the direction of the current mouse movement. This does not yield pixel-level precision, partly because doing so would require taking into account the resolution of the monitor, the model and sensitivity of the mouse as well as many other environmental factors.

All ML models are trained on a Dell laptop with Intel(R) Core(TM) i7-12700H 2.30 GHz, and NVIDIA GeForce RTX 3050, with 16 GB of RAM.

3.1 Mouse and Mouse Pad

In this study, the *'mouse pad'* is the mouse movement area, either an exterior mouse pad or the flat surface near the laptop, upon which the latter rests. Initially we experimented with various mouse pads, discovering that, if audio signals are "audible" by the microphone it yields the same experimental results. Similarly, the experimental mouse is a HP X500 mouse, which is a typical commodity office-style model. For the sake of consistency, the same mouse and mouse pad are used for all experiments.

We use a coordinate system on the mouse pad to categorize mouse movements. Directional movements are defined relative to specific points on the pad: 'T' for top, 'B' for bottom, 'L' for left, and 'R' for right. Also, 'M' represents the middle along the y-axis, while 'C' indicates the center along the x-axis. Therefore, a movement labelled 'TL $\rightarrow$ BR' means a diagonal trajectory from the top-left to the bottom-right corner of the pad.

3.2 Audio Recording Methods

Each experiment uses different recording methods based on the objective. Key parameters of audio recordings, such as sample rate, bit depth, channels, and file format, play a crucial role in recorded data. Sample rate, typically measured in kHz, defines the number of samples captured per second, where higher rates allow for a more accurate representation of the original sound [7]. Moreover, according to [7], bit depth determines the resolution of the amplitude of each audio sample. This influences the noise level of the recording. The number of channels affects the spatial representation of the sound: mono contains only one waveform, while stereo contains two.

Our experiments use the Waveform Audio File Format (WAV), due to being uncompressed, thus ensuring high fidelity of recordings [7]. We record in mono for experiments utilizing a single microphone since the sound comes from a single source, switching to a stereo configuration in scenarios where a second microphone is used. Bit depth selected for each experiment varies from 16 to 32 bits. Sample rate ranges from 44.1 kHz to 48 kHz in the smartphone range.

In most of our experiments, the microphone is directly connected to the computer. To record with a second microphone, using a smartphone, we use an Android application 'AudioRec' (available in the Google PlayStore[1]).

4 Acoustic Analysis of Mouse Movements

Recall that our goal is to find a correlation between audio signals and mouse movements. While existing literature studied the inference of clicking and scrolling events via acoustic signals, the potential for inferring mouse movements from these emissions remains unexplored. This section addresses the central research question:

> *Can the sounds generated by the movements of a computer mouse be used to infer its trajectory?*

We initially conduct experiments using a single microphone, focusing on identifying and categorizing distinct movement patterns. Next, we integrate a second microphone to further expand the potential of the attack while maintaining a viable attack model.

4.1 Single Microphone Analysis

Audio Capture via Processing Software. In the first phase, we assess whether an acoustic leakage model could distinguish between four basic directional movements of a mouse: up, down, left, and right. Specifically, the experiment is to determine the feasibility of detecting mouse movements along four directions using acoustic signals. A microphone is positioned on the left side of a right-handed mouse pad. The mouse is then moved repetitively along the 'X' and 'Y' axes. The movements are captured using experimental Java Processing Sketch software designed to record these events. As seen in Fig. 1, the starting point (in green) triggers the beginning of the recording process, while the

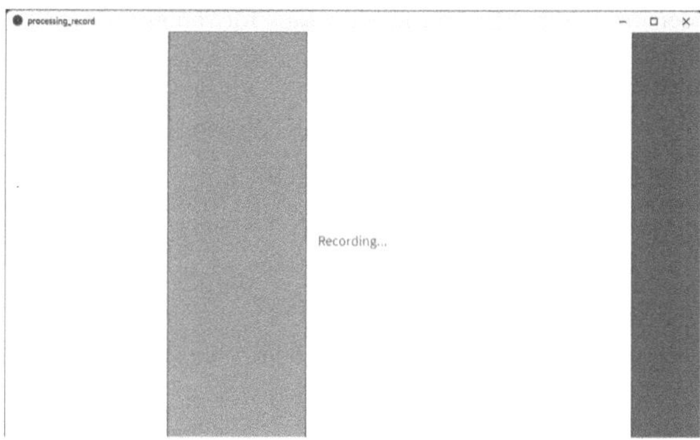

Fig. 1. Recording phase of Left to Right movement. The user has to move the cursor from the green area to the red one. (Color figure online)

red-colored end-point stops it. As a result, the Java Processing Sketch returns acoustic representations of mouse movements in various directions.

We collected 6,000 samples for each direction, yielding a total of 24,000 samples. We note that acoustic signal data-frames, due to the characteristics of the recording procedure, are not of a uniform size. Samples are pre-processed and cleaned for a Convolutional Neural Network (CNN) [17] ML model. We process the waveform of noises to obtain MFCCs using the *Sliding Windows* method with a size of 36 milliseconds (ms), which lead to the best result. At this point, we end up with two lists: (1) one of MFCCs and (2) the other – of encoded labels from 0 to 3 corresponding to each direction. The list is split into training and testing samples, with the testing dataset taking up 35% of the original dataset size. The data are entered in a CNN model, compiled with the default Adam optimizer and categorical cross-entropy as a loss function using ten epochs.

While this model may appear complex, the experiment shows no signs of over-fitting [20], as confirmed by the loss functions in Fig. 2. We also verified lack over-fitting with a second validation dataset split before any computations. After several checks, we deem the results of this experiment as reliable and representative. This model demonstrates a high level of confidence, achieving a 98% classification accuracy rate, with F1-scores ranging between 97% and 99% for each category. These performance metrics are reflected in Table 1.

This study highlights the viability of ASCA. While the current experiment is limited due to simple and constrained movements, it still serves as a proof-of-concept, motivating further research.

Continuous Audio Capture During Mouse Movements. The following experiment we remove the recording software of the previous experiment and use continuous audio capture corresponding to a more realistic setup. This means that the mouse moves on the desktop without predefined patterns and the audio

[1] https://play.google.com/store/apps/details?id=com.audioRec&hl=en_US&gl=US.

Table 1. Classification Report for classifying four movement categories using Processing Record script.

Class	Precision	Recall	F1-Score	Support
Up	0.97	0.97	0.97	2093
Left	0.99	0.99	0.99	2084
Down	0.99	0.99	0.99	2144
Right	0.97	0.97	0.97	2079
Accuracy			0.98	8400

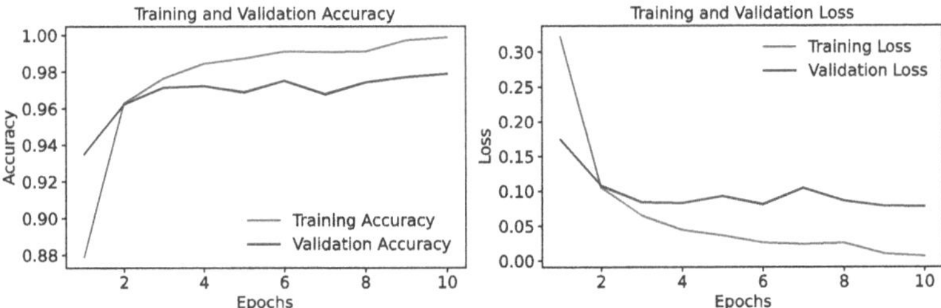

Fig. 2. Accuracy and loss graph for classifying four movement categories using Processing Record script.

is recorded simultaneously. In order to set a direction vector and a movement angle corresponding to movements in an audio frame, we synchronized the background recording with the mouse movements.

The experiment consists of repeating mouse movements in cardinal directions. The mouse coordinates are then converted to direction vectors and associated with acoustic signal frames. The size of each acoustic signal chunk is formed by 8,192 samples with a frame rate of 44.1 kHz, resulting in approximately 186 milliseconds. For classification, the angles are categorized into four (cardinal) directions using the angle representation of the direction vector, meaning that 25% of the circle spatial dimension was associated to a single label. For instance, with a clockwise circle with 0° at the top (north), the angles between −45° and 45° corresponded to the up (or north) direction category.

Computation of MFCCs and their normalization are the only pre-processing steps required and are computed using the same methods as in the previous experiment. Using ten epochs with the same CNN model, we observe a plateau of the validation loss, which leads us to assume that the ML model stops improving. Based on the batch size of 32 data points and the *"sum_over_batch_size"* reduction method during loss calculation, we obtain a loss value of 0.8, which is higher than the average loss for each data point. With this setting, we achieved the overall accuracy of 74%. The classification report can be seen in Table 2, along with the accuracy and loss function graph in Fig. 3.

Table 2. Classification Report for four area classification using background recording.

Class	Precision	Recall	F1-Score	Support
Up	0.69	0.74	0.72	755
Left	0.80	0.68	0.73	852
Down	0.77	0.79	0.78	698
Right	0.72	0.78	0.75	765
Accuracy			0.74	3070

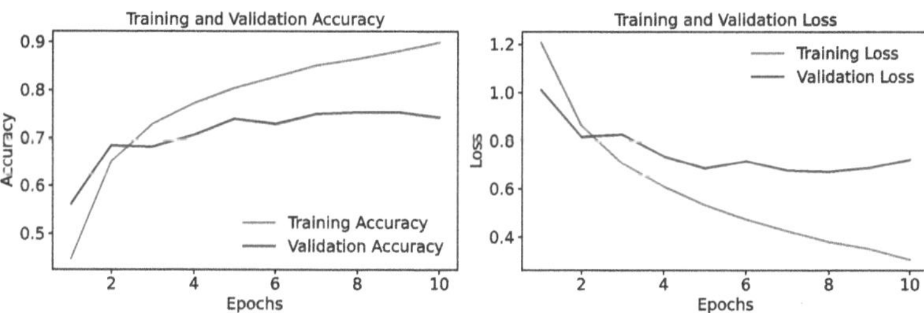

Fig. 3. Accuracy and loss graph for four area classification using background recording.

The accuracy is lower than that obtained with the previous experiment since the recording setup has a greater degree of freedom. With 74% accuracy we can state with confidence that movements in four directions leak enough audio information to determine some user activities. Using a pipeline of a Standard Scaler to normalize, Principal Component Analysis (PCA) and a Support Vector Machine (SVM), we obtain the accuracy of 83%. Consequently, this section answers the leakage model research question: There is indeed an acoustic leakage model for SCA, and it is possible to use audio to infer mouse movements.

Limitation of Single Microphone Approach. Previous experiments show acoustic leakage that can be split into four categories, However, it does not represent realistic mouse movements. Of course, the idea is to increase the level of granularity in order to understand what angles are discernible with a single microphone and the ML algorithm.

We now evaluate our model to distinguish between eight directions simultaneously, doubling the bins of the previous classifier (classes from 0 to 360° with a step of 45°). The classification report shows that the model has an accuracy of at most 15%. Despite employing the ML framework similar to prior successful models, the algorithm's performance with the eight different categories is sub-optimal.

With different ML models and multiple datasets that cover various environments, the results are still falling short: from Support Vector Regression (SVR) to more complex CNNs. (These experiments are not described due to space limitations).

No model could differentiate between eight movement angle categories. It thus becomes clear that a singular microphone is a limiting factor in capturing similar mouse movements. This prompts us to explore other methods.

Rather than classifying angles into bins, we attempt to apply a regression learning algorithm to predict the angle continuously. The data points are recorded using a similar approach adopted in Sect. 4.1, though using a circular pattern. The software, shown in Fig. 4, initiates the recording when the mouse reaches the middle of the screen (i.e., a red square) and stops it when the green target box is touched by the cursor. The green boxes are always placed on top of the circumference depending on a certain angle with respect to the starting point, i.e., the circle center.

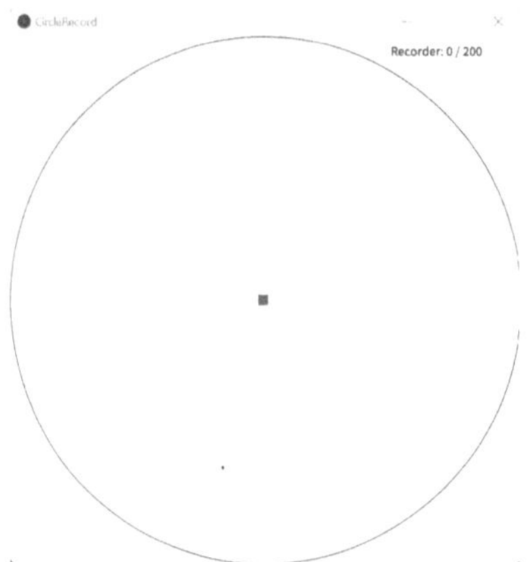

(a) Move cursor towards red center to start the recording.

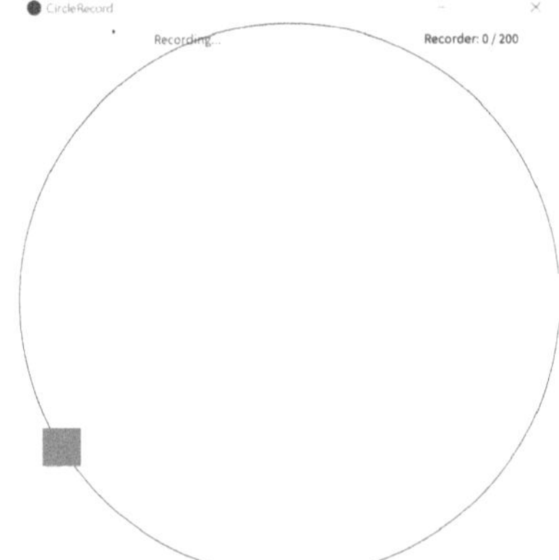

(b) Move cursor towards green box to stop the recording.

Fig. 4. Circular recording method. (Color figure online)

The CNN model comprises convolutional layers and fully connected layers to map to the target angle prediction. We use a L1 loss function, which simply calculates the mean absolute difference, to approach the target angle as closely as possible. The dataset is composed of ~21,000 samples of audio MFCCs and mouse movement angles. The results clearly show randomness in the angle predictions. Thus, we believe that predicting angles from 0 to 360 degrees is complicated due to the circular nature of angle measurements. For instance, angles close to 0 degrees and 360 degrees are nearly identical in orientation but are numerically distant. This discontinuity, also known as "Angle Periodicity" [13], can complicate regression models. To address the issue, we conduct a new experiment where angles are represented using their cosine and sine values [13]. However, the average angle difference between the model output and the real angle is 88.79°, which is close to the random angle choice.

To rule out the possibility that the size of the dataset compromises the outcome of the regression, we test data augmentation by adding white noise. However, the model clearly shows signs of over-fitting on randomized augmented data.

The series of regression experiments conducted using a single microphone yield insights into the possibility of accurately predicting continuous mouse movement angles. However, a single microphone is insufficient to regress the angle of mouse movements.

4.2 Analyzing Proximity Through Sound Amplitude Variations

The methodology for analyzing acoustic signals requires a new approach. To this end, we examine the amplitude of sound waves to infer the proximity of sound-generating movements to a microphone. Sounds produced nearer the microphone register with higher amplitudes than those originating at a greater distance. This variance in amplitude could provide a metric for determining the relative distance of the sound source along an axis.

The test involves a controlled series of bidirectional movements along the microphone's directional orientation axis with ten samples. During this process, we segment the audio recordings into two categories: from left to right to represent the mouse trajectory away from the microphone and from right to left to denote its approach towards the microphone. Next, we plot the audio amplitude over time. To do so, we segment the sound into discrete windows and calculate the amplitude, which are then be compared and plotted to visualize waveform differences in time. Afterwards, we apply a linear regression analysis to amplitude data points for a better graphical interpretation. This facilitates a general trend which reveals whether there is a consistent increase or decrease in amplitude. The slope of the regression line serves as the indicator of this relationship.

To highlight the outcomes, we incorporate three distinct auditory markers, consisting of three circles done with the mouse at the start and at the end of the recording. This modification aims to extend the time frame in which the microphone captures the audio.

The experimental data in Figs. 5a and 5b illustrate the subtle, yet consistent, trends in sound amplitude corresponding to the mouse movement near the microphone. We obtain a negative slope angle value of $-0.0000099°$ decrement in amplitude when the mouse moves away from the microphone, and a positive slope angle value of $0.0000068°$ in the opposite direction. Although minimal, the differences are consistent across different tests. These outcomes demonstrate the possibility to visually determine the direction of a mouse movement, validating the hypothesis that a distinctive mouse movement along a single axis can be accurately inferred.

4.3 Dual-Microphone Approach for Two-Dimensional Amplitude Analysis

We now introduce a second microphone. This is achieved using two microphones of a smartphone to provide a two-dimensional measurement plane, as shown in Fig. 6.

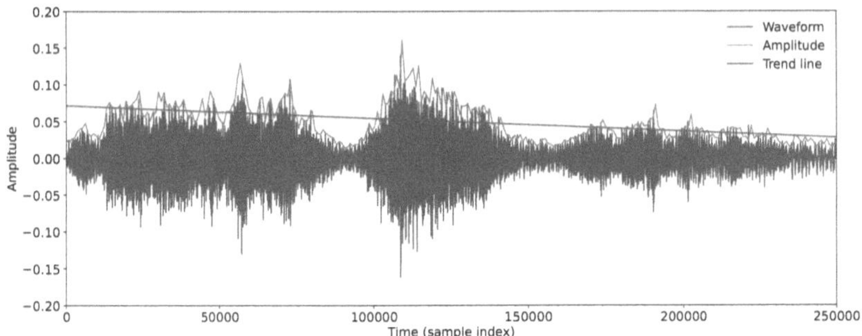

(a) Waveform of the sound moving away from the microphone (descending amplitude).

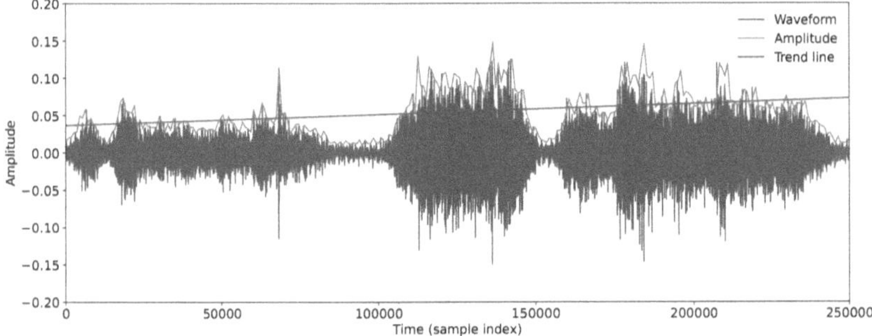

(b) Waveform of the sound moving towards the microphone (ascending amplitude).

Fig. 5. Audio waveforms for the mouse moving towards and away from the microphone.

The recording movement pattern involves traversing the length of the mouse pad, moving from top to bottom and back, and then horizontally from a location near the phone to the far edge and back. Figure 7 shows the resulting graph of two out of four cardinal directions. Each graph shows a plot of the two microphone amplitudes and the amplitude difference. Also, regression lines for the microphone amplitudes are plotted to show general trends. As can be seen in Fig. 7a, representing moves up the mouse pad, a criss-cross effect is evident, indicating one microphone "hearing" louder than the other one at certain times. Furthermore, Fig. 7b shows a negative slope trend in the regression line when moving away from the microphone. Even though graphs are not included in Fig. 7 for the sake of simplicity, we observed coherent behaviour while moving the mouse towards or down with respect to the smartphone.

Since the experiment yield visual representations that clearly distinguish various directional movements, we extend it by attempting to distinguish among ten movement patterns. We record and segment approximately 250 samples for each directional shift using Audacity[2]. The ten categories are a combination of

[2] https://www.audacityteam.org/.

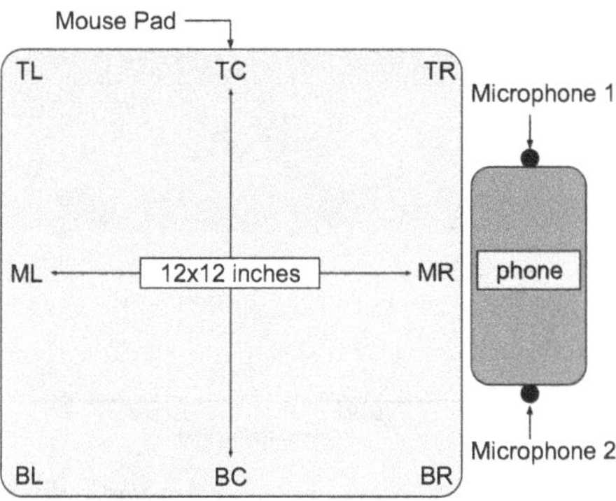

Fig. 6. Dual-Microphone experimental setup.

movements with specific start and ending points of the mouse pad, where each point is defined by a vertical and horizontal location. These points are shown in Fig. 6.

Table 3. Classification report on ten movements with dual-microphone setup.

Class	Precision	Recall	F1-Score	Support
BC→TC	0.91	0.94	0.92	52
BL→BR	1.00	1.00	1.00	50
BR→BL	1.00	0.96	0.98	50
BR→TR	1.00	0.97	0.99	39
ML→MR	0.83	0.88	0.85	49
MR→ML	0.88	0.86	0.87	49
TC→BC	0.98	0.98	0.98	52
TL→TR	0.98	1.00	0.99	49
TR→BR	0.97	0.92	0.95	39
TR→TL	1.00	1.00	1.00	49
Accuracy			0.95	478

Using a Random Forest (RF) Classifier with 100 estimators, the model achieves a high accuracy of 95.19%, indicating good performance across various movement categories. The classification report in Table 3 provides further insights into the model's efficacy.

In summary, this experiment demonstrates the ability to infer ten distinct two-dimensional movements, thus exposing the ability to understand the leakage model of a mouse using a smartphone.

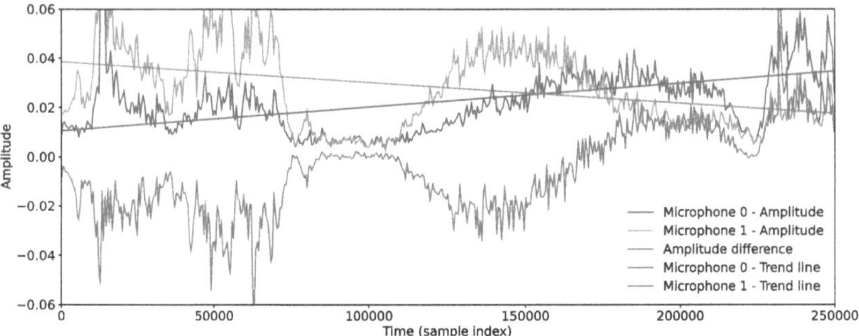

(a) Amplitude difference and trend when moving vertically up the table.

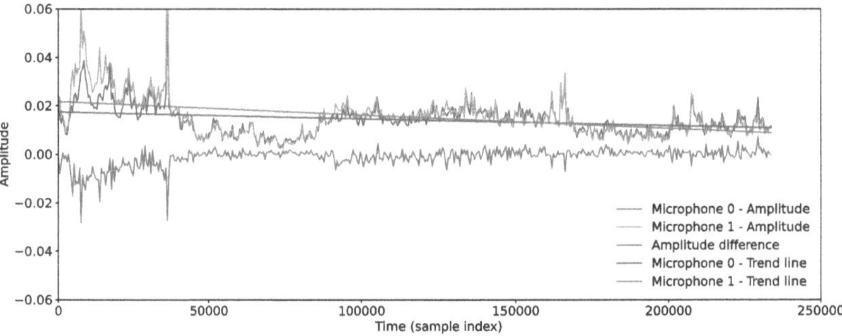

(b) Amplitude trend with a negative slope, indicating movement away from the phone.

Fig. 7. Graphs showing amplitude differences and trends based on the mouse's movement.

5 Real-World Implications and Security Risks

We now consider practical implications of our results. A typical attack environment could be a shared office (e.g., a cubicle farm), a cafeteria, or a library. A simple smartphone represents the attack vector. Modern smartphones now feature dual microphones, mainly to reduce ambient noise during calls. The smartphone is a perfect attack vector since it can safely be placed near the victim. It requires no interaction in the recording phase and can remain with its screen off the entire time, so as not to arouse suspicion. With this scenario in mind, we now describe experiments with six participants who volunteered to be recorded while performing predefined mouse movements. Based on this, we examine real-world implications of such attacks.

5.1 Experiment with Other Participants

The next phase involves extending the experiments to include a multiple participants. This helps to determine the generalizability of the model and its applicability to various environments. The new experiment involves six participants performing a series of mouse movements on a mouse pad. This setup is identical to the one described in Sect. 4.3.

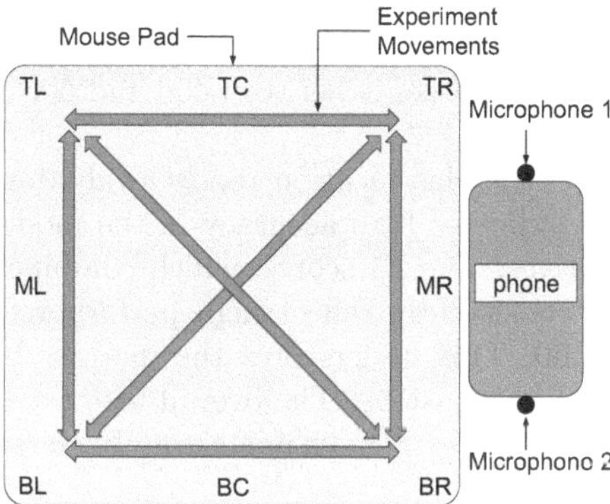

Fig. 8. Mouse pad with mouse movements in the experiment.

Participants are instructed to perform distinct back-and-forth movements along both axes within a 12-by-12 mouse pad placed on an office table. The experiment details specific movement patterns, shown in Fig. 8. Participants are asked to repeat each back-and-forth movement for one to two minutes, resulting in a total experiment time ranging from 6 to 12 minutes. To help with the segmentation during the pre-processing phase, participants are also asked to clap before and after recording each axis to delineate the start and end.

The six participants show variability in their mouse usage methods. For example, some execute the movements rapidly, resulting in a higher frequency of actions, while others move more slowly. In addition, the amount of pressure applied to the mouse differs among participants, producing distinct acoustic signatures. These speed, frequency, and pressure differences are integral to the experiment's design. They provide a diverse dataset that is ideal for testing the generalizability of the ML model, since it reflects a more realistic range of user behavior.

We specifically isolate the segments corresponding to mouse movements using a thresholding technique to distinguish them from background noise, followed by a manual verification process to ensure accuracy. The waveforms from both channels are split into 50 discrete windows and then we calculate their amplitude. Thus, we extract the same number of data points for each movement acoustic signal regardless of different recording methods used by the participants.

Our analysis also involves calculating the difference in amplitudes between the top and the bottom microphone recordings for each segment. Consequently, we generate a "difference amplitude line" of 50 data points representing the differential amplitude for each audio sample. These data points serving as learning values are the 'X' values. The 'Y' values, or target labels, are assigned based on the type of directional movement recorded. This structured dataset, with its distinct X and Y values, allows us to train and test the ML model to classify the directional movements based on acoustic signatures.

This experimental setup is designed to record six types of back-and-forth mouse movements, resulting in 12 distinct directional classes. In the course of the experiments, a total of 5,507 samples are collected. The distribution of these samples across the different classes is between 406 and 506 samples each, with a mean of 456 samples per class.

As shown in Table 4, the classification report evaluates performance of the ML model. The report indicates high accuracy in the model's predictions, with a low rate of false positives. The F1-scores, which combine precision and recall into a single measure, consistently reflect high performance, predominantly in the range of 0.93 to 1.00. This underscores the model's balanced accuracy in both precision and recall dimensions. The overall test accuracy of the model is 96%, indicating the model's effectiveness in accurately classifying the directional movements.

Table 4. Classification Report for a Random Forest Classifier with 100 estimators.

Direction	Precision	Recall	F1-Score	Support
BL→BR	0.98	0.99	0.98	81
BL→TL	0.91	0.98	0.95	97
BL→TR	0.96	0.90	0.93	91
BR→BL	0.99	0.93	0.96	82
BR→TL	0.94	0.99	0.96	91
BR→TR	1.00	0.99	0.99	89
TL→BL	0.92	0.97	0.94	96
TL→BR	0.98	0.93	0.96	92
TL→TR	0.99	1.00	1.00	101
TR→BL	0.93	0.95	0.94	91
TR→BR	0.98	0.93	0.95	90
TR→TL	0.99	0.99	0.99	101
Accuracy			0.96	1102

5.2 Inferring Realistic Mouse Movements

To demonstrate the feasibility of such attacks and help in assessing their potential impact on real-world scenarios, we shift towards a more practical approach. The next experiment aims to determine whether it is possible to detect when a user clicks the 'close' button, typically found at the top right of a Windows laptop screen. While closing a window may not directly reveal sensitive security information, the ability to discern precise user interactions like button clicks might have broader implications for user privacy.

The experiment is structured around two recording sessions of five minutes with a single participant. It is performed on a traditional mouse setting using a Windows laptop and an entry level office mouse pad. In the first session, the participants are instructed to move the cursor from random points on the screen to the top-right corner and click the red 'X', a standard action for closing a window (positive class for the classification). The second session involves recording mouse movements followed by clicks at various random points on the screen – a pattern different from the first recording session (negative class).

We focus on a time window around click events in the analysis phase. This approach allows us to compare acoustic characteristics of mouse movements and clicking sounds associated with both positive and negative classes.

The recordings are first analyzed to determine the clicking events through quantile thresholding methods. To do so, the waveforms are converted to a mono channel to calculate the absolute value of the waveform. This threshold can be modified based on the visual representation and manually adjusted.

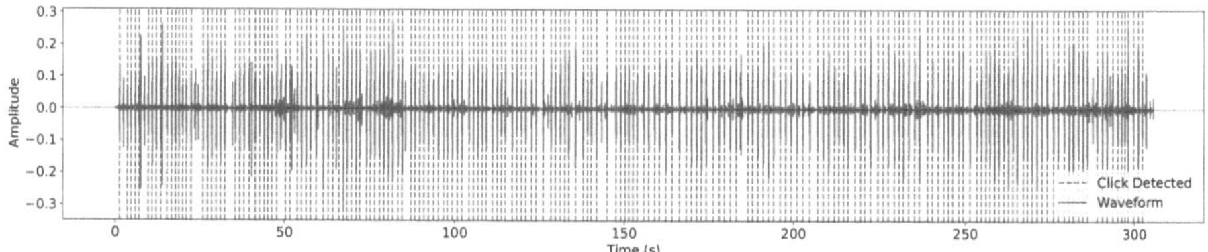

Fig. 9. Detected clicks out of mouse events. (Color figure online)

With these clicking events, the goal is to find the two consecutive peaks corresponding to the mouse press and release events during a mouse click. To do so, we use a minimum and maximum distances between detected peaks to isolate the events that seem to correspond to mouse-clicking events. These distances are again estimated via empirical values, in this case, from 10 ms to 200 ms. Figure 9 shows the waveform with red vertical lines representing detected clicks.

As we iterate through each identified mouse click event, a specific time window surrounding each event is extracted to prepare the data for the ML pipeline. Next, we apply an MFCC transformation to each window. These transformed data points are fed into a ML pipeline utilizing a binary RF Classifier with 50 estimators.

The classification report shown in Table 5 reflects the performance of the binary classification model. The model demonstrates a high level F1-score, with class *'General Click'* reaching 0.92 and class *'Closing Click'* at 0.89. These scores indicate a strong balance between precision and recall across both classes. The overall accuracy of the model stands at 91% for the 136 samples tested, further reinforcing the model's effectiveness.

This experiment successfully demonstrates the ability to distinguish between various mouse movements and their associated clicks. Using controlled recordings, we train a RF Classifier to recognize these specific activities.

Table 5. Classification Report for the Binary Classification Model.

Class	Precision	Recall	F1-Score	Support
General Click	0.89	0.96	0.92	77
Closing Click	0.94	0.85	0.89	59
Accuracy			0.91	136

The goal of the next phase is to extend this approach to more natural computer usage scenarios. We record the authors' typical computer activities, including working on a project in Overleaf, ending with the action of closing a window. These recordings are then processed using the steps previously described. Next, we apply the pre-trained RF Classifier to the isolated mouse click events and infer the type and nature of mouse movements and clicks. This step marks a significant advance in applying our model to real-world scenarios, thus bridging the gap between controlled experimental conditions and everyday computer usage.

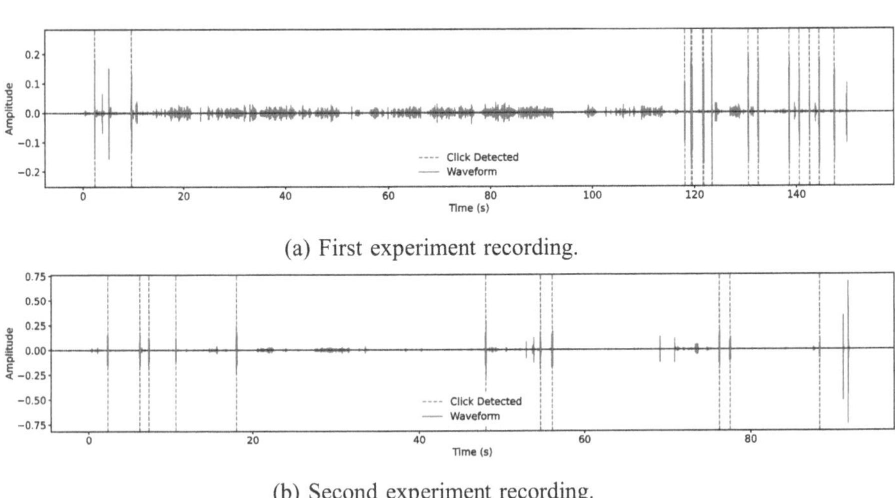

(a) First experiment recording.

(b) Second experiment recording.

Fig. 10. Comparison of waveform detection in two iterations of the experiment. (Color figure online)

We experiment by dividing it into two iterations. In the first iteration, the red 'X' is clicked to close the window, together with other two additional clicks. However, due to the thresholding method employed, the final click is not detected, as per Fig. 10a. In the second iteration, the red 'X' to close the window is clicked and followed with two hand claps. This means that the final detected mouse click is the one of interest, as showin in Fig. 10b. Looking at the waveforms, we observe background audio that includes random clicks and typing, creating a consistent hum of ambient noise.

In the first recording, among 15 clicks detected, the next-to-last sample is accurately identified as corresponding to the 'clicking red X' event. Similarly, in the second recording, out of 11 notable events, the model correctly infers the last one as the 'click red X' event. These results demonstrate the trained model's ability to accurately identify window-closing events, pointing at the feasibility of employing such a method in an ASCA.

5.3 Security Risks of ASCAs Based on Mouse Movement Inference

Ourfindings illustrate the feasibility of developing a generalizable model applicable to various users and the capability to accurately identify specific mouse movements in realistic settings. Looking ahead, these insights open the door to potential future threats where ASCAs might be combined with keyboard inference techniques, further reconstructing a victim's computer activities. For instance, one plausible scenario involves targeting users filling out sensitive online forms, such as tax documents. In such cases, the model could be fine-tuned to recognize distinctive patterns associated with clicking specific buttons or entering information into text fields. The attack does not aim to recognize by the number of pixels across which the user moved the moue and whether, at any given time, the cursor is precisely over a specific button. However, pattern recognition in mouse movements could lead to a comparable result. This highlights the importance of further research into ASCAs.

6 Ethical Concerns

Our institutions do not require any formal IRB approval to perform the experiments described above. Nonetheless, all experiments and corresponding evaluations were performed in accordance with the guidelines of the Menlo report [1]. We preemptively informed all voluntary participants about intended usage of their data. We obtained their informed consent before the recording process. We anonymously recorded only audio samples produced by a mouse and explicitly asked all participants to avoid speaking during recording sessions. We used audio samples for research purposes only.

7 Conclusions

This paper explored mouse-based ASCAs. It analyzed whether audio emanations from mouse movements reveal any sensitive information. The initial experiments show that it is possible to differentiate among four basic mouse movements. We then looked into granular measurements in order to predict the direction angles of a moving mouse, which shows some predictive complexities.

Next, we switched to a stereo recording method using two microphones on a smartphone, This allowed us to distinguish between ten two-dimensional movements on a mouse pad.

We then considered potential real-world scenarios where mouse-based ASCAs pose a security risk. To this end, we trained a model – using six participants – that predicts 12 distinct two-dimensional movements. This shows the ability to generalize the model and attack multiple users in one recording environment. Furthermore, we experimented with detection of a specific user actions, such as closing a Window, which demonstrated the efficacy of our approach in a realistic experimental setup. Consequently, we showed that the acoustic leakage model of mouse activity poses a real security risk.

Since this is only the first attempt to experiment with mouse-based ASCAs we identify directions for future work. Applying regression algorithms to stereo acoustic data would be interesting, since all regression tests so far are performed using one microphone. Moreover, the experiment in Sect. 5.2 could be extended by finding other situations where mouse movements reveal sensitive information. Finally, mouse-based ASCAs can be conducted in tandem with keyboard-based ones. The combination of the two might yield even more leakage.

References

1. Bailey, M., et al.: The Menlo report. IEEE Secur. Priv. **10**(2), 71–75 (2012)
2. Balagani, K., et al.: We can hear your PIN drop: an acoustic sidechannel attack on ATM PIN pads. In: Atluri, V., Di Pietro, R., Jensen, C.D., Meng, W. (eds.) ESORICS 2022. LNCS, vol. 13554, pp. 633–652. Springer, Cham (2022). https://doi.org/10.1007/978-3-031-17140-6_31
3. Cecconello, S., et al.: Skype & type: keyboard eavesdropping in voiceover- IP. ACM Trans. Priv. Secur. (TOPS) **22**(4), 1–34 (2019)
4. Chabanne, H., et al.: Side channel attacks for architecture extraction of neural networks. CAAI Trans. Intell. Technol. **6**(1), 3–16 (2021)
5. Chen, M., et al.: Behavicker: eavesdropping computer-usage activities through acoustic side channel. Wirel. Commun. Mob. Comput. **2022** (2022)
6. Cheng, P., et al.: SonarSnoop: active acoustic side-channel attacks. Int. J. Inf. Secur. **19**, 213–228 (2020)
7. Ciesla, R.: Bits, sample rates, and other fundamentals of digital audio. In: Sound and Music for Games: The Basics of Digital Audio for Video Games, pp. 1–24. Apress, Berkeley (2022)
8. Edfors, O., et al.: An introduction to orthogonal frequency-division multiplexing (1996)
9. Fang, Y., et al.: Eavesdrop with PoKeMon: position free keystroke monitoring using acoustic data. Futur. Gener. Comput. Syst. **87**, 704–711 (2018)
10. Genkin, D., Pattani, M., Schuster, R., Tromer, E.: Synesthesia: detecting screen content via remote acoustic side channels. In: 2019 IEEE Symposium on Security and Privacy (SP), pp. 853–869. IEEE (2019)
11. Gupta, H., Sural, S., Atluri, V., Vaidya, J.: Deciphering text from touchscreen key taps. In: Ranise, S., Swarup, V. (eds.) DBSec 2016. LNCS, vol. 9766, pp. 3–18. Springer, Cham (2016). https://doi.org/10.1007/978-3-319-41483-6_1
12. Kogler, A., et al.: Collide+power: leaking inaccessible data with software-based power side channels. In: 32nd USENIX Security Symposium (USENIX Security 2023), pp. 7285–7302 (2023)

13. Li, H., et al.: Deep learning methods for protein torsion angle prediction. BMC Bioinform. **18**(1), 1–13 (2017)
14. Lipp, M., et al.: PLATYPUS: software-based power side-channel attacks on x86. In: 2021 IEEE Symposium on Security and Privacy (SP), pp. 355–371. IEEE (2021)
15. Liu, J., et al.: Snooping keystrokes with mm-level audio ranging on a single phone. In: Proceedings of the 21st Annual International Conference on Mobile Computing and Networking, pp. 142–154 (2015)
16. Nandakumar, R., et al.: FingerIO: using active sonar for finegrained finger tracking. In: Proceedings of the 2016 CHI Conference on Human Factors in Computing Systems, pp. 1515–1525 (2016)
17. O'Shea, K., Nash, R.: An introduction to convolutional neural networks. In: arXiv preprint arXiv:1511.08458 (2015)
18. Spolaor, R., et al.: Plug and power: fingerprinting USB powered peripherals via power side-channel. In: IEEE INFOCOM 2023-IEEE Conference on Computer Communications, pp. 1–10. IEEE (2023)
19. Yilmaz, B.B., Prvulovic, M., Zajić, A.: Electromagnetic side channel information leakage created by execution of series of instructions in a computer processor. IEEE Trans. Inf. Forensics Secur. **15**, 776–789 (2019)
20. Ying, X.: An overview of overfitting and its solutions. J. Phys.: Conf. Ser. **1168**, 022022 (2019)
21. Zeng, Y., et al.: Spectrogram based multi-task audio classification. Multimed. Tools Appl. **78**, 3705–3722 (2019)
22. Zhu, T., et al.: Context-free attacks using keyboard acoustic emanations. In: Proceedings of the 2014 ACM SIGSAC Conference on Computer and Communications Security, pp. 453–464 (2014)

Using Semgrep OSS to Find OWASP Top 10 Weaknesses in PHP Applications: A Case Study

Lukas Kree(✉), René Helmke, and Eugen Winter

Fraunhofer FKIE, Bonn, Germany
{lukas.kree,rene.helmke,eugen.winter}@fkie.fraunhofer.de

Abstract. Given PHP's continuous success, it remains an important task to ensure security in its applications. While code reviews are a common measure to catch bugs during development, they lack scalability, are error-prone, and time-consuming [2,38,41]. Thus, static analysis tools like Semgrep emerged to provide programmatic feedback on code. But static analyses often show low precision, which can jeopardize utility.

In this case study, we investigate precision rates for Semgrep OSS for common web weaknesses from the OWASP Top 10 [35]. We explore method and tool limitations in weakness detection, OWASP classes, and Semgrep's public PHP rule set. We apply the latter to 300 open source applications, invest 34 h in manual sample validation, and derive precision rates for each OWASP class.

Our validation shows that the rules correctly detected weaknesses for seven OWASP classes with 86% precision, demonstrating the tool's utility. Yet, we estimate that most findings are not exploitable (81%). Thus, there is still considerable assessment overhead for users. Our work further highlights that only a subset of weaknesses are detectable, as dimensions such as runtime context and insecure design remain hidden. Finally, we advise practitioners to not exclusively rely on public rules, as translating application-specific business logic and design choices may open up to the detection of previously uncovered weaknesses.

1 Introduction

W3Techs approximates that 76% of web pages use PHP, including popular services as Facebook, Wikipedia, and WordPress [49]. Given its continuing relevance in web development, it remains important to detect security bugs in PHP code.

Human error is an intuitive source of security issues. During development, code reviews reduce risks, because they offer protection against *undesirable* code changes [2]. Yet, identifying insecure code needs expert knowledge [30,38,42] and bugs may slip through the cracks as reviews are error-prone and hard to scale [2,38,41]. Thus, novel Static Application Security Testing (SAST) tools with PHP support emerged to automate fault detection.

F. Maggi et al. (Eds.): DIMVA 2024, LNCS 14828, pp. 64–83, 2024.
https://doi.org/10.1007/978-3-031-64171-8_4

Yet, some static methods show low precision, which adds validation overhead and jeopardizes utility [19,30]. Furthermore, result expectations must be adjusted to method limitations: Some bugs manifest in dimensions invisible to the methods in use, e.g., runtime context [10,23,40]. Thus, academia explores approaches to improve applicability and effectiveness [3,7,8,27].

There are many commercial and open source tools, e.g., Semgrep [43], SonarQube [46], or PHPStan [31]. Their technical maturity makes them a prevalent choice for developers and, thus, highly relevant in practice. In contrary to academic tools like RIPS [7], which is now part of the commercial SonarQube [46], we found few performance data for prevalent tools on real world PHP data sets. Synthetic benchmarks such as SARD [32] exist, but they can only partially model the real world. Thus, it remains unknown how well some of these tools perform.

In this paper, we perform a case study on real PHP applications to collect performance data on the detectability of prevalent web weaknesses for an open source SAST tool. As this involves the manual verification of roughly hundred findings, we limit the scope to one tool and focus on precision rates. We chose the rule-based Semgrep OSS [43]. It is freely available, supports PHP, and already maps its public rule set to the latest revision of the OWASP Top 10 ranking, which helps us to classify prevalent web weaknesses for our case study. It has a broad user base, is the youngest of the commercially available SAST tools, and we did not find any performance data for it on PHP code in related work.

We approach the topic in two steps: The first step investigates which prevalent web weaknesses Semgrep OSS is able to detect and points out its possible limitations (Sect. 2). We provide insights on the OWASP Top 10 2021 to show the taxonomy's impact on our results. As Semgrep OSS can only detect OWASP classes for which there are rules, we investigate the used rule set's distribution across the taxonomy (Sect. 3). The case study is in part two (Sect. 4). Through this top-down approach, we establish a more comprehensive, fair, and transparent perspective on the performance results we are presenting.

Our contributions are summarized as follows:

1. We analyze the OWASP Top 10 2021 to estimate which prevalent weaknesses are hard to discover for Semgrep OSS due to method limitations. For this purpose, we introduce a scoring system that pairs weakness properties with method-specific limitations and expert knowledge.
2. We analyze Semgrep's public PHP rule set regarding its distribution across the OWASP Top 10 2021 to provide further insights on detectability.
3. We apply Semgrep to 300 carefully chosen open source PHP projects from GitHub and gather 9,301 findings. We invest approximately 34 h for manual validation of 98 sample findings and derive the tool's precision on PHP code for each OWASP class.

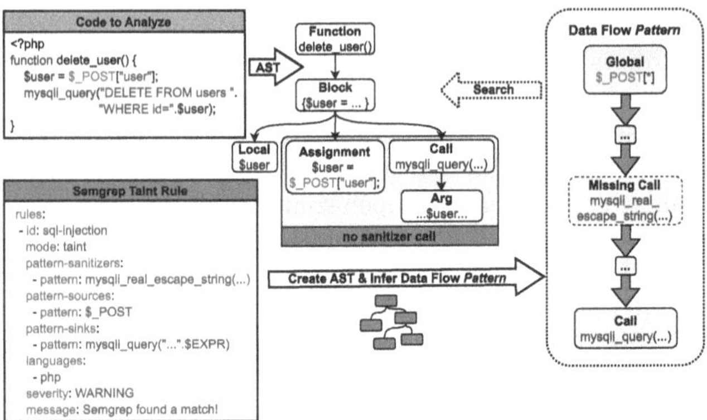

Fig. 1. High-level example of Semgrep's taint mode to find SQL injections [43]. The tool creates an AST from the code to analyze. Similarly, it translates the rule patterns to another AST, which is used to infer a data flow signature. The tool searches for the signature in the target code. The code AST is truncated for illustrative purposes.

2 Exploring Method Limitations in Detecting Web Weaknesses from the OWASP Top 10 2021

Using automation to detect security issues is a difficult task [44]. Weaknesses have varying properties and reside in dimensions which might be invisible to the methods used. As the goal of this case study is to estimate precision rates for Semgrep OSS, we analyze its detection capabilities for weaknesses to establish result transparency. We discuss the tool's fundamental methods and identify three weakness properties that help to estimate limitations. We also introduce the OWASP Top 10 2021 [35] as empirical ground truth for web weakness prevalence. We use a scoring system to rate their detectability using Semgrep OSS.

2.1 Semgrep OSS, Method Limitations, and Weakness Properties

In the following, we summarize Semgrep OSS's static analysis methods and then discuss their boundaries to identify limiting weakness properties.

Analysis Methods. Semgrep OSS [43] is a static analysis tool that uses rules to specify and detect patterns in source code. Patterns resemble code snippets from the target's language to describe semantic properties of potentially dangerous constructs. Semgrep OSS builds Abstract Syntax Trees (ASTs) for both target code and rule. These are directed graphs that describe code hierarchically: Nodes decompose code semantics, with increasing granularity down the hierarchy. This traversable representation helps to apply rules in two different modes [43]:

1. As patterns are AST sub-graphs, it can search for **semantic properties**, such as calls to known vulnerable functions or missing return checks.

2. It can infer data flows from the target's AST and then use **taint analysis**.

The latter tracks taint data propagation, such as user inputs, from predefined code sources to its sinks, e.g., functions [16,30]. This is done by interpreting the AST and inferring all potential data paths in the code [43]. Thus, Semgrep OSS can find *taint-style* issues, which result from missing sanitization of input reaching sensible routines, e.g., SQL injections. Figure 1 illustrates the analysis steps to detect SQL injections with Semgrep OSS and exemplifies the rule format.

Method Limitations. In the context of our case study, which uses a predefined rule set, we identify four bug discovery limitations (L1-L4) across two categories for Semgrep OSS: First, there are tool-specific limitations that are special to the analysis engine. Second, there are limitations inherent to rule-based semantic and taint analyses. Both are independent of the underlying program language.

As for tool-specific limitations[1], Semgrep OSS does neither support inter-file nor inter-procedural analyses (L1). Thus, it cannot track data flows or find patterns that span across multiple functions or files. This is a significant limitation because bugs often affect multiple components. For example, the injection in Fig. 1 remains undetected if we wrap the call to `mysqli_query` into another function - a false negative. Conversely, if we sanitize the input in a nested function call, the tool yields a false positive.

Regarding method limitations, Semgrep OSS is limited to code and cannot infer certain information: As it does not execute code and has no mechanism to calculate states, e.g., via symbolic execution, it can neither analyze nor estimate runtime behavior (L2). Also, code design is dictated by human decisions, which are not always inferable by semantics or taint patterns. Examination needs an entity, human or artificial, that provides context beyond the expressiveness of code. There are two cases we distinguish using an access control mechanism example: First, if a human specifies an access control matrix that does not fully or incorrectly accounts for critical code paths, Semgrep OSS cannot be taught to find these design flaws (L3). Second, assuming that the access control matrix is sound, we must correctly translate it to Semgrep OSS rules. Otherwise, it cannot detect incorrect implementations. It is generally possible to implement such rules, but infeasible to provide the application-specific knowledge for each target in our study. Thus, this negatively impacts detection capabilities (L4).

Limiting Weakness Properties. We identify three weakness properties that can hinder detectability in our case study and map them to L1 to L4. We derive all properties from the second category of limitations, because they are conceptual and, thus, mappable to CWEs. Tool-specific limitations also affect detectability, but describe implementation-specific conditions, e.g., if the flaw spans across multiple functions in code (L1). Thus, the properties are:

---

[1] https://semgrep.dev/docs/semgrep-code/overview/#oss-versus-pro-engine.

- **R) Runtime-Behavior or -Context.** Semgrep OSS has no behavioral analysis to probe runtime contexts or states (L2): There is no execution environment that might influence system behavior. Thus, it cannot detect weaknesses that depend on these conditions [12]. For example, attacks like CVE-2022-40482 use observable variations in function execution time.
- **D) Insecure System Design.** We observe that vulnerabilities caused by insecure system design emerge from fundamental, conceptual problems [37]. These problems cannot be fixed by a perfect implementation – they are inherently insecure [36]. Therefore, it is infeasible for the analysis to detect such weaknesses (L3): There are no detectable semantic or data flow patterns in the code, and the required knowledge base that could be used to learn the nature of such flaws is unknown to Semgrep OSS. Instead, detection requires a holistic system view and thorough evaluation requiring cognitive abilities that are possibly exclusive to humans.
- **B) Business Logic.** In contrary to insecure system design, this property focuses on faulty implementations of sound design. These usually implement application-specific (or business) logic. As a specification is present, which must not be the case for property D), there is the possibility that we *can* find and implement patterns for Semgrep OSS. This implies that the property is not necessarily a limitation for weakness detectability in general. However, L4 applies to our case study as it is infeasible to formulate rules that detect such application-specific implementation flaws for hundreds of projects. Thus, it is a limiting weakness property within the scope of this paper.

We do not claim that this list is complete. However, as we will see in Sect. 2.4, all properties show applicability to each OWASP Top 10 2021 class.

2.2 The OWASP Top 10 2021 Web Application Security Risks

Given the identified properties, we require an up to date catalogue of prevalent weaknesses to check against. The Open Worldwide Application Security Project (OWASP) maintains a list of the Top 10 Web Application Security Risks, with the latest iteration being from 2021 [35]. The project gathers and evaluates empirical vulnerability data from over 500,000 real-world web applications. Vulnerabilities are associated with CWEs and are ranked in a system of risk metrics such as incidence rates in vulnerability databases, and exploitability and severity scores. The resulting list clusters empirical vulnerability data from 193 CWEs in ten prevalence-ranked classes[2]. Table 1 lists the classes from the 2021 iteration.

In Fig. 2, we visualize all CWEs referenced by each OWASP Top 10 class in a network. Each color encodes an OWASP class and the node shapes represent different CWE types. These express different weakness abstraction levels and range from explicit implementation flaws to abstract concepts that reference other, more specific CWEs. Thus, the edges represent parent-child-relations between CWEs. We followed these relations across classes (n = 1) and added the corresponding edges between Common Weakness Enumerations (CWEs).

[2] https://cwe.mitre.org/data/definitions/1344.html, Accessed: 2024-04-15.

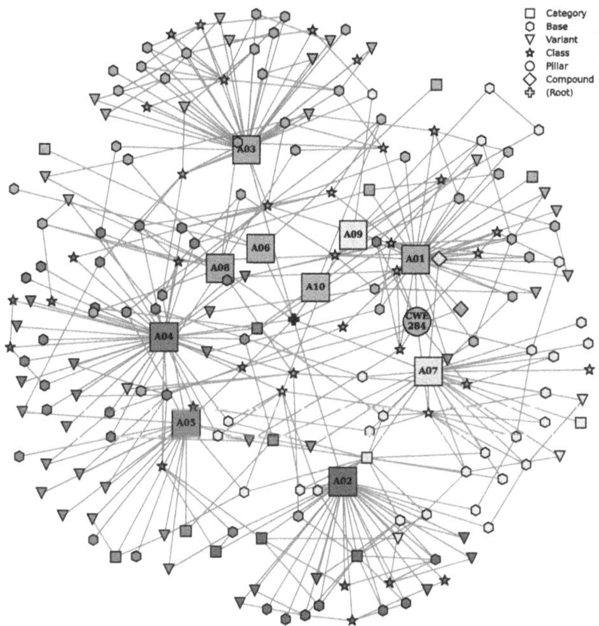

Fig. 2. Visualization of relationships between all CWEs mapped to OWASP Top 10 2021 classes with recursion depth one. We observe strongly intertwined connections between classes and their weaknesses. Combined with the varying abstraction levels of CWEs, borders between OWASP classes become imprecise and blurry.

As Fig. 2 shows, the OWASP classes are deeply intertwined and their CWEs vary greatly in abstraction. For example, there is the abstract OWASP class A04) Insecure Design. In contrary, there is A10) Server Side Request Forgery, which is also classified as A03) Injection. The injection class also references CWE-284, which describes Improper Access Control (A01). The longer we search for relations between classes, the blurrier the lines become.

This imprecision complicates the detectability scoring of common web weaknesses in Sect. 2.3. Yet, we choose to stick with the OWASP Top 10 since it is the de facto standard reference in the web security industry and largest source of empirical data on web weakness prevalence.

2.3 Scoring System for Weakness Detectability per OWASP Class

We now propose a scoring system to estimate the detection capabilities of Semgrep OSS for all OWASP Top 10 2021 classes. Consider the three previously discussed weakness properties Business Logic (**B**), Insecure System Design (**D**) and Runtime-Behavior or -Context (**R**) from Sect. 2.1. Ideally, there is an objective metric to rate the 193 associated CWEs across all properties. Unfortunately, our observations in Sect. 2.2 let us conclude that this is infeasible due to the imprecisions. Thus, the authors use their security expertise to analyze each OWASP

Top 10 2021 class regarding presence of the weakness attributes **B**, **D**, and **R** in referenced CWEs. When entering the blurry lines between OWASP classes, they apply their web security knowledge and discard or clearly assign a CWE to a single class. They independently rate the degree to which a property can negatively affect the detectability of an OWASP class on a scale of four values:

1. ● (1): A property (almost) never affects the OWASP class.
2. ◉ ($\frac{2}{3}$): A property affects the class in some, but not the majority of cases.
3. ⊙ ($\frac{1}{3}$): A property affects a class in the majority, but not all cases.
4. ○ (0): A property (almost) always affects the OWASP class.

Then, they determine a final score for each property and class through majority voting. Finally, the arithmetic mean $\bar{x}_P$, where $p \in \{B, D, R\}$, is being calculated for each OWASP class, resulting in one of three categories:

1. **High coverage.** The majority of weaknesses within a class should be detectable by Semgrep OSS. ($1 \geq \bar{x}_P > \frac{2}{3}$)
2. **Medium coverage.** Semgrep OSS should be able to identify several, but not all, of the included weaknesses. ($\frac{2}{3} \geq \bar{x}_P > \frac{1}{3}$)
3. **Low coverage.** Only a few weaknesses within a class should be detectable using Semgrep OSS. ($\frac{1}{3} \geq \bar{x}_P \geq 0$)

If one property is mapped to ○, the class has low coverage: If, e.g., the weaknesses in a class can only be detected at runtime, Semgrep OSS won't be able to detect any, regardless of the properties. This avoids coverage overestimations.

2.4 Summary of Scoring Results

The scoring results in Table 1 suggest that Semgrep OSS only covers a subset of the OWASP Top 10. Yet, there is high confidence that most weaknesses categorized under A03 and A10 are detectable. They can typically be discovered by analyzing the flow of user-controllable input using taint analysis.

Classes A02, A05, A06, and A08 obtained medium coverage. While Semgrep OSS can find known vulnerable software or functions in the code, it cannot detect missing protection, requiring knowledge of which data needs safeguarding. Also, it cannot determine which source is allowed to manage data without insights on security rules like access control matrices. Developers must also maintain a comprehensive list of used software and known vulnerabilities, adding complexity.

Classes A01, A07, A09 and A04 have lower coverage. We assume that detecting weaknesses within them requires a deeper understanding of application-specific logic, such as permissions and roles, or relies on runtime behavior. Also, A04 weaknesses mostly do not surface in specific coding patterns and, therefore, cannot be detected by Semgrep OSS.

This scoring serves as an initial expectation of Semgrep OSS's detection capabilities. It derives from discussions regarding weakness detection using Semgrep OSS for each OWASP class, which are provided separately[3].

[3] https://github.com/fkie-cad/2024-paper-owasp-weaknesses-in-php.

Table 1. Expected detection capabilities of Semgrep OSS on all OWASP Top 10 2021 classes. Scoring is based on three weakness properties: (B: Does not require business logic; R: Does not require runtime-behavior or -context; D: Is not caused by insecure system design). Their applicability determines the overall detection coverage $\bar{\mathbf{x}}_{\mathbf{P}}$.

OWASP Top 10 2021 Class	Property Score			
	B	R	D	Coverage ($\bar{x}_P$)
A01) Broken Access Control	⊙	⊙	⊙	low
A02) Cryptographic Failures	⊙	●	◉	medium
A03) Injection	◉	●	●	high
A04) Insecure Design	⊙	◉	○	low
A05) Security Misconfiguration	⊙	◉	●	medium
A06) Vulnerable and Outdated Components	⊙	◉	●	medium
A07) Identification and Authentication Failures	⊙	⊙	⊙	low
A08) Software and Data Integrity Failures	⊙	◉	⊙	medium
A09) Security Logging and Monitoring Failures	○	⊙	●	low
A10) Server-Side Request Forgery	◉	●	●	high

3 Exploring Semgrep's Public PHP Rule Set

Intuitively, Semgrep OSS can only detect weaknesses from the OWASP Top 10 2021 when rules cover its classes. For our case study, we chose to use the public PHP rule set[4], because it is maintained by the tool's developers, which we argue ensures rule quality. It also shows practical relevance, as it ships with the tool itself. The maintainers mapped each rule to the OWASP Top 10 2021, which allows us to analyze the distribution across all classes.

The set contains 57 rules, with 30 of them covering native PHP code. The remaining 27 are framework-specific. They address Doctrine (two rules), Laravel (eleven), Symfony (three), and WordPress (eleven).

The inclusion of highly popular [47] frameworks provides valuable insights regarding general detectability: Uncustomized rule sets can hardly account for application-specific logic or implementation. Yet, as frameworks provide a fixed ecosystem of functions, we can detect errors in these standardized but non-native functionalities in a scalable way. If there is no generalizable pattern, detection fails for all non-standardized implementations. Class A01 is an example, as an authentication matrix and its enforcement is application-specific, but gets dictated by the framework's ecosystem. Furthermore, rules for the same weakness but different frameworks exist, which inflates rule counts. Thus, we cannot assume there is a correlation between rule counts and general OWASP class detectability.

[4] https://github.com/semgrep/semgrep-rules.

Table 2. Distribution of Semgrep's public PHP rules across the OWASP Top 10 2021.

	A01	A02	A03	A04	A05	A06	A07	A08	A09	A10	No Class	$\sum$
# Rules	11	6	21	0	6	0	3	5	0	3	2	57
%	19.3	8.8	36.8	0	10.5	0	7	8.8	0	5.3	3.5	100

Table 2 provides an overview of the rule distribution across the OWASP Top 10 2021. Classes A03) Injection and its variant A10) Server-Side Request Forgery (SSRF) achieve high coverage and have a share of 42%. Additionally, classes A02, A05, and A08, categorized as having medium coverage, each encompass about 10% of the rules. There are no rules mapped to A04 and A09. The low coverage category A01 represents roughly 19% of the rules. Here, eight out of the eleven rules are framework-specific, exemplifying our observation discussed above. The related class A07 has three rules: Two rules check for overly permissive Cross-Origin Resource Sharing (CORS), and the other detects weak password usage for LDAP services. A06) Vulnerable and Outdated Components has no rules. However, around half of the rules target frameworks, which can be considered as third-party components. Thus, many of the included rules can be indirectly mapped to A06. This observation aligns with our findings in Sect. 2.2 regarding the imprecision of OWASP classes. Therefore, mappings lack explicitness; rules can be associated not only with the targeted weakness, but also with the broader concept of being part of a vulnerable component. Finally, two rules are not mapped to any OWASP class.

4 A Case Study on the Precision of Semgrep Findings

With the composition of the OWASP Top 10, the tool limitations, and the rule distribution across the OWASP Top 10 in mind, we now conduct an empirical case study on real-world PHP applications. Figure 3 illustrates our approach. It is a three-step process consisting of target projects sampling, the tool application to all included PHP files, and the analysis and validation of tool findings.

Target Selection. We create a corpus of real-world PHP applications for analysis. Ideally, it has the following properties to attain representativeness:

1. The corpus shall contain *heterogeneous* applications with different use cases, code quality, popularity, and project sizes. A mix of framework-based and plain PHP projects is desirable.
2. The heterogeneous properties should be *uniformly distributed* to avoid data skew from dominating clusters, e.g., projects with overly high or low quality.
3. The corpus should be *large* enough to derive robust precision rates.

We scrape open source PHP repositories from GitHub to establish *heterogeneity*: There are many available applications, ranging from amateur websites

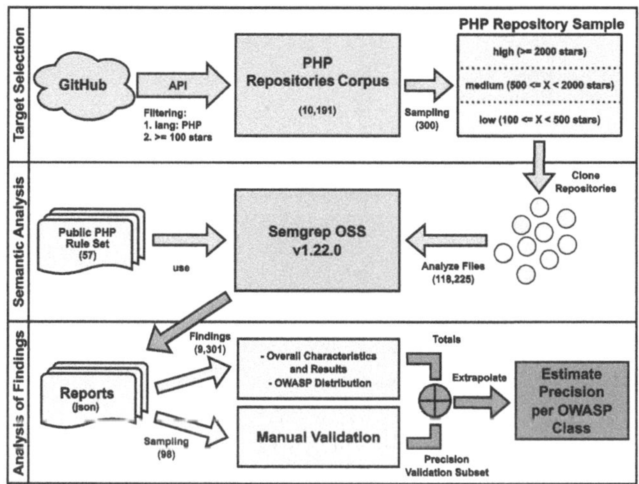

Fig. 3. High-level overview of the evaluation methods applied in our case study. It describes a multi-phase process with the three steps *Target Selection*, *Semantic Analysis*, and *Analysis of Findings*.

to professional software. We also annotate our data with GitHub *stars*: They are an effective way to measure project popularity [5,6,18]. We assume that it is a proxy metric for code quality, as projects must adhere to the standards of their community to remain popular. This may imply a certain degree of security awareness, safe coding practices, and quality control.

On 2023-05-04, we collected 10,191 PHP repositories using GitHub's API. A lower bound of 100 stars filters out projects with little to no engagement. In the raw data, we found that the majority of projects have less than 500 stars. We also noticed a drop in project frequency above 2,000 stars. To preserve a heterogeneous star distribution, we divide the set into three bins: high ($\geq$ 2,000 stars), medium (2,000 $>$ X $\geq$ 500 stars), and low (500 $>$ X $\geq$ 100 stars). This resulted in 629 high-star, 1,825 medium-star, and 7,737 low-star repositories. Figure 4 visualizes the resulting star distribution across the bins.

A sampling process uniformly draws 100 repositories from each bin:

1. Sort each bin by star count.
2. Set the sample size to $n = 100$.
3. Calculate step size $S_{\text{bin}} = \frac{U_{\text{bin}} - L_{\text{bin}}}{n}$ for each bin. U_{bin} describes the highest and L_{bin} the lowest star counts per repository in a bin.
4. Starting with the lowest star count per set, sweep over it with step size S_{bin}.
5. For each step, select the sample repository that is in closest star proximity.

The final corpus contains 300 repositories with varying star counts and differing code quality, and is uniformly distributed within each bin.

Semantic Analysis. We pull the main branches of all 300 projects and collect 118,225 PHP files. We pass each file to Semgrep OSS v1.22.0 for analysis and collect the results in JSON format. The used rule set is described in Sect. 3. The used rule commit hash is `e32bb84`.

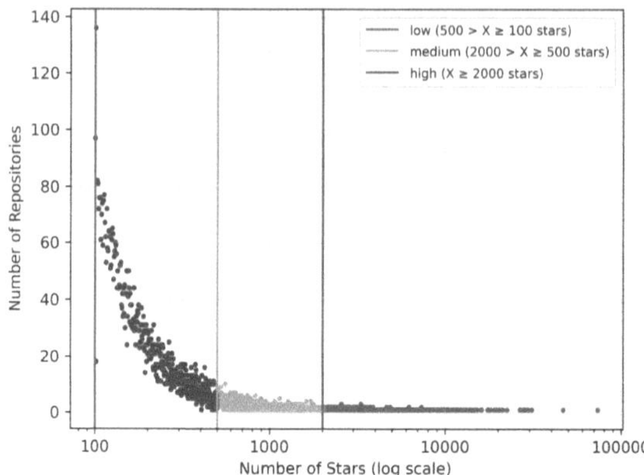

Fig. 4. Distribution of stars of the PHP repository corpus collected from GitHub (≥100 stars). Using the star metric, we classify repositories into the groups *low*, *medium*, *high*.

Analysis of Findings. We analyze the findings, present overall match numbers, and their distribution across the OWASP Top 10 2021. We also inspect data artifacts to gain insights on detection challenges per class. Then, we derive precision rates of correctly identified weaknesses through result validation: Five findings per star bin in each OWASP class are randomly picked, resulting in a maximum of 15 data points per class. We aimed for an even rule distribution to cover as much of the applied rule set as possible.

Analysis is limited to 20 min per finding, leading to roughly 34 h of manual validation. Each sample is classified in two dimensions: First, we assess if the rule correctly identified the targeted weakness. When a finding is a true positive, it indicates a security *risk*. Yet, it is unclear if the risk is a *vulnerability* with true impact, which can be exploited through external data, e.g., user input. Thus, we classify findings as exploitable (E) or not exploitable ($\bar{\text{E}}$). If the 20 min are exceeded, findings are marked as not determinable (N/A).

For example, consider Semgrep OSS's *exec-use* rule to detect non-constant shell commands: It correctly matches the `passthru` function, but also any function that starts with said name, e.g. `passthru_sanitize`; a false positive in identification. In another case, validation confirms non-constant use of `passthru`, but users cannot control the input; a correct finding, but not exploitable.

4.1 Semgrep Analysis Results

Semgrep OSS generated 9,301 findings in 118,225 files across 196 repositories. The other 104 had no matches. Figure 5 shows the distribution of findings across the OWASP Top 10 2021. The tool detected 1,940 potential A01 weaknesses, representing approximately 20.86% of all findings. Out of the ten rules that produced matches, eight targeted PHP frameworks. However, they only account for 14.22% of all A01 findings. Notably, the *unlink-use* rule, which detects unsafe

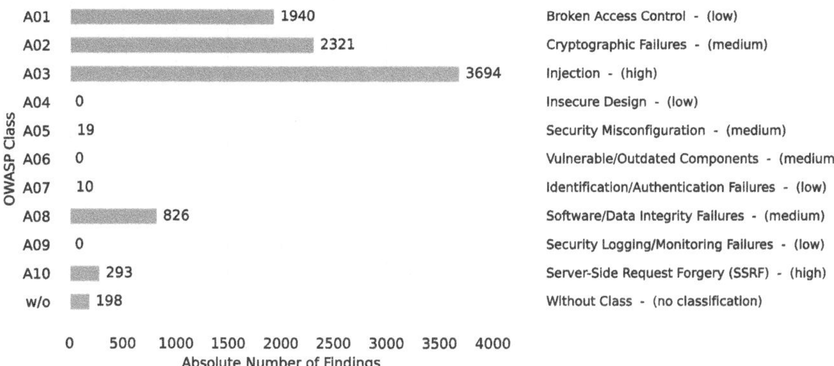

Fig. 5. Absolute number of findings per OWASP class using Semgrep's publicly available PHP rule set. Our classification of Semgrep's coverage is indicated in parentheses.

`unlink`[5] use, contributes to 84.79% of A01 matches. This provides evidence for the A01 detection limitations discussed in Sect. 2.1, as most findings disclose sensitive function use without addressing explicit access control paradigms.

For A02, Semgrep OSS found 2,321 potentially vulnerable patterns, accounting for 24.95% of all matches. Here, the *weak-crypto* rule is responsible for over 90% of findings. This rule detects weak hash functions based on a wordlist. An additional 8% of A02 findings are mapped to the *ftp-use* rule, which identifies the usage of the unencrypted File Transfer Protocol (FTP) protocol. Thus, most A02 findings relate to detecting known weak or insecure functions, not incorrect implementation, which we assumed to be challenging to detect.

The rules for A03) Injection and its variant A10) SSRF collectively generated 3,987 (42.9%) findings. Out of 24 rules, 15 produced matches, six of which used Semgrep's `taint` analysis mode. The tool found most types of injection weaknesses: file injection (133), SSRF (293), XSS (595), SQL injection (630), and command injection (801). The *non-literal-header* rule, which detects issues leading to XSS or SSRF, identified the remaining 1,535 matches. This aligns with our estimation regarding injection detectability.

Six rules target A05 weaknesses, which only generated 22 (0.2%) matches. This outcome could have several reasons, as all of them are framework-specific: First, frameworks may provide secure default configurations, e.g., Laravel sets the 'HTTP' attribute for cookies to true[6]. Second, not all projects use frameworks. Lastly, A05 weaknesses may only occur once during configuration.

Three A07 rules generated ten (0.11%) findings. Two rules aim to uncover overly permissive CORS settings, allowing untrusted entities to send requests and access target resources. As CORS headers are unset by default, one must explicitly modify them. This is only relevant when multiple domains are

[5] https://www.php.net/manual/de/function.unlink.php.
[6] https://github.com/laravel/laravel/blob/10.x/config/session.php.

Table 3. Outcomes of our result validation for all 98 sampled matches, sorted by OWASP Top 10 2021 weakness classes. We classified each finding in correct identification, assessing if the rule correctly matched the weakness it was designed to find, and their actual exploitability. *Note: We do not consider the two rules mapped to no class.

ID	#	Correct Identification			Exploitability			
	Findings	TP	FP	Precision	E	$\bar{E}$	N/A	E. Rate
A01	1,940	10	5	66.7%	1	12	2	7.7%
A02	2,321	14	1	93.3%	7	8	0	46.7%
A03	3,694	14	1	93.3%	4	8	3	33.3%
A04	0	–	–	–	–	–	–	–
A05	19	8	6	57.1%	0	14	0	0%
A06	0	–	–	–	–	–	–	–
A07	10	9	0	100%	0	7	2	0%
A08	826	15	0	100%	1	10	4	9.1%
A09	0	–	–	–	–	–	–	–
A10	293	15	0	100%	3	9	3	25%
–	198*	–	–	–	–	–	–	–
$\sum$	9,103*	85	13	86.7%	16	68	14	19%

involved. Thus, the such modifications may be less frequent. The last rule detects password-less connections to LDAP. They occur less often than subsequent connection use. Therefore, there might be fewer matches.

The A08 rules generated 826 (8.88%) findings. Semgrep's *unserialize-use* rule detected 653 of them, matching untrusted `unserialize`[7] input. The remaining findings were identified by the *laravel-dangerous-model-construction* rule, which detects best practice violations for model creation in the framework.

Since there are no rules for classes A04, A09, and A06, there are no findings.

4.2 Result Validation

Table 3 summarizes our manual validation results. We documented the selection process at in the beginning of this chapter; our goal was to sample five matches for each OWASP class and star bin combination. Yet, as there are not enough results for all combinations of OWASP class and star bin, we drew 98 samples.

Regarding correct identification, 85 findings were classified as true and 13 as false positives, resulting in an 86.7% precision rate. Of the five incorrect A01 identifications, three were caused by misidentifying a Symfony function name used by another framework. Framework-specific mismatching also applies to eight other instances: The remaining two A01 false positives stemmed from false detection of missing Laravel CSRF mitigations. One resulted from tool limitations, as

[7] https://www.php.net/manual/en/function.unserialize.php.

Table 4. Characterization of all 16 findings deemed exploitable. Projects are considered active, if their last commit was less than one year ago. Respective OWASP classes and CWEs are provided by Semgrep.

#	Star Bin	Active	OWASP	CWE	Description
1	low	No	A02	CWE-319	Cleartext Transmission
2	low	No	A02	CWE-319	Cleartext Transmission
3	low	No	A02	CWE-319	Cleartext Transmission
4	low	No	A03	CWE-94	Code Injection
5	low	No	A03	CWE-98	PHP Remote File Inclusion
6	low	No	A08	CWE-502	Deserialization of Untrusted Data
7	medium	Yes	A01	CWE-1275	Improper Cookie Attributes
8	medium	Yes	A02	CWE-319	Cleartext Transmission
9	medium	No	A02	CWE-319	Cleartext Transmission
10	medium	No	A02	CWE-319	Cleartext Transmission
11	medium	Yes	A03	CWE-79	Cross-site Scripting
12	medium	Yes	A03	CWE-98	PHP Remote File Inclusion
13	medium	Yes	A10	CWE-918	Server-Side Request Forgery
14	medium	Yes	A10	CWE-918	Server-Side Request Forgery
15	medium	Yes	A10	CWE-918	Server-Side Request Forgery
16	high	Yes	A02	CWE-328	Use of Weak Hash

safeguards were implemented in a separate function. The other was due to Semgrep misinterpreting if-statements within Laravel Blade templates. Six A05 false positives came from the *laravel-cookie-secure-set* rule, which incorrectly flagged cookie settings. Here, newer Laravel versions set a secure attribute differently than specified in the pattern. A03 yields one false positive sample: The *exec-use* rule matches any function starting with `exec`. Thus, it also matched an `execute` helper function for Laravel Commands, which dispatches application tasks. The last false positive was produced by the *md5-used-as-password* rule incorrectly matching MD5 being used for a pseudo-random salt, rather than a password.

Overall, we conclude that Semgrep identifies most issues with good precision. Misidentification was mainly due to pattern quality rather than tool limitations.

When assessing exploitability, 14 samples could not definitely be classified. The remaining 84 samples yielded 16 likely exploitable findings, which are listed in Table 4. Regarding star bin distribution, six (38%) potential vulnerabilities reside in low, nine (56%) in medium, and one (6%) in high star repositories. We contacted the corresponding projects for further assessment. Looking at OWASP class distribution, approximately 43% of the potentially exploitable findings relate to injection vulnerabilities, including code, file, command, and SSRF injections. Generally, such vulnerabilities come with high impact scores, but without further assessment by vendors, we refrain from unverified claims. Additionally, 43% of potential vulnerabilities involve insecure transmission of

Table 5. Related Work: SAST tools, system papers, and case and performance studies.

Comm. & OSS Tools		SAST System Papers		Case & Perf. Studies	
Tool	*Release*	*Paper*	*Year*	*Paper*	*Year*
Semgrep OSS [43]	2020	Luo et al. [27]	2022	Etemadi et al. [11]	2023
PHPStan [31]	2016	Li et al. [24]	2021	Li et al. [25]	2023
Psalm [48]	2016	Backes et al. [3]	2017	Lipp et al. [26]	2022
Phan [22]	2015	Olivo et al. [34]	2015	Pierce et al. [39]	2022
SonarQube [46]	2007	Nunes et al. [33]	2015	Kluban et al. [20]	2022
CodeQL [14]	2006	Dahse et al. [7]	2014	Bandara et al. [4]	2020
		Dahse et al. [9]	2014	Lenarduzzi et al. [21]	2020
		Dahse et al. [8]	2014	Florin et al. [13]	2020
		Medeiros et al. [29]	2014	Marcilio et al. [28]	2019
				Algaith et al. [1]	2018

sensitive data (A02), covering improper certificate verification using cURL and utilizing unencrypted FTP protocols. The remaining samples regard improper cookie configurations (A01) and the deserialization of user data (A08).

The remaining 68 matches (81%) were deemed non-exploitable. In total, 21 matches were marked as such because of custom sanitization or mitigation routines in place, which are impossible to address with generalized rules. Additionally, there were eleven non-constant calls to sensitive functions, yet the data passed to them could not be modified. Another seven instances identified weaknesses within code not exposed to or reachable for users. The remaining matches were classified as non-exploitable for various reasons: For instance, rules matched non-exploitable code within dummy or test files (5), detected deviations from best practices rather than explicit vulnerabilities (6), or matched weak patterns, such as the use of MD5 for temporary file names, in non-cryptographic contexts.

Ultimately, this implies that although Semgrep effectively detected potential weaknesses, substantial manual overhead is required to determine exploitability.

4.3 Threats to Validity

This case study is subject to the following threats to validity: First, we assumed no prior knowledge of each application, except for the source code itself. This allows for a large analysis corpus but poses challenges in granularity, as we miss application-specific findings and rule development. Especially the latter could positively affect precision. Also, manual validation introduces human error and subjective judgment. The author's practical knowledge of web security influences results. Despite efforts to ensure objectivity, individual expertise may introduce variability. A more objective approach would involve the recruitment of external parties, which was infeasible due to the limited resources for this case study. This also applies to our scoring system in Sect. 2.4. Lastly, our evaluation does not account for false negatives, which are missed weaknesses that may exist but were not detected by Semgrep. Therefore, this case study is not a complete assessment of Semgrep OSS's capabilities and limitations in detecting OWASP weakness classes, but rather provides first insights and data on this matter.

5 Related Work

Table 5 provides an overview of related work in the SAST domain. We categorize by Commercial & OSS Tools, SAST System Papers, and Case & Performance Studies. To the best of our knowledge, our large-scale empirical case study is the first to explore Semgrep OSS precision on real-world PHP applications.

Commercial & OSS Tools. Aside of Semgrep OSS [43] (cf., Sect. 2), there are other commercial and open source SAST tools for PHP in active development. All of them use the AST, but only partially overlap in purpose and functionality. PHPStan [31], e.g., is rule-based and retrofits PHP with static type checker functionality to find bugs. In contrast to Semgrep, type annotations must be added to the code and it does not support taint analysis. The Phan [22] static code analyzer has similar functionality. Psalm [48] is another tool of the same category, implementing type and scope checks for annotated PHP code. There is rudimentary taint analysis support for a set of predefined sources and sinks. They can be extended by adding specific annotations to the code. SonarQube [46] is a commercial tool that is closely related to Semgrep [43]: It has a rule-based format that defines semantics for potentially dangerous or unwanted code patterns, and also provides a taint analysis engine. The latter originates from the RIPS tool by Dahse et al. [7] and supports inter-procedural analyses, but it is not part of the open source version. Finally, there is CodeQL [14], which creates a database of semantic code properties. It introduces a query language similar to SQL, which allows security analysts and developers to query the AST for complex structures, including bug patterns. There is a taint analysis engine, but CodeQL does not support PHP.

SAST System Papers. In the past years, academia explored several SAST techniques for PHP: Dahse et al. present three papers revolving around their prototype called RIPS and its extensions [7–9]. When RIPS was introduced in 2014 [7], it combined intra- and inter-procedural approaches with the creation of block and function summaries for the incremental semantic analysis of blocks and efficient backwards-directed taint analysis. Dahse et al. then extend RIPS to detect second-order vulnerabilities [8], which use persistent and global data stores as intermediate step in the data flow and are, thus, harder to detect for taint analyses. Another extension [9] introduces PHP object injection detection. Olivo et al. [34] focus on second-order Denial of Service (DoS) bugs. They use a 2-phase taint analysis that separates the tracking of writing from reading data flows to identify vulnerable database queries. Subsequently, they use symbolic execution to propose candidate attack vectors. Nunes et al. [33] use inter-procedural taint analysis to detect XSS and SQLi bugs in 35 WordPress plugins and show that their tool achieves higher precision rates than RIPS. Luo et al. [27] observe that existing approaches perform worse on object-oriented PHP code. Therefore, they model several relational structures and object-oriented programming features before performing an inter-procedural taint analysis. They show that their tool outperforms the state of the art in finding SQLi, XSS, and DoS vulnerabilities. Li et al. [24] focus on loose comparison bugs inferred by weak typing. They combine their taint analysis with a novel type inference algorithm that tracks type reassignments in the data flow as well. Backes et al. [3] are the first to use code property graphs and introduce a framework that allows complex graph traversal queries on PHP, similar to CodeQL [14]. Analysts can then formulate queries on, e.g., the AST or CFG to specify signatures for vulnerabilities. Finally, Medeiros et al. [29] propose a taint analysis algorithm that is validated by a machine learning approach based on data mining to detect false-positives and, thus, improve precision rates.

Case & Performance Studies Using Commercial or Open Source SAST Tools. Lipp et al. [26] evaluate the effectiveness of static C analyzers, including CodeQL. They use Magma [17] and then compare found ground truth bugs by each SAST tool, which yields mixed results for all of them. Bandara et al. [4] use CodeQL to perform bug remediation analyses on real projects. They find that many bugs stem from commits intended to fix previous issues. Most could have been avoided through SAST. Pierce et al. [39] assess if GitHub Copilot proposes insecure code snippets to its users. Through manual and CodeQL-based assessments, they approximate that 40% of all tested programs are vulnerable. For SonarQube, Etemadi et al. [11] fixes tool findings by creating automatic patch suggestions for selected rules. Marcilio et al. [28] mine project issues and find that the resolution rates of tickets generated by SonarQube warnings is 13%. Lenarduzzi et al. [21] analyze SonarQube rules for fault-proneness in Java and conclude that only few rules are robust. Florin et al. [13] use SonarQube to find bugs in domain-specific software, but do not further validate results. Kluban et al. [20] use Semgrep with rules derived from Snyk [45] and VulnCode [15] to detect prototype pollution and DoS in JavaScript. Li et al. [25] search for taint-style vulnerabilities in Lua web interfaces using Semgrep. Tests on real-world firmware yield true positive rates of 42.5%. In 2018, Algaith et al. [1] compare the performance of five SAST tools on real-world PHP code. They measure true-positives and -negatives for WordPress PHP plugins to test the effectiveness of tool combinations. They find that specific combinations boost effectiveness, but did not include Semgrep, as it was released two years later.

6 Conclusion and Future Work

We performed a case study on the precision of Semgrep OSS, an open source semantic analysis tool, in detecting OWASP Top 10 2021 weaknesses. After introducing Semgrep and its method and tool limitations, we identified three weakness properties hindering detectability in our scenario: 1) Insecure System Design, 2) Runtime-Behavior or -Context, and 3) Business Logic. We then applied these properties to all OWASP classes and provide insights into their detection.

Ultimately, we conduct a case study to evaluate Semgrep's precision. During analysis, the tool consolidated 9,301 findings across 300 carefully collected GitHub PHP repositories, of which we manually validated a subset:

Semgrep correctly found weaknesses across all OWASP classes that are included in the rule set with a precision of 86.7%. As these findings pose potential security risks, the tool shows utility in real-world scenarios. However, we determined that roughly 81% of samples are unexploitable. Nonetheless, as unexploitable findings may become exploitable with ongoing code changes, we argue that most true positives are worth of further investigation. Yet, we express concerns that these unexploitable issues may be erroneously assessed as false positives by developers, which can skew the perceived utility of the tool [19]. We contacted projects affected by likely exploitable matches for further assessment.

For future work, the inclusion of custom rules that consider application-specific logic and design choices might yield new findings. Furthermore, experiments on data sets with known vulnerability ground truth, such as SARD [32], would provide interesting insights on false negative rates.

Finally, we gave insights on the vagueness of the OWASP Top 10. It would be highly beneficial to further investigate this aspect, especially regarding the impact on the utility of such taxonomies for scientific purposes, e.g., performance evaluations. We provide an anonymized set of our raw data and results: https://github.com/fkie-cad/2024-paper-owasp-weaknesses-in-php.

References

1. Algaith, A., Nunes, P., Jose, F., Gashi, I., Vieira, M.: Finding SQL injection and cross site scripting vulnerabilities with diverse static analysis tools. In: Proceedings of the 2018 14th European Dependable Computing Conference (EDCC), pp. 57–64. IEEE (2018)
2. Bacchelli, A., Bird, C.: Expectations, outcomes, and challenges of modern code review. In: 2013 35th International Conference on Software Engineering (ICSE), pp. 712–721 (2013)
3. Backes, M., Rieck, K., Skoruppa, M., Stock, B., Yamaguchi, F.: Efficient and flexible discovery of PHP application vulnerabilities. In: Proceedings of the 2017 IEEE European Symposium on Security and Privacy (Euro S&P), pp. 334–349. IEEE (2017)
4. Bandara, V., et al.: Fix that fix commit: a real-world remediation analysis of Javascript projects. In: 2020 IEEE 20th International Working Conference on Source Code Analysis and Manipulation (SCAM), pp. 198–202 (2020)
5. Borges, H., Tulio Valente, M.: What's in a GitHub star? Understanding repository starring practices in a social coding platform. J. Syst. Softw. **146**, 112–129 (2018)
6. Coelho, J., Valente, M.T., Milen, L., Silva, L.L.: Is this GitHub project maintained? Measuring the level of maintenance activity of open-source projects. Inf. Softw. Technol. **122**, 106274 (2020)
7. Dahse, J., Holz, T.: Simulation of built-in PHP features for precise static code analysis. In: NDSS, vol. 14, pp. 23–26 (2014)
8. Dahse, J., Holz, T.: Static detection of second-order vulnerabilities in web applications. In: Proceedings of the 23rd USENIX Security Symposium (USENIX Security 2014), pp. 989–1003 (2014)
9. Dahse, J., Krein, N., Holz, T.: Code reuse attacks in PHP: automated pop chain generation. In: Proceedings of the 2014 ACM SIGSAC Conference on Computer and Communications Security, pp. 42–53 (2014)
10. Deepa, G., Thilagam, P.S.: Securing web applications from injection and logic vulnerabilities: approaches and challenges. Inf. Softw. Technol. **74**, 160–180 (2016)

11. Etemadi, K., et al.: Sorald: automatic patch suggestions for sonarqube static analysis violations. IEEE Trans. Dependable Secure Comput. **20**(4), 2794–2810 (2023)
12. Ferrara, P., Mandal, A.K., Cortesi, A., Spoto, F.: Static analysis for discovering IoT vulnerabilities. Int. J. Softw. Tools Technol. Transf. **23**, 71–88 (2021)
13. Florin, I.L., Bălan, T.: Vulnerability remediation in ICS infrastructure based on source code analysis. In: 2020 19th RoEduNet Conference: Networking in Education and Research (RoEduNet), pp. 1–6 (2020)
14. GitHub: Codeql (2023). https://codeql.github.com
15. Google: The vulnerable code database (vulncode-db) (2019). https://www.vulncode-db.com/
16. Hauzar, D., Kofron, J.: Framework for static analysis of PHP applications. In: Boyland, J.T. (ed.) 29th European Conference on Object-Oriented Programming (ECOOP 2015). Leibniz Int. Proc. in Informatics (LIPIcs), vol. 37, pp. 689–711. Schloss Dagstuhl–Leibniz-Zentrum fuer Informatik, Dagstuhl (2015)
17. Hazimeh, A., Herrera, A., Payer, M.: Magma: a ground-truth fuzzing benchmark. Proc. ACM Meas. Anal. Comput. Syst. **4**(3), 1–29 (2020)
18. Hu, Y., Zhang, J., Bai, X., Yu, S., Yang, Z.: Influence analysis of GitHub repositories. SpringerPlus **5**(1), 1268 (2016)
19. Johnson, B., Song, Y., Murphy-Hill, E., Bowdidge, R.: Why don't software developers use static analysis tools to find bugs? In: 2013 35th International Conference on Software Engineering (ICSE), pp. 672–681 (2013)
20. Kluban, M., Mannan, M., Youssef, A.: On measuring vulnerable Javascript functions in the wild. In: 2022 ACM Asia Conference on Computer and Communications Security (ASIA CCS 2022), ASIA CCS 2022, pp. 917–930. Association for Computing Machinery, New York (2022). https://doi.org/10.1145/3488932.3497769
21. Lenarduzzi, V., Lomio, F., Huttunen, H., Taibi, D.: Are sonarqube rules inducing bugs? In: 2020 IEEE 27th International Conference on Software Analysis, Evolution and Reengineering (SANER), pp. 501–511 (2020)
22. Lerdorf, R., Morrison, A.: Phan. https://github.com/phan/phan
23. Li, J.: Vulnerabilities mapping based on OWASP-SANS: a survey for static application security testing (SAST). Ann. Emerg. Technol. Comput. **4** (2020)
24. Li, P., Meng, W.: LChecker: detecting loose comparison bugs in PHP. In: Proceedings of the 30th International Conference on World Wide Web (WWW 2021), pp. 2721–2732 (2021)
25. Li, X., Wei, Q., Wu, Z., Guo, W.: Finding taint-style vulnerabilities in lua application of iot firmware with progressive static analysis. Appl. Sci. **13**(17) (2023)
26. Lipp, S., Banescu, S., Pretschner, A.: An empirical study on the effectiveness of static C code analyzers for vulnerability detection. In: Proceedings of the 31st ACM SIGSOFT International Symposium on Software Testing and Analysis, ISSTA 2022, pp. 544–555. Association for Computing Machinery, New York (2022)
27. Luo, C., Li, P., Meng, W.: TChecker: precise static inter-procedural analysis for detecting taint-style vulnerabilities in PHP applications. In: Proceedings of the 2022 ACM SIGSAC Conference on Computer and Communications Security, pp. 2175–2188 (2022)
28. Marcilio, D., Bonifácio, R., Monteiro, E., Canedo, E., Luz, W., Pinto, G.: Are static analysis violations really fixed? A closer look at realistic usage of sonarqube. In: 2019 IEEE/ACM 27th Int. Conference on Program Comprehension (ICPC), pp. 209–219 (2019). https://doi.org/10.1109/ICPC.2019.00040

29. Medeiros, I., Neves, N.F., Correia, M.: Automatic detection and correction of web application vulnerabilities using data mining to predict false positives. In: Proceedings of the 23rd International Conference on World Wide Web (WWW 2014), pp. 63–74 (2014)
30. Medeiros, I., Neves, N., Correia, M.: Detecting and removing web application vulnerabilities with static analysis and data mining. IEEE Trans. Reliab. **65**(1), 54–69 (2016)
31. Mirtes, O.: PHPStan. https://phpstan.org/
32. NIST: Software assurance and reference dataset. https://samate.nist.gov/SARD/
33. Nunes, P.J.C., Fonseca, J., Vieira, M.: phpSAFE: a security analysis tool for OOP web application plugins. In: Proceedings of the 45th IEEE/IFIP International Conference on Dependable Systems and Networks, pp. 299–306. IEEE (2015)
34. Olivo, O., Dillig, I., Lin, C.: Detecting and exploiting second-order denial-of-service vulnerabilities in web applications. In: Proceedings of the 22nd ACM SIGSAC Conference on Computer and Communications Security, pp. 616–628 (2015)
35. OWASP Foundation: OWASP top 10 2021 (2021). https://owasp.org/www-project-top-ten/
36. OWASP Foundation: OWASP top 10 2021: A04:2021 - insecure design (2021). https://owasp.org/Top10/A04_2021-Insecure_Design/
37. OWASP Foundation: How to use the OWASP top 10 as a standard (2023). https://owasp.org/Top10/A00_2021_How_to_use_the_OWASP_Top_10_as_a_standard/
38. Paul, R., Turzo, A.K., Bosu, A.: Why security defects go unnoticed during code reviews? A case-control study of the chromium OS project. In: Proceedings of the 43rd International Conference on Software Engineering, ICSE 2021, pp. 1373–1385. IEEE Press (2021)
39. Pearce, H., Ahmad, B., Tan, B., Dolan-Gavitt, B., Karri, R.: Asleep at the keyboard? Assessing the security of GitHub copilot's code contributions. In: 2022 IEEE Symposium on Security and Privacy (SP), pp. 754–768 (2022)
40. Dewhurst, R.: Static code analysis (2023). https://owasp.org/www-community/controls/Static_Code_Analysis
41. dos Santos, E.W., Nunes, I.: Investigating the effectiveness of peer code review in distributed software development. In: Proceedings of the XXXI Brazilian Symposium on Software Engineering, SBES 2017, pp. 84–93. Association for Computing Machinery, New York (2017)
42. Seixas, N., Fonseca, J., Vieira, M., Madeira, H.: Looking at web security vulnerabilities from the programming language perspective: a field study. In: 2009 20th International Symposium on Software Reliability Engineering, pp. 129–135 (2009)
43. Semgrep: Project page (2023). https://semgrep.dev
44. Shahriar, H., Zulkernine, M.: Mitigating program security vulnerabilities: approaches and challenges. ACM Comput. Surv. **44**(3) (2012)
45. Snyk.io: Snyk Vulnerability DB. https://security.snyk.io/
46. Sonar: Sonarqube (2023). https://www.sonarsource.com/products/sonarqube
47. Statista: Most used web frameworks among developers worldwide, as of 2022 (2023). https://www.statista.com/statistics/1124699/worldwide-developer-survey-most-used-frameworks-web/
48. Vimeo: Psalm. https://psalm.dev/
49. W3Techs: Usage statistics of server-side programming languages for websites (2023). https://w3techs.com/technologies/overview/programming_language

Modularizing Directed Greybox Fuzzing for Binaries over Multiple CPU Architectures

Sofiane Benahmed, Abdullah Qasem, Anis Lounis(✉), and Mourad Debbabi

Concordia University, 1515 Ste. Catherine St. W., Montreal, QC, Canada
{sofiane.benahmed,abdullah.qasem,anis.lounis}@mail.concordia.ca,
mourad.debbabi@concordia.ca

Abstract. Directed Greybox Fuzzing has proven effective in vulnerability detection areas such as bug reproduction and patch testing. However, existing directed fuzzers are often difficult to customize, lack modularity and have limited binary support. This constrains their usability on complex software or when the source code is unavailable; a challenge encountered when fuzzing embedded systems. This article addresses these limitations by introducing the Directed Fuzzing Toolkit (DRIFT) as a platform for directed fuzzing within the modular framework LibAFL. DRIFT modularizes techniques from the state-of-the-art directed fuzzer AFLGo and adapts them for binary applications thereby augmenting LibAFL's highly customizable fuzzers with directed fuzzing capabilities. Additionally, by leveraging Ghidra's analysis, DRIFT achieves architecture agnostic static analysis, opening doors for DGF to tackle previously challenging scenarios. Our evaluation of DRIFT shows a 90% correlation in static analysis metrics over binary compared to its source-code counterpart. Fuzzing performance was also notable despite operating over emulation. In benchmarks, DRIFT's performance exceeds the original fuzzer with up to doubled bug discovery rates and 9–40× faster exploitation times of target bugs. These results are attributed to the toolkit's modular design and its integration with LibAFL. Additionally, DRIFT includes a profiling platform for DGF metrics and is incorporated with the Magma benchmark. Together, these features position DRIFT as a practical advancement in directed fuzzing within LibAFL.

Keywords: Fuzzing · Vulnerability Discovery · Dynamic Analysis · LibAFL · Greybox Fuzzing · Multiple CPU Architectures

1 Introduction

The proliferation of Internet of Things (IoT) devices, expected to reach 29 billion by 2030 [25], has brought unparalleled connectivity and convenience to daily life. However, this widespread integration is accompanied by a surge in cybersecurity risk akin to the recent increases in vulnerabilities [20]. In response to this

F. Maggi et al. (Eds.): DIMVA 2024, LNCS 14828, pp. 84–103, 2024.
https://doi.org/10.1007/978-3-031-64171-8_5

heightened threat landscape, vulnerability detection has evolved significantly in the past decade. In particular, fuzzing, a prominent method for identifying software vulnerabilities, has gained traction, including in the IoT domain [7]. Fuzzing involves providing pseudo-random input to a program and closely monitoring its behaviour to uncover anomalies. This technique has proven remarkably effective, as demonstrated by Google's OSS-Fuzz[1], which unearthed over 10,000 vulnerabilities in open-source software as of the current writing. Researchers have significantly advanced fuzzing since its introduction by incorporating new techniques leading to a more effective and efficient fuzzing process. One particularly novel technique deserving attention is directed greybox fuzzing (DGF) [1,5]. Introduced by Böhme *et al.*, AFLGo [1] propelled directed greybox fuzzing into the spotlight [5,11,13,21,28,30,32]. This technique deviates from conventional coverage-guided fuzzing by prioritizing the exploration and testing of specific code areas, as opposed to maximizing overall program coverage. Showcasing recent advancements in DGF, notable progress is exemplified by works such as Constraint Guided Directed Fuzzing (CAFL) [16], Monte Carlo Counting (MC2) [24], and Titan [11]. With a proven track record, these techniques can offer valuable applications in vulnerability detection including in the domain of IoT in areas such as patch testing [27]. However, the applicability of these techniques is limited by their implementations. Applying DGF to complex closed-source programs meets considerable challenges:

1. **Lack of Modularity:** Most directed fuzzers are forks of AFL or similar tools [26] which inherently lack modularity. Despite shared high-level components, each implementation becomes specific to its fuzzer, hindering integration across works. This trend, previously noted by LibAFL's authors [8], complicates the customization of the fuzzers which is often required when dealing with complex software.
2. **Limited Binary Support:** Directed fuzzers primarily target source code [26]. While some fuzzer tackled binary-only fuzzing (e.g., UAFuzz [21]), the overall focus on source code limits the applicability of DGF in situations where the source code is unavailable, especially for binaries compiled for less common architectures.
3. **Complexity and Maintainability:** AFL-like fuzzers are complex software. While extending them with heterogenous tools greatly improves their performance (e.g., FirmAFL [31]), it also introduces challenges like instability, maintenance burden, and difficulty integrating techniques like DGF.

These challenges underscore the need for improvements in modularity, binary application support, and user-friendly design. Within the state-of-the-art coverage-based fuzzers, LibAFL [8] addressed these challenges by standardizing and modularizing the fuzzing process. Recognizing the potential applicability of this approach to binary directed fuzzing, our primary objective of this research is to enhance the adaptability of DGF techniques specifically for binary-only

[1] https://github.com/google/oss-fuzz.

fuzzing in a modular manner, while ensuring compatibility across multiple architectures. This approach aims to move towards a seamless application of DGF in previously challenging scenarios, such as IoT binary fuzzing. Our research fills the gap in the existing literature by extending LibAFL's methodology to encompass directed greybox fuzzing with native support for cross-architecture binary applications. Emphasizing native integration into the LibAFL ecosystem has resulted in the creation of the Directed Fuzzing Toolkit (DRIFT). The latter is a library that introduces practical features, including the integration of a binary-only version of AFLGo [1] into LibAFL. Our implementation serves two purposes. Firstly, it provides researchers with a starting point for directed fuzzing, allowing them to engage in this area without the burden of re-implementing or navigating a steep learning curve associated with complex systems. Secondly, it offers practitioners an easily customizable directed fuzzer that can be tailored to suit specific targets leveraging the benefits of LibAFL. Our experiments show a noteworthy enhancement in vulnerability discovery performance, with a doubling effect and a significant acceleration, ranging from 9–40×, in target bug exploitation times. These improvements were achieved by incorporating five DRIFT components into a standard LibAFL fuzzer. During the design of this toolkit, we prioritized its application to binaries such as the ones present in IoT devices by ensuring at a minimum inherent support for multiple CPU architectures. Additionally, the framework includes a profiling platform for directed fuzzing metrics and is integrated with the de-facto benchmark Magma [9], simplifying directed fuzzing technique development and evaluation. Together, these features position DRIFT as a practical advancement in directed fuzzing within LibAFL, offering a versatile toolkit focused on adaptability and native binary support. We make the following contributions:

- Streamlining research on directed fuzzing by providing a novel DGF toolset to LibAFL based on the popular directed fuzzer, AFLGo.
- Enhancing the applicability of DGF for binaries across architectures.
- Assisting researchers in developing novel directed fuzzing techniques with real-time insight into DGF metrics via a profiling module.
- Providing integration into an adapted version of the Magma benchmark for binary directed fuzzers.
- Demonstrating the toolkit's benefits by outperforming AFLGo in binary mode through a novel combination of DRIFT and LibAFL components.

The remainder of this paper is organized as follows: Sect. 2 elaborates on the foundational concepts of DGF and modular fuzzers. In Sect. 3, we depict the challenges specific to binary applications of DGF, elucidating the motivations and design principles guiding the development of DRIFT. Section 4 details DRIFT's system design and implementation. Sections 5 and Sect. 6 present evaluations of our system in static analysis and fuzzing, respectively. Section 7 provides an overview of the related work, exploring potential integration points with DRIFT. Finally, in Sect. 8, we conclude with the contributions and insights of this paper.

2 Background

This section explores the research foundation of DRIFT, beginning with Directed Greybox Fuzzing followed by the recent trend towards modular fuzzer design.

2.1 Directed Greybox Fuzzing

The field of fuzzing categorizes tools based on their understanding of a program's internals [18]. Three main categories emerge white-box fuzzers, requiring extensive program understanding; black-box fuzzers, operating without prior knowledge; and our focus, grey-box fuzzers, acquiring insights through analysis. Our research relates to DGF, a branch of greybox fuzzing. AFLGo [1], a pioneering DGF tool, introduced this concept, departing from conventional coverage-based methods. DGF targets specific program areas, guided by a distance metric, representing proximity to one or more targets. The fuzzer evaluates this metric for each input execution and adjusts the mutations based on its value. By iterating through this process, the fuzzer minimizes the distance between input executions and targets thereby *directing* the fuzzing towards those areas. Subsequent works improved on the distance metric [5], its use in specific use cases [21,28], its impact on the fuzzing process [16,24,30] and handling of multiple target locations [10,11]. This targeted approach proves valuable in bug reproduction, patch testing, and scenarios requiring focused Program Under Test fuzzing (PUT) [26].

2.2 Modularity in Fuzzing

Despite significant progress in the field, practical adoption of fuzzing remains challenging [22]. Fioraldi *et al.*'s work on LibAFL [8] highlighted the difficulties arising from the common trend of innovators implementing their ideas as forks of established monolithic (non-modular) platforms like AFL [29], resulting in numerous independently maintained versions. This renders customization and integration across versions difficult and requires a thorough understanding of both the base fuzzer's internals and the variations between different versions. Employing these versions in real-world scenarios with complex programs proves even more challenging due to their limited accessibility, extensibility, and compartmentalization [8]. These challenges extend beyond coverage fuzzers, as many directed fuzzers adopt a similar approach [26]. Modular fuzzing frameworks, exemplified by LibAFL [8], address these challenges. They abstract the fuzzing process into abstract entities to foster reusability and customization in a modular manner. These traits can prove invaluable to both IoT fuzzing [7], and for the potential of integration across various fuzzing techniques [8].

3 Binary DGF Model Motivation

Upon examination of DGF literature [26], a prevailing trend emerges, wherein fuzzers primarily focus on source code rather than binary. This limitation significantly restricts the technique's applicability in scenarios where the code is inaccessible, despite operational commonalities between the source code and binary fuzzing. AFLGo, initially tailored for source code, suggests the potential for binary applications [1], a notion subsequently realized by Nguyen *et al.* [21] highlighting the technique's applicability to binary and its minimal overhead. In an effort to modularize and extend this approach across various architectures, we introduce changes to the two phases of directed fuzzing shown in Fig. 1:

1. **Static Analysis:** In AFLGo, the static analysis phase heavily depends on source code instrumentation accomplished through two compilations for structural information extraction and runtime distance metric accumulation. Unlike source-code methods, binary instrumentation does not rely on compilers; instead, execution information is obtained at runtime. Similarly to previous works [21,28], we replace the source code instrumentation through the use of binary decompilation and lightweight static analysis. The recent advances in decompilation tools [19] make them a viable approach.
2. **Fuzzing:** Directed fuzzers are often derived from coverage fuzzers like AFL [29] and VUzzer [26] which are augmented by incorporating the distance metrics in various fuzzing phases [16,24], such as *power scheduling* [1]. These modifications apply similarly to both binary and source code versions. The key distinction lies in program execution and program monitoring. Common techniques in this regard include emulation using tools like QEMU(Quick Emulator)[2] and dynamic instrumentation. Emulation is particularly interesting given its native support for multiple architectures.

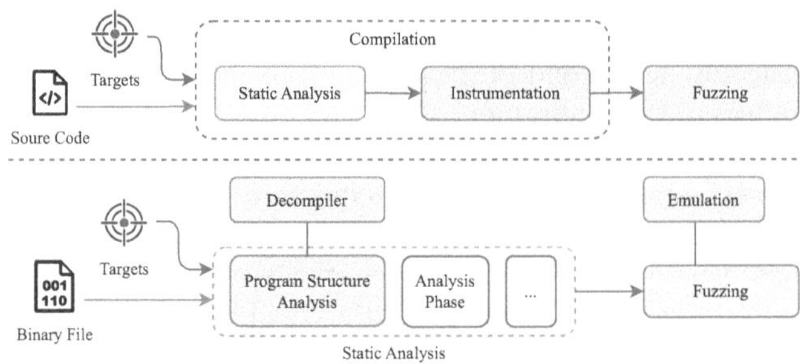

Fig. 1. Directed Greybox Fuzzing for Source Code (top) vs. Binary (bottom)

4 System Design and Implementation

In this section, we describe the concretization of our proposed model. Following the design principles established for this work, we detail the architecture and components of DRIFT.

[2] https://www.qemu.org/.

4.1 Design Principals

The design of the DRIFT prioritizes two key principles: modularity and binary compatibility. To minimize development effort, the system is built with swappable components, allowing for independent use and customization. This modularity applies to both the core fuzzing functionalities (based on LibAFL [8]) and the custom static analysis. Additionally, the model adopts a "binary-first" approach to work across different architectures. This is achieved by utilizing AFLGo's DGF techniques, known for their source-and-binary support [1,21,28], along with a carefully chosen decompiler with a powerful intermediate representation and LibAFL's QEMU extension for cross-architecture execution.

4.2 System Architecture

Following the model presented in Sect. 3, our system comprises two parts:

- **Static Analysis Engine (SAE):** This module in binary-only fuzzing is responsible for analyzing the PUT's binary. It extracts relevant program graphs, identifies target locations, and computes the necessary distance metrics used to direct the fuzzer toward those targets using AFLGo's algorithms.
- **Fuzzing Engine (FE):** Comprising a set of LibAFL-compatible components, this module provides directed fuzzing capabilities based on the AFLGo power schedule. The toolset includes auxiliary modules for monitoring runtime distance metrics and providing components for easier harnessing of Command Line Interface (CLI) and Common Gateway Interface (CGI) binaries, both often found in IoT firmwares.

The system architecture is visually represented in Fig. 2, illustrating these two modules as well as their internal components.

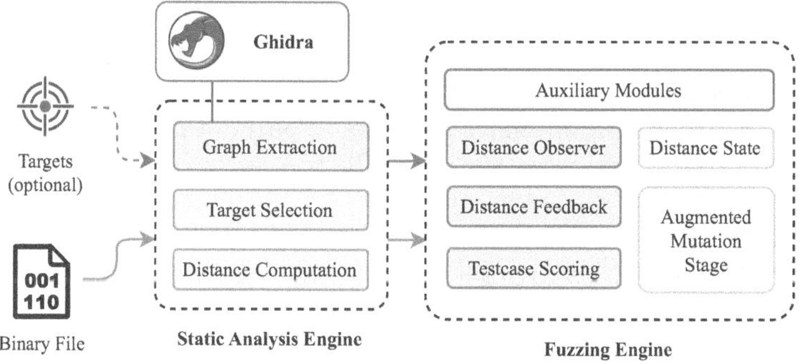

Fig. 2. High-level view of the system architecture and internal components

4.3 Static Analysis Engine

In Sect. 2.1, we emphasize the role of the static analysis phase in applying DGF to binaries. The module responsible for it serves two roles: structural information extraction and distance computation. While the distance computation closely mirrors its source code counterpart, the extraction of structural information from binaries is different. Specifically, SAE needs to extract the call graphs and control flow graphs directly from the binary to compute the AFLGo [1] distance metric. To this end, our system requires the use of a decompiler. We select the software reverse engineering suite Ghidra[3] for three key reasons:

1. Ghidra's free and open-source nature ensures platform accessibility and provides advanced users with greater control over binary analysis.
2. PCode, Ghidra's lifted language, provides CPU architecture-agnosticism, crucial for supporting diverse architectures with minimal overhead.
3. Ghidra's API facilitates scripting for extracting the supplementary data required by advanced distance algorithms [16,21] and dynamic target selection.

Following the integration of Ghidra, we divide the static analysis engine into three modules: Graph Extraction, Target Selection and Distance Computation.

Graph Extraction. The main role of this module is the extraction of the call graphs and intraprocedural control flow graphs needed by the AFLGo distance computation algorithm. It achieves this by leveraging the Rust library CWE Checker[4] which interfaces natively with Ghidra through a custom wrapper that also provides pre-analysis caching optimization. In addition to the required graphs, we extend the graph extraction's functionality to include the information on indirect calls and the program's interprocedural control flow graph, also represented in *high* PCode, Ghidra's intermediate language. We anticipate that subsequent works can benefit from an extensive set of pre-extracted graphs.

Target Selection. The design of the SAE incorporates flexibility by supporting two modes of target selection. The first aligns with approaches observed in AFLGo [1], Hawkeye [5], and other general-purpose DGFs. In this mode, a predetermined set of fixed targets is specified. Analogous to using source code line numbers, we rely on entry addresses of basic blocks. During the targeting phase, the module validates these addresses as legitimate basic blocks and extracts corresponding function information for distance computation. Inspired by UAFuzz [21], the second mode is more suitable for vulnerability discovery. It bypasses pre-defined targets, instead dynamically selecting them via program analysis. This analysis could involve leveraging the already integrated CWE Checker or even machine learning on software metrics extracted from Ghidra, similar to LEOPARD [6]. Figure 3 illustrates this integration within SAE.

[3] https://ghidra-sre.org/.

[4] https://github.com/fkie-cad/cwe_checker.

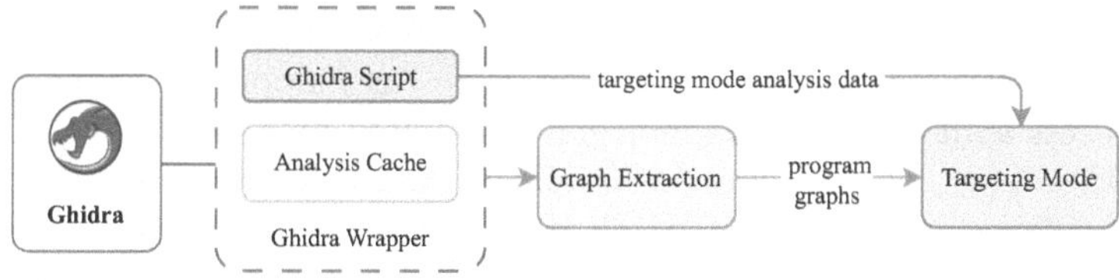

Fig. 3. High-level view of the dynamic target selection system integration

Distance Computation. This module is responsible for computing the distance metric—a crucial metadata used by the fuzzer to guide execution towards specified targets. In this initial iteration, we have implemented the distance computation algorithm of AFLGo [1]. This implementation lays the groundwork for the development of more sophisticated algorithms. The accuracy of this version is evaluated against the original, and the results are detailed in Sect. 5. After computing the distance metric for each basic block, this module outputs key-value pairs, representing annotated basic block addresses and their corresponding distance metrics.

4.4 Fuzzing Engine

DRIFT's fuzzing engine isn't a single module, but rather a concept for fuzzers built using LibAFL that leverage DRIFT for directed fuzzing. FE provides a DGF library integrating with LibAFL through four key components: Emulation Helpers, Observers, Feedback, and Scheduling. As seen in Listing 1, these components enable the creation of fuzzer configurations used in Sect. 6.

Emulation Helpers. As part of the binary-only support offered by LibAFL, the `libafl_qemu` crate introduces the concept of QEMU Helpers. They provide additional logic applied to the emulator before, during, and after the execution of an input. They prove particularly useful for extracting emulator-specific information during execution or customizing the emulator's behaviour. In our case, replicating the behaviour of distance accumulation typically done via instrumentation (mentioned in Sect. 2.1) requires tracking the addresses of the executed basic blocks for each input. To accomplish this, we developed the *Block Execution Tracking Helper*, which compiles a list of executed basic block addresses and provides them to the *Distance Observer*.

Observers. In LibAFL, observers record execution data, such as edge coverage, which later determines the input's *"interestingness"* [8] to guide processes like mutation and scheduling. Unlike AFLGo we do not instrument the source code to compute the distance metric hence our need to observe the execution of the program and compute it on the fuzzer side. In LibAFL terms, this translates to the creation of the *Distance Observer* for seed distance calculation.

Let s be an input (or seed). $V(s)$ is the set of basic block addresses visited by s, collected by the emulation helper, and $M(b)$ the mapping between the basic block address b and its distance metric as computed by SAE. Equation 1 shows the function $D(s)$, which, using $M(b)$, allows the observer to compute the seed distance metric for s. Note that $|V(s)|$ represents the cardinality of $V(s)$ where $M(b)$ is defined. With this equation, we replicate the logic performed by the instrumentation code added by AFLGo to a program at runtime.

$$D(s) = \frac{\sum_{b \in V(s)} M(b)}{|V(s)|} \tag{1}$$

Feedback. Feedback mechanisms interpret fuzzing information by contextualizing input observations against the campaign records. In the context of distance evaluation, understanding the metric's evolution requires interpreting it against previously computed values. To facilitate this, DRIFT introduces the *Distance Feedback*. This component serves two primary roles. Firstly, akin to AFLGo, it normalizes seed distance against recorded maximum and minimum values. The normalized distance is then stored in the corresponding Testcase object for scheduling. The second role of this component is to influence whether an input is retained for further mutations based on the newly computed seed distance being a new minimum. Diverging from AFLGo, this adaptation is LibAFL-specific and ensures the retention of seeds closest to targets, irrespective of coverage.

Scheduling. AFLGo directs the fuzzing process by adjusting AFL's power scheduling mechanism [1]. Specifically, it modifies the *performance score* calculated by AFL's *Explore* scheduler [2] through the integration of a *power factor*, computed via a Simulated Annealing algorithm applied to the normalized distance metric [15]. By incorporating this factor with AFL's performance score, AFLGo exerts influence over subsequent scheduling steps. To replicate this functionality, an examination of LibAFL's power scheduling components is necessary. These components include the Schedulers, responsible for gathering execution metadata; the *Corpus Power Testcase Score* (CPTS), which computes a performance score akin to AFL's; and the *Power Mutational Stage*, which utilizes the score to influence mutations.

Using this mechanism, DRIFT injects AFLGo's power factor by adjusting the original score computed by the CPTS. This adjustment is done through the custom *AFLGo Testcase Score*. To enable this score to influence scheduling, this score is integrated into the Power Mutation Stage from LibAFL creating the *AFLGo Power Mutation Stage*. Figure 4 provides a visual representation of this process and the aforementioned components.

4.5 Auxiliary Components

In addition to the aforementioned modules, DRIFT includes auxiliary LibAFL-compatible components for binary directed fuzzing:

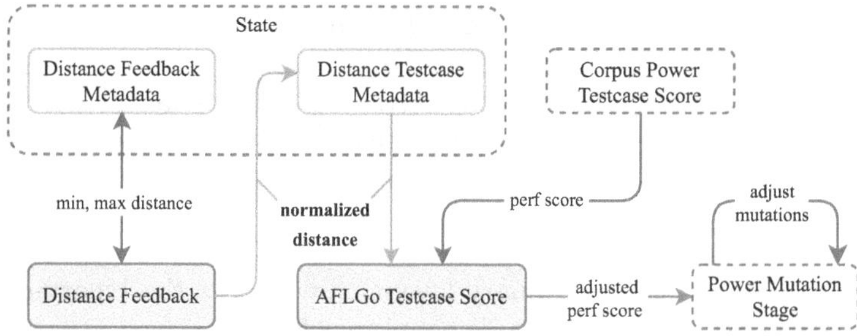

Fig. 4. Diagram illustrating the Power Scheduling components and their interactions

Distance Monitoring. When investigating new distance metrics, visualizing the distance metric's evolution and the paths taken by inputs provides valuable insights. Accordingly, we incorporate a module that exports this data at runtime, complemented by an interface for real-time profiling.

Harnessing Modules. To provide a base for harnessing IoT binaries, we designed two modules:

- **File Input Helper:** Inspired by contributions from the LibAFL community[5], this module handles input passing through files. This is done by hooking file-based system calls such as `open`, `read`, `write`, `lseek` and `stat` using the emulator and passing the data through the return values.
- **CGI Input Helper:** The second more IoT-specific provides the ability to harness CGI binaries. These binaries take inputs via environment variables. To pass this data, the fuzzer preloads a library provided with DRIFT that takes the inputs via a buffer in memory. At runtime, the module places inputs in this buffer via the emulator application programming interface.

4.6 Benchmark Integration

Magma [9] is a distinguished benchmarking tool and dataset for fuzzers, emphasizing vulnerability detection over metrics like coverage. While existing implementations support directed fuzzing in Magma, our version goes a step further by adding enhanced binary support. The extension facilitates the selection of target Bug IDs[6] during the campaign configuration and integrates a Ghidra-powered analysis tool to translate these IDs into their corresponding basic block addresses within the binaries. This process is achieved by locating the corresponding magma bug patches [10]. We incorporate our fuzzing engine into Magma to improve the developer experience and allow for rapid testing of prototyping of directed fuzzers on real-world programs.

5 https://epi052.gitlab.io/notes-to-self/tags/libafl.

6 An identifier defined by Magma for each CVE/Bug retrofitted to the programs.

5 Static Analysis Engine Evaluation

This section assesses the fidelity of the static analysis engine. Fidelity refers to how closely SAE's distance metrics, calculated directly on the binary code, align with those produced by AFLGo which were computed on the source code. The evaluation also investigates how factors like program size (affecting basic block matching), optimization levels (potentially disrupting block alignment), and CPU architecture (influencing binary layout) can impact the binary-computed distance compared to the source code counterpart.

5.1 Basline Evaluation

The baseline evaluation uses three programs originally examined by AFLGo: `LRZip`, `ObjDump`, and `XMLlint`. We compute the distance metric for the specified targets within each program and extend the analysis to the compiled programs using SAE. This evaluation requires comparing distance metrics calculated by SAE (basic block analysis) and AFLGo (source code analysis). To bridge this gap, a mapping between basic blocks and their corresponding source code elements was established. The **`readelf`** tool was employed to extract debugging information `.debug_line` that provides basic block addresses for each code block. The code was compiled with the `-O0` flag (no optimizations) to minimize compiler optimizations that could disrupt the mapping. Finally, leveraging these matched blocks, a correlation analysis was performed on the distance metrics calculated by both tools. Table 1 presents the results of our experiments, detailing the number of aligned blocks (Block Count) for each program, along with the corresponding distance metric correlation and the average delta between the two metrics. These results reveal a strong correlation ranging between 87% and 95% across the three programs, indicating a robust alignment between both tools.

Table 1. Distance Metrics Correlation, Delta and Count for Matched Blocks in LRZip, ObjDump, and XMLLint Programs.

Program	LRZip	ObjDump	XMLLint
Block Count	46	473	4536
Correlation	0.9467	0.8684	0.9455
Delta	5.6147	7.6465	5.5167

5.2 Block Count Sensitivity

Given the variability in the number of matched blocks across different programs (e.g., `LRZip` with 46 exact matches compared to `XMLLint`'s 4,500), we conduct additional analysis to assess the correlation's sensitivity to the number of blocks.

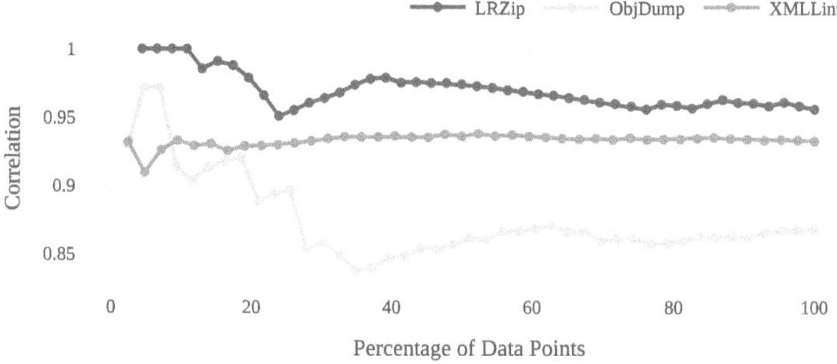

Fig. 5. Variation in distance metric correlation across different data points

We randomly select subsets of blocks from each program and compute the correlation for each subset. Results are shown in Fig. 5. Despite fluctuations in the correlation observed in all programs, we notice it stabilizes beyond 40%, consistently remaining at a notably high level above 85%. The correlation is not absolute because of various factors, such as compiler transformations, programming language syntax, and other factors that contribute to disparities between the original code and its disassembled representation [19].

5.3 Optimization Level Evaluation

Compiler optimizations can alter the program structure during compilation [23], potentially affecting SAE's accuracy. To assess this impact, we compiled the previous three programs with GCC[7] at four optimization levels (-O0 to -O3) and compared the resulting binaries with source code metrics. Figure 6 shows the correlation for each program across optimization levels.

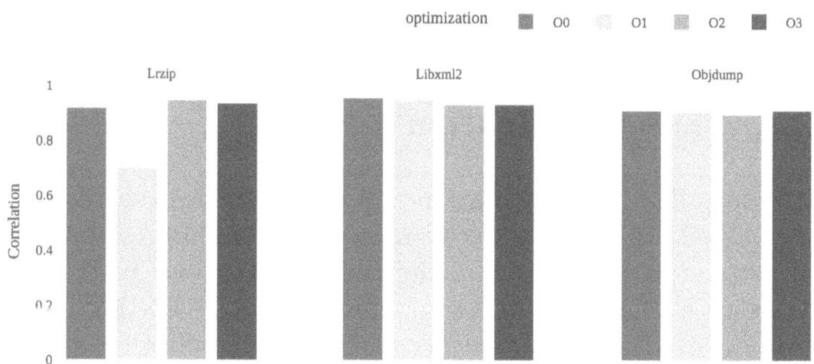

Fig. 6. Distance correlation for all programs across optimization levels

[7] https://gcc.gnu.org/.

Results showed a high correlation of around 85% that dipped slightly with higher optimization levels due to code changes for performance. An outlier was `LRZip` with `-O1`. Here, optimizations reduced the number of matched basic blocks, affecting the source-to-binary correlation. `LRZip` already had few matches at `-O0`, and `-O1` reduces it to 18 matches. Conversely, programs with many matched blocks (300–4000) maintained high correlations of around 90%.

5.4 Cross Architecture Evaluation

Cross architecture applicability was assessed using `XMLLint` due to its high matched blocks. We evaluate the correlation across four common IoT architectures (`ARMv7`, `ARM64`, `MIPS`, `X86_64`). Results as shown in Fig. 7 indicate a high correlation of more than 90% for all architectures. Architectures with similar register sizes (`ARMv7/MIPS` and `X86_64/ARM64`) had slightly closer correlations, suggesting register size might influence the source-to-binary mapping on the PCode level.

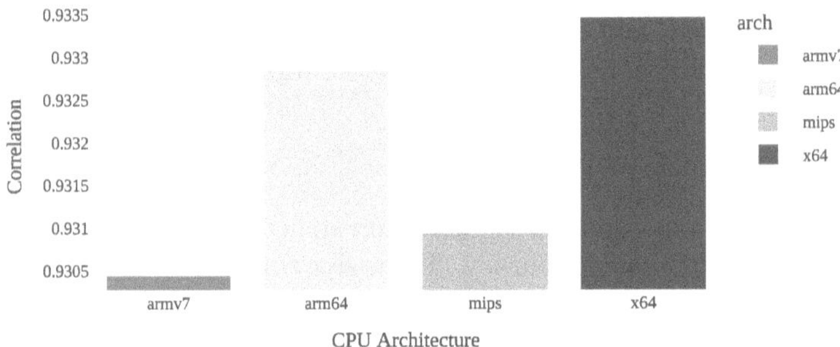

Fig. 7. Distance correlation for XMLLint across CPU architectures

6 Fuzzing Engine Evaluation

To facilitate controlled experimentation and isolate the impact of DRIFT components, we build DRIFT AFL, a coverage fuzzer based on LibAFL's sample binary-only fuzzer `fuzzbench_qemu`. Prioritizing control over peak performance, DRIFT AFL serves as a baseline for our directed fuzzer DRIFT AFLGo. This fuzzer incorporates DRIFT components while keeping all other elements identical. This approach allows for a direct comparison of the added components' performance against unmodified DRIFT AFL, AFLGo, and vanilla AFL using the modified Magma benchmark detailed in Sect. 4.6.

```
// Initialize distance observer using SAE-generated files
let observer = DistanceObserver::from_files(distances, targets);
// Update hooks with block execution tracking and other QEMU helpers
let mut hooks = QemuHooks::new(&emu, tuple_list!(
    BlockExecutionTrackingHelper::from_observer(&observer),
    // Other QEMU helpers ...
));
// Add distance feedback connected to the observer
let mut feedback = feedback_or!(
    DistanceFeedback::with_observer(&observer),
    // Other feedbacks ...
);
// Add AFLGo scheduler metadata to the global state
state.add_metadata(AFLGoSchedulerMetadata::new(
    SACoolingSchedule::EXP, exploration_time,
));
// Substitute StdPowerMutationalStage with AFLGo version
let power = AFLGoPowerMutationalStage::new(mutator);
```

Listing 1: Snippets showing the integrated DRIFT modules in LibAFL Fuzzer.

6.1 Fuzzer Configuration

- **DRIFT AFL:** Building on `fuzzbench_qemu`, we deliberately made trade-offs to avoid overemphasizing coverage. Specifically, we exclude features like `Cmp Logging` and `MOpt` mutations [17] and select the `AflMap` feedback over the `MaxMap` one. These choices, based on experimentation, ensure that the distance feedback, once integrated, can have a significant impact on the fuzzer's scheduling, guiding it toward more effective exploration paths.
- **DRIFT AFLGo:** This fuzzer is the directed counterpart of DRIFT AFL. It integrates the directed components from Sect. 4.4. The benefits of modularity are showcased in this extension which only requires four additional components and minimal adjustments of the base fuzzer as seen in Listing 1.

6.2 Benchmark Configuration

We compare the effectiveness of our two fuzzers against their source-code counterparts using a Magma benchmark campaign. The campaign includes five programs: `LibPNG`, `LibXML`, `LibTiff`, `LibSNDFile`, and `Poppler`. Other programs are excluded due to Ghidra decompilation issues. For each program, we run a 24-hour benchmark with three repetitions using one executable per program. The campaign consists of two phases:

- **Phase 1:** We evaluate the coverage fuzzers AFL and DRIFT AFL. From their results, we chose one high Time-To-Exploit (TTE) bug per program. This ensures exploitability within the campaign timeframe and provides ideal test cases for directed fuzzers.

– **Phase 2:** We evaluate the directed fuzzers AFLGo and DRIFT AFLGo using the selected bugs[8] `PNG001`, `XML009`, `TIF014`, `SND006`, and `PDF010`.

Both phases ran on the same Linux server: an Intel Xeon Gold 6130 CPU (2.10 GHz) with 500 GB of RAM, with one pinned core per fuzzer/target pair.

Table 2. Table displaying the restricted mean survival time for bug triggering, calculated using the Kaplan-Meier estimator. The 'Time Out' (T.O.) cells indicate untriggered bugs. The Ratio and p-values represent are compared to DRIFT AFLGo

Bugs	driftaflgo	afl			aflgo			driftafl		
	Time	Time	Ratio	P	Time	Ratio	P	Time	Ratio	P
XML009	1321.67	60818.33	46.02	—	59546.67	45.05	—	32186.67	24.35	—
TIF014	4881.67	37611.67	7.70	—	45805.00	9.38	—	37770.00	7.74	—
PDF010	8911.67	180.00	0.02	0.01	T.O	—	—	1898.33	0.21	0.02
SND006	2051.67	T.O	—	—	T.O	—	—	15353.33	7.48	0.28
PNG001	T.O	T.O	—	—	T.O	—	—	58013.33	—	—
PDF016	3998.33	313.33	0.08	0.02	315.00	0.08	0.02	866.67	0.22	0.03
PNG003	16.67	15.00	0.90	0.42	15.00	0.90	0.42	35.00	2.10	0.37
PNG007	14086.67	73690.00	5.23	—	32678.33	2.32	—	T.O	—	—
SND001	168.33	T.O	—	—	T.O	—	—	965.00	5.73	0.27
SND005	29351.67	1136.67	0.04	—	961.67	0.03	—	52168.33	1.78	—
SND007	860.00	T.O	—	—	T.O	—	—	683.33	0.79	0.03
SND017	54843.33	5288.33	0.10	—	4235.00	0.08	—	T.O	—	—
SND024	886.67	T.O	—	—	T.O	—	—	331.67	0.37	0.04
TIF007	386.67	3823.33	9.89	<0.01	3880.00	10.03	<0.01	63.33	0.16	0.10
TIF012	38813.33	8915.00	0.23	0.32	8645.00	0.22	0.33	72283.33	1.86	—
XML017	33.33	20.00	0.60	0.02	20.00	0.60	0.02	15.00	0.45	0.01

6.3 Results

In this section, we delve into the results of the Magma benchmark, employing the Kaplan-Meier restricted mean estimator [12] to assess the TTE. Presented in Table 2, the analysis focuses on bugs triggered by at least one fuzzer. For each fuzzer, we compute performance ratios and associated p-values relative to DRIFT AFLGo. It is important to note that the reported metrics, sourced from Magma, do not factor in emulation overhead which can be substantial, potentially ranging from 2–22× [21]. In analyzing these results we attempt to answer the following questions:

What Impact Did DRIFT Have on Target Bugs TTE? Our primary focus is to assess the impact of our directed fuzzing components on overall performance. We compare results from DRIFT AFLGo and DRIFT, revealing significant speed improvements, notably in `LibXML`, `LibTiff`, and `LibSNDFile`, with ratios ranging from 7–24× as shown in Table 2. While DRIFT AFLGo fell short in triggering the bug `PNG001` and underperformed with `Poppler`[9], the overall noticeable performance gain underscores the effectiveness of directed fuzzing.

[8] Full list of Magma bugs available at https://hexhive.epfl.ch/magma/.

[9] Poppler BugIDs are prefixed with `PDF`.

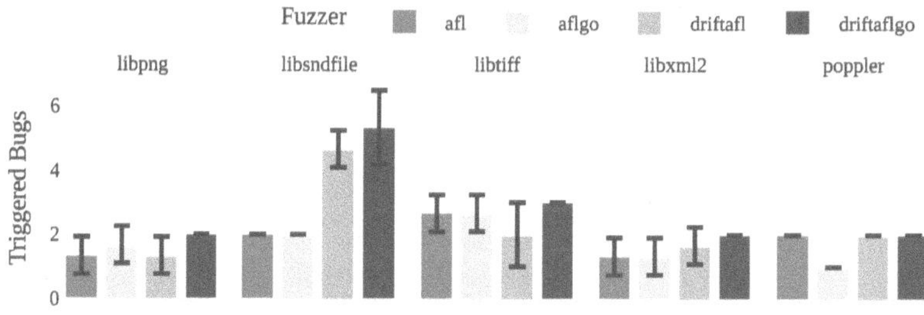

Fig. 8. Number of bugs triggered by each fuzzer, and the deviation between campaigns

How Does DRIFT AFLGo Compare to AFLGo? Examining the survival probability graphs for target bugs in Fig. 9, it is evident that DRIFT AFLGo consistently outperforms AFLGo in 4 out of 5 experiments. Table 2 provides further insights, indicating that it successfully exploited twice the number of bugs in our experiment while triggering the targetted ones significantly faster. Non-target bugs were, however, exploited more quickly by AFLGo. This experiment suggests two points: (1) our directed fuzzing implementation is valid and was able to compete against the original and (2) the advanced features of LibAFL and the modifications made by DRIFT help overcome the emulation overhead making directed fuzzing on binary viable.

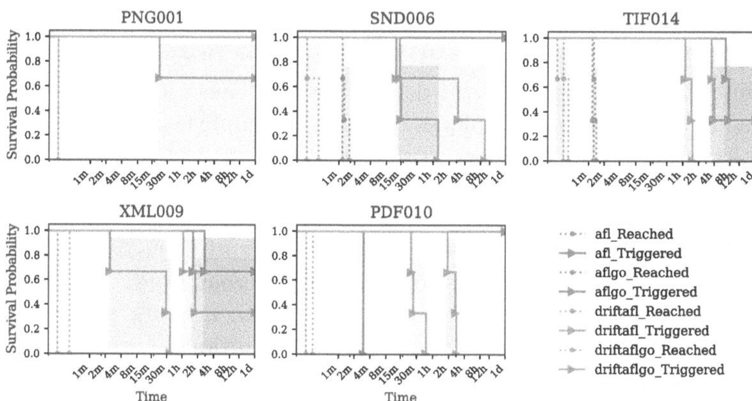

Fig. 9. Kaplan-Meier survival curves for the five selected target bugs.

How Did DRIFT Influence Overall Bug Discovery? Our exploration into the impact on bug discovery reveals that fuzzers based on LibAFL, overall, demonstrate similar or superior performance compared to their AFL-based counterparts with both DRIFT AFL and DRIFT AFLGo equaling and in cases doubling the number of bugs triggered. This is likely attributed to the enhanced scheduling and feedback mechanisms provided by LibAFL. Figure 8 highlights this difference by showing the number of bugs triggered by each fuzzer. When we compare LibAFL-based fuzzers among themselves we see that DRIFT AFLGo consistently triggers as many bugs or more than DRIFT, affirming that our directed fuzzing approach not only did not impede bug discovery but also exhibited notable performance gains in bug coverage.

7 Related Work

Recent developments in fuzzing were discussed in Sect. 2, laying the foundation for an exploration into related works and their potential integration into DRIFT. Starting with DGF, a significant focus lies on enhancing the Static Analysis phase, where various contributions have augmented distance computation by assimilating additional information from the PUT. Hawkeye [5] concentrates on call sequences necessary to trigger a target, UAFuzz [21] delves into the sequence order of memory-management functions pertaining to User-After-Free bugs, CAFL [16] considers data and order constraints on targets, and Titan [11] addresses the synergy between targets. These advancements can be integrated into SAE as new distance computation modules, accompanied by corresponding target selection strategies. In some cases like with CAFL [16], custom Ghidra scripts may be required to extract additional information from the PUT. We notice a trend in input prioritization strategies for enhancing the fuzzing phase of DGF. For instance, CAFL [16] introduces an input scoring algorithm incorporating target constraints. MC2 [24] utilizes a Monte Carlo counting model on execution information to prioritize inputs with lower difficulty reaching the targets. FishFuzz [30] improves multi-target support through dynamic ranking and culling of seeds based on their distance score. These strategies, along with other queue optimization techniques [3], can be implemented using a combination of custom testcase scores, as demonstrated in Sect. 4.4, and schedulers akin to the Minimizer and Probabilistic Sampling Schedulers offered by LibAFL [8]. Recent techniques from the state-of-the-art in IoT binary fuzzing such as mocking `libnvram` calls proposed by FirmAE [14] and Firmadyne [4] could be combined with DRIFT for better IoT support. Recent work, such as Butterfly[10], which explores the applicability of LibAFL in network protocol fuzzing could also be combined with DRIFT to support additional harnessing methods for directed fuzzing. Experimenting novel combinations from the state-of-the-art works is facilitated through DRIFT, thanks to both its modular architecture for quick prototyping and its integration with Magama for out-of-the-box benchmarking.

8 Conclusion

The culmination of this work is DRIFT, a modular binary-only directed fuzzing extension designed for seamless integration with LibAFL. Our approach involved presenting a high-level model of directed fuzzing on binary code, emphasizing

[10] https://github.com/fkie-cad/butterfly.

its distinctive requirements compared to its source code counterpart. Following this model, we successfully adapted directed fuzzing techniques from AFLGo to binary ensuring cross-architecture support by design through the use of emulation and Ghidra's lifted language. Evaluating our approach shows a high correlation in distance computation despite utilizing decompilation instead of source code. The two consistently correlated with rates of 85% on average. The results remained consistent across consistent across architectures and optimization levels. When tested against the original fuzzer on the Magma benchmark, our evaluation fuzzer yielded promising results with up to a doubling in bug discovery rates and speed benefits ranging from 9–40x in target bugs compared to AFLGo all without considering emulation overhead. It is fair to say that these benefits stem from the modularity of our system and its integration with the well-designed and highly performant LibAFL. We designed DRIFT intending to make it applicable to IoT software. While we recognize that IoT fuzzing is a complex challenge, DRIFT is a step towards addressing it, leveraging its support for various architectures and its highly customizable design. In future works, we aim to enhance our system support for IoT fuzzing and conduct evaluations across all major CPU architectures. Additionally, we plan to implement additional state-of-the-art techniques and assess the potential benefits of their combined use. We believe that the resulting toolkit not only offers a practical solution to current modularity challenges in directed fuzzing but also holds the potential for widespread reuse, contributing to improved performance and enhanced security across various platforms, including the intricate landscape of IoT.

References

1. Böhme, M., Pham, V., Nguyen, M., Roychoudhury, A.: Directed greybox fuzzing. In: Thuraisingham, B., Evans, D., Malkin, T., Xu, D. (eds.) Proceedings of the 2017 ACM SIGSAC Conference on Computer and Communications Security, CCS 2017, Dallas, TX, USA, 30 October–03 November 2017, pp. 2329–2344. ACM (2017)
2. Böhme, M., Pham, V., Roychoudhury, A.: Coverage-based greybox fuzzing as Markov chain. IEEE Trans. Softw. Eng. **45**(5), 489–506 (2019)
3. Canakci, S., Matyunin, N., Graffi, K., Joshi, A., Egele, M.: TargetFuzz: using darts to guide directed greybox fuzzers. In: Suga, Y., Sakurai, K., Ding, X., Sako, K. (eds.) ASIA CCS 2022: ACM Asia Conference on Computer and Communications Security, Nagasaki, Japan, 30 May–3 June 2022, pp. 561–573. ACM (2022)
4. Chen, D.D., Woo, M., Brumley, D., Egele, M.: Towards automated dynamic analysis for Linux-based embedded firmware. In: 23rd Annual Network and Distributed System Security Symposium, NDSS 2016, San Diego, California, USA, 21–24 February 2016. The Internet Society (2016)
5. Chen, H., et al.: Hawkeye: towards a desired directed grey-box fuzzer. In: Lie, D., Mannan, M., Backes, M., Wang, X. (eds.) Proceedings of the 2018 ACM SIGSAC Conference on Computer and Communications Security, CCS 2018, Toronto, ON, Canada, 15–19 October 2018, pp. 2095–2108. ACM (2018)
6. Du, X., et al.: Leopard: identifying vulnerable code for vulnerability assessment through program metrics. In: Atlee, J.M., Bultan, T., Whittle, J. (eds.) Proceedings of the 41st International Conference on Software Engineering, ICSE 2019, Montreal, QC, Canada, 25–31 May 2019, pp. 60–71. IEEE/ACM (2019)

7. Eceiza, M., Flores, J.L., Iturbe, M.: Fuzzing the internet of things: a review on the techniques and challenges for efficient vulnerability discovery in embedded systems. IEEE Internet Things J. **8**(13), 10390–10411 (2021)
8. Fioraldi, A., Maier, D.C., Zhang, D., Balzarotti, D.: LibAFL: a framework to build modular and reusable fuzzers. In: Yin, H., Stavrou, A., Cremers, C., Shi, E. (eds.) Proceedings of the 2022 ACM SIGSAC Conference on Computer and Communications Security, CCS 2022, Los Angeles, CA, USA, 7–11 November 2022, pp. 1051–1065. ACM (2022)
9. Hazimeh, A., Herrera, A., Payer, M.: Magma: a ground-truth fuzzing benchmark. In: Huang, L., Gandhi, A., Kiyavash, N., Wang, J. (eds.) SIGMETRICS 2021: ACM SIGMETRICS/International Conference on Measurement and Modeling of Computer Systems, Virtual Event, China, 14–18 June 2021, pp. 81–82. ACM (2021)
10. He, Y., Zhu, Y.: RLTG: multi-targets directed greybox fuzzing. PLOS ONE **18**(4), 1–23 (2023)
11. Huang, H., Yao, P., Hung-Chun, C., Guo, Y., Zhang, C.: Titan: efficient multi-target directed greybox fuzzing. In: 2024 IEEE Symposium on Security and Privacy (SP), p. 59. IEEE Computer Society (2024)
12. Kaplan, E.L., Meier, P.: Nonparametric estimation from incomplete observations. J. Am. Stat. Assoc. **53**(282), 457–481 (1958)
13. Kim, J., Yun, J.: Poster: directed hybrid fuzzing on binary code. In: Cavallaro, L., Kinder, J., Wang, X., Katz, J. (eds.) Proceedings of the 2019 ACM SIGSAC Conference on Computer and Communications Security, CCS 2019, London, UK, 11–15 November 2019, pp. 2637–2639. ACM (2019)
14. Kim, M., Kim, D., Kim, E., Kim, S., Jang, Y., Kim, Y.: FirmAE: towards large-scale emulation of IoT firmware for dynamic analysis. In: ACSAC 2020: Annual Computer Security Applications Conference, Virtual Event/Austin, TX, USA, 7–11 December 2020, pp. 733–745. ACM (2020)
15. Kirkpatrick, S., Gelatt, C.D., Jr., Vecchi, M.P.: Optimization by simulated annealing. Science **220**(4598), 671–680 (1983)
16. Lee, G., Shim, W., Lee, B.: Constraint-guided directed greybox fuzzing. In: Bailey, M.D., Greenstadt, R. (eds.) 30th USENIX Security Symposium, USENIX Security 2021, 11–13 August 2021, pp. 3559–3576. USENIX Association (2021)
17. Lyu, C., et al.: MOPT: optimized mutation scheduling for fuzzers. In: Heninger, N., Traynor, P. (eds.) 28th USENIX Security Symposium, USENIX Security 2019, Santa Clara, CA, USA, 14–16 August 2019, pp. 1949–1966. USENIX Association (2019)
18. Manès, V.J.M., et al.: The art, science, and engineering of fuzzing: a survey. IEEE Trans. Softw. Eng. **47**(11), 2312–2331 (2021)
19. Mantovani, A., Compagna, L., Shoshitaishvili, Y., Balzarotti, D.: The convergence of source code and binary vulnerability discovery - a case study. In: Suga, Y., Sakurai, K., Ding, X., Sako, K. (eds.) ASIA CCS '22: ACM Asia Conference on Computer and Communications Security, Nagasaki, Japan, 30 May–3 June 2022, pp. 602–615. ACM (2022)
20. National Institute of Standards and Technology (NIST): NVD - CVSS Severity Distribution Over Time. https://nvd.nist.gov/general/visualizations/vulnerability-visualizations/cvss-severity-distribution-over-time. Accessed 12 May 2023
21. Nguyen, M., Bardin, S., Bonichon, R., Groz, R., Lemerre, M.: Binary-level directed fuzzing for use-after-free vulnerabilities. In: Egele, M., Bilge, L. (eds.) 23rd International Symposium on Research in Attacks, Intrusions and Defenses, RAID 2020, San Sebastian, Spain, 14–15 October 2020, pp. 47–62. USENIX Association (2020)

22. Nourry, O., Kashiwa, Y., Lin, B., Bavota, G., Lanza, M., Kamei, Y.: The human side of fuzzing: Challenges faced by developers during fuzzing activities. ACM Trans. Softw. Eng. Methodol. **33**(1) (2023)
23. Ren, X., Ho, M., Ming, J., Lei, Y., Li, L.: Unleashing the hidden power of compiler optimization on binary code difference: an empirical study. In: Proceedings of the 42nd ACM SIGPLAN International Conference on Programming Language Design and Implementation, pp. 142–157 (2021)
24. Shah, A., She, D., Sadhu, S., Singal, K., Coffman, P., Jana, S.: MC2: rigorous and efficient directed greybox fuzzing. In: Yin, H., Stavrou, A., Cremers, C., Shi, E. (eds.) Proceedings of the 2022 ACM SIGSAC Conference on Computer and Communications Security, CCS 2022, Los Angeles, CA, USA, 7–11 November 2022, pp. 2595–2609. ACM (2022)
25. Vailshery, L.S.: IoT connected devices worldwide 2019-2030 | statista (2023). https://www.statista.com/statistics/1183457/iot-connected-devices-worldwide/. Accessed 15 Nov 2023
26. Wang, P., Zhou, X.: SoK: the progress, challenges, and perspectives of directed greybox fuzzing. CoRR abs/2005.11907 (2020). https://arxiv.org/abs/2005.11907
27. Xiao, F., Sha, L., Yuan, Z., Wang, R.: VulHunter: a discovery for unknown bugs based on analysis for known patches in industry internet of things. IEEE Trans. Emerg. Top. Comput. **8**(2), 267–279 (2020)
28. Yang, S., et al.: 1dFuzz: reproduce 1-day vulnerabilities with directed differential fuzzing. In: Proceedings of the 32nd ACM SIGSOFT International Symposium on Software Testing and Analysis, ISSTA 2023, pp. 867–879. Association for Computing Machinery, New York (2023)
29. Zalewski, M.: American fuzzy lop - a security-oriented fuzzer. https://github.com/google/AFL (2014). http://lcamtuf.coredump.cx/afl. Accessed 2023
30. Zheng, H., et al.: FISHFUZZ: catch deeper bugs by throwing larger nets. In: Calandrino, J.A., Troncoso, C. (eds.) 32nd USENIX Security Symposium, USENIX Security 2023, Anaheim, CA, USA, 9–11 August 2023, pp. 1343–1360. USENIX Association (2023)
31. Zheng, Y., Davanian, A., Yin, H., Song, C., Zhu, H., Sun, L.: FIRM-AFL: high-throughput greybox fuzzing of IoT firmware via augmented process emulation. In: Heninger, N., Traynor, P. (eds.) 28th USENIX Security Symposium, USENIX Security 2019, Santa Clara, CA, USA, 14–16 August 2019, pp. 1099–1114. USENIX Association (2019)
32. Zhu, X., Böhme, M.: Regression greybox fuzzing. In: Kim, Y., Kim, J., Vigna, G., Shi, E. (eds.) CCS 2021: 2021 ACM SIGSAC Conference on Computer and Communications Security, Virtual Event, Republic of Korea, 15–19 November 2021, pp. 2169–2182. ACM (2021)

Malware and Threats

Constructs of Deceit: Exploring Nuances in Modern Social Engineering Attacks

Mohammad Ali Tofighi, Behzad Ousat(✉), Javad Zandi, Esteban Schafir, and Amin Kharraz

Florida International University, Miami, USA
bousa001@fiu.edu

Abstract. Despite the increasing effort in the defense community to develop robust security solutions, social engineering attacks are getting more prevalent every year. Detecting fraudulent websites has been a concurrent task of both academia and industry in combating this type of attack. A common approach is to use supervised methods and labeled data to locate suspicious cases. In this paper, we evaluate a set of more common features related to the development and deployment aspects of websites that have been widely used in detecting scam and phishing websites over the years. As threat actors and the defense community are in a cat-and-mouse game, we aim to investigate whether such features are still prevalent or how to move forward in determining signs of malice when looking at the problem space at scale. Our study challenges the efficacy of deployment-based features, such as infrastructure providers or certificate issuers, in detecting fraudulent websites. Additionally, we perform an empirical analysis of the development aspects of websites that can be utilized in the detection pipeline.

1 Introduction

Social engineering attacks persist as a significant security threat. The impact of these attacks is often deep and consequential. Modern social engineering attacks have evolved to deliver different classes of malicious code while collecting extensive financial and personal information [1–5]. Moreover, these attacks result in significant collateral damage by harming the reputation and necessitating substantial effort to mitigate abuse. Over the years the security defense community has developed various methods and tools to fight against social engineering attacks [6–10]. The core insight in a large number of prior works is that the majority of adversaries behind social engineering attacks are *cost sensitive*. That is, adversaries aim to minimize their costs to develop and distribute their social engineering web attacks. This observation has been translated into several detection heuristics on the defense side. For instance, prior work incorporated features that extracted the type of domain name or the network address to which they were resolved based on the intuition that adversaries are more likely cheaper domain names or network addresses or that they would use free services such as Let's Encrypt [11] more frequently to develop realistic websites.

F. Maggi et al. (Eds.): DIMVA 2024, LNCS 14828, pp. 107–127, 2024.
https://doi.org/10.1007/978-3-031-64171-8_6

Given that adversaries are continuously adapting their techniques to evade detection, a research question that arises is to what degree these features are still relevant in today's threat landscape. This paper aims to revisit the corresponding insights, aiming to answer what measures are still effective and can be insightful, and what measures are losing their effectiveness in this evolving landscape. Our work is guided by three primary research questions. First, we aim to investigate what has changed in network infrastructure to host fraudulent websites and if there still exists any distinguishing patterns in the usage of network addresses. Second, we investigate how premium certificate services, as a key component of the deployment mechanism, are being used in fraudulent website development. Lastly, we evaluate the adoption of software development techniques in the design and development of fraudulent websites and how they can be translated into defense mechanisms. To answer these questions, we partnered with a well-known security company and received daily access to their URL seeds for six months – from August 2022 to February 2023. We crawled these websites using an instrumented browser, collecting development and deployment-related information when visiting these websites. In the following, we describe the main findings of the experiments by analyzing 9.5 TBs of data collected over this period. It is also notable that to reduce false positives and evade websites, we removed all scans that resulted in an HTTP redirection to another domain. We also collected a list of the top 1 million websites from Chrome's UX Report (CrUX) and scanned them using our crawler, collecting the same set of artifacts.

The Underlying Infrastructure of Fraudulent Websites are Getting More Diverse and Distributed. We identified 1,995 infrastructure providers utilized in fraudulent websites. 1,764 (89%) of the providers are being also used in the top legitimate websites. We observed that 20 ASNs were present in over 1,000 cases on both fraudulent and legitimate websites. Notably, *Cloudflare* was the top infrastructure provider in both categories with 461,913 (51.22%) instances in legitimate URLs and 66,258 (30.96%) times observed for the fraudulent group. The second common provider was *Amazon* with 279,135 (30.95%) and 31,043 (14.5%) instances in legitimate and fraudulent groups respectively.

The Adoption of Certificate Services in Adversarial Settings is Getting Significantly More Diverse. While we observed that Let's Encrypt, a free certificate authority, is the most prevalent certificate issuer for both legitimate and fraudulent websites (38% and 36% respectively), premium SSL certificate services are also prevalent in the fraudulent websites. In particular, we observed that over 84,289 (40%) of the certificates used in fraudulent websites belong to premium services such as DigiCert, GoDaddy, and Sectigo.

The Usage of Major Web Technologies is Often Very Similar Among Fraudulent and Legitimate Websites. Similar to legitimate websites, modern fraudulent websites also use popular Javascript libraries (e.g., core-js, lodash), or development frameworks (e.g., bootstrap, animate-ccs). However, the difference is visible for technologies from advertising and analytics categories. Specifically, we observe that over 600,000 (67%) of legitimate websites utilize *analytics* tools but the usage is diminished to nearly 24,000 (11%) on fraudulent webpages. We used Wappalyzer to detect web technologies used in websites; based on the version we used in our analysis, we were able to detect more than 4,500 technologies across 102 categories. Among other findings, perhaps the main takeaway of this paper is to provide practical evidence in modern social engineering attacks. The features that were historically being used in classifications are not likely to be effective for detecting and attributing cost-sensitive attacks anymore. We claim that the definition of cost-sensitive attacks needs to be adapted to stay effective, given the dynamic changes that are taking place in the adversarial landscape.

This paper makes the following contributions:

- We built a dataset containing various artifacts, including the run-time behavior artifacts, web technologies being used, certificate details, and underlying network infrastructure for 213,958 fraudulent and 901,817 legitimate websites over 12 months.
- We performed a large-scale longitudinal analysis of fraudulent websites, including phishing and scam websites, on how they differ from legitimate websites in development and deployment
- We provided two case studies using the collected dataset. The dataset is accessible upon request containing over 9TBs of artifacts for legitimate and fraudulent websites.

2 Background and Related Work

A substantial volume of research has been dedicated to the analysis and identification of fraudulent websites. We specifically focus on supervised approaches, wherein diverse features are extracted from both legitimate and fraudulent websites. Typically, a machine learning model is trained to classify between the two categories effectively. To comprehensively analyze the related studies, we categorize the features into two groups: development and deployment. Development features are the ones that are determined during the development of the website such as the structure of the page or the utilized technologies. On the other hand, deployment features are considered the ones that come from the deployment decisions such as the infrastructure provider or SSL certificate. Table 1 includes a summary of common features in each of these categories and shows whether a related study has used them in their classifier or not. In the following, we briefly describe the features and the related works that have utilized them for detection purposes. Despite the merits associated with each of these approaches, they all exhibit vulnerability to evasion tactics employed by attackers

Page DOM. DOM of the Page can play a crucial role in fraud detection as it provides insights into the structure and content of a page. Pan et al. [12] utilized features related to a web page's identity, including the page title, destination of HTTP requests, anchor elements, and content of the HTML body. Similarly, Rosiello et al. [13] focused on the graph representation of the page DOM to extract features related to the number and structure of HTML tags. Other studies such as [15,17,20,21] also leveraged page DOM features in their classifiers.

Table 1. Related Studies on Fraudulent Website Detection

Study	Venue	Year	Selected Features								
			Development						Deployment		
			Page DOM	Remote Resources	Technologies	HTML Content	CSS	Script Tags	SSL Cert	ASN/IP Address	Page URL
[12]	ACSAC	2006	✓	✓		✓					
[13]	SecureComm	2007	✓								
[14]	NDSS	2010	✓	✓						✓	✓
[15]	ACM TOPS	2011	✓	✓							✓
[16]	NDSS	2011								✓	
[17]	INCoS	2013	✓				✓				
[18]	APWG eCrime	2015							✓		
[19]	USEC	2016	✓								
[20]	ESORICS	2017	✓	✓		✓	✓				✓
[21]	ASIA CCS	2017	✓	✓							
[22]	IEEE ISI	2018			✓			✓	✓		
[23]	AISec	2018							✓		
[24]	IEEE S&P	2018	✓	✓					✓		
[25]	EURASIP	2019					✓				
[26]	ESWA	2022	✓	✓	✓						✓
[27]	IEEE S&P	2023	✓	✓						✓	
[28]	ISCC	2023	✓								

Remote Resources. The utilization of remote resources is another aspect considered in fraud detection models. Whittaker et al. [14] discussed Google's ML-powered classifier, analyzing the URL of the website and considering external links and images included in the web page to make a decision. Studies like [15,26] also explored remote resource features in their detection mechanisms.

Web Technologies. The technologies employed in a website can be indicative of its legitimacy or fraudulent intent. Studies including [26] considered the presence of specific web technologies in their classifiers. Niakanlahiji et al. [22] detected web technologies used in the website using HTTP headers and used them to detect the targets.

HTML Content. Features extracted from the HTML content of websites provide valuable information for detection. Pan et al. [12] and Corona et al. [20] utilized features related to HTML content to distinguish between legitimate and fraudulent websites. These features include the content of the HTML body, page title, and stylesheet-related elements in the DOM.

CSS Features. The CSS of a page can also serve as a discriminative feature in classification. Studies such as [17,20,25] extracted features from the CSS of web pages to build their classifiers.

Script Tags. The presence and characteristics of script tags in the HTML code were considered by related studies. In [22], authors check the code complexity of the included JavaScript as features for detecting fraudulent websites.

SSL Certificate. SSL certificate-related features have been utilized in scam detection. Torroledo et al. [23] and Dong et al. [18] based their work solely on SSL certificates, using various metadata for detection.

ASN/IP Address. Analyzing ASN/IP addresses is another aspect considered in scam detection. Whittaker et al. [14] used the ASN/IP address as one of the features in their model. Additionally, Bilge et al. [16] focused on DNS traffic and their resolved IP addresses.

Page URL. In some cases, the URL of the website is used to decide without actually visiting the website. Studies including [15,26,27], incorporated features derived from the page's URL in their classifiers.

Despite all of the advancements in the last two decades, the core techniques used by the defense community to detect malicious entities on the Web remain the same. This paper aims to see an adaptation of adversaries to these approaches and whether they are still good indicators or whether there is a need for novel data points to distinguish between the two groups.

3 Methodology

In this section, we introduce the three research questions we seek to answer. We then elaborate on how we conducted our research to collect forensically relevant artifacts to answer these questions. We also explain our approach to clean the dataset before running the large-scale analysis.

3.1 Research Questions

Our research seeks to answer the following research questions:

RQ1: How do adversaries abuse network infrastructure? There have been several studies on the use of free hosting services or abusing legitimate infrastructure to publish fraudulent websites. This research question aims to answer how the usage of underlying infrastructure has changed over time and if there is a distinguishing pattern in the underlying network structure in legitimate and fraudulent websites in modern social engineering attacks.

RQ2: How do adversarial campaigns make use of premium certificate services? There is no lack of evidence that adversaries have been using certificates to look legitimate when developing their fraudulent websites. The question we seek to answer is to investigate the distribution of legitimate/premium certificate providers in fraudulent web development and how the distribution differs between legitimate and fraudulent websites.

RQ3: How do fraudulent websites differ from legitimate websites when considering web development aspects of the websites? Our intuition is that, unlike legitimate and high-profile websites, adversaries are less likely to prioritize best practices and software development details of their fraudulent websites. We take a deeper look on if this hypothesis holds and how it would be different among different forms of websites.

Scope of the Study. There have been several work on different classes of social engineering attacks. Scareware [29,30], Survey and technical scams [24,31], themed-based attacks [32] are just a few examples of social engineering attacks.

The term fraudulent website we are using in this paper covers websites such as phishing as well as those types of scam websites designed to trick incoming users and incorporate certain elements of social engineering such as deception to deliver the attack. That said, the term does not cover other forms of social engineering attacks that do not require development of a website like common forms of spear-phishing or technical scams [31].

3.2 Datasets

Fraudulent Websites. We collaborated with a security company and subscribed to their premium URL seed, which exclusively provides URLs confirmed to direct users to fraudulent web pages. We collected the longitudinal list of 313,110 fraudulent websites over the period of six months from August 2022 to February 2023. This dataset was continuously updated at five-minute intervals, initiating scanning immediately upon capture. Throughout the scanning process, we encountered certain failures, primarily stemming from two key sources. The first category of failed scans arose from non-responsive URLs, which often result from the short lifespan of phishing websites. The second category encompassed scans that appeared to evade the scanner by redirecting to known legitimate web pages. After removing the failed scans including non-responsive URLs, possible evasions, and duplicate URLs, we ended up with 213,958 successful scans of phishing, scams, and other types of fraudulent websites. It is noteworthy to mention that the fraudulent dataset covers malicious websites in 50 different languages which shows the diversity of the targets.

Legitimate Websites. Our scanning pipeline is not limited to fraudulent websites. To have a more comprehensive overview, we incorporated the top 1 Million URLs included in Chrome's User Experience Report [33]. The complete dataset which is publicly accessible includes over 15M URLs. The dataset provides many attributes including the origin (URL) of the pages per month [34]. We used the data from February 2023 in our crawling pipeline. An overview of the datasets and the number of successful scans on each of them is presented in Table 2.

Table 2. URLs Scanned Using Different Data Sources

Type	Legitimate	Fraudulent
Source	CrUX Top 1M	Daily Seed
Date	Feb 2023	Aug 2022 - Feb 2023
Scans	901,817	213,958

3.3 Collected Artifacts

We built the forensics layer on top of the Chrome Devtools Protocol [35] to minimize the source code modifications and avoid possible crashes. The Chrome Debugging Protocol offers programmatic access to the browser engine and allows

Table 3. Top infrastructure providers used in legitimate and fraudulent websites. 153,286 (71.64%) of the fraudulent websites share the same infrastructure provider to deliver their page.

#	Legitimate Websites			Fraudulent Websites		
	Name	Presence		Name	Presence	
		#	%		#	%
1	Cloudflare	461,913	51.22	Cloudflare	66,258	30.96
2	Amazon	279,135	30.95	Weebly	38,356	17.92
3	Google	32,950	36.53	Amazon	31,043	14.50
4	Microsoft	31,975	35.45	Fastly	22,652	10.58
5	Akamai	31,450	34.87	Google	16,335	7.63
6	OVH	31,226	34.62	Microsoft	10,527	4.92
7	Fastly	29,614	32.83	Unifiedlayer	9,092	4.24
8	Hetzner	23,852	26.44	Dedipath	8,121	3.79
9	DigitalOcean	15,269	16.93	Namecheap	6,862	3.20
10	Incapsula	7,845	8.69	Network Solutions	5,355	2.50
11	Azion Technologies	6,832	7.57	DigitalOcean	3,796	1.77
12	XServer	6,689	7.41	Protocol	2,713	1.26
13	Hostinger	5,829	6.46	OVH	2,675	1.25
14	Ionos	5,186	5.75	Quantline Networks	2,597	1.21
15	Korea Telecom	4,737	5.25	Quadranet Global	2,093	0.97

the code to attach to open windows and interact with the loaded JavaScript and the DOM tree created for the window. We used Chrome's Lighthouse [36] which is an open-source tool to monitor the performance, quality, and correctness of web applications. This tool is offered as a library that includes all of the extracted data from the website. We based our crawler on this tool and took advantage of custom data gatherers to collect additional data from different aspects of each page. These custom gatherers include but are not limited to *DOM Graph Gatherer* which adds the graph representation of the DOM of the page to be used in graph analysis; The *Network Metadata Gatherer*, which captures all DNS A records that the domain resolves to, along with ASN information of the associated IP addresses for the website's domain. Additionally, we implemented the *SSL Certificate Gatherer*, which captures the SSL certificate of the domain and stores it in a PEM file for future use. Furthermore, we incorporated a fingerprinting technique to detect used libraries in the target web applications [37]. While Lighthouse includes some JavaScript libraries in the result, the service does not cover a large list of web technologies used at the internet scale. In each scan, we collect the source code of the page, the loaded JavaScript files and their execution, resources and technologies, the entire HTTP request and responses, and the certificate information loaded into the browser while visiting the target website. The result of each scan is a JSON object that is archived based on the date of the scan. A sample of the collected data for both legitimate and fraudulent websites is publicly available in a GitHub repository[1]. Details to access the full dataset are provided in the repository.

[1] https://github.com/phishvsbenign/phishvsbenign.

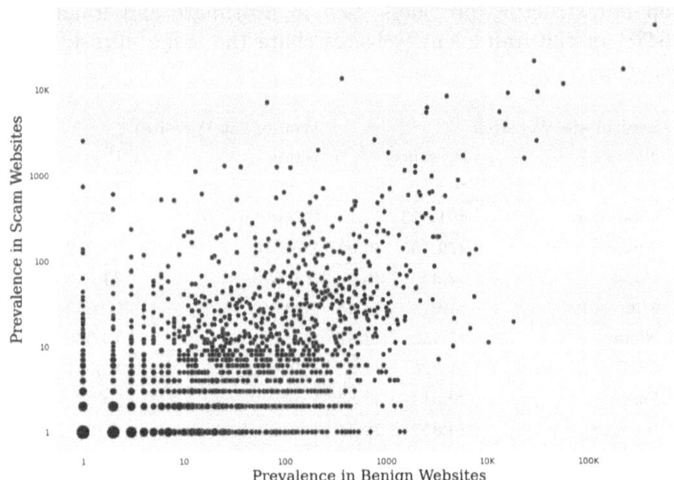

Fig. 1. The use of ASNs in legitimate and fraudulent websites. 1,764 (89%) ASNs observed in the fraudulent websites are present on the legitimate side as well. X and Y axes are log-scaled for visualization.

4 Deployment Aspects of Modern Web Ecosystem

In this section, we empirically study the deployment of fraudulent websites and compare them with legitimate ones by looking at the implementation details in three different ways: (1) the underlying infrastructure used to deploy fraudulent websites, (2) the certificate services used to issue certificates for fraudulent pages, (3) the common web technologies used to deliver these websites while comparing them with legitimate websites.

4.1 Deployment Distribution

To gain insights into the hosting environment of both fraudulent and legitimate websites, we analyzed the deployment of these websites in the wild by studying their network addresses. Our methodology involved identifying the top ASNs associated with fraudulent websites and examining their prevalence in the benign context. The results of this investigation are depicted in Fig. 1, which illustrates the frequency of each target ASN appearing in both fraudulent and legitimate websites. The analysis shows that 1,764 of 1,995 (89%) ASNs observed in the fraudulent websites deployment are being used in both fraudulent and legitimate websites. Table 3 shows the top cloud service providers observed in our analysis. Nearly 72% of the fraudulent websites share the same infrastructure services (e.g., Cloudflare, Amazon, Google, Fastly, and OVH) being used in legitimate websites. This observation suggests that modern social engineering attacks are increasingly using infrastructures that were historically serving legitimate services – influencing the effectiveness of security mechanisms that monitor the hosting reputation.

Table 4. Top certificate issuers used in legitimate and fraudulent websites. 175,363 (81.96 %) of the fraudulent websites use the same top certificate authorities that legitimate websites use.

Legitimate Websites						Fraudulent Websites					
Name	Presence		Validity Days			Name	Presence		Validity Days		
	#	%	min	max	mean		#	%	min	max	mean
Let's Encrypt	328,207	38.3	90	90	90	Let's Encrypt	78,855	36.8	90	90	90
Cloudflare	140,651	16.4	15	366	365	DigiCert	27,662	12.9	91	825	371
DigiCert	90,507	10.5	19	826	365	Google Trust Services	17,306	8.0	84	91	88
Sectigo Limited	68,790	8.0	26	397	376	cPanel	13,344	6.2	91	366	92
Amazon	49,396	5.7	141	396	394	Cloudflare	12,322	5.7	365	366	365
GlobalSign	41,796	4.8	22	825	372	Sectigo Limited	9,767	4.5	91	397	376
Google Trust Services	25,698	3.0	31	91	88	GoDaddy.com	3,888	1.8	39	398	380
GoDaddy.com	23,705	2.7	26	398	380	Amazon	3,842	1.7	338	396	394
cPanel	22,611	2.6	91	366	92	ZeroSSL	2,717	1.2	91	366	114
Entrust, Inc.	7,650	0.8	79	397	376	Entrust, Inc.	2,138	0.9	277	397	376
Unizeto Technologies	5,670	0.6	48	395	364	Microsoft Corp.	1,563	0.7	182	366	347
ZeroSSL	5,599	0.6	91	366	114	GlobalSign nv-sa	1,462	0.6	34	826	372
Starfield Technologies	4,898	0.5	28	398	374	Starfield Technologies	497	0.2	187	398	374
GEANT Vereniging	4,759	0.5	117	396	365	D-Trust GmbH	198	0.0	233	394	366
SECOM Trust Systems	2,692	0.2	131	397	385	TrustAsia	178	0.0	91	397	359

4.2 Certificate Analysis

In addition to the deployment infrastructure, we analyzed how certificate authorities are being abused in fraudulent websites. In particular, we analyzed the certificate issuers and certificate validity periods in the collected datasets and compared them with each other. Aligned with prior work [38–40], usage of SSL certificates by phishing and scam websites has become increasingly prevalent to imitate. We observed that 176,397 (82.4%) of the fraudulent websites are using SSL certificates to deliver their content. This number for the legitimate websites was 856,144 (94.9%) which shows a very narrow gap between the two.

Certificate Authorities. In our datasets, we identified a total of 147 certificate authorities involved in issuing SSL certificates. From these, 48 (33%) of the CAs had been abused in generating at least one certificate for a malicious website. Table 4 shows the top certificate issuers in legitimate as well as fraudulent websites. While the analysis shows that fraudulent websites use a smaller set of certificate issuers to generate certificates compared to legitimate websites, the number of certificates and frequency at which we observed in the legitimate dataset do not reveal any evident patterns, That said, it is not likely that certificate information, including certificate issuance, lifespan helps to distinguish these adversarial cases compared to legitimate websites.

Certificate Validity Periods. Additionally, we analyzed the validity period by calculating the number of days between issuance and expiration dates. The results of this analysis are presented in Table 4, revealing analogous trends in both scenarios, where the average duration for both categories of websites exhibits a high degree of similarity.

5 Development Practices in Fraudulent Websites

In this section, we present an analysis of the development aspects of fraudulent websites, from the complexity of their resulting HTML pages to UX and security practices that are expected to be seen in professionally developed legitimate websites. Specifically, we analyze the attributes of the Document Object Model (DOM), the prevalence of distinct HTML tags, the utilization of CSS/JS properties, two of the most common security practices, and web technologies used in the websites. This comparative study aims to highlight distinctive patterns and characteristics that can aid in the identification and differentiation of fraudulent websites from their counterparts.

5.1 DOM Stats Analysis

Analysis of a small set of data showed us that if we create the DOM graph of the page, the size of the graph is significantly different for legitimate versus fraudulent pages. Figure 2 presents a sample login form and its corresponding DOM graph using the graph drawing tool available at [41]. The Figure illustrates the elements of the page as well as width and height.

We investigated the idea by extracting the DOM statistics from the collected data and comparing the results for each category. Figure 3 showcases the CDF graph of DOM statistics for legitimate and fraudulent scans. The number of total body elements has a significant difference in the two cases. The mentioned reasons made us believe that DOM stats may be a useful feature to include in our analysis. We aim to identify distinguishing metrics that could effectively

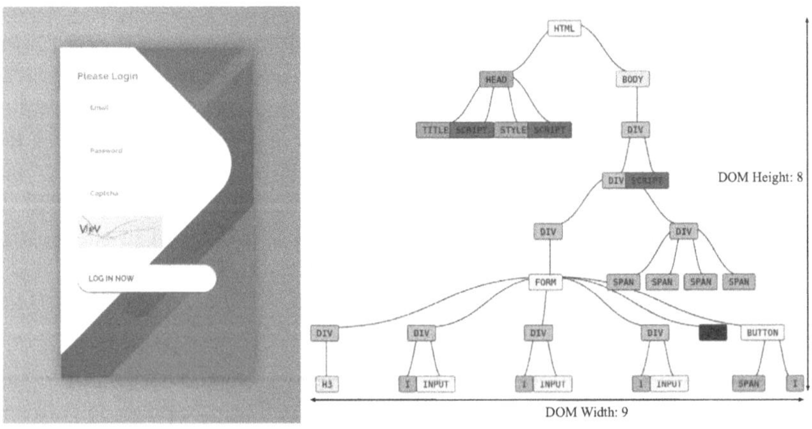

Fig. 2. Sample of a login form and its corresponding DOM graph

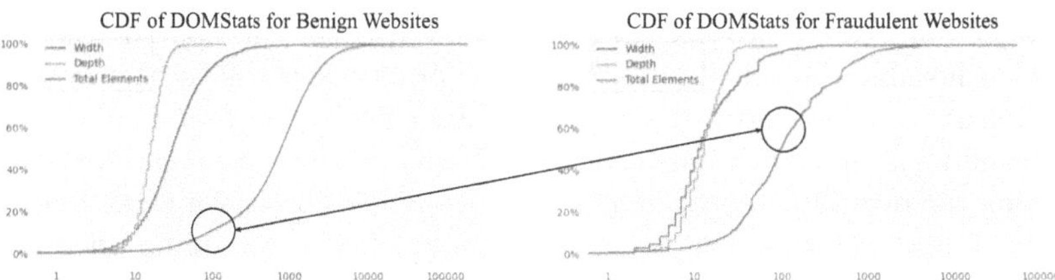

Fig. 3. DOM Statistic Comparison Between Legitimate and Fraudulent Websites. 60% of fraudulent websites have total elements of only 100 or less.

differentiate between the two by fingerprinting the usage of different resources in the document and the statistics of DOM elements in the pages. By examining these aspects, we try to uncover potential patterns that could aid in accurately distinguishing between the two types of websites. Figure 3 showcases the CDF graph of DOM statistics for legitimate and fraudulent scans. The number of total body elements has a significant difference in the two cases. We can see that a larger percentage of fraudulent websites have a low number of elements. For example, 60% of fraudulent websites have total elements of only 100 or less.

5.2 Resource Inclusions

The hypothesis is that developers of scam and phishing websites focus mainly on the look and feel of their target websites to defraud victims and focus less on common programming best practices that a legitimate website would consider. This part of the analysis aims to see whether fraudulent websites are less complex than legitimate ones in terms of included resources, or if they do they are most likely come with errors or not used at all when observing at the run-time behavior of the website. Figure 4 shows the comparison between the two.

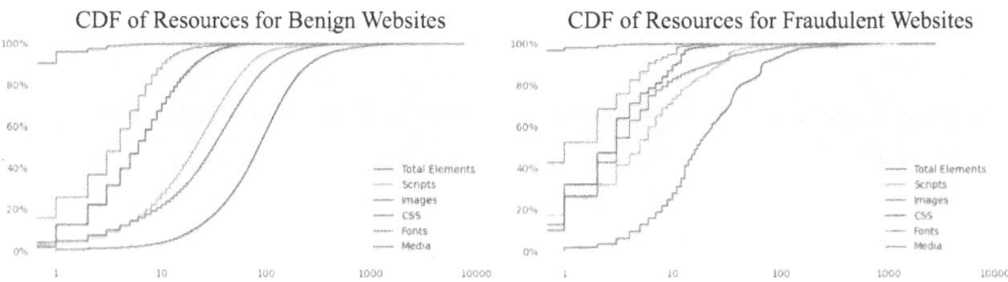

Fig. 4. Resource inclusion comparison between legitimate and fraudulent websites. Legitimate websites are more complex in terms of each resource type.

5.3 Scripts Analysis

In our investigation, we analyzed the sourced JavaScript in the collected websites and focused on evaluating the implementation of best practices, such as minification and the removal of unused JavaScript. As anticipated, the results revealed a notable discrepancy between the practices employed by developers of fraudulent websites and legitimate ones. Specifically, we found that developers of fraudulent websites tend to neglect these optimization techniques. When comparing the overall percentage of potential savings (based on total JavaScript size), legitimate websites demonstrated a modest 0.82% for unmodified JS and a more substantial 14.17% for unused JS. In contrast, these figures increased significantly for fraudulent websites, reaching 3.59% for unmodified JS and an impressive 29.15% for unused JS. Our findings underscore the importance of adhering to best practices in JavaScript development, as they not only optimize website performance but also serve as potential indicators for distinguishing between legitimate and fraudulent websites. By recognizing and addressing these disparities, we can gain better insights for developing protection solutions against social engineering attacks.

5.4 Security Mechanisms

One of the hypotheses we explored in our research was the disparity in security focus between fraudulent websites and legitimate ones. To investigate this, we employed two essential criteria: (1) Content Security Policy (CSP) headers, and (2) The usage of HTTPS. Upon analyzing the data, striking differences emerged between fraudulent and legitimate websites. We found that only 17% of legitimate websites had enforced CSP policies, while this number drastically dropped to less than 8% for fraudulent websites.

In terms of HTTPS analysis, we observed that the majority (82.4%) of the fraudulent websites were equipped with an SSL certificate. However, the main question would be to check whether these websites exclusively operated under HTTPS or not. This entails determining if plain-text HTTP requests are made while the primary page is presented through HTTPS, or if the website effectively redirects all HTTP traffic to HTTPS. We observed that 11% of legitimate websites lacked full HTTPS support, but this figure skyrocketed to 45% for fraudulent websites, which shows that while they are paying attention to obvious notations to use an SSL certificate, they do not care about the follow-up security practices. It is crucial to note that while these factors alone may not be conclusive evidence of a website being fraudulent or legitimate, incorporating them alongside other relevant indicators can significantly improve the accuracy of detection. By considering these features in combination with other factors, we can enhance our ability to identify and combat potential phishing and scam threats more effectively.

5.5 Technologies Analysis

One of the crucial aspects of analyzing websites from a development perspective is examining the employed technologies. We investigated the adoption of software packages on fraudulent and legitimate websites. Specifically, we queried the category of the observed packages and marked the top five most widely utilized ones

Table 5. Top Five Mostly Observed Technologies in Different Categories for Legitimate and Fraudulent Websites

Category	Rank	Legitimate Websites			Fraudulent Websites		
		Name	Presence		Name	Presence	
			#	%		#	%
JavaScript Libraries	1	jquery	682,617	75.69%	jquery	110,275	51.54%
	2	core-js	352,214	39.05%	core-js	33,726	15.76%
	3	lodash	102,086	11.32%	lodash	15,907	7.43%
	4	lazysizes	82,983	9.20%	fancybox	9,886	4.62%
Programming Languages	1	php	435,297	48.26%	php	77,904	36.41%
	2	java	41,587	4.61%	node-js	6,352	2.96%
	3	node-js	34,504	3.82%	typescript	4,503	2.10%
	4	python	14,836	1.64%	java	4,036	1.88%
Analytics	1	google-analytics	599,485	66.47%	google-analytics	24,906	11.64%
	2	facebook-pixel	192,854	21.38%	snowplow-analytics	19,505	9.11%
	3	hotjar	48,276	5.35%	datadog	9,249	4.32%
	4	yandex-metrika	35,223	3.90%	ads-conversion-tracking	3,424	1.60%
Font Scripts	1	google-font-api	314,761	34.90%	google-font-api	34,935	16.32%
	2	font-awesome	189,689	21.03%	font-awesome	25,024	11.69%
	3	twitter-emoji-twemoji	91,451	10.14%	typekit	1,682	0.78%
	4	typekit	18,170	2.01%	twitter-emoji-twemoji	1,209	0.56%
Video Players	1	youtube	50,279	5.57%	videojs	9,820	4.58%
	2	mediaelement-js	12,320	1.36%	vimeo	9,493	4.43%
	3	vimeo	11,599	1.28%	mediaelement-js	9,488	4.43%
	4	videojs	8,082	0.89%	youtube	1,521	0.71%
Tag Managers	1	google-tag-manager	497,073	55.11%	google-tag-manager	15,744	7.35%
	2	adobe-launch	7,321	0.81%	adobe-launch	820	0.38%
	3	tealium	3,747	0.41%	ensighten	543	0.25%
	4	matomo-tag-manager	1,363	0.15%	tealium	528	0.24%
Advertising	1	google-adsense	102,534	11.36%	microsoft-advertising	1,143	0.53%
	2	google-publisher-tag	48,804	5.41%	33across	1,053	0.49%
	3	twitter-ads	45,319	5.02%	twitter-ads	1,016	0.47%
	4	microsoft-advertising	39,519	4.38%	dtscout	910	0.42%
Web Servers	1	nginx	225,808	25.03%	apache	63,116	29.49%
	2	apache	203,441	22.55%	nginx	53,948	25.21%
	3	iis	88,416	9.80%	litespeed	13,056	6.10%
	4	litespeed	30,493	3.38%	openresty	8,144	3.80%
CMS	1	wordpress	184,452	20.45%	weebly	19,175	8.96%
	2	drupal	18,580	2.06%	wordpress	5,478	2.56%
	3	joomla	8,182	0.90%	godaddy-website-builder	2,836	1.32%
	4	1c-bitrix	5,970	0.66%	adobe-experience-manager	2,134	0.99%
Frontend	1	bootstrap	293,081	32.49%	bootstrap	51,112	23.88%
	2	animate-css	47,785	5.59%	tailwind-css	5,120	2.39%
	3	zurb-foundation	16,606	1.84%	animate-css	3,442	1.60%
	4	tailwind-css	16,167	1.79%	marko	2,417	1.12%

for each category. Several categories, including JS libraries, programming languages, and web servers, emerged as consistent front-runners within both groups. Yet, a noteworthy observation emerged-categories like advertising and analytics exhibited a pronounced prevalence primarily within legitimate websites. We present the outcomes of our investigation into the utilized technologies in each website group. Our findings indicate that the technologies used in the development of fraudulent websites differ from those utilized in legitimate websites. While there are certain similarities in the technology stacks of both categories, we observe that scam websites are less likely to employ technologies from *analytics, advertising,* and *tag managers* categories. Table 5 provides a comprehensive overview of the top 10 categories observed in fraudulent and legitimate websites, showcasing the prevalence of top packages in each category. These disparities in technology adoption between fraudulent and legitimate websites serve as valuable indicators that can aid in distinguishing between the two and contribute to more effective identification and mitigation of social engineering attacks.

6 Case Studies

Throughout our analysis of different features between legitimate and fraudulent websites, we found interesting cases revealing possible future avenues of analysis in the fraudulent websites. The first case study discusses how utilizing development best practices on the defense side can potentially lead to more robust detection mechanisms where analyzing visual representation, certificate details, and infrastructure may not be sufficient. The second case study relies on the idea that cost-sensitive adversaries rely on code reuse practices, similar to legitimate developers, to reduce the cost of development. We investigate how this intuition can be translated into threat detection and attribution.

6.1 Case Study 1: Cloning Legitimate Websites for Fraud

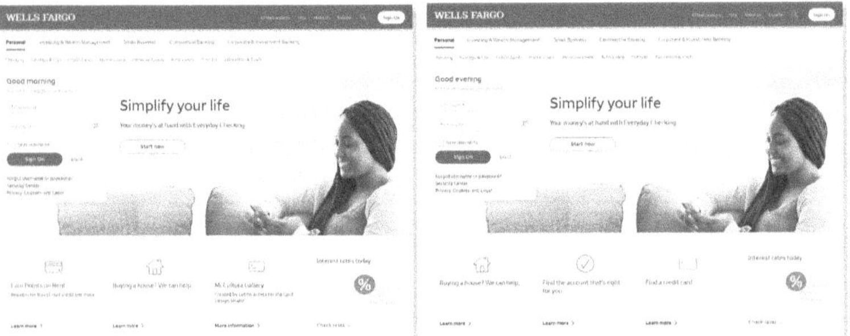

(a) Original Wellsfargo Website (b) Fake Wellsfargo Website

Fig. 5. Example of scam page developed by duplicating the target legitimate counterpart. The two pages look identical and use the same set of technologies.

In numerous instances, we observed a disconcerting trend where fraudulent/phishing websites closely mimic their legitimate counterparts. Our investigation suggests that these deceptive pages are crafted by duplicating the entire target page and hosting it on a different domain with minimal alterations. Consequently, many of the criteria discussed in Sect. 5 prove inadequate in differentiating between authentic and fraudulent web pages. Figure 5 exemplifies one of these cases, showcasing two nearly identical pages of *Wellsfargo Bank*. Besides the visual similarity, we verified that the identified technologies were too similar to be reliable discriminators. However, a more in-depth analysis revealed a crucial distinction- the prevalence of errors, such as console messages, during the loading process of the page. Notably, the authentic Wells Fargo website exhibited zero console messages, whereas the fraudulent counterpart generated 12 network and JavaScript errors resulting from failures in accessing the original Wells Fargo servers. This is due to the fact that the scam page includes scripts that directly interact with the original servers. The act of duplicating these scripts without modification led to the observed errors. The detailed breakdown of errors for this specific case is available in Table 6. It is essential to note that the infrastructure features, including domain, certificate, and ASN, differ between the two websites; however, they cannot be deemed as dependable identification factors. As outlined in Sect. 4.1, scam pages strategically leverage the same deployment environment as authentic websites. Consequently, it becomes feasible to migrate the exemplified fraudulent website to the identical environment as the original target page, effectively obfuscating any differentiating elements in the infrastructure analysis. This case study underscores that despite adversaries duplicating their target websites to manipulate end-users, there are discernible differences that can be identified. With the right tools and perspectives, the defense side can effectively safeguard users from falling victim to such manipulative tactics.

Table 6. Console Errors Generated from Scam Wellsfargo Website

Error Source	Error Text
javascript (*6)	Access to XMLHttpRequest at https://connect.secure.wellsfargo.com/... from origin www–wellsfargo–com–... has been blocked by CORS policy: The 'Access-Control-Allow-Origin' header contains the invalid value 'connect.secure.wellsfargo.com'.
network (*6)	Failed to load resource: net::ERR_FAILED https://connect.secure.wellsfargo.com/.../...638a.js

6.2 Case Study 2: Code Reuse for Development

Among our cases, we found fraudulent pages that were identical in several aspects. One of these cases is presented in Fig. 6 where the three pages have the same DOM structure with minimal changes in the logo image. Table 7 presents the comparison of these three pages based on their domain, technologies, DOM stats, certificate issuer, and IP address. This particular case highlights the tendency of adversaries, much like legitimate developers, to employ code reuse, facilitating the widespread deployment of their fraudulent pages. An application of the aforementioned features discussed earlier in this manuscript is the potential to identify such scam campaigns, which may originate from the same source. We believe this is an interesting opportunity for defenders to formulate robust defense mechanisms for similarity testing. That is, customized similarity hashing mechanism similar to prior work in malware analysis research [42] can assist us to better catalog similar fraudulent websites without relying on the visual representation, infrastructure, and certificate which can be evaded.

Table 7. Comparison between a set of visually-identical fraudulent websites

Domain	*creditinternationalbank.com*	*wealthpathbank.com*	*moonstandcu.com*
Technologies	jQuery, jQuery-Fast, FastClick, Drupal		
DOM Stats	DOM Height: 20, DOM Width: 31, DOM Elements: 409		
Certificate	cPanel Inc.	Let's Encrypt	cPanel Inc.
IP Address	82.163.176.106	192.185.46.77	162.241.176.106

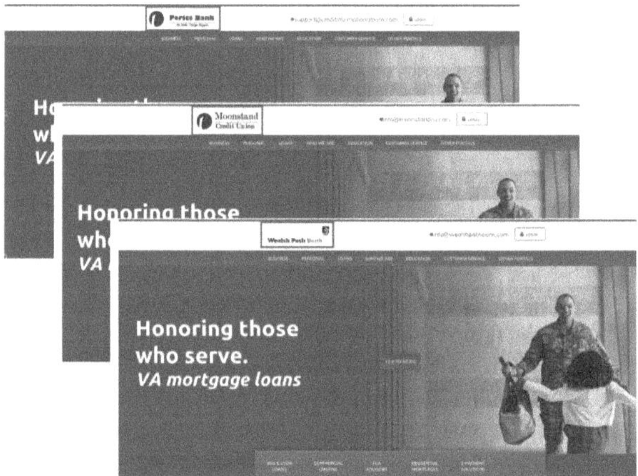

Fig. 6. Screenshots of different domains using the same design, but changing the bank name and logo.

7 Discussion and Future Works

In this section, we discuss the key findings derived from various analyses conducted during this research. Additionally, we put forth prospective avenues for further exploration and investigation.

Detecting Fraudulent Websites. Across various sections, we delved into multiple feature categories obtained from analyzing fraudulent web pages. Our observations indicate that although some of these features have been employed in prior phishing and scam detection efforts, their reliability may not be consistent across all cases. Notably, our analysis unveiled similarities between the deployment environment and certificate issuers of the fraudulent and legitimate websites. While past studies have found success in leveraging these features for phishing and scam detection, we believe that depending exclusively on such features diminishes the accuracy of the approach when applied in real-world scenarios. Additionally, our examination extended to the HTML features of the target pages, revealing a notable difference in complexity between scam pages and legitimate websites. Primarily, fraudulent pages exhibit a significantly reduced usage of fonts, images, and scripts, along with a smaller total count of DOM elements. This disparity can be attributed to the fact that adversaries prioritize functionality over aesthetics, concentrating solely on obtaining visitor credentials with minimal effort, thereby ensuring the scalability of their deceptive sites. Furthermore, we analyzed the development aspects of the pages to compare the utilized technologies and best practices. We observed differences in the usage of specific technology categories and libraries. Overall, we conclude that effective phishing and scam detection solutions necessitate a comprehensive approach, considering various factors and analyzing multiple facets of the target website to arrive at a reliable decision.

Runtime Features Analysis. In our investigations, we encountered instances where the static features of fraudulent pages closely mirrored those of their legitimate counterparts. This resemblance is especially prevalent in phishing websites, which can adeptly clone the front-end of authentic pages. Section 6.1 provides a detailed exploration of one such case. However, despite the visual similarity, we identified that the generated error messages on the scam page can serve as a robust indicator for detection. This insight underscores the potential effectiveness of analyzing runtime features in distinguishing between fraudulent and legitimate pages. The generated error messages represent a subset of other runtime features, encompassing console messages, network traces, and logs from development tools. Each of these features includes valuable information that holds promise for enhancing fraud detection. We leave the comprehensive analysis of runtime features to future research endeavors.

Campaign Analysis. Throughout our research, our focus was on identifying distinctive features of fraudulent and legitimate websites. During this process, we encountered instances where the same scam page was found across different domains, maintaining an identical design while altering the name and logo of the bank. Intriguingly, our brief analysis illuminated that the clusters of websites were operational under disparate domain names and distinct networks, characterized by distinct IP addresses and ASNs. While a subset of these clusters displayed visual resemblances, it was fascinating to find groups that exhibited entirely different designs while focusing on the same target group. These examples underscore the significant influence that integrating multiple page attributes can wield in enhancing the detection of phishing websites and conducting thorough campaign analysis. Discovering campaigns can help find scam pages at a higher rate and even prevent similar scam pages from being deployed from scratch which helps the defense side in the arm-race against adversaries.

8 Conclusion

In this study, we present an overview of techniques for detecting fraudulent websites. To investigate the details of the features prevalent in the contemporary fraud ecosystem, we collected artifacts by crawling a live feed of fraudulent URLs over six months. Our experiments indicate that the deployment aspects of fraudulent websites closely resemble those of legitimate ones, making them an unreliable source for distinction. However, observable differences emerge in the development features of the two groups. Notably, we found that fraudulent pages exhibit significantly lower complexity compared to legitimate ones. Furthermore, the adoption of best practices is largely overlooked in the fraud ecosystem. Additionally, we showcase case studies on how the collected features can serve purposes beyond mere detection, emphasizing the broader utility of our findings.

Acknowledgments. We would like to thank the anonymous reviewers for their thoughtful feedback. This project was supported by Microsoft AI Security and CITES/IUCRC Grant No. 2113880. Opinions, findings, conclusions, or recommendations expressed in this material are those of the author(s) and do not necessarily reflect the views of Microsoft Incorporation or NSF.

References

1. Business Insurance. Ransomware victims paid $18 billion ransom in 2020 (2021). https://www.businessinsurance.com/article/20210429/STORY/912341506/Ransomware-victims-paid-$18-billion-ransom-in-2020
2. PurpleSec Inc. 2021 Ransomware Statistics, Data, & Trends (2021). https://purplesec.us/resources/cyber-security-statistics/ransomware/
3. Venkatesha, S., Reddy, K.R., Chandavarkar, B.R.: Social engineering attacks during the COVID-19 pandemic. SN Comput. Sci. **2**(2), 1–9 (2021)
4. Verizon Inc. Data Breach Investigation Report 2020 (2020). https://enterprise.verizon.com/resources/reports/2020-data-breach-investigations-report.pdf
5. Security Magazine Inc. Microsoft warns of Russian Nobelium phishing campaign (2020). securitymagazine.com/articles/95328-microsoft-warns-of-russian-nobelium-phishing-campaign
6. Stringhini, G., Thonnard, O.: That ain't you: blocking spearphishing through behavioral modelling. In: Almgren, M., Gulisano, V., Maggi, F. (eds.) DIMVA 2015. LNCS, vol. 9148, pp. 78–97. Springer, Cham (2015). https://doi.org/10.1007/978-3-319-20550-2_5

7. Bhardwaj, T., Sharma, T.K., Pandit, M.R.: Social engineering prevention by detecting malicious URLs using artificial bee colony algorithm. In: Pant, M., Deep, K., Nagar, A., Bansal, J.C. (eds.) Proceedings of the Third International Conference on Soft Computing for Problem Solving. AISC, vol. 258, pp. 355–363. Springer, New Delhi (2014). https://doi.org/10.1007/978-81-322-1771-8_31
8. Neupane, A., Saxena, N., Kuruvilla, K., Georgescu, M., Kana, R.K.: Neural signatures of user-centered security: an fMRI study of phishing, and malware warnings. In: NDSS (2014)
9. Abeywardana, K.Y., Pfluegel, E., Tunnicliffe, M.J.: A layered defense mechanism for a social engineering aware perimeter. In: 2016 SAI Computing Conference (SAI), pp. 1054–1062. IEEE (2016)
10. Wen, Z.A., Lin, Z., Chen, R., Andersen, E.: What. Hack: engaging anti-phishing training through a role-playing phishing simulation game. In: Proceedings of the 2019 CHI Conference on Human Factors in Computing Systems, pp. 1–12 (2019)
11. Let's Encrypt. Let's Encrypt Certificate Provider (2024). https://letsencrypt.org/
12. Pan, Y., Ding, X.: Anomaly based web phishing page detection. In: 2006 22nd Annual Computer Security Applications Conference (ACSAC 2006), pp. 381–392 (2006)
13. Rosiello, A.P.E., Kirda, E., Kruegel, C., Ferrandi, F.: A layout-similarity-based approach for detecting phishing pages. In: 2007 Third International Conference on Security and Privacy in Communications Networks and the Workshops - SecureComm 2007, pp. 454–463 (2007)
14. Whittaker, C., Ryner, B., Nazif, M.: Large-scale automatic classification of phishing pages. In: NDSS 2010 (2010)
15. Xiang, G., Hong, J., Rose, C.P., Cranor, L.: Cantina+: a feature-rich machine learning framework for detecting phishing web sites. ACM Trans. Inf. Syst. Secur. **14**(2) (2011)
16. Bilge, L., Kirda, E., Kruegel, C., Balduzzi, M.: Exposure: finding malicious domains using passive DNS analysis. In: NDSS, pp. 1–17 (2011)
17. Mao, J., Li, P., Li, K., Wei, T., Liang, Z.: BaitAlarm: detecting phishing sites using similarity in fundamental visual features. In: 2013 5th International Conference on Intelligent Networking and Collaborative Systems, pp. 790–795 (2013)
18. Dong, Z., Kapadia, A., Blythe, J., Camp, L.J.: Beyond the lock icon: real-time detection of phishing websites using public key certificates. In: 2015 APWG Symposium on Electronic Crime Research (eCrime), pp. 1–12 (2015)
19. Ardi, C., Heidemann, J.: AuntieTuna: personalized content-based phishing detection. In: NDSS Usable Security Workshop (USEC) (2016)
20. Corona, I., et al.: DeltaPhish: detecting phishing webpages in compromised websites. In: Foley, S.N., Gollmann, D., Snekkenes, E. (eds.) ESORICS 2017. LNCS, vol. 10492, pp. 370–388. Springer, Cham (2017). https://doi.org/10.1007/978-3-319-66402-6_22
21. Rao, R.S., Pais, A.R.: Detecting phishing websites using automation of human behavior. In: Proceedings of the 3rd ACM Workshop on Cyber-Physical System Security, CPSS 2017, pp. 33–42. Association for Computing Machinery, New York (2017)
22. Niakanlahiji, A., Chu, B.-T., Al-Shaer, E.: PhishMon: a machine learning framework for detecting phishing webpages. In: 2018 IEEE International Conference on Intelligence and Security Informatics (ISI), pp. 220–225 (2018)

23. Torroledo, I., Camacho, L.D., Bahnsen, A.C.: Hunting malicious TLS certificates with deep neural networks. In: Proceedings of the 11th ACM Workshop on Artificial Intelligence and Security, AISec 2018, pp. 64–73. Association for Computing Machinery, New York (2018)
24. Kharraz, A., Robertson, W., Kirda, E.: SurveyLance: automatically detecting online survey scams. In: 2018 IEEE Symposium on Security and Privacy (SP), pp. 70–86. IEEE (2018)
25. Mao, J., et al.: Phishing page detection via learning classifiers from page layout feature. EURASIP J. Wirel. Commun. Netw. **1**, 2019 (2019)
26. Sánchez-Paniagua, M., Fidalgo, E., Alegre, E., Alaiz-Rodríguez, R.: Phishing websites detection using a novel multipurpose dataset and web technologies features. Expert Syst. Appl. **207**, 118010 (2022)
27. Bitaab, M., et al.: Beyond phish: toward detecting fraudulent e-commerce websites at scale. In: 2023 IEEE Symposium on Security and Privacy (SP), pp. 2566–2583 (2023)
28. Zhang, Y., Fu, X., Yang, R., Li, Y.: DRSDetector: detecting gambling websites by multi-level feature fusion. In: 2023 IEEE Symposium on Computers and Communications (ISCC), pp. 1441–1447 (2023)
29. Kharraz, A., Robertson, W., Balzarotti, D., Bilge, L., Kirda, E.: Cutting the gordian knot: a look under the hood of ransomware attacks. In: Almgren, M., Gulisano, V., Maggi, F. (eds.) DIMVA 2015. LNCS, vol. 9148, pp. 3–24. Springer, Cham (2015). https://doi.org/10.1007/978-3-319-20550-2_1
30. Kharraz, A., Arshad, S., Mulliner, C., Robertson, W., Kirda, E.: UNVEIL: a large-scale, automated approach to detecting ransomware. In: 25th USENIX Security Symposium (2016)
31. Miramirkhani, N., Starov, O., Nikiforakis, N.: Dial one for scam: analyzing and detecting technical support scams. In: 22nd Annual Network and Distributed System Security Symposium (NDSS 2016). NDSS (2016)
32. Ousat, B., Tofighi, M.A., Kharraz, A.: An end-to-end analysis of COVID-themed scams in the wild. In: Proceedings of the 2023 ACM Asia Conference on Computer and Communications Security, ASIA CCS 2023, pp. 509–523. Association for Computing Machinery, New York (2023)
33. Google. Chrome UX Report (2023). https://developer.chrome.com/docs/crux/
34. Durumeric, Z.: Cached chrome top million websites (2023). https://github.com/zakird/crux-top-lists
35. Google. Chrome Devtools Protocol (2024). https://chromedevtools.github.io/devtools-protocol/
36. Google. Chrome's Lighthouse (2024). https://developer.chrome.com/docs/-lighthouse/overview/
37. Wappalyzer. Technology Profiler (2024). https://www.wappalyzer.com/
38. Drury, V., Meyer, U.: Certified phishing: taking a look at public key certificates of phishing websites. In: Fifteenth Symposium on Usable Privacy and Security (SOUPS 2019), Santa Clara, CA, pp. 211–223. USENIX Association (2019)
39. Aas, J., et al.: Let's encrypt: an automated certificate authority to encrypt the entire web. In: Proceedings of the 2019 ACM SIGSAC Conference on Computer and Communications Security, CCS 2019, pp. 2473–2487. Association for Computing Machinery, New York (2019)
40. Drichel, A., Drury, V., von Brandt, J., Meyer, U.: Finding phish in a haystack: A pipeline for phishing classification on certificate transparency logs. In: Proceedings of the 16th International Conference on Availability, Reliability and Security, ARES 2021. Association for Computing Machinery, New York (2021)

41. Edward. Dom visualizer (2024). https://0xedward.github.io/dom-visualizer/
42. Mirzaei, O., Vasilenko, R., Kirda, E., Lu, L., Kharraz, A.: SCRUTINIZER: detecting code reuse in malware via decompilation and machine learning. In: Bilge, L., Cavallaro, L., Pellegrino, G., Neves, N. (eds.) DIMVA 2021. LNCS, vol. 12756, pp. 130–150. Springer, Cham (2021). https://doi.org/10.1007/978-3-030-80825-9_7

Tarallo: Evading Behavioral Malware Detectors in the Problem Space

Gabriele Digregorio[1(✉)], Salvatore Maccarrone[1], Mario D'Onghia[1], Luigi Gallo[2], Michele Carminati[1], Mario Polino[1], and Stefano Zanero[1]

[1] Politecnico di Milano, Milan, Italy
{gabriele.digregorio,mario.donghia,michele.carminati, mario.polino,stefano.zanero}@polimi.it, salvatore.maccarrone@mail.polimi.it

[2] Cybersecurity Lab, TIM S.p.A., Turin, Italy
luigi1.gallo@telecomitalia.it

Abstract. Machine learning algorithms can effectively classify malware through dynamic behavior but are susceptible to *adversarial attacks*. Existing attacks, however, often fail to find an effective solution in *both* the feature and problem spaces. This issue arises from not addressing the intrinsic nondeterministic nature of malware, namely executing the same sample multiple times may yield significantly different behaviors. Hence, the perturbations computed for a specific behavior may be ineffective for others observed in subsequent executions. In this paper, we show how an attacker can augment their chance of success by leveraging a new and more efficient feature space algorithm for sequential data, which we have named *Position Sensitive - Fast Gradient Sign Method*, and by adopting two problem space strategies specially tailored to address nondeterminism in the problem space. We implement our novel algorithm and attack strategies in *Tarallo*, an end-to-end adversarial framework that significantly outperforms previous works in both white and black-box scenarios. Our preliminary analysis in a sandboxed environment and against two Recurrent Neural Network (RNN)-based malware detectors, shows that Tarallo achieves a success rate up to 99% on both feature and problem space attacks while significantly minimizing the number of modifications required for misclassification.

Keywords: adversarial machine learning · dynamic analysis · malware detection

1 Introduction

Malware detectors rely on *software analysis techniques* to identify malicious programs. These techniques are grouped into two main categories: *static* and *dynamic*. Static analysis techniques examine the code without running the program, such as strings or IP addresses within the code [20]. While providing a fast way to analyze software, static analysis falls short against obfuscated samples [11,16,48,53]. On the other hand, *dynamic analysis* gathers information

F. Maggi et al. (Eds.): DIMVA 2024, LNCS 14828, pp. 128–149, 2024.
https://doi.org/10.1007/978-3-031-64171-8_7

while the program is running, capturing its *dynamic behavior*. This includes system calls, network traffic, and memory usage. One commonly employed dynamic feature is the sequence of executed Application Programming Interface (API) calls, which can effectively represent the program functions and goals.

Machine Learning (ML), and particularly Deep Learning (DL), can be employed to detect or classify malware, leveraging both static [4,5,38] and dynamic [32,39,54] features. They represent a more effective alternative to traditional approaches focusing on pattern identification (e.g., byte signatures) or heuristics due to their well-understood generalization capabilities [7,31]. Nonetheless, ML algorithms are susceptible to *adversarial* attacks, which consist in forcing an ML model to misclassify specially *perturbed* samples [18]. These attacks may heavily impact the reliability of ML for cybersecurity applications, forcing their advocates to preemptively research and address vulnerabilities. A substantial body of work in adversarial attacks against static analysis and ML-based malware detectors exists in the literature [10,23,24,27,45]. Fewer works have addressed the security of classifiers employing dynamically extracted features. Of great relevance are the works by Rosenberg et al. [42,43], which introduced an adversarial framework that can target a variety of ML-based malware detectors that employ API call sequences. The authors also discussed and showed how to modify a program's dynamic behavior without *breaking* its functionality, an essential requirement for the attack to be considered successful, formally known as preserved semantics [35]. However, these works do not explicitly discuss or evaluate the real impact of their attacks in the *problem space*. Our preliminary experiments show that only "attacking" the final ML classifier does not suffice to evade the *whole* detection system. This is due to the nondeterminism and probabilistic nature of malware behavior. Hence, computing an adversarial perturbation after observing a specific behavior may not be enough to guarantee a successful attack in subsequent executions. Furthermore, as shown in Sect. 5, the original feature space attack proposed by Rosenberg et al. [43] requires a very large number of additional API calls to evade newer and more sophisticated classifiers such as [25,54]. This may affect both the stealthiness of the attack as well as its capacity to deliver its original and intended behavior.

Our main contribution consists in designing an Adversarial Machine Learning (AML) attack that is effective both in the feature *and* problem spaces while targeting systems that employ sophisticated DL algorithms such as [25,54]. First, we propose *Position Sensitive - Fast Gradient Sign Method*, a new adversarial attack specific to sequential data, able to produce evasive behaviors with minimal perturbations. Then, we address the problem space limitations through two variants: with the first, the attacker executes several times the malware sample, recording all behaviors, and then selects the longest. The second approach consists of performing the attack while considering multiple behaviors at once.

We preliminary evaluate our attacks through an array of experiments against two state-of-the-art Recurrent Neural Networks (RNNs)-based malware detectors, measuring the attack success in the *problem space* [35] in both white- and black-box scenarios. We confirm that after modification and re-execution in the sandbox, the adversarial samples can still evade detection. Our more stringent criterion contrasts with those adopted in previous works [42,43], which solely focused on evaluating the evasion rate in the feature space. Tarallo achieved a success rate of 99% in the feature space. Additionally, it significantly reduced the number of injections required to achieve misclassification, with an average of 27 additional API calls required, a value significantly lower than what is required by current state-of-the-art approaches against weaker models. In the problem space experiments, our method attains a success rate of up to 99%. While securing high evasion rates, our approach also maintains the original malware functionality. In 89% of the cases, the modified malware preserves its original behavior, thus ensuring its capability to execute its original malicious functions while evading detection. Overall, the results from our experimental evaluations highlight the effectiveness and efficiency of our approach, particularly when compared to state-of-the-art attacks.

The main contributions of this research are summarized below:

- We introduce *Position Sensitive - Fast Gradient Sign Method*, a novel AML attack algorithm that targets discrete sequences, which, against RNN models, proves to be more effective and efficient than state-of-the-art approaches.
- We design two strategies to compute an optimal API call sequence that aims to evade detection in the problem space, addressing the nondeterminism between executions.
- We propose a less invasive code-modification strategy that preserves the original functionality of modified malware samples.
- We present Tarallo, an end-to-end framework that modifies the apparent dynamic behavior of malware, revealing vulnerabilities in state-of-the-art RNN-based detection systems. We make Tarallo publicly available to ensure reproducibility and encourage further research on the topic[1].

2 Related Work

Mimicry Attacks. In the original formulation [50], mimicry attacks aimed to camouflage malicious activities by emulating legitimate goodware behavior. A follow-up work refined this strategy, enabling to fool Intrusion Detection Systems (IDSs) that monitor system calls by adding ones that have no effect or swapping some for equivalent system calls [51]. Nevertheless, mimicry attacks, as well as the related countermeasures (e.g., [17]), targeted IDSs relying on simpler anomaly detection techniques (e.g., by comparing recorded behaviors against known benign ones and modeling behaviors as automata [44,46,52]) than the one based on ML and DL targeted in this work.

DL-Based Behavioral Malware Detection. This section presents a brief discussion of DL algorithms for malware detection employing dynamically extracted API call sequences as features. In particular, we focus on the use of RNNs to

[1] https://github.com/necst/Tarallo.

process long sequences of API calls, as those are the targets of the attacks presented in this paper. We refer the readers interested in a more in-depth analysis of the topic, including the use of traditional ML as in [13,47,49], to [6].

In [54], the authors presented a feature representation for arguments of API calls, in contrast with most existing works that focus primarily on API names. They designed a Deep Neural Network (DNN) architecture tailored to process these features, which are encoded using a hashing trick method. Samples are executed in a sandbox that records the sequences of API calls. To capture the relationships between these API calls, the model incorporates gated-Convolutional Neural Networks (CNNs), batch normalization layers, a bidirectional Long Short-Term Memory (LSTM) [3], a global max-pooling layer, and a dense layer. This configuration achieved an accuracy of $\approx$95% and a recall of $\approx$71%.

In [25], the authors introduced a DNN for dynamic malware detection based on intrinsic features extracted from API call sequences. Similar to [54], each malware sample is executed in a sandbox, where all the API call sequences are recorded. However, instead of using arguments to encapsulate the semantic information of each API call, their proposed model considers three distinct attributes: the category, the action, and the operation object associated with it. Let us consider the API *RegCreateKeyExW*: here, *Create* is the action, *RegKeyEx* the operation object, and *registry* the category. The architecture of the model consists of embedding layers, multi-layer CNNs, a bidirectional LSTM, and dense layers. This architecture achieved an accuracy and recall of 97% and 98%, respectively.

Adversarial Machine Learning. Research in AML has focused on inducing misclassification in ML classifiers at inference time, particularly in the image domain [12,18,28]. However, traditional AML attacks for images (e.g., FGSM [18]) cannot be directly applied to malware detectors due to specific problem-space constraints (i.e., functionality preservation) and the discrete and sequential nature of API call sequences.

There is a growing body of research on perturbing Windows Portable Executable (PE) files; in particular, attacks on static features have been extensively explored in the literature. These attacks involve modifying the headers [40,41] or the raw bytes of PE files [10,23,27,45]. Fewer works have instead targeted dynamic and machine learning-based malware detectors [21,22,30,42,43]. While Ming et al. focused on evading a particular detection strategy based on the *system call dependency graph* [30], we tackle a more general and powerful detection approach that looks at the actual dynamic behaviors (namely, the API call sequence) as discriminating features.

Most closely related to ours are the works by Rosenberg et al. [42,43], which propose an adversarial framework for deceiving ML detectors that work on sequences of dynamically recorded API calls, including RNNs and feed-forward DNNs. Furthermore, they included a tool that modifies malware samples to reflect the required changes in their dynamic behavior. In contrast to [43], our experimental evaluation considers state-of-the-art DL-based behavioral malware detectors. We show that the method from [43] is less effective and efficient against these advanced models, while our strategy proves both highly effective and efficient in evading them. Additionally, Rosenberg et al. [43] assess their attack effectiveness solely in the feature space, deeming it successful if the adversarial algorithm can transform a malicious API call sequence into a variant misclassified as benign by the target detector. In contrast, our study adopts a more complex and realistic approach. An attack is deemed successful if it can alter a malicious sample in a manner that, *upon re-execution*, it produces an API call sequence misclassified as benign by the target model. Our method addresses the significant challenge of varying API call sequences that result from different executions of the same malware sample, an aspect overlooked in [43]. Lastly, in [43] the authors suggest using a wrapper as proxy code, serving as an intermediary between the malware and the DLLs executing the API calls. Our method directly modifies the malware bytecode to intercept API calls. This alteration of the bytecode itself removes the necessity for an extra proxy layer, making the attack more realistic and potentially lowering the likelihood of detecting the modification. Rosenberg et al. also introduced a more efficient black-box attack, necessitating fewer queries to the target model [42]. This method employs a Generative Adversarial Network [9] to mimic genuine benign API call sequences. The authors evaluated the efficacy of this attack with varying query budgets. While it outperforms their earlier method [43] under a limited query budget, their findings also indicate that a gradient-based attack becomes more effective with a larger budget. In contrast, our adversarial algorithm not only surpasses the gradient-based attack from [43] in efficiency and query requirement but also remains effective even with minimal or no knowledge of the target model.

3 Background and Threat Modeling

Adversarial Malware. Based on the work by Pierazzi et al. [35], we formally define the notions employed in this paper. Consider a set of executable programs denoted as E. Let F represent the feature space in which our model operates. We define g as the feature extraction function that maps an executable to a specific point within the feature space, i.e., $g : E \to F$. The readers must notice that g is generally not a surjective function, as it may be generally not possible to identify an executable corresponding to a certain point in the feature space [18]. Let M represent a machine learning classifier, which accepts as input a point in the feature space representing a sequence of API calls. M outputs a prediction from the predefined label set $L = \{malware, goodware\}$, i.e., $M : F \to L$. Given an executable e belonging to the set E and its corresponding representation in the feature space $x_e \in F$, we define the adversarial sample $\hat{x_e}$ for the model M as $\hat{x_e} \triangleq x_e + \delta$ where $\delta \in F$ is the adversarial perturbation, such that M classifies correctly x_e while misclassifies $\hat{x_e}$.

Crafting adversarial samples in the malware problem space presents significant challenges. In contrast to computer vision, where attackers aim to introduce subtle and imperceptible perturbations to images, adversarial malware must be valid and executable programs. This requirement limits the range of possible points in the feature space that can be considered, as not all points correspond to valid and executable programs. Furthermore, the original functionality of the program must be preserved for the attack to be considered successful. This further restricts the extent of possible modifications that an attacker can apply. An attacker can only add extra API calls to the original sequence; in fact, deleting or modifying existing ones would compromise its original functionality.

Threat Model. We examine two different scenarios based on the attacker's knowledge of the target model. In the white-box scenario, the attacker has complete access to its internal parameters. In this case, the model used as the target in the white-box attack directly serves as an oracle. In the black-box scenario, the attacker has no access to the target model architecture or parameters. Instead, the attacker relies on a surrogate oracle, which may have a different structure than the target detector. The only assumption in our experimental evaluation is that the target model employs API call sequences as features.

4 Tarallo

The Tarallo framework operates in an end-to-end manner, to modify the apparent dynamic behavior of malware. It comprises three key components, the first being a Cuckoo sandbox [1] controller that automates the submission of malware samples and the retrieval of the corresponding reports. The second component is responsible for computing the *adversarial* API call sequence and consists of the *Position Sensitive - Fast Gradient Sign Method* (PS-FGSM) algorithm and the two problem space strategies to maximize the evasion probability. Lastly, it comprises a *PE patcher* that parses and modifies a PE file to make it reflect the evasive behavior computed by the second component. Figure 1 depicts Tarallo's workflow. To transform a given malware sample into a functioning adversarial sample, it performs the following steps: 1. It executes the sample multiple times in a controlled environment (sandbox) and records the dynamic behavior of each execution as a sequence of API calls. 2. It runs the PS-FGSM to manipulate the original API call sequence and generate a modified version that can evade the target machine learning model. 3. It modifies the original malware sample to alter its apparent dynamic behavior so that it closely resembles the evasive behavior produced by the adversarial machine learning attack. 4. It executes the modified sample multiple times in the sandbox, recording its dynamic behavior. 5. It lastly evaluates the attack success by examining the effect on the target machine learning model, while also ensuring the preservation of its functionality.

4.1 Position Sensitive - Fast Gradient Sign Method

The main contribution of this work is a novel AML attack designed to target ML models that rely on discrete and sequential data. We call this attack *Position Sensitive - Fast Gradient Sign Method* as it builds upon the *Fast Gradient Sign Method* introduced by Goodfellow et al. [18]. Specific to malware detection, we employ PS-FGSM against models that employ API call sequences as input

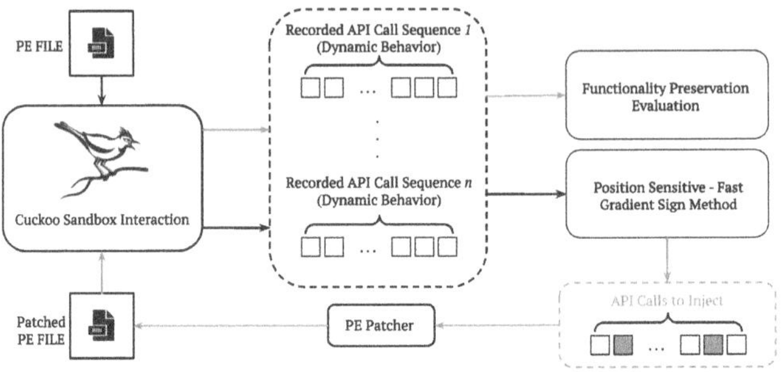

Fig. 1. The workflow of the Tarallo framework.

features. The algorithm relies on an oracle with known architecture and parameters. It can either work in a white-box mode, in which the oracle is the target model, or in a full black-box manner. As a variation of the Fast Gradient Sign Method [18,33], the algorithm we present in this work relies on the *sign* of the *inverse* Jacobian of the model's output with respect to its input to minimize the probability of the sample to be classified as belonging to its real class, in this case "malware." As a reminder, the Jacobian matrix represents all the partial derivatives of a function. Since we are considering the *inverse* Jacobian, we are computing the *gradients* of the model from the output to the input.

The PS-FGSM repetitively selects the *best position* within a given subsequence of consecutive API calls (henceforth called *windows*) and adds an extra API call at that position. The algorithm iteratively attacks windows instead of whole sequences because the targeted models are most often RNNs which process one window per time. We take inspiration for the PS-FGSM from the algorithm originally introduced by Rosenberg et al. [43], significantly improving its effectiveness. In particular, we found a way to explicitly compute the position that most affects the output of the model at each iteration, significantly augmenting the impact of each single injection. The rationale behind choosing a specific position within the current window—rather than just focusing on the type of API—is that single API calls cannot be strictly classified as malicious or benign. Therefore, the impact of injecting an API call may vary significantly depending on its position within the original sequence. Indeed, some positions may disrupt "malicious" windows, while others may have little to no effect on the final classification. Furthermore, as the attacker is only allowed to *add* API calls, it intuitively follows that the algorithm must work by displacing a certain API within a sequence, rather than eliminating it.

In Algorithm 1, which showcases the PS-FGSM algorithm for API sequence-based malware detectors, $\perp$ represents the concatenation operation. In its basic implementation, the *stop_condition* is triggered either when the oracle is evaded or when the maximum number of injected API calls is reached. At each iteration, the algorithm processes a window of API calls of size n, which is constrained by the accepted input sequence length of the oracle model. Sequences shorter than n

Algorithm 1. *Position Sensitive - Fast Gradient Sign Method*

```
Require: f (target model), x (malicious sequence to perturb), n (window size)
procedure AdversarialAttack(f, x, n)
    x_new ← [ ]
    r ← 0
    for window w_j of size n in x do
        while stop_condition do
            J_j ← J_f(w_j)[f(w_j)]
            P* ← ComputeBestPosition(J_j, r)
            w_j[P*] ← arg min_a ||sign(w_j[0 : P* − 1] ⊥ a ⊥ w_j[P* : n − 1]) − sign(J_j)||
            r ← r + 1
        end while
        x_new ← x_new ⊥ w_j
    end for
    return x_new
end procedure
procedure ComputeBestPosition(J_j, r)
    if r mod 4 == 0 then
        norm ← ||J_j||_1
    else
        norm ← ||J_j||_−1
    end if
    P* ← argmax(norm)
    return P*
end procedure
```

are padded, whereas sequences longer than n are truncated. Moreover, our approach includes a mechanism to shift the last API call from the current window to the subsequent one each time a new API call is injected. The position where we inject the additional API call is selected by looking at the norms of the Jacobian matrix. Being API calls discrete entities, these are generally embedded into multidimensional tensors of order k by the classifier. Hence, to evaluate the impact of each API call, we compute their norms along the second-last dimension of their embedding (i.e., assuming k is the number of dimensions of the embedding, we only consider the d_{k-1}^{th} dimension). In this way, the algorithm will produce a matrix of norms with one column for each API call. To select the best candidate in the window, we then select the position with the *greatest smallest norm*. In other words, we collect the smallest norm for each API in that window and then select the greatest. We have empirically verified that this works generally better than selecting the position corresponding to the *greatest absolute norm*. Nonetheless, selecting the *greatest smallest norm* may lead the solution to stagnate; we combine the two norms by adopting the *greatest absolute norm* every c iteration, where c is a small integer (4 in our experiments). Given $U_i \in \mathbb{R}^{d \times n}$ and $U_i := [u_1, \ldots, u_{n-1}, u_n]$ s.t. $u_j = [1, 1, \ldots, 1]$ iff $j = i$ else $u_j = [0, 0, \ldots, 0]$, this process is defined as:

$$\mathbf{N}^{\mathbf{w},\mathbf{r}} = \sum_{i}^{n} \ell_1 \left(\mathbf{J}_{emb\_api_i} \right) * U_i$$

$$\mathbf{P}_{\mathbf{1}}^{\mathbf{w},\mathbf{r},*} := \arg \max_{i}^{n} \left(\min \mathbf{N}_i^{w,r} \right), \mathbf{P}_{\mathbf{2}}^{\mathbf{w},\mathbf{r},*} := \arg \max_{i}^{n} \left(\max \mathbf{N}_i^{w,r} \right)$$

In the above equations, $\mathbf{N}^{\mathbf{w},\mathbf{r}}$ is the norm matrix computed for the $\mathbf{w}^{th}$ API call window at the $\mathbf{r}^{th}$ iteration. $\mathbf{P}_{\mathbf{1}}^{\mathbf{w},\mathbf{r},*}$ and $\mathbf{P}_{\mathbf{2}}^{\mathbf{w},\mathbf{r},*}$ both compute the best position in the $\mathbf{w}^{th}$ window at the $\mathbf{r}^{th}$ iteration, with $\mathbf{P}_1^{\mathbf{w},\mathbf{r},*}$ selecting the one corresponding to the *greatest smallest norm* whereas $\mathbf{P}_2^{\mathbf{w},\mathbf{r},*}$ selects the position where the *greatest absolute norm* is found.

After determining the optimal position, the PS-FGSM selects the API call for injection among those that do not disrupt the original functionality of the program. Moreover, the set is restricted to API calls that are both accurately tracked by Cuckoo and employed as features by the oracle. We denote the ordered set of available API calls as A. Adding API calls not belonging to A would not affect the final classification, as they would be filtered out before classification in the problem space.

Given a window of API calls w and a best position $\mathbf{P}^*$ with respect to w, PS-FGSM selects the API call $a_{inj} \in A$ as:

$$a_{inj} = \arg\min_{a \in A} ||sign(w\,[0 : \mathbf{P}^* - 1] \perp a \perp w\,[\mathbf{P}^* : n - 1]) - sign(\mathbf{J})||$$

In the above equation, we compute the ℓ_1 norm between the sign of the modified window $(w\,[0 : \mathbf{P}^* - 1] \perp a \perp w\,[\mathbf{P}^* : n - 1])$ and the Jacobian of the original one. In other words, we select the API that most closely draws the modified window in the direction of the gradient, expressed by the Jacobian matrix $\mathbf{J}$ (similarly to [43]). Despite the constraints posed by the limited set of API calls—a detail frequently overlooked in previous research [21,22,30,42, 43]—Tarallo still manages to evade detection using a particularly small number of injected API calls, especially when compared to Rosenberg et al. [43].

Dealing with Nondeterminism. As a premise for this work, we stated that state-of-the-art attacks do not address the nondeterminism typical of real-world malware. This characteristic results in differences among behaviors observed across multiple executions, which in turn strongly impacts the effectiveness of the attack in the problem space. Several factors contribute to nondeterminism, including the system's state at the time of execution (e.g., available resources and network status), operating system scheduling combined with hardware and software multithreading [26], interactions with other executing processes, and evasion techniques like probabilistic control flow approaches [34] and dormant periods [8]. These factors affect the API call sequences observed during different executions of the same malware, a phenomenon that is known in the literature [37] and we encountered in our experiments. While nondeterminism has been studied in the context of dynamic analysis, to the best of our knowledge, this work is the first to address this problem in the context of evading behavioral malware detectors that rely on API call sequences.

We propose two possible strategies to address this issue: Longest Known Behavior (LKB) and Behavior Cascade Optimization (BCO), which maximize the attack success in case the exhibited behavior differs from the one observed by the adversary. Both variations build upon the intuition that the success probability will increase as the confidence score associated with the malicious class decreases. Therefore, the general PS-FGSM is modified to not interrupt the optimization process for the current window when this is classified as benign by the oracle. Instead, the window is optimized until all n_A allowed API calls are injected. The algorithm stores every prediction score obtained after each injection, along with the partial API call sequence composed so far. After all n_A sequences are generated, the algorithm then selects the one that obtained the lowest score by the oracle (i.e., the closest to the goodware label 0). The LKB variant runs the same sample b times, obtaining this way up to b different behaviors. It then selects the behavior that contains the most information, namely the longest sequence of API calls. It lastly performs the variation of the PS-FGSM. Conversely, the BCO builds upon the intuition that the probability of success will increase as the number of behaviors *simultaneously* considered in the feature space grows: i.e., the more behaviors the attacker considers when running the PS-FGSM, the more successful the attack will be. Specifically, the attacker considers b behaviors collected from b executions of the same sample. They sort them out in descending order by looking at the classification scores given by the oracle. Then they attack the most "malicious" sequence and propagate the solution found to all the other behaviors. Suppose the solution found by the PS-FGSM for the first sequence is to inject API calls $\mathbf{a_r}$ before the first a_j occurrence and $\mathbf{a_b}$ before the first a_v, so that $s_1 = |a_o|a_p|a_b|a_j|a_b|a_z|a_v|$ becomes $s_1^* = |a_o|a_p|a_b|\mathbf{a_r}|a_j|a_b|a_z|\mathbf{a_b}|a_v|$. Now suppose $s2 = |a_a|a_b|a_j|a_d|$ and $s_3 = |a_i|a_j|a_k|a_z|a_v|$; before computing s_2^* and s_3^*, the solution found for s_1 is propagated to the two sequences, obtaining in this way: $s_2' = |a_a|a_b|\mathbf{a_r}|a_j|a_d|$ and $s_3' = |a_i|\mathbf{a_r}|a_j|a_k|a_z|\mathbf{a_b}|a_v|$. The attacker then computes the solution $s_2'^*$ from s_2' and propagates the results to s_1^* and s_3'. The algorithm repeats this procedure until all the modified sequences are *simultaneously* classified as benign by the oracle.

4.2 PE Patcher

Lastly, the PE patcher component is responsible for altering the dynamic behavior of malware samples to resemble the adversarial sequence generated by the PS-FGSM. The PE patcher does not require access to the malware source code but is able to preserve its functionality. It supports both x86 and x86-64 PEs. The PE patcher identifies the assembly instructions corresponding to calls to imported APIs through heuristics (e.g., matching call and jump assembly instruction opcodes). It then modifies the arguments of those instructions to redirect them to a code snippet that implements the *hijacking* logic and that was previously injected into the PE. Indeed, our PE patcher directly modifies the malware bytecode, in contrast to state-of-the-art solutions that rely on a wrapper as an intermediary between the malware and the DLLs. This approach makes the PE patcher arguably stealthier. The PE patcher can selectively choose which assembly calls to hijack and modify, while leaving other API calls of the same type unaltered. This level of granularity and flexibility enables to replicate the desired API call sequences with high precision.

Hijacking Logic. When hijacking many API calls, allocating a separate handler for each is inefficient. Such an approach would require modifying extensively the

PE, compromising the attack's stealthiness. Instead, the PE patcher creates a single, adaptable code segment capable of handling the different injections. The underlying logic operates following the static directives produced by the PS-FGSM. These directives specify which API calls to hijack and inject. The code segment produced by the PE patcher consists of two modules. The first one is a jump table, which acts as a trampoline for the second module. This second module contains the actual code responsible for performing both the injected and hijacked API calls. Upon hijacking an API call, its assembly call instruction is modified to point to a specific entry in the jump table. Each entry in this table sets up the stack with relevant information about the current execution context, such as the details of the specific API call that has been hijacked. The entire hijacking logic is placed in an additional section added at the end of the PE.

Functionality Preservation. One of the main goals of the Tarallo framework is to preserve the functionality of the malware sample while altering its apparent dynamic behavior. However, formally proving that the modified program behaves exactly as the original is impossible as it can be reduced to the halting problem. Instead, Tarallo adopts a dual strategy.

To avoid changes in its functionality, the PE patcher ensures that each API call performed by the original malware is also executed by the modified one with the same memory and register state. The values of caller-saved registers are stored before calling any injected API, to later enable the reconstruction of the original stack state. In this way, Tarallo ensures that the execution of the original API call produces the intended output, prevents subsequent injected API calls from interfering with one another, and, therefore, does not alter the original functionality. To prevent any disruption in the process flow, the PE patcher ensures that all injected API calls are performed with valid parameters, and conform to the guidelines outlined by the Microsoft documentation [29]. The PE patcher achieves this by passing the correct parameters[2] via the stack or registers, adhering to the calling convention specified by the executable. When the API call needs a pointer as an argument, a pointer within the section added by the PE patcher is provided. If the argument is used only in read mode, the PE patcher fills the location pointed to by that pointer, with null characters or a random string of the required size [29]. Conversely, if the pointer is used in write mode, a pointer to a sufficiently large area of the section, not used for other operations, is provided. Additionally, before executing any injected API call, the stack is aligned to avoid potential issues that could arise from SIMD (Single Instruction, Multiple Data) operations [15].

To demonstrate the preservation of functionality between the original and modified versions, the PE patcher employs an empirical approach. The framework runs both the original and modified versions, recording their API call sequences. It then compares these sequences to determine if the modified version maintains the original behavior. If executed under the same conditions, the sequences should match, except for injected API calls in the modified version.

[2] The list of parameters can be found at: https://github.com/necst/Tarallo/blob/main/ChainFramework/config/config_api_args.py.

However, replicating identical execution conditions is not always feasible due to nondeterminism. To address this issue, each malware sample is executed multiple times, both before and after the attack. From each run, Tarallo extracts a set of the executed API calls and their parameters. It then computes an invariant that represents the pre-modification behavior. This is done by computing intersections among sets of non-empty API call sequences from the original malware executions. The invariant serves as a reference point to compare against post-modification behaviors, which are derived through a set union operation of API call sequences from different executions of the modified malware. If the invariant (i.e., the original behavior) is a subset of the post-modification behavior, then Tarallo preserves the program functionality.

5 Experiments

The experiments aim to establish whether the PS-FGSM is effective in deceiving state-of-the-art detectors [25,54] both in the feature and problem spaces, whether the resulting adversarial malware samples preserve their original functionality, and to measure the number of injections required to evade detection (i.e. attack overhead). The feature space evaluation quantifies the attack ability to defeat *only* the DL classifier, which means that the object of the evaluation is solely the API call sequence modified by the PS-FGSM. On the other hand, the problem space evaluation concerns the ability to modify a malware sample that, upon re-execution, will display a behavior able to evade the target DL classifier. In brief, the following experimental evaluation aims to answer four research questions:

RQ1. Is the original functionality of the malware sample preserved after applying the changes to make it evasive? (*Functionality preservation*)

RQ2. Can PS-FGSM generate API call sequences from previously detected malware that can evade state-of-the-art detectors? (*Feature space experiment*).

RQ3. Can PS-FGSM combined with either the LKB or the BCO effectively fool a detection system in an end-to-end manner? (*White- and black-box problem space experiments*).

RQ4. Does the attack transfer to different RNN models employing a different encoding for the API call sequences? (*Black-box problem space experiment*).

Sandbox Setup. We employ 35 Oracle VirtualBox (v5.1.38) virtual machines running Windows 7. Each sample is run with a time limit of 300 s [16,36], a duration greater than in [19,42,43]. We disabled the Cuckoo options for simulating random human interactions to ensure a controlled analysis environment and enabled deterministic human interactions, like button-clicking operations, to manage software requiring human interactions before exhibiting their behavior. However, these options cannot encompass all scenarios, leading to software that exhibits limited behaviors in the sandbox environments. This is a recognized challenge in dynamic analysis, part of the Reverse Turing Test [2].

Datasets. Throughout the experiments, we employ different datasets. The first dataset, presented in [25], includes 2000 malware samples first observed in 2019

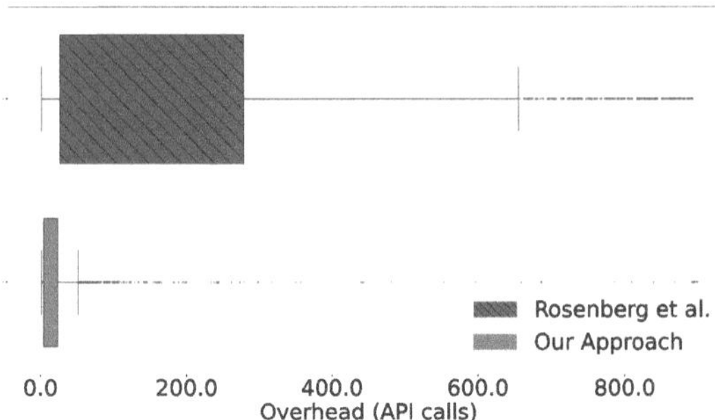

Fig. 2. Box plot of the overhead distribution for the feature-level attack against Li et al. [25], using both the proposed PS-FGSM and Rosenberg et al. [43].

Table 1. Effectiveness scores for the **feature-level** attack against Li et al. [25] using the proposed PS-FGSM, Rosenberg et al. [43], and a random approach, evaluated with both the VirusShare Dataset No. 375 and the Li et al. Dataset [25].

Overhead limit	VirusShare Dataset n.375			Li et al. Dataset [25]		
	20%	50%	70%	20%	50%	70%
PS-FGSM	**0.9742**	**0.9841**	**0.9863**	**0.9815**	**0.9920**	**0.9925**
Rosenberg et al. [43]	0.6492	0.8556	0.9261	0.5696	0.8178	0.9044
Random Approach	0.2670	0.3674	0.7020	0.2638	0.3634	0.7192

and 2000 benign executables. The malware was sourced from VirusShare [14], and the benign samples were obtained from popular free software sources like Softonic, SourceForge, and Portableapps. Another dataset, referenced in [54], encompasses the features extracted from 15931 malicious samples detected in April 2017 and an additional 11856 samples from May 2017. This dataset also contains 11417 benign samples gathered in April 2017 and a further 21983 samples collected in May 2017. We also employ Dataset n.375 from VirusShare, comprising 5474 PE files. Of these, 4917 are 32-bit PEs, and 557 are x86-64 PE files. Finally, Dataset n.290 from VirusShare is used, consisting of 7651 PE files. Among these, 7636 are 32-bit and 15 are 64-bit, all collected in May 2017.

Functionality Preservation (RQ1). We executed the original and the modified malware five times in the Cuckoo sandbox. Comparing their behaviors, we observed functionality preservation in 89% of the cases, slightly lower than Rosenberg et al. [43] where all modified malware maintain their functionality. However, 89% is a lower bound. Some differences might result from nondeterminism rather than actual disruptions in functionality, a factor not treated by previous works that do not execute the original and modified malware multiple times. Additionally, Tarallo modifies the malware's bytecode directly instead of using wrappers, increasing realism but also implementation complexity.

5.1 Feature Space Experiment (RQ2)

The feature space experiment evaluates the performance of the PS-FGSM against both state-of-the-art and random algorithms in deceiving the malware detector presented by Li et al. [25]. This evaluation is conducted in a white-box setting, where both the oracle and the target model are the same as Li et al. [25]. Each adversarial sequence is generated from a malicious API call sequence extracted from the execution of a real malware sample. Each test is deemed successful if the adversarial sequence is misclassified as benign by the target detector. Additionally, we measure the number of injected API calls required to achieve misclassification. We refer to this number as the "attack overhead," and we use this as an indicator of the attack efficiency.

Evaluation Result. In this experiment, we employ the VirusShare dataset n.375 and the one used in [25]. The first one is the closest temporally to the one used to train the oracle. Using datasets from different time periods would result in a weak performance by the target detector, which would in turn introduce a bias in favor of the adversarial attack success. We also filter out all the malware samples not detected by the target model, ending up with 2708 malware samples from the VirusShare dataset and 2000 from the one in [25].

In Table 1, we present the results obtained with our PS-FGSM, the approach devised in [43], and an attack consisting in randomly injecting API calls. Both PS-FGSM and the approach in [43] demonstrate almost 100% effectiveness when not imposing strict limits on the attack overhead. However, when we consider a more reasonable attack overhead, the PS-FGSM significantly outperforms the approach in [43]. The box plot in Fig. 2 provides a visual representation of the attack overhead by PS-FGSM and the approach in [43]. In the VirusShare dataset, the average overhead for PS-FGSM is 27, indicating that, on average, we have to inject only 27 additional API calls to evade detection. In contrast, the approach in [43] has an average overhead of 202. The same happens for dataset in [25], where the average overhead for the PS-FGSM is 30, while for [43] is 252. Furthermore, when looking at the median overhead, we observe that half of the evaluated samples require 12 or fewer API calls to achieve misclassification when employing the PS-FGSM. In contrast, the median overhead for the approach in [43] is 110, indicating that the majority of samples require a much higher number of injected API calls. Similarly, the median overhead of PS-FGSM is 13 for the dataset in [25], while the approach in [43] has an average overhead of 165.

5.2 White-Box Problem Space Experiment (RQ3)

This experiment shows the capability of Tarallo to produce a modified malware with an evasive dynamic behavior. These settings present additional challenges compared to the feature space attack; e.g., the solution computed in the feature space may include calls to API that are not imported by the original malware. Moreover, it explicitly addresses the nondeterminism in the execution of malware samples in a sandbox, which results in different API call sequences during

Table 2. Longest Known Behavior (LKB) and Behavior Cascade Optimization (BCO) **white-box** attacks overhead and effectiveness in **problem space** against Li et al. [25].

	Overhead Limit	5%			20%			120%
	Arsenal size	(2%, 3%]	(2%, 6%]	Any	(2%, 3%]	(2%, 6%]	Any	Any
LKB	Total Executions	73	165	416	115	267	800	1414
	Evasive Executions	72	129	249	98	189	455	806
	Avg Injected APIs	≈19	≈19	≈22	≈56	≈56	≈74	≈406
	Attack Effectiveness	**0.9863**	0.7818	0.5986	0.8522	0.7079	0.5688	0.5700
BCO	Total Executions	90	219	723	100	278	1120	1571
	Evasive Executions	80	196	629	90	252	837	1018
	Avg Injected APIs	≈15	≈17	≈18	≈25	≈40	≈56	≈219
	Attack Effectiveness	0.8889	**0.8950**	**0.8700**	**0.9000**	**0.9065**	**0.7473**	**0.6480**

different executions of the same malware sample, a problem not addressed in previous works such as [43]. To explicitly tackle the complexity of the problem space settings, we employ the LKB and the BCO strategies, both based on the PS-FGSM. We consider the attack successful if and only if re-executing the malicious sample produces an API call sequence that is misclassified as benign by the target model. In the white-box tests, we use the malware detector described by Li et al. in [25] as both the oracle and the target model.

Evaluation Result. The VirusShare dataset n.375 is employed in this experiment; in particular, we keep only executables classified as malicious by the target model. Moreover, we exclude all malware with API call sequences shorter than a threshold set to 15 APIs, ensuring that only malware with significant recorded dynamic behavior is considered for modification. For each malware sample, we define an "arsenal" of API calls available for injection, which comprises the intersection of the APIs imported by the malware, the APIs considered safe to inject without disrupting its normal execution, and the APIs recorded by Cuckoo. The results are evaluated using various arsenal size ranges to measure their impact on the attack performance. To evaluate the attack's effectiveness, we run each malware sample five times in Cuckoo. Then, after applying the LKB and BCO strategies, each modified sample is executed five more times per strategy. The resulting API call sequences are submitted for evaluation. In Table 2, we present the results of this evaluation, detailing different arsenal ranges and overhead limits. The arsenal ranges are expressed as the percentage of the total number of API calls recorded by Cuckoo, which is just over 300. For example, a 2% value corresponds to approximately 6 API calls. We compute the attack effectiveness as $EvasiveExecutions/TotalExecutions$.

The problem space constraints have an impact on the attack effectiveness to some extent. However, despite the complex settings, the modified malware still manages to exhibit an evasive apparent dynamic behavior that successfully deceives the white-box target model. For the LKB strategy, we find that the attack effectiveness score decreases as the arsenal size and the overhead limit increase. A larger arsenal size implies a greater variety of API calls available to the malware, which implies that the problem space is broader and, therefore, more challenging from the attacker's perspective, as it amplifies the impact of nondeterminism on the attack performance. As a result, when the malware is re-executed post-attack, the resulting API call sequence may significantly deviate from the original, including differences in API calls used. Moreover, increased overhead generally results in a more fragile attack. Specifically, a high number of injected API calls has an increased chance that variations in the sequence of API calls due to nondeterminism will shift these injections, rendering them ineffective. The BCO strategy shows a more stable attack effectiveness than the LKB strategy, underscoring its superior ability to navigate the breadth of the problem space and handle variations in executed behaviors due to nondeterminism. Moreover, the BCO strategy generally outperforms the LKB strategy, supporting the soundness of its underlying rationale. The only instance where the LKB strategy surpasses BCO is in scenarios with the most limited overhead and arsenal size. This is expected as it is the scenario with a narrower problem space, where nondeterminism has minimal impact. This observation further reinforces the link between the BCO strategy's effectiveness and nondeterminism.

Additionally, we analyze the ability of the BCO and LKB strategies to generate adversarial API call sequences in a constrained environment. We use a subset of the VirusShare dataset no. 375, limiting to a maximum of 800 injected API calls per window and an overall overhead of 120%. Interestingly, the BCO and LKB strategies are successful with different malware samples. In our tests, only about 15% of the samples with an adversarial sequence are targeted by both strategies. Furthermore, the malware samples effectively attacked by the BCO strategy have an average pre-attack sequence length that is roughly 25% shorter than those targeted by the LKB strategy. The LKB strategy succeeds with the most informative (i.e., longest) sequence across different executions of the same malware sample, showing greater effectiveness in creating adversarial sequences with longer available sequences. Conversely, the BCO strategy derives information from all executions of a sample. Consequently, shorter sequences in each run, being less informative, fit within the set overhead limit, making BCO effective in these cases. Notably, unlike the LKB strategy, the BCO strategy's suggested injections are meant for different executions of the same sample, allowing for higher overhead limits compared to LKB. With increased overhead, BCO can attack more samples, including some previously only vulnerable to LKB, although these do not completely overlap.

5.3 Black-Box Problem Space Experiment (RQ3, RQ4)

The black-box, problem space test follows a procedure similar to the white-box one, with a key difference. In this test, the final API call sequences extracted from the modified malware are not evaluated against the model by Li et al. [25], which serves as the oracle for the PS-FGSM attack. Instead, these sequences are tested against a different model, unknown to the attacker. The target model for this black-box test is the DL framework proposed by Zhang et al. [54]. This

Table 3. Effectiveness scores of Longest Known Behavior (LKB) and Behavior Cascade Optimization (BCO) approaches in **black-box, problem space** settings against Zhang et al. [54]-evaluated across different prediction thresholds, True Positive Rate (TPR), and False Positive Rate (FPR)-using the model by Li et al. [25] as an oracle.

Threshold	TPR	FPR	Available Executions		Evading Executions		Attack Effectiveness	
			LKB	BCO	LKB	BCO	LKB	BCO
0.9683	0.992	0.325	309	161	154	82	0.4984	**0.5093**
0.9776	0.989	0.300	263	146	130	71	**0.4943**	0.4863
0.9868	0.986	0.275	225	131	115	75	0.5111	**0.5725**
0.9951	0.978	0.225	162	86	79	51	0.4877	**0.5930**
0.9981	0.971	0.178	141	82	69	52	0.4894	**0.6341**
0.9988	0.968	0.150	132	68	67	39	0.5076	**0.5735**
0.9996	0.958	0.100	115	57	54	33	0.4696	**0.5789**

model presents a significant challenge as it takes into account the arguments of API calls, which are not explicitly manipulated by the AML attack. Instead, the arguments for the injected API calls are defined by the PE patcher to be valid and ensure they do not disrupt the malware's normal execution.

Evaluation Result. We train the black-box model using the dataset and code provided in [54]. We follow the recommendation from the original work to set the prediction threshold corresponding to a FPR of 0.1%. However, we encountered challenges with this configuration as the model is unable to accurately identify the original malware samples from both dataset n.375 and dataset n.290 from VirusShare. This suggests that the chosen threshold and configuration may not be suitable for our specific purposes and dataset. Hence, we select threshold values that correspond to high TPRs, rather than focusing on a specific FPR. Although this strategy results in higher FPRs, it allows us to assess the effectiveness of the Tarallo framework in fooling robust machine learning systems. By testing our method with different threshold values, we can demonstrate how this parameter influences the effectiveness of our attack and provide a comprehensive validation of the experimental evaluation. In our tests, we run the original malware samples from the VirusShare dataset n.375 through the Cuckoo sandbox, recording their API call sequences. We repeat this process five times to collect a sufficient number of executions. For each threshold value considered during the test, we select only the malware samples that are classified as malicious by the black-box model in *all* the executions. We attack each sample with both the LKB and BCO attack strategies and then run the modified versions of these malware executables through the Cuckoo sandbox five times for each strategy, recording their evading API call sequences. The results of these experiments are summarized in Table 3. The attack effectiveness is instead computed as $EvadingExecutions/TotalExecutions$. Restricting the target model to a lower FPR -though still above the value reported in the original work- by increasing the classification threshold, leads to fewer correctly classified malware samples (i.e., a lower TPR), particularly those the model is most confident about. This makes attacks on these samples harder for both the LKB and BCO, reducing the effectiveness score compared to other threshold settings. Lowering the threshold results in an increase in the FPR, which may be less realistic, but it also enhances the TPR. This improvement boosts the model's malware detection capabilities, more accurately mirroring real-world conditions. Under these conditions, both attacks tend to exhibit improved performance. However, when we further increase both the TPR and the FPR, the attack effectiveness score starts to decline again. This decrease occurs because the model begins to incorrectly identify a large number of benign behaviors as malicious, leading to a misclassification of the injection in the original API call sequence as harmful. The tests underscore the need to balance TPR and FPR in detection system design, requiring careful domain-specific consideration to optimize accuracy while reducing evasion attack risks. Additionally, the consistently better performance of the BCO strategy confirms that considering multiple sequences during the attack results in more resilient and general adversarial sequences.

6 Conclusions

In this paper, we introduced Tarallo, an end-to-end framework designed to modify the apparent dynamic behavior of malware to deceive ML detectors that employ API call sequences as features. Our first major contribution is a novel AML attack that targets discrete and sequential data. Moreover, we explicitly addressed the problem of nondeterministic execution in a sandbox, which makes the design of problem space attacks particularly challenging. Lastly, we introduced a new approach for modifying the run-time behavior of malicious software that does not require access to the source code. The preliminary experimental evaluation in a sandboxed environment and against two RNN-based malware detectors, showed that our novel AML algorithm, namely the PS-FGSM, outperforms current state-of-the-art algorithms in terms of both effectiveness and efficiency. In both feature and problem spaces, our algorithm achieves up to 99% effectiveness, while also minimizing overhead as measured by the number of injected APIs. Although we showed that obfuscating the behavior of existing malware is a nontrivial problem and, therefore, that current antimalware solutions are very robust, we also demonstrated that such attacks are in fact possible.

Future work will focus on assessing the effectiveness of Tarallo against different malware detectors, especially those not based on RNN. Moreover, we will explore methodologies that allow defenders to remove non-operational APIs from API call sequences, enhancing the overall performance of the detection systems. We also intend to employ the PS-FGSM to build an adversarial training module to strengthen state-of-the-art DL-based detectors.

Acknowledgements. This study was carried out within the MICS (Made in Italy - Circular and Sustainable) Extended Partnership and received funding from Next-Generation EU (Italian PNRR - M4 C2, Invest 1.3 - D.D. 1551.11-10-2022, PE00000004). CUP MICS D43C22003120001. Mario D'Onghia acknowledges support from TIM S.p.A. through the PhD scholarship.

References

1. Cuckoo (2024). https://github.com/cuckoosandbox/cuckoo
2. Afianian, A., Niksefat, S., Sadeghiyan, B., Baptiste, D.: Malware dynamic analysis evasion techniques: a survey. ACM Comput. Surv. **52**(6) (2019). https://doi.org/10.1145/3365001
3. Agrawal, R., Stokes, J.W., Marinescu, M., Selvaraj, K.: Neural sequential malware detection with parameters. In: 2018 IEEE International Conference on Acoustics, Speech and Signal Processing (ICASSP), pp. 2656–2660. IEEE (2018)
4. Anderson, H.S., Roth, P.: Ember: an open dataset for training static PE malware machine learning models. arXiv preprint arXiv:1804.04637 (2018)
5. Arp, D., Spreitzenbarth, M., Hubner, M., Gascon, H., Rieck, K., Siemens, C.: Drebin: effective and explainable detection of Android malware in your pocket. In: NDSS, vol. 14, pp. 23–26 (2014)
6. Berman, D.S., Buczak, A.L., Chavis, J.S., Corbett, C.L.: A survey of deep learning methods for cyber security. Information **10**(4), 122 (2019)
7. Catak, F.O., Yazı, A.F., Elezaj, O., Ahmed, J.: Deep learning based sequential model for malware analysis using windows exe API calls. PeerJ Comput. Sci. **6**, e285 (2020)
8. Comparetti, P.M., Salvaneschi, G., Kirda, E., Kolbitsch, C., Kruegel, C., Zanero, S.: Identifying dormant functionality in malware programs. In: 2010 IEEE Symposium on Security and Privacy, pp. 61–76. IEEE (2010)
9. Creswell, A., White, T., Dumoulin, V., Arulkumaran, K., Sengupta, B., Bharath, A.A.: Generative adversarial networks: an overview. IEEE Signal Process. Mag. **35**(1), 53–65 (2018)
10. D'Onghia, M., Di Cesare, F., Gallo, L., Carminati, M., Polino, M., Zanero, S.: Lookin'out my backdoor! investigating backdooring attacks against DL-driven malware detectors. In: Proceedings of the 16th ACM Workshop on Artificial Intelligence and Security, pp. 209–220 (2023)
11. D'Onghia, M., Salvadore, M., Nespoli, B.M., Carminati, M., Polino, M., Zanero, S.: Apícula: static detection of API calls in generic streams of bytes. Comput. Secur. **119**, 102775 (2022)
12. Eykholt, K., et al.: Robust physical-world attacks on deep learning visual classification. In: Proceedings of the IEEE Conference on Computer Vision and Pattern Recognition, pp. 1625–1634 (2018)
13. Fang, Y., et al.: A new malware classification approach based on malware dynamic analysis. In: Pieprzyk, J., Suriadi, S. (eds.) ACISP 2017. LNCS, vol. 10343, pp. 173–189. Springer, Cham (2017). https://doi.org/10.1007/978-3-319-59870-3_10
14. Forensics, C.: Virusshare (2023). http://virusshare.com/
15. Furht, B.: SIMD Single Instruction Multiple Data Processing. In: Furht, B. (ed.) Encyclopedia of Multimedia, pp. 817–819. Springer, Boston (2008). https://doi.org/10.1007/978-0-387-78414-4_220
16. Galloro, N., Polino, M., Carminati, M., Continella, A., Zanero, S.: A systematical and longitudinal study of evasive behaviors in windows malware. Comput. Secur. **113**, 102550 (2022)
17. Giffin, J.T., Jha, S., Miller, B.P.: Automated discovery of mimicry attacks. In: Zamboni, D., Kruegel, C. (eds.) RAID 2006. LNCS, vol. 4219, pp. 41–60. Springer, Heidelberg (2006). https://doi.org/10.1007/11856214_3
18. Goodfellow, I.J., Shlens, J., Szegedy, C.: Explaining and harnessing adversarial examples. arXiv preprint arXiv:1412.6572 (2014)

19. Hariom, Handa, A., Kumar, N., Kumar Shukla, S.: Adversaries strike hard: adversarial attacks against malware classifiers using dynamic API calls as features. In: Dolev, S., Margalit, O., Pinkas, B., Schwarzmann, A. (eds.) CSCML 2021. LNCS, vol. 12716, pp. 20–37. Springer, Cham (2021). https://doi.org/10.1007/978-3-030-78086-9_2
20. Hassen, M., Carvalho, M.M., Chan, P.K.: Malware classification using static analysis based features. In: 2017 IEEE Symposium Series on Computational Intelligence (SSCI), pp. 1–7. IEEE (2017)
21. Hu, W., Tan, Y.: Black-box attacks against RNN based malware detection algorithms. arXiv preprint arXiv:1705.08131 (2017)
22. Hu, W., Tan, Y.: Generating adversarial malware examples for black-box attacks based on GAN. In: Tan, Y., Shi, Y. (eds.) DMBD 2022. CCIS, vol. 1745, pp. 409–423. Springer, Singapore (2022). https://doi.org/10.1007/978-981-19-8991-9_29
23. Kolosnjaji, B., et al.: Adversarial malware binaries: evading deep learning for malware detection in executables. In: 2018 26th European Signal Processing Conference (EUSIPCO), pp. 533–537. IEEE (2018)
24. Kreuk, F., Barak, A., Aviv-Reuven, S., Baruch, M., Pinkas, B., Keshet, J.: Deceiving end-to-end deep learning malware detectors using adversarial examples. arXiv preprint arXiv:1802.04528 (2018)
25. Li, C., Lv, Q., Li, N., Wang, Y., Sun, D., Qiao, Y.: A novel deep framework for dynamic malware detection based on API sequence intrinsic features. Comput. Secur. **116**, 102686 (2022)
26. Liu, T., Curtsinger, C., Berger, E.D.: DTHREADS: efficient deterministic multithreading. In: Proceedings of the Twenty-Third ACM Symposium on Operating Systems Principles, pp. 327–336 (2011)
27. Lucas, K., Sharif, M., Bauer, L., Reiter, M.K., Shintre, S.: Malware makeover: breaking ML-based static analysis by modifying executable bytes. In: Proceedings of the 2021 ACM Asia Conference on Computer and Communications Security, pp. 744–758 (2021)
28. Machado, G.R., Silva, E., Goldschmidt, R.R.: Adversarial machine learning in image classification: a survey toward the defender's perspective. ACM Comput. Surv. (CSUR) **55**(1), 1–38 (2021)
29. Microsoft: Programming reference for the win32 API (2024). https://learn.microsoft.com/en-us/windows/win32/api/
30. Ming, J., Xin, Z., Lan, P., Wu, D., Liu, P., Mao, B.: Impeding behavior-based malware analysis via replacement attacks to malware specifications. J. Comput. Virol. Hacking Tech. **13**, 193–207 (2017)
31. Moser, A., Kruegel, C., Kirda, E.: Limits of static analysis for malware detection. In: Twenty-Third Annual Computer Security Applications Conference (ACSAC 2007), pp. 421–430. IEEE (2007)
32. Or-Meir, O., Nissim, N., Elovici, Y., Rokach, L.: Dynamic malware analysis in the modern era-a state of the art survey. ACM Comput. Surv. (CSUR) **52**(5), 1–48 (2019)
33. Papernot, N., McDaniel, P., Swami, A., Harang, R.: Crafting adversarial input sequences for recurrent neural networks. In: 2016 IEEE Military Communications Conference, MILCOM 2016, pp. 49–54. IEEE (2016)
34. Pawlowski, A., Contag, M., Holz, T.: Probfuscation: an obfuscation approach using probabilistic control flows. In: Caballero, J., Zurutuza, U., Rodríguez, R.J. (eds.) DIMVA 2016. LNCS, vol. 9721, pp. 165–185. Springer, Cham (2016). https://doi.org/10.1007/978-3-319-40667-1_9

35. Pierazzi, F., Pendlebury, F., Cortellazzi, J., Cavallaro, L.: Intriguing properties of adversarial ml attacks in the problem space. In: 2020 IEEE Symposium on Security and Privacy (SP), pp. 1332–1349. IEEE (2020)
36. Polino, M., et al.: Measuring and defeating anti-instrumentation-equipped malware. In: Polychronakis, M., Meier, M. (eds.) DIMVA 2017. LNCS, vol. 10327, pp. 73–96. Springer, Cham (2017). https://doi.org/10.1007/978-3-319-60876-1_4
37. Polino, M., Scorti, A., Maggi, F., Zanero, S.: Jackdaw: towards automatic reverse engineering of large datasets of binaries. In: Almgren, M., Gulisano, V., Maggi, F. (eds.) DIMVA 2015. LNCS, vol. 9148, pp. 121–143. Springer, Cham (2015). https://doi.org/10.1007/978-3-319-20550-2_7
38. Raff, E., Barker, J., Sylvester, J., Brandon, R., Catanzaro, B., Nicholas, C.K.: Malware detection by eating a whole exe. In: Workshops at the Thirty-Second AAAI Conference on Artificial Intelligence (2018)
39. Rieck, K., Trinius, P., Willems, C., Holz, T.: Automatic analysis of malware behavior using machine learning. J. Comput. Secur. **19**(4), 639–668 (2011)
40. Rosenberg, I., Meir, S.: Bypassing NGAV for Fun and Pro t (2020)
41. Rosenberg, I., Meir, S., Berrebi, J., Gordon, I., Sicard, G., David, E.O.: Generating end-to-end adversarial examples for malware classifiers using explainability. In: 2020 International Joint Conference on Neural Networks (IJCNN), pp. 1–10. IEEE (2020)
42. Rosenberg, I., Shabtai, A., Elovici, Y., Rokach, L.: Query-efficient black-box attack against sequence-based malware classifiers. In: Annual Computer Security Applications Conference, pp. 611–626 (2020)
43. Rosenberg, I., Shabtai, A., Rokach, L., Elovici, Y.: Generic black-box end-to-end attack against state of the art API call based malware classifiers. In: Bailey, M., Holz, T., Stamatogiannakis, M., Ioannidis, S. (eds.) RAID 2018. LNCS, vol. 11050, pp. 490–510. Springer, Cham (2018). https://doi.org/10.1007/978-3-030-00470-5_23
44. Somayaji, A., Forrest, S.: Automated response using {System-Call} delay. In: 9th USENIX Security Symposium (USENIX Security 2000) (2000)
45. Suciu, O., Coull, S.E., Johns, J.: Exploring adversarial examples in malware detection. In: 2019 IEEE Security and Privacy Workshops (SPW). IEEE (2019)
46. Tan, K., Maxion, R.: "Why 6?" defining the operational limits of stide, an anomaly-based intrusion detector. In: Proceedings 2002 IEEE Symposium on Security and Privacy, pp. 188–201 (2002). https://doi.org/10.1109/SECPRI.2002.1004371
47. Tian, R., Islam, R., Batten, L., Versteeg, S.: Differentiating malware from cleanware using behavioural analysis. In: 2010 5th International Conference on Malicious and Unwanted Software, pp. 23–30. IEEE (2010)
48. Ucci, D., Aniello, L., Baldoni, R.: Survey of machine learning techniques for malware analysis. Comput. Secur. **81**, 123–147 (2019)
49. Uppal, D., Sinha, R., Mehra, V., Jain, V.: Malware detection and classification based on extraction of API sequences. In: 2014 International Conference on Advances in Computing, Communications and Informatics (ICACCI). IEEE (2014)
50. Wagner, D., Dean, R.: Intrusion detection via static analysis. In: Proceedings 2001 IEEE Symposium on Security and Privacy, S&P 2001, pp. 156–168. IEEE (2000)
51. Wagner, D., Soto, P.: Mimicry attacks on host-based intrusion detection systems. In: Proceedings of the 9th ACM Conference on Computer and Communications Security, pp. 255–264 (2002)
52. Warrender, C., Forrest, S., Pearlmutter, B.: Detecting intrusions using system calls: alternative data models. In: Proceedings of the 1999 IEEE Symposium on Security and Privacy (Cat. No. 99CB36344), pp. 133–145. IEEE (1999)

53. You, I., Yim, K.: Malware obfuscation techniques: a brief survey. In: 2010 International Conference on Broadband, Wireless Computing, Communication and Applications, pp. 297–300 (2010). https://doi.org/10.1109/BWCCA.2010.85
54. Zhang, Z., Qi, P., Wang, W.: Dynamic malware analysis with feature engineering and feature learning. In: Proceedings of the AAAI Conference on Artificial Intelligence, vol. 34, pp. 1210–1217 (2020)

Evading Userland API Hooking, Again: Novel Attacks and a Principled Defense Method

Cristian Assaiante(✉), Simone Nicchi, Daniele Cono D'Elia, and Leonardo Querzoni

Sapienza University of Rome, Rome, Italy
{assaiante,delia,querzoni}@diag.uniroma1.it, simone.nicchi@gmail.com

Abstract. Monitoring how a program utilizes userland APIs is behind much dependability and security research. To intercept and study their invocations, the established practice targets the prologue of API implementations for inserting hooks. This paper questions the validity of this design for security uses by examining completeness and correctness attacks to it. We first show how evasions that jump across the hook instrumentation are practical and can reach places much deeper than those we currently find in executables in the wild. Next, we propose and demonstrate TOCTTOU attacks that lead monitoring systems to observe false indicators for the argument values that a program uses for API calls. To mitigate both threats, we design a static analysis to identify vantage points for effective hook placement in API code, supporting both reliable call recording and accurate argument extraction. We use this analysis to implement an open-source prototype API monitor, Toxotidae, that we evaluate against adversarial and benign executables for Windows.

1 Introduction

Programmers largely resort to userland Application Programming Interfaces (APIs) to carry heterogeneous tasks such as file and string manipulation, process management, network communication, and many others. For this reason, monitoring what APIs a program invokes represents a common strategy in much dependability and security research to study the externally observable behavior of a program [12]. This practice is also known colloquially as "API hooking" due to the probes (*hooks*) that a monitoring system inserts in the execution runtime to intercept API calls. This paper focuses on Windows programs accessing APIs from userland libraries on x86/x64 architectures, but the general concepts we present in it may be relevant for other platforms, too.

As typically with dynamic analysis endeavors, adversarial programs may attempt to resist monitoring attempts by employing API call obfuscation techniques [12,19]. Some research has explored defensive methods that increase the robustness of the monitoring through transparent instrumentation [12] or taint

F. Maggi et al. (Eds.): DIMVA 2024, LNCS 14828, pp. 150–173, 2024.
https://doi.org/10.1007/978-3-031-64171-8_8

tracking [19]. Nonetheless, an enduring weakness that we find in current API monitoring embodiments [12,16,27,28,31] is that they assume API calls to follow the calling convention of the target platform.

As a result, they record API identity and input arguments once execution reaches an API prologue, and output arguments and return value upon returning to the caller. However, as we show in this paper, both the completeness and the correctness of the tracing are affected if adversarial code invokes APIs in unexpected ways.

Completeness attacks aim to hide calls from monitoring systems. Prior research [19] describes *stolen code* attacks where an adversary emulates instructions from the API prologue and then jumps into API code with a function-boundary unaligned control transfer. While currently known stolen attacks are limited to a few instructions, in this paper we generalize them to much deeper points with modest effort.

Correctness attacks aim to provide fake indicators to monitoring systems and, to the best of our knowledge, are unexplored for userland malware. In this paper, we propose and demonstrate the feasibility of a time-of-check to time-of-use (TOCTTOU) attack on userland Windows APIs, exhibiting benign argument values upon entering the API and replacing them with rogue ones before API code uses them.

To explore the potential extent of both attacks and counter them, we propose a static program analysis technique that tracks how argument values flow through the storage locations in use to API code. Our goal is to intercept call arguments right before API code uses them: this information can guide us in inserting *deep hooks* in a principled way that effectively counters both types of attacks.

We extensively evaluate the attacks and the proposed defense on popular Windows APIs. First, we show that existing solutions are vulnerable to both attacks and that the hooking locations our method identifies are deeper than where either attack can reach. Then, we build an open-source[1] API monitoring system prototype, Toxotidae, using our deep hooks. We test its efficacy against real-world samples that use classic stolen code techniques and against proof-of-concept implementations of our attacks. Following recent literature [12], we extensively test its performance using the Wine test suite: compared to a traditional design, Toxotidae adds an average 7.62% slowdown and recovers API argument values with 99.40% accuracy.

2 Background and Motivation

Windows APIs. Windows offers programmers an extensive collection of APIs through its DLL (Dynamic Link Library) files. A compiled binary can expose a list of APIs it seeks to use and the Windows loader will fill an array, called Import Address Table (IAT), with their run-time addresses. Alternatively, a

[1] Our artifacts are available at: https://github.com/cristianassaiante/toxotidae.

program may look up an API address at run-time among currently loaded code modules or after loading a new DLL.

Compiled code accesses Windows APIs through well-defined calling conventions for argument passing. On 32-bit systems, all arguments are placed on the stack, typically in 4-byte slots, following the `stdcall` convention. On 64-bit systems, the first four arguments are passed through registers, whereas any following argument is placed on the stack, typically in 8-byte slots. Arguments come with markers (*modifiers*) that identify an input (IN) and/or an output (OUT) value, where an OUT argument is typically a pointer to a storage location for saving a result from the API [12].

Windows API code shows by design several regularities [25]. Functions come with a prologue and an epilogue block. The prologue allocates a fixed-size space[2], which we term **local area**, for hosting local variables and spilling callee-saved registers; on 64-bit API code, the allocation includes also the space needed to call functions internally. The prologue supplies the area size to an instruction (`sub` or `lea`) that updates the stack pointer accordingly. If the function needs to allocate stack space dynamically (`alloca`), the prologue must set a frame pointer register to mark the base of the fixed part of the stack. The epilogue simply restores the initial value of the stack pointer.

A peculiarity of 64-bit API code is that the caller reserves four stack slots before the return address (and thus the local area). We refer to this space as the **register area**. The called function may use it, if needed, for spilling registers.

Pitfalls of Hooking. To interpose on API calls, current monitoring solutions target for hook insertion either the IAT by replacing its contents or the prologue of API implementations by adding trampolines. The first strategy is an easy prey of adversaries that look up functions at run-time [12]. As for trampolines, adversaries can detect them [12] or, as with attacks to anti-virus userland agents [1], even remove them by restoring the original instructions. Recent research [12] recommends using more reliable and transparent instrumentation mechanisms, such as dynamic binary instrumentation [9] or hypervisor-based invisible breakpoints [22], to mitigate these threats.

However, a natural question is whether intercepting relevant program behavior at system-call level would sidestep the pitfalls of userland API hooking. A common belief is that kernel code is a harder target for attackers to tamper with (but also for defenders to instrument) and that the behaviors of interest will show up in low-level system call activity. When only a coarse-grained characterization of a program is needed, this may be a reasonable compromise: for example, malware sandboxes can instrument system calls to reliably capture in one place behaviors that can be exercised by multiple APIs [11], for example when involving the registry or files.

[2] With the exception of leaf functions [25], which cannot modify the stack pointer and other callee-save registers. Due to their simple nature, they are marginal for this work.

However, this does not hold for general uses, as well as for other important malware analysis scenarios. One prominent example are anti-virus products: since the introduction of PatchGuard in 2005, they can no longer insert hooks in Windows kernel code, and many rely on userland API hooks as a primary source of information[3]. Another relevant use case is reverse engineering: when attempting, for example, manual analysis of an untrusted object or efforts like malware lineage [17], determining how a program enacts a behavior can be just as valuable as intercepting it in a coarse-grained way.

Semantic loss can also be a problem. For example, network APIs typically see their inputs lowered to a flat data representation for the device driver associated to the connection. This makes, for instance, a `gethostbyname` operation virtually indistinguishable from an `InternetConnectA` one. Other APIs show generic system call activity, or even none: this is the case, e.g., with `CryptBinaryToString`, `GetModuleHandle`, `GetProcAddress`, and other helpers often used by malware authors.

We conclude that userland API hooking remains necessary for general monitoring.

3 Proposed Attacks

This section discusses two attacks to the completeness and correctness of current API hooking systems: an extended form of stolen code and a novel TOCTTOU attack.

As **threat model** and defense capabilities, we make the following assumptions. The adversary knows the Windows build installed on the victim or prepares a payload for multiple versions. Attacks to the instrumentation technique are out of scope. If attacking one property, the other is not a primary concern for the adversary: for instance, they are not interested in hiding calls to the API undergoing a TOCTTOU attack. Defenses that operate outside userland (e.g., hypervisor-based system call monitoring, ETW-based monitoring) are out of scope and attackers may deploy existing evasion methods if needed: this paper focuses on userland API hooking and its applications.

3.1 Stolen Code Attack to Monitoring Completeness

Kawakoya et al. in [19] report on *stolen code* attacks observed in malware samples in the wild. In such an attack, the adversary copies at run-time[4] some instructions from the prologue of an API to a memory area owned by the sample. Then, when the sample has to call the API, it first executes the copied instructions

[3] At times alongside Event Tracing for Windows (ETW), for which several evasions exist.

[4] In an attack variant dubbed *sliding call*, the sample contains such instructions already in its compiled body. The distinction is irrelevant for the techniques presented in this paper.

and then jumps to the instruction in the original API code next to the copied ones.

Stolen code instances reported in literature [8,21,29] or that we found in real-world malware and executable protectors (ASProtect, Enigma, Obsidium, PELock, Themida, VMProtect) span the first few instructions (1–3) of an API[5]. Those showcase helpful regularities among the vast majority of APIs. For example, in 32-bit Windows APIs the first instruction is always a `mov edi, edi` that the compiler emits for hot-patching purposes [7]; then, the next two instructions are typically `push ebp; mov ebp, esp`. Thanks to such regularities, implementing an unaligned jump becomes straightforward and already suffices to defeat existing API monitoring solutions.

One could argue that defenders should move their hooks to later instructions. However, two problems arise: how deep a stolen sequence can be and if the API argument values are still visible by then. In this paper, we study how deep such attacks can go with only a modest effort for the attacker. Hence, with longer sequences, one can realize cases where the argument is no longer visible at the original location. Even worse, also branching decisions may become part of the stolen code, potentially requiring monitoring systems to deploy path-sensitive hooks to intercept execution.

Ultimately, an attacker may decide to push the attack to go as deep as possible. We believe a reasonable **boundary** is when the API makes a call. The target can be a helper (typically out-of-scope for hooking [12] as uninteresting) or another API, which the attacker may or may not decide to attack recursively[6]. We remark that stealing an entire function is possible but, in our opinion, hardly profitable. The presence of internal API calls entails a recursive stealing process, with potential dependencies on the Windows version in use and the DLLs involved across calls. As for self-contained APIs, it would be simpler to use an own re-implementation of the functionality.

Instantiating the Attack. We implemented a generalized version of stolen code that targets the call boundary. The workflow is as follows. The attacker prepares a memory region for hosting the API code to steal and locates the run-time address of the API. Then, the attacker copies the entire API body or, more surgically (using a disassembler or precomputed information), just the bytes needed to cover the basic blocks that execution can traverse before reaching an internal call. At each place where a call occurs, the attacker overwrites the opcode with a gadget that, by reading the target from an attacker-configured location, redirects execution as a tail jump to the call instruction in the original API. The adversarial program can call the attacked API with a normal call to where the stolen code resides, and pass the parameters according to the calling convention.

5 The most complex stolen code attack that we observed in the wild is from the Obsidium protector and encompasses the first 6 instructions of an API.

6 As advanced monitors filter out internal calls due to their sheer number [12].

As we show in Sect. 5, our extended attack enables the stealing of dozens of instructions. At such depths, API arguments may no longer be visible at the original locations. Furthermore, a deferred single hook may not suffice to intercept API execution due to argument-dependent control-flow choices taken in the stolen code. For example, on the DLL collection that we study in Sect. 5, our attack copies path-sensitive code for about 53.77% of the APIs in Windows 10 64-bit.

3.2 TOCTTOU Attack to Monitoring Correctness

Time-to-check to time-to-use (TOCTTOU) race conditions create a window of opportunity for an attacker when a data item is accessed a first time for checking purposes and then a second time for modifying execution state. During this window, the attacker may deliberately alter the data and subvert the semantics of the initial check. TOCTTOU attacks are well documented in UNIX-like systems, notoriously affecting file systems [33] and the userland/kernel interface [4].

In our context, we propose and demonstrate the feasibility of a TOCTTOU attack on userland API monitoring, targeting the window between call argument inspection at the API prologue and when API code later acts according to such values.

Main Idea. As discussed before, current systems retrieve API argument values right before executing the API prologue, inspecting the appropriate registers and stack locations according to the calling convention and the API prototype. To hamper monitoring correctness, an attacker may invoke the API with misleading (e.g., innocent-looking) argument values to deceive the logging or security policy in place and eventually replace them with the intended ones.

A vulnerable time window opens up between the value logging and the first use of the argument value in API code, where the use is identified after excluding trivial copies (e.g., for register spilling). From a separate thread, the attacker targets the location from which the first use (and subsequent ones, if any) takes place and updates the value with the intended one. As we will see, due to internal data movements, such location often differs from the one where the API initially received the value.

In the following, we assume to have an oracle for determining the visibility and location of API argument values at different points in API code and where their first uses occur. Section 4 will present program analysis techniques supporting these tasks.

Vulnerable Locations. As the attacker issues the API call, the stack pointer value upon entering the API is known. The attacker can use a relative displacement to write the desired value to the location from which API code reads the value for its first use.

The attack we propose is not limited to the argument locations dictated by the calling convention. It also benefits from data movements that API code makes when copying an argument to a spilling slot or to a local variable that will hold it for the remainder of the computation. The oracle informs the attacker on the location of such copies and where the first "real" use of an argument value happens, either as an operand of a non data-movement instruction or if passed as argument to an internal call.

We identify three types of potentially vulnerable locations:

❶ *On-entry slot.* The attacker overwrites the stack location where the caller passes the argument (Sect. 2). A shortcoming of this case is that, on 64-bit code, the first four arguments of any API are out of reach as they reside in registers.

❷ *Local-area slot.* The attacker overwrites a copy of the argument value that API code makes to a location in the *local area* (Sect. 2) of its stack frame. This happens, for example, for register spilling purposes. Eventually, the program accesses this copy to read the argument value and make use of it.

❸ *Register-area slot.* This case is analogous to the previous one, with the copy being located in one of the slots of the *register area* (Sect. 2) that the caller reserves when invoking a 64-bit API. The API code may spill there a copy of an argument (typically one passed via register, but not necessarily so) and eventually re-read it.

Figure 1a shows the distribution of such vulnerable locations across the APIs available in popular DLLs on selected versions of Microsoft Windows. While ❶ is the most frequent case, on the 64-bit build types ❷ and ❸ account for the majority of locations.

Timing the Attack. For a successful attack, the attacker should overwrite a target memory location between the check (i.e., when the API hooking system logs argument values) and the first use from API code. As we will measure in Sect. 5, this window typically encompasses a dozen or more instructions, and in our experiments we were able to attack windows of even just two instructions with good success.

As the attacker controls when the call is issued, locations of type ❷ or ❸ can be attacked in a straightforward and reliable way. As they are not monitored, an attacker may set a loop that overwrites the location even before the API call takes place[7].

For locations of type ❶, the difficulty for a successful TOCTTOU attack is to time accurately when to start the overwriting. This should happen only after the API monitoring system has logged the initial fake value (or it would log the rogue one) and before its first use. Furthermore, the latency of the API call processing is system-dependent and, in the general case, unknown. For this task, we envision two avenues.

For high reliability, the attacker may build a deterministic method to intercept when execution reaches a certain API instruction and then overwrite the

[7] The attentive reader may argue that this holds only if a slot is not reused with different variables: in this case, the overwriting should be timed (e.g., alike to what we do for ❶) or correctness may be affected. However, in our tests, slot reuse appeared quite rarely.

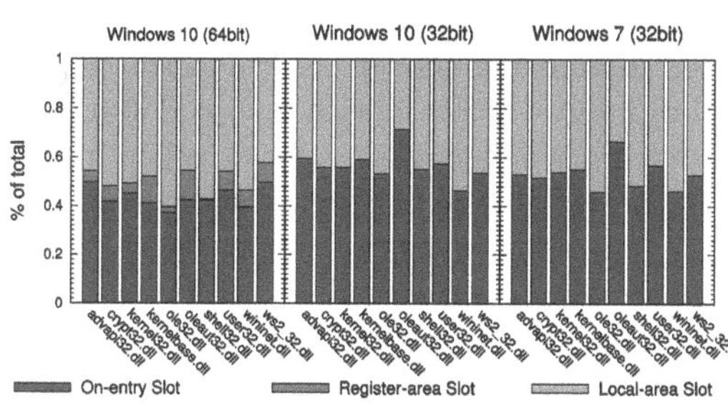

(a) Type distribution for vulnerable locations

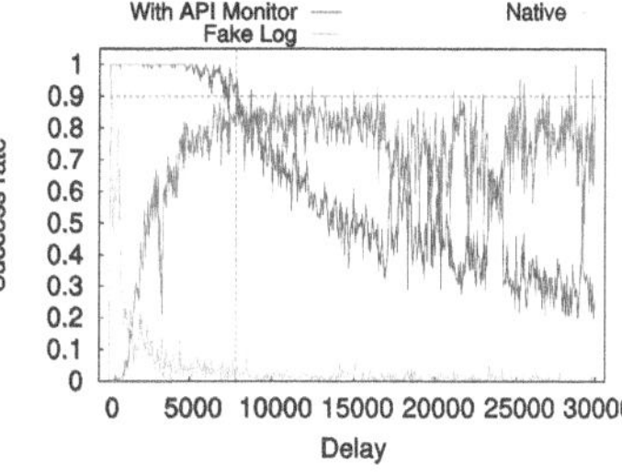

(b) Delay calibration: the value where the two lines intersect is a good pick.

Fig. 1. Breakdown of TOCTTOU-attackable locations and feasibility in hardest setting.

location. As one possible embodiment, the attacker sets an exception handler and then a hardware breakpoint at the desired API instruction. We find this a safe evasion option when the goal is to impede reverse engineering (Sect. 2), but it may be realistic also against security solutions, as shown by recent attacks that disable security products [18] or the Windows Anti-Malware Scanning Interface (AMSI) using hardware breakpoints. Less refined variants, for example based on software breakpoints, are also possible.

Alternatively, the attacker may be willing to accept occasional failures in exchange for simpler deployment and maximum stealth. For this scenario, we came up with three techniques to determine with high probability that the monitoring has ended. The first two are surgical and mainly reliable: to time the overwriting, they wait for some sentinel value to appear on the stack as part of API code execution. The third is more general and attempts multiple TOCTTOU attacks with a benign value on an API to indirectly measure the delay introduced by the monitoring.

Sentinels. One sentinel kind are *stack immediates*: i.e., immediate values that API code pushes to fixed stack locations. A typical example in 32-bit API code is the size pushed on stack as argument when calling internally `__SEH_prolog()` to configure Structured Exception Handling. The attacker implements a busy-waiting loop that inspects the stack memory until the known constant appears: when it does, execution has left the monitor, and the attacker can start overwriting the target location.

Another sentinel kind are *spilled registers*: i.e., registers that API code spills to fixed stack locations and whose content is known to the attacker. The typical case are callee-saved registers, which the attacker can arbitrarily set before calling the API, but also API argument values and even the base pointer value (from the `push ebp` in the API prologue) can serve the purpose in a busy loop designed as above.

Delay Loop. When no sentinel is available, the attacker may exploit TOCTTOU race conditions also to guess when the monitoring is over. The attacker chooses two non-suspicious values for an API of interest, calls it with one, and starts the overwriting operation with the other with a delay. Then, the attacker inspect the result of the API (or its effects on the system) to determine which value was used.

The attacker employs a *delay calibration loop* to gradually vary the delay and measure the stability of the result on multiple calls. The goal is to find a delay sufficiently high such that, if increased further, the overwriting will likely happen after the use; aiming for a high delay provides the attacker with confidence that the monitoring has, by then, ended. From then on, to call the API with conspicuous arguments, the attacker issues only a single call and adds the measured delay to the thread carrying the overwriting presented for the main attack. The value can be reused to attack other APIs with a similar expected logging time (e.g., for argument count and types).

Figure 1b plots the effects of the delay, measured as the trip count of a `loop` assembly instruction for busy-waiting, in a native execution and under a popular [12] hooking tool, *API Monitor*. The attack targets the `WSAStartup` API. In the Windows 10 build used for the test, the API offers a TOCTTOU window size of 21 instructions. The attacker identifies a high delay value that results in the API executing with the rogue argument with, at least, the desired probability. As we inspected the logs of this and similar experiments, we saw that a very high success rate in timely argument overwriting is associated with the API monitoring system logging the fake initial values. In the figure, the point where the dashed lines insersect shows a 91% success rate if the attacker waits for the corresponding delay before starting the attack.

Discussion. The TOCTTOU attack we propose is the first targeting API hooking correctness and, to the best of our knowledge, also the first entirely within userland. We target a design weakness of current hooking systems, exposing fake indicators for argument values and then overwrite them as late as possible, either in their original location or in later copies depending on the opportunities that the victim API offers. We remark that the attack applies also to OUT arguments, as the attacker can put a fake result in the original buffer and have the API write the result to a new buffer.

Our goal is to show that TOCTTOU attacks are feasible. We note that many targets (type ❷ and ❸) are straightforward and we propose alternatives to comprehensively treat locations of type ❶ with different degrees of reliability and complexity.

The next section presents techniques that examine API code for identifying what are the instances of an argument value and where the code uses them. While we will use them for defensive purposes (including the defeat of our attack), they can also measure to what extent Windows APIs may be vulnerable to TOCTTOU races.

An orthogonal aspect worth mentioning is that the delay loop technique may be used to detect if a hooking system is present. In particular, we noted that the distribution of the delay value compared to its effects on the API outcome changes significantly between native and monitored executions, with the latter needing much higher values.

4 Towards a Principled Defense

This section proposes countermeasures for more reliable API hooking designs. In the following, we present program analysis techniques to track the propagation of the values provided as call arguments throughout API code and study their liveness, which is important in turn in order to identify what value copies may be accessed.

4.1 Overview

A clear weakness of current API monitoring systems is their on-entry hooking policy. We argue that placing probes in deeper spots would significantly reduce, if not close, the attack surface for completeness and correctness attacks as those from Sect. 3.

To this end, the main obstacle to overcome is that, for deeper spots, defenders should determine until when argument values are still visible, including in the reasoning the locations where they may have flown through data movements. For example, compilers perform a register allocation process to control which intermediate values and expressions reside in registers, and when a register must be temporarily freed for other computations, the compiler spills its value on the stack [6]. Also, API code may read argument values and host them in temporary, internal variables before using them. Even debug information available for Windows DLLs as PDB files is insufficient to track these movements. Therefore, we designed a binary-level static analysis to track the propagation of API call argument values through the control-flow graph (CFG).

Main Idea. We want to determine the deepest point(s) in the CFG of the API where one can place hooks and still be able to determine not only that an API call took place, but also which were the parameters supplied for it. We aim for a data-flow analysis that can trace parameter value propagation from the entry point of the API. More precisely, we want to see how the values they contain flow through API code and where they end up stored (in spilling slots or copied to internal variables hosted on stack or in registers) for eventually being read again for computations that make use the value (e.g., branches, arithmetic manipulations, supplying an argument to an internal call).

The most convenient place where a parameter can be retrieved, either in its original location or through a copy, is where API code uses its value last. This scenario covers both unaligned invocations (as with stolen code schemes) and TOCTTOU attacks[8].

The data-flow analysis we design computes for every basic block in the API an input list describing which parameters are visible and their current location(s) when execution enters the block. Then, we evaluate how instructions in the block can affect such list (for instance, by polluting a register containing a dead value), and produce an output list for the block. For any block, the input list is given by a merge operation where we intersect the output lists of the blocks with an outgoing edge to it.

The input and output lists that we build for the data-flow analysis also allow us, through further refinements, to precisely locate when the API code accesses a parameter value, regardless of its availability for inspection. Identifying where uses happen is important because an attacker may apply the TOCTTOU idea to overwrite the location after the use and mislead a hooking system that inspects a value "too late".

Finally, relying on topological properties of the CFG for efficient insertion, we identify suitable places for placing hooks to log each API parameter, so as to cover all the alternative execution paths in the CFG while being at the deepest possible places.

4.2 Tracking Parameter Value Propagation

We model the control-flow graph as a directed graph $G = (V,\ E)$ with a set of vertices V and edges E. The API prologue corresponds to vertex v_0. We assume there is a single vertex with no incoming edges and that there are no isolated vertices.

Algorithm 1 is designed to work on compiler-generated code that is well-behaved [23] in terms of stack manipulations: by that, we mean that every basic block can alter the stack pointer register only by a predetermined quantity and that CFG join points (i.e., at basic blocks with multiple predecessors) see an identical stack pointer variation across all the incoming paths compared when execution reached the prologue v_0. We note that Windows API code naturally meets these characteristics (Sect. 2).

For each vertex $v \in V$, the algorithm maintains two sets $PAR_{in}[v]$ and $PAR_{out}[v]$ that it updates using a fixed-point approach. Initially, $PAR_{in}[v]$ and $PAR_{out}[v]$ are not instantiated (we use $\perp$), with the exception of $PAR_{in}[v_0]$ that we populate with the set of pairs for each argument and its location at the prologue according to the calling convention. The algorithm uses a worklist W with initially $W = \langle v_0 \rangle$.

The algorithm pops a vertex v from the worklist, and if $PAR_{in}[v]$ is not initialized, it computes it by merging $PAR_{out}[v_{pred}]$ for each predecessor v_{pred} of v. Informally, all the results from the computation of the predecessors of vertex

[8] In case of multiple uses, the attacker has to overwrite the value already before the first use.

Algorithm 1. Proposed Algorithm for Tracking Parameter Value Propagation

$\forall v \in V - \{v_0\} : PAR_{in}[v] = \bot$; $\forall v \in V : PAR_{out}[v] = \bot$
$PAR_{in}[v_0] = \{(arg_i,\ loc_i) \mid$ value of arg_i is at loc_i by calling convention$\}$
$W = \langle v_0 \rangle$
while not $W.empty()$ **do**
    $v = W.pop()$
    $I = \{PAR_{out}[v_{pred}] \mid (v_{pred},\ v) \in E\ \wedge\ PAR_{out}[v_{pred}] \neq \bot\}$
    if $I \neq \emptyset$ **then**
        $new\_in = \mathrm{merge}(I)$
        if $PAR_{in}[v] == new\_in$ **then** *continue*
        $PAR_{in}[v] = new\_in$
    $new\_out = \mathrm{compute}(v,\ PAR_{in}[\mathrm{v}])$
    if $PAR_{out}[v] \neq new\_out$ **then**
        $PAR_{out}[v] = new\_out$
        for each v_{succ} $s.t.$ $(v,\ v_{succ}) \in E$ **do**
            $W.push(v_{succ})$

v are merged together. After the initialization of $PAR_{in}[v]$, the algorithm computes statically the effects on memory and registers of the instructions in the basic block and computes the output state. Mainly, the static analysis recognizes the introduction of value copies to new locations through data movement instructions, while it discards a currently known location when its contents are overwritten. If the output state differs from prior iterations, $PAR_{out}[v]$ is updated and all the successors of v enter the worklist.

We treat information from multiple predecessor blocks via set intersection. In particular, the merge() operator takes all the pairs (arg_i, loc_i) such that $(arg_i, loc_i) \in PAR_{out}[v_{pred}]$ for all the predecessors v_{pred} of the current vertex v. Informally, an argument-location pair is added to the result set iff. it is in the PAR_{out} set of each predecessor. This approach is the most conservative possible for value correctness, as it discards information for locations whose visibility is potentially path-sensitive.

For the static analysis of compute(), we track registers, memory contents, and related aliases; we include provisions for more complex cases of pointer expressions on structure fields and nested dereferencing. For instructions that modify operands for purposes other than data movements, we clear from the set all the arguments that see their location altered by the instruction. This choice is conservative too, as compilers may realize data movements also through arithmetic/logic instructions: however, in our experience, such cases are rare in Windows API code; therefore, we opted not to include heuristics for them, avoid the risk of introducing unnecessary imprecision.

4.3 Identifying Live Copies of Argument Values

The conservative approach taken at merging data-flows in Algorithm 1 is adequate for retrieving call argument value against traditional API obfuscation attacks. That is, when the analysis reports a location for a parameter value at a certain program point from API code, the value will be identical to the one visible upon entering the prologue. However, if the adversary attempts TOCTTOU attacks like the one we described in Sect. 3.2, this property no longer holds.

Consider a scenario, rather frequent in our observations, where API code maintains two copies of an argument value. At some point, API code no longer uses the first copy, and does not reuse its storage location: therefore, the copy remains visible at later basic blocks, ideally even until the API epilogue. From that point forward, API code accesses instead the other copy and, for example, issues an internal API call using it or makes other program state changes. If the storage location of this other copy is vulnerable to a TOCTTOU attack, an API monitoring system that logs only the first copy (for example, just because it is visible longer) will see only the initial parameter value, missing the value that the adversary sets instead via the attack.

Therefore, at every program point we should look for what parameter value instance(s) API code may access and use in the remainder of the execution. This problem is similar to what in compiler theory goes by live variable (or liveness) analysis [26]. In particular, a variable is said to be live at a program point if the program may access it in the remainder of the computation without reassigning it first. Alternatively, this can also be modeled as a reaching definition problem [26]. Techniques for conducting either analysis on source code are well established.

To cope with our scenario, we first extend Algorithm 1 to compute PAR_{in} and PAR_{out} sets at the granularity of individual instructions. Then, we adapt the function compute() to construct the standard $LIVE_{in}$ and $LIVE_{out}$ sets of a standard backward liveness analysis, again at the granularity of individual instructions. Therefore, $LIVE_{out}[i]$ contains the locations that the program may reference in the reminder of the execution past i (i.e., the locations that are live after i), and $LIVE_{in}[i]$ contains the locations that are live at i. At control-flow join points, the backward analysis assigns $LIVE_{out}[i]$ as the union of the sets $LIVE_{in}[j]$ where j is an immediate successor of i.

Therefore, to solve our problem, we can intersect the locations from $LIVE_{in}[i]$ with the locations that appears in the pairs at $PAR_{in}[i]$ as hosting a parameter value. This will remove stale copies of parameters values from the results of Algorithm 1, exposing the sole instance(s) that a monitoring system should consider as valid.

4.4 Hook Placement

Knowing which copies of parameter values are live at each instruction in API code provides sufficient information for an API monitoring implementation to reliably trace API call parameter values. The last element of our approach to counter completeness and correctness as in Sect. 3 is to determine what are the most convenient places for hook insertion to combine efficacy (i.e., to defeat the attacks) with performance.

In general, aiming for the deepest places is a good strategy as long as all the alternative execution paths throughout the API traverse them. Unfortunately, not all paths are similarly deep. This implies that some paths will need an early logging, whereas for deeper paths we may have to re-inspect the value against TOCTTOU attacks.

We can address this problem by building dominator information on the CFG and using it as follows. For each leaf node in the dominator tree and for each API parameter, we see if a live copy of the parameter value is visible at any of its instructions, starting from the last in the basic block. If found, we mark the instruction as a place for hook insertion. Otherwise, we inspect its ancestors until a copy becomes available and we hook it. The intuition behind the criterion is that we want to defer logging values unless their visibility becomes path-sensitive.

4.5 Implementation

We implement the analysis techniques from the previous sections in Python in about 2200 LOC. For disassembly and CFG reconstruction, we use off-the-shelf tools: in particular, we opt for IDA Pro [13] as it also natively integrates with the PDB debug information files that Microsoft makes available for Windows DLLs. These files report the stack pointer variation at each instruction compared to its value at the prologue. We retrieve this information directly from IDA, including the variations induced by internal function calls when they clean up the stack upon return. For API prototypes, we resort to existing efforts [12] based on automatic extraction from Windows header files. Finally, to capture the semantics of instructions, we use the Capstone [5] framework to recover operands (including the individual components of complex memory addressing expressions) and memory access kinds (read, write, or both).

We then implement TOXOTIDAE, a prototype API tracer based on the design we argue for in this paper. For its realization, similarly as in the most recent work in the area [12], we use Pin [24] for dynamic binary instrumentation. As we only need to hook specific addresses, we remark that also implementations based on virtualization-assisted hardware breakpoints or even user-space inline hooks remain possible. We provide TOXOTIDAE with precomputed program points for hook insertion (currently for selected Windows 7, 10, and 11 builds, but adding more is straightforward) and with prototypes for all the APIs from the DLLs of interest for monitoring.

4.6 Discussion

The program analysis techniques we presented allow us to study the propagation and life span of parameter values throughout the locations in use to a function. We use this information to revisit the canonical design of API hooking solutions, identifying multiple spots for hook insertion that account for the plurality of alternative paths and the decaying visibility of value copies. Each parameter sees the insertion of dedicated hooks; as we measure in Sect. 5, their number is small in practice.

The output of Algorithm 1 also provides an obvious way for computing vulnerable locations for TOCTTOU attacks according to the patterns ❶-❷-❸ we showed in Sect. 3.2. In Sect. 5, we will experimentally measure that our defensive hook are deeper than the reach of both our extended stolen code attack and the TOCTTOU one.

Finally, our analysis techniques may be of general interest. For example, we believe they may be helpful to amend debug information that optimizers discard inadvertently [2] or emit incorrectly [10] due to inaccurate implementations, by comparing it with the storage locations we track. We leave this investigation to future work.

5 Evaluation

For evaluating the attacks and the defense methodology presented in the paper, we seek to answer the following research questions:

RQ1: How are common Windows APIs vulnerable to our attacks?
RQ2: How many hooks are needed for the proposed defense? How deep are they?
RQ3: What are the accuracy and performance costs of the proposed defense?

We conducted our evaluation mainly on two virtual machines running Windows 7 SP1 32-bit (build 7601) and Windows 10 64-bit (version 1803). The host machine has an Intel Core i7-11800H CPU @ 2.30 GHz with 8 physical cores (16 threads) and 32 GB of RAM. Each VM received 4 virtual CPU cores.

Table 1 shows the 10 DLLs we select for the study, following the choices of the authors of the SNIPER tracer [12]. These DLLs feature heterogeneous APIs used both in malware and goodware, covering disparate tasks from file manipulation to network communication and cryptography.

We analyze only APIs that take arguments. Then, we further check if they have at least one argument that is live at the prologue, labeling them *parametric*. APIs not in this group are deprecated or dismissed APIs or return constant values.

For comparison, we consider three popular monitoring tools (Rohitab API Monitor, SpyStudio, WinAPIOverride) and a state-of-the-art academic system (SNIPER [12]).

Attacks. To answer **RQ1**, we first study in Fig. 2 how deep that the extended stolen code and the TOCTTOU attack can go. For a meaningful comparison, we consider for the chart only the functions that are amenable to both. Provided as a reference, the average function depth of functions is measured as the length of the maximal acyclic path in the CFG in terms of traversed instructions.

We recall that TOCTTOU attacks on 32-bit APIs are always possible thanks to pattern ❶, hence the distinction is relevant only for 64-bit API. For the other attack, to keep the analysis realistic, we consider only functions with at least 10 instructions and that are not stubs that jump into another API (or an attacker would just attack it).

The data for stolen code on the 64-bit APIs not amenable to TOCTTOU attacks do not show relevant differences, hence we omit them for brevity. For

Table 1. DLLs and analyzed APIs (H: Hooks, P: Parameters, A: API).

DLL	Windows 10 (64bit)				Windows 10 (32bit)				Windows 7 (32bit)			
	Analyzed	Parametric	H/P	H/A	Analyzed	Parametric	H/P	H/A	Analyzed	Parametric	H/P	H/A
advapi32.dll	626	612	1.092	3.6127	643	632	1.0513	3.6345	680	669	1.1936	4.142
crypt32.dll	258	258	2.8088	10.6279	284	280	2.464	9.5643	284	280	2.1117	7.8143
kernel32.dll	1152	1147	1.076	2.476	1158	1150	1.0477	2.6296	1109	1097	1.3872	3.6737
kernelbase.dll	1315	1309	2.1804	5.8892	1352	1347	2.0753	6.0371	504	502	2.0277	5.9801
ole32.dll	239	228	2.2024	5.7061	218	211	2.1315	5.6161	315	305	2.3748	5.9607
oleaut32.dll	407	406	3.024	7.8818	395	395	2.7983	7.3671	392	392	2.4534	6.5434
shell32.dll	330	300	1.6542	4.3633	379	354	1.5549	4.1695	404	392	1.728	4.8673
user32.dll	679	658	1.7289	4.114	676	670	1.5849	4.0	672	665	1.3846	3.588
wininet.dll	275	264	2.8932	10.4735	298	276	2.3741	8.5688	295	274	2.6854	9.6131
ws2_32.dll	161	161	2.7717	9.8944	149	149	2.9597	10.1946	147	147	2.288	7.585

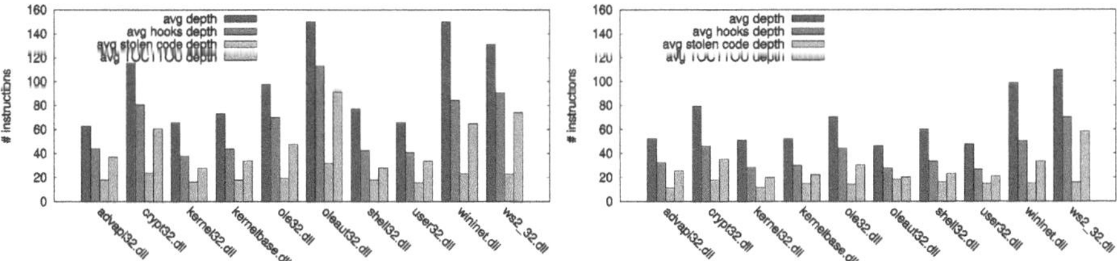

Fig. 2. Average depths. Left: Windows 10 (64-bit). Right: Windows 10 (32-bit).

similar reasons, we omit details on the 32-bits DLLs of Windows 7 and refer for them to the Windows 10 counterparts. The average function depth differs between the 32-bit and 64-bit APIs mainly because vulnerable 32-bit APIs are a superset of vulnerable 64-bit APIs.

Extended Stolen Code Attack. We can immediately see from Fig. 2 that depths of about 1–2 dozens of instructions are possible. Figure 6 (Appendix A) provides additional data points, including the median absolute deviation. These depths are significantly larger than in currently documented stolen code instances (Sect. 3.1).

Also, they back our claim that current on-entry hooking designs cannot be amended by simply pushing their analysis a few instructions deeper. Not only argument locations may change (especially with 64-bit code), but further analyses we carried on these Windows 10 64-bit APIs showed that about 53.77% of the code sequences our method can steal include path-sensitive code (which requires multiple hooks for interposition).

To show the feasibility of the method, we implemented a script (Sect. 3.1) that automatically and surgically steals API code until the call boundary. We use it to shield synthetic test programs, as well as several calls in the code of Al-Khaser, a popular utility for stressing malware analysis environments. For the latter, we test 21 APIs (such as `AdjustTokenPrivileges`, `CheckRemoteDebuggerPresent`, `EnumProcesses`, `SleepEx`, and `VirtualProtect`) with heterogeneous depths. All protected calls operate correctly both in a native execution and when under the four monitoring systems we chose for the evaluation, going unnoticed by all of them.

DLL	Vulnerable APIs (%)	Redirecting	Parametric
advapi32.dll	147 (49.00)	312	612
crypt32.dll	122 (51.19)	23	258
kernel32.dll	128 (32.82)	757	1147
kernelbase.dll	432 (36.12)	113	1308
ole32.dll	40 (24.69)	66	228
oleaut32.dll	41 (11.61)	53	406
shell32.dll	55 (26.83)	95	300
user32.dll	127 (32.15)	263	657
wininet.dll	104 (40.94)	10	264
ws2_32.dll	67 (43.23)	6	159

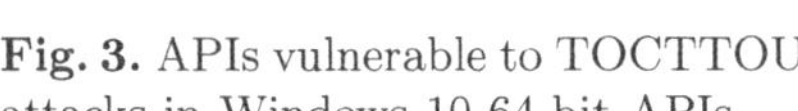

Fig. 3. APIs vulnerable to TOCTTOU attacks in Windows 10 64-bit APIs.

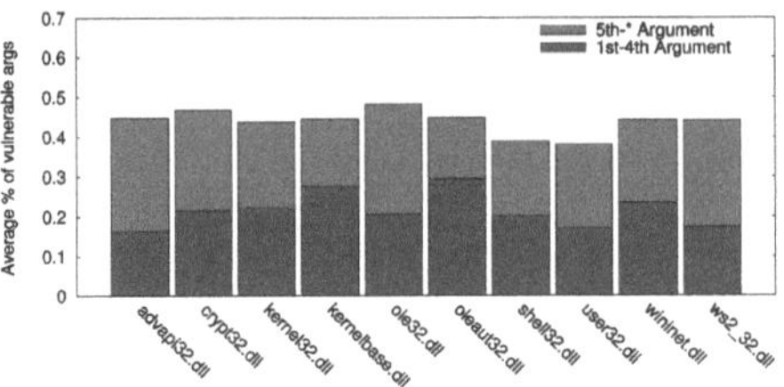

Fig. 4. Arguments vulnerable to TOCTTOU attacks in Windows 10 64-bit APIs.

TOCTTOU Attack. The average depth of vulnerable locations reported in Fig. 2 represents the average size, in terms of instructions, of the TOCTTOU window that an attacker can target before API code uses a parameter value (for purposes other than making a local copy). On 64-bit code, these windows are deeper: this is due to the fact that, as we discuss next, fewer arguments see their values being stored in vulnerable locations—whereas all APIs and all arguments are attackable on 32-bit DLLs.

Figure 3 shows how many 64-bit parametric APIs from Table 1 are vulnerable. We further refine parametric APIs by discarding redirection stubs (similarly as with stolen code), as for an attacker it would be more convenient to attack their targets directly. Therefore, we report the number of APIs that we can attack in one or more argument and express them also in terms of percentage of non-redirecting parametric APIs.

Figure 4 shows how frequently arguments are attackable: the total amount is computed as the distinct vulnerable arguments among those actually used by each vulnerable API inside the DLL. Then, we further distinguish the identity of arguments between those passed by registers (from the first up to the fourth) and those passed via memory. Overall, across all vulnerable APIs, we may attack nearly half of their arguments.

Finally, regarding vulnerable storage locations, we already reported the distribution of locations types ❶-❷-❸ in Fig. 1 when presenting the attack in Sect. 3.2. Here, to complete our investigation on the potential attack surface of Windows APIs, we provide additional insights on how we can attack locations of type ❶. Figure 5 analyzes all vulnerable locations and shows as "none" those that we classify as ❷ or ❸ (which are straightforward to attack, Sect. 3.2). Then, it divides locations of type ❶ according to the method needed to attack them. To recap, either the attacker may benefit from sentinel values or they have to explicitly determine (with a delay calibration loop or with a breakpoint-like mechanism) when the on-entry monitoring is over. We remark that the latter is the hardest but also, by far, the least frequent case in practice.

Summarizing, TOCTTOU attacks are possible and require sophistication in the overwriting step only for a limited fraction of parameters. 32-bit APIs are naturally more vulnerable due to stack-passing for all arguments, whereas for

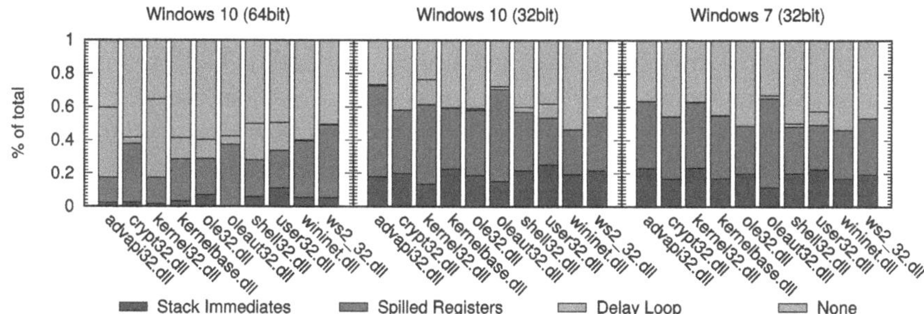

Fig. 5. Deferred overwriting strategy needed at TOCTTOU-vulnerable locations.

64-bit (with the exception of the APIs from `oleaut32.dll`) the DLLs in our study show that 11.61 to 51.19% of non-redirecting APIs are vulnerable in one or more of their arguments.

As for feasibility, we conducted a preliminary investigation with several proof-of-concept implementations (PoCs) on 12 machines featuring different CPU models (9 Intel Core models from 5 generations and 3 AMD Ryzen from 2 generations) and Windows versions (Windows 11 Home builds 22631 and 22621, Windows 10 Home 19045, Windows 10 Education 17134). The PoCs cover multiple APIs (e.g., `CryptBinaryToStringA`, `ShellExecuteA`, `RegOpenKeyA`, `WSAStartup`, `inet_ntop`) and all the location types and overwriting strategies presented in the paper. Each PoC attempts 20 executions of the API and checks whether it executed using the initial or the overwritten parameter(s). Across our tests, we obtained a success rate of 97.06%. When inspecting the failures, we noticed that they originated from a fragile monitoring choice on one stack-immediate sentinel value, which was selected too close to the instruction accessing the parameter location under attack.

After excluding this program, we repeated these tests on the machine used for the main evaluation using the four API monitoring tools chosen for the study. None of the tools was able to log any rogue argument value in the 20 trials.

However, while the others operated correctly, API Monitor resulted in several crashes on 64-bit code. We noticed that it alters the stack layout significantly: its hooking stub uses the register area that the program sets up for the API (Sect. 2) to spill register-passed arguments, copies stack-passed arguments lower in the stack, and restores register-passed arguments before issuing the call to the API. This has two drawbacks: transparency, as its hooking is straightforward to detect by monitoring the original register area contents, and security, as now every arguments becomes a type ❷ location and we can attack it reliably by just adding a predetermined offset.

Defenses. Table 1 and Fig. 2 contain relevant information for discussing the effectiveness of the proposed defense (**RQ2**). The key insight from Fig. 2 is that the hooks we can place thanks to our value tracking analysis are appreciably deeper than the reach of both extended stolen code and TOCTTOU attacks.

Besides average values, we checked that no attack path is deeper than where we put our deepest hooks. When this hold, the hooks we insert counter completeness and correctness attacks by construction. Across all the APIs we tested, the only failure was one stolen code case surpassing the last point where our analysis deems an argument visible (hence we hook it). A bug in the analysis had made us miss a copy available within a struct, despite the analysis tracked the storage of the struct correctly.

The path-sensitive insertion policy guided by dominator information that we use for hook insertion results in a number of hooks that is rather limited on average. In Table 1, we plot as H/P the average number of hooks that we need for each parameter of the API throughout the CFG. The average is computed on a per-API basis and then aggregated as a geometric mean across APIs. Across the three DLL collections we study, we find that this number is in the range $1.0477 - 3.024$. Then, we plot as H/A the average number of hooks that we insert for each API. This number is in the range $2.476 - 10.4735$. While H/P reflects the complexity of the interactions in API code over argument values, H/A reflects the opportunity that we take for optimizing hook insertion, grouping per-parameter hooks when at identical program points. Furthermore, in practice, almost the entirety of these hooks execute only once per API call, as for the DLLs that we study they are inserted within loops only rarely.

Testing. To test the efficacy of our TOXOTIDAE system, we successfully run the PoCs that we synthesized for the extended stolen code and TOCTTOU attacks. We also test real-world instances of traditional stolen code attacks, in particular executables shielded with protectors (Obsidium, Themida, VMProtect) and 10 malware samples (Table 2). We obtained the samples thanks to the authors of API Chaser [19], who shared with us the 2 samples used for its evaluation, and to a professional malware analyst that spotted them during their daily job; the 8 samples are mainly from 2020 and 2021. As expected, all these executables successfully evade the four existing API hooking tools we tested, whereas TOXOTIDAE traces their calls correctly.

An interesting sample is `051...dc4`, which invokes the `LoadLibraryA` API normally excepts when it loads, using stolen code, `cmcfg32.dll` for configuring the Microsoft Connection Manager for remote accesses. We also find it interesting to report on `08c...4b1`, a sample presumably protected with Obsidium. After invoking in a normal way typical APIs for code unpacking, it performs evasive sequences around debugger and virtualization artifacts (e.g., `CreateFileW` on `\??\VBOXGUEST`) using stolen code. We believe this is done to frustrate analysts that try to analyze the sample using manual debugging tools that do not automatically counter these tactics.

Performance. Finally, we tested accuracy and overheads of TOXOTIDAE using the test suite of the Wine emulator for the DLLs selected for our study. This suite, used also in the evaluation of SNIPER [12], stresses different implementation

Table 2. Analyzed malware samples. `5fd..` and `e09..` are from API Chaser [19].

Hashes (md5)	
9407345f8a1d891624fcba99d20a8b33	61f5e1d9bdc23b0328ef7e874314e3a7
886a19aa14a41386dfb695ad908a1999	08cfaf120cd12bafe53c2fbb50c204b1
33fb8185801d229c91d4aef1efba941d	de71c06b30caa7488f6f42380316ef18
f1cfdb7530169e9398bcc78095ecdcf3	0519f94bf975fdc6b92b40f282853dc4
5fd727d3c11c66583f92970e4c0ec197	e0974042a67ad3db9042e16e4dcb0465

aspects of APIs; also, being deterministic and more API call-intensive than real-world applications, it enables reliable accuracy tests and simulates worst-case overhead scenarios. While the authors of SNIPER picked one workload per DLL, we used all the 130 available for our DLLs.

For accuracy, we configure TOXOTIDAE to log parameters also upon entering an API and systematically compare their values against those from our deeper hooks. Across all tests, our hooks retrieve correct values in 99.40% of all invocations of 64-bit APIs and 99.82% on 32-bit APIs. We find this result encouraging for a static analysis.

The main cause of inconsistency in our results are partial uses of argument locations that our static analysis does not model correctly: for example, a `BOOL` type on Windows is sized as an `int` but most APIs only use its least significant byte.

For hook coverage, we compute the geometric mean of the distinct hooks hit across all tests among those we set for an API: we measure a hit ratio of 86.66% on 32-bit APIs and 85.32% on 64-bit APIs, reflecting the good heterogeneity of the test suite. The average depth of hooks hit is of 25.73 instructions on 32-bit APIs and 23.29 on 64-bit APIs.

For overheads, we compare the running time of our approach against TOXOTIDAE configured to log arguments like traditional hooking tools do. Building both tools with `-O2` in MSVC and the optimized buffering of SNIPER, we measure an average slowdown of 4.66% (std. dev. $\sigma = 0.0399$) on 64-bit tests and of 7.62% ($\sigma = 0.0765$) on 32-bit tests. Hence, the increased number of hooks results in modest run-time overhead increases on call-intensive tests. This answers **RQ3** and concludes our evaluation.

6 Other Related Works

Binary Analysis. In the vast universe of binary-level program analysis techniques proposed over the years, our work shares loose analogies with techniques that have been explored, among others, for pointer analysis [20], pinpointing memory vulnerabilities in unknown layouts [32], or recovering symbolic information from stripped binaries [3]. However, we are unaware of techniques directly applicable to the tasks of Sect. 4.

Stolen Code. Countermeasures for stolen code techniques are given in API Chaser [19]. The authors propose techniques based on pre-boot disk tainting and code taint propagation to tag instructions according to their source: *benign* for system code and known applications, *api* for API code, and *malware* if from the untrusted program under analysis. The method intercepts stolen code sequences as the stolen code will have a benign tag and the control transfer to it will have a malicious tag. Documented ways to hinder this approach include issuing calls via code reuse gadgets, as they would be tagged as from a benign location, and imprecisions of taint analysis with implicit flows. API Chaser comes with important setup costs and overheads due to the taint analysis. Our approach is more efficient in both respects. API Chaser would handle our stolen code extension but be defeated by the TOCTTOU attack.

TOCTTOU Attacks. These attacks are well-known to users of UNIX-like system. For example, Wei et al. [33] exploit a race condition between the code responsible for checking file existence and permissions and the code for opening a file, resulting in gaining access to files restricted to superusers only. Two recent industry presentations [14,15] leverage a race condition between a system call invocation and the Linux kernel when copying memory content from user space to kernel space. An attacker may use the condition to hide the parameters for a few system calls. Our attack works instead in userland, systematically identifying races and categorizing them for exploitability.

Schwarz et al. [30] propose microarchitectural techniques to detect double-fetch operations and a fuzzer to narrow down the exploitable cases, attacking trusted execution environments and system calls. Our approach is different, as we study code statically and do not need a double-fetch vulnerability to attack a location, but it may be interesting to look into similar exploitation techniques for generalizations of our attack.

7 Limitations and Future Works

We acknowledge the following limitations for the research presented in this paper.

As we rely on static analysis for analyzing Windows DLLs, we may miss locations or model wrong ones due to imperfection in the CFG reconstruction phase, when modeling the effects of specific instructions in compute(), or when PDB files do not contain prototype information for internal calls to helper functions. We discussed an example of modeling imprecision in Sect. 5; more may occur when considering other libraries or, more generally, code generated for other platforms or systems.

For stolen code attacks, we left out cases of complete code copies, considering them inconvenient for an attacker. In particular, recursive stealing of internal calls is necessary for stealth (or those would show as from program code and hooking tools would log them) and comes with issues that we discussed in Sect. 3.1. Nonetheless, complementary measures are a possibility, such as using the code tainting of API Chaser [19] or shepherding memory and file content transfers involving Windows DLLs.

For TOCTTOU attacks, one aspect that we did not cover in the presentation is when an API monitoring solution may re-inspect the initial storage of parameters upon API return. This may expose attacks to locations of type ❶ at API return time. Therefore, the attacker would need to restore the initial fake values at such locations. While for API entry events we came up with sentinel values or delay calibration loops, the only strategy that we foresee for an attacker is to halt execution with a breakpoint on an instruction within the API epilogue or one nearby that dominates it.

More evaluation and calibration is needed for threat actors to deploy our TOCTTOU attack at a scale, also depending on their reluctancy to occasionally fail and reveal the intended API arguments. This observation mainly applies to type-❶ locations when the attacker prefers not to use a deterministic execution interception mechanism and no sentinel value is available, leaving the delay calibration loop as only viable option. Nonetheless, the goal of this paper is showing that these attacks are possible and that, as we showed, can be countered in a principled way.

Finally, we are hopeful that the techniques presented in this paper may be helpful for other tasks. We hinted at a potential application for debug information recovery in Sect. 4.6. Another interesting opportunity could be to study how our deep hooks may be used to raise the bar to unhooking attacks to antivirus and EDR products and to similar approaches that disable key Windows components like AMSI and ETW.

Acknowledgements. We thank our anonymous reviewers for their comments. We also thank Federico Palmaro and Franco Gioia for their feedback on our design. This work was supported by the Italian MUR National Recovery and Resilience Plan funded by the European Union - NextGenerationEU through project SERICS (PE00000014).

A Appendix

Figure 6 shows average stolen code depths and their median absolute deviation computed for all APIs in a DLL, excluding cases of trivial redirections, as opposed to the analysis of Fig. 2 that features only TOCTTOU-vulnerable APIs.

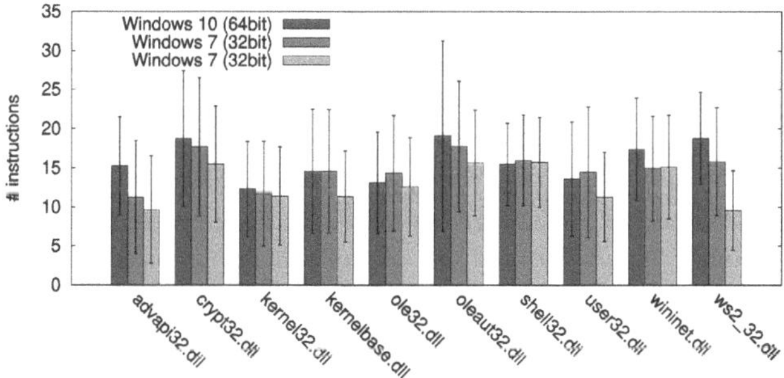

Fig. 6. Stolen code depth with mean absolute deviation bars.

References

1. Apostolopoulos, T., Katos, V., Choo, K.K.R., Patsakis, C.: Resurrecting anti-virtualization and anti-debugging: unhooking your hooks. Futur. Gener. Comput. Syst. **116**, 393–405 (2021)
2. Assaiante, C., D'Elia, D.C., Di Luna, G.A., Querzoni, L.: Where did my variable go? Poking holes in incomplete debug information. In: ASPLOS 2023, pp. 935–947. ACM (2023)
3. Balakrishnan, G., Reps, T.: Analyzing memory accesses in x86 executables. In: Duesterwald, E. (ed.) CC 2004. LNCS, vol. 2985, pp. 5–23. Springer, Heidelberg (2004). https://doi.org/10.1007/978-3-540-24723-4_2
4. Bhattacharyya, A., Tesic, U., Payer, M.: Midas: systematic kernel TOCTTOU protection. In: 31st USENIX Security Symposium (USENIX Security 2022), pp. 107–124. USENIX Association (2022)
5. Capstone Engine: Capstone, The Ultimate Disassembler. https://www.capstone-engine.org/
6. Chaitin, G.J.: Register allocation & spilling via graph coloring. In: Proceedings of the 1982 SIGPLAN Symposium on Compiler Construction, SIGPLAN 1982, pp. 98–105. ACM (1982)
7. Chen, R.: Why do Windows functions all begin with a pointless MOV EDI, EDI instruction? https://devblogs.microsoft.com/oldnewthing/20221109-00/?p=107373
8. Cheng, B., et al.: Towards paving the way for large-scale windows malware analysis: generic binary unpacking with orders-of-magnitude performance boost. In: CCS 2018, pp. 395–411. ACM (2018)
9. D'Elia, D.C., Coppa, E., Nicchi, S., Palmaro, F., Cavallaro, L.: SoK: using dynamic binary instrumentation for security (and how you may get caught red handed). In: Asia CCS 2019, pp. 15–27. ACM (2019)
10. Di Luna, G.A., Italiano, D., Massarelli, L., Österlund, S., Giuffrida, C., Querzoni, L.: Who's debugging the debuggers? Exposing debug information bugs in optimized binaries. In: ASPLOS 2021, pp. 1034–1045. ACM (2021)
11. D'Elia, D.C., Coppa, E., Palmaro, F., Cavallaro, L.: On the dissection of evasive malware. IEEE Trans. Inf. Forensics Secur. **15**, 2750–2765 (2020)
12. D'Elia, D.C., Nicchi, S., Mariani, M., Marini, M., Palmaro, F.: Designing robust API monitoring solutions. IEEE Trans. Dependable Secure Comput. **20**(1), 392–406 (2023)
13. Guilfanov, I.: IDA Pro. https://hex-rays.com/ida-pro
14. Guo, R., Zeng, J.: Phantom attack: evading system call monitoring (2021). https://www.youtube.com/watch?v=yaAdM8pWKG8, DEFCON 29
15. Guo, R., Zeng, J.: Trace me if you can: bypassing Linux syscall tracing (2022). https://www.youtube.com/watch?v=yFl_ScKA300, DEFCON 30
16. Hasherezade: Tiny tracer. https://github.com/hasherezade/tiny_tracer
17. Iwamoto, K., Wasaki, K.: Malware classification based on extracted API sequences using static analysis. In: Proceedings of the 8th Asian Internet Engineering Conference, AINTEC 2012, pp. 31–38. ACM (2012)
18. Kalendarov, I.: Blindside: A New Technique for EDR Evasion with Hardware Breakpoints. https://cymulate.com/blog/blindside-a-new-technique-for-edr-evasion-with-hardware-breakpoints
19. Kawakoya, Y., Iwamura, M., Shioji, E., Hariu, T.: API chaser: anti-analysis resistant malware analyzer. In: Stolfo, S.J., Stavrou, A., Wright, C.V. (eds.) RAID

2013. LNCS, vol. 8145, pp. 123–143. Springer, Heidelberg (2013). https://doi.org/10.1007/978-3-642-41284-4_7

20. Kim, S.H., Zeng, D., Sun, C., Tan, G.: BinPointer: towards precise, sound, and scalable binary-level pointer analysis. In: CC 2022, pp. 169–180. ACM (2022)
21. Lee, J., Han, J., Lee, M., Choi, J., Baek, H., Lee, S.J.: A study on API wrapping in Themida and unpacking technique. J. Korea Inst. Inf. Secur. Cryptol. **27**(1), 67–77 (2017)
22. Lengyel, T.K., Maresca, S., Payne, B.D., Webster, G.D., Vogl, S., Kiayias, A.: Scalability, fidelity and stealth in the DRAKVUF dynamic malware analysis system. In: ACSAC 2014, pp. 386–395. ACM (2014)
23. Linn, C., Debray, S., Andrews, G.: Stack analysis of x86 executables. Technical report, Department of Computer Science, University of Arizona (1212)
24. Luk, C.K., et al.: Pin: building customized program analysis tools with dynamic instrumentation. In: PLDI 2005, pp. 190–200. ACM (2005)
25. Microsoft: API stack usage. https://learn.microsoft.com/en-us/cpp/build/stack-usage?view=msvc-170
26. Muchnick, S.S.: Advanced Compiler Design and Implementation. Morgan Kaufmann Publishers Inc. (1998)
27. Nektra: Spystudio. https://www.nektra.com/products/spystudio-api-monitor/
28. Rohitab: API monitor. http://www.rohitab.com/apimonitor
29. Roundy, K.A., Miller, B.P.: Binary-code obfuscations in prevalent packer tools. ACM Comput. Surv. **46**(1) (2013)
30. Schwarz, M., et al.: Automated detection, exploitation, and elimination of double-fetch bugs using modern CPU features. In: ASIACCS 2018, pp. 587–600. ACM (2018)
31. Shudrak, M., Bruening, D., Testa, J.: Drltrace. https://github.com/mxmssh/drltrace
32. Wang, H., et al.: Locating vulnerabilities in binaries via memory layout recovering. In: ESEC/FSE 2019, pp. 718–728. ACM (2019)
33. Wei, J., Pu, C.: TOCTOU vulnerabilities in UNIX-Style file systems: an anatomical study. In: 4th USENIX Conference on File and Storage Technologies (FAST 2005). USENIX Association (2005)

Extended Abstract: Evading Packing Detection: Breaking Heuristic-Based Static Detectors

Alexandre D'Hondt(✉), Charles Henry Bertrand Van Ouytsel(✉), and Axel Legay(✉)

Université Catholique de Louvain, Rue Archimede 1, Louvain-la-Neuve, Belgium
{alexandre.dhondt,charles-henry.bertrand,axel.legay}@uclouvain.be

Abstract. Nowadays, executable packing remains an open issue in its detection especially when it comes to static analysis. Packing is significantly used in malware to hide malicious code from detection systems. These last years, many studies about static packing detection addressed this problem with heuristics and machine learning, considering different ad hoc techniques, algorithms and feature sets but very few addressed it from the adversarial point of view, that is, how to fool heuristics by altering samples with targeted modifications. The objective of this work is to study to what extent it is easy to evade detection by open source static detectors that are commonly used by the community by applying alterations on packed samples, which require only slight adaptations of the related packers, resulting in evasion. An adversarial setting from the problem-space perspective is addressed by using realistic modifications of binary samples that target common significant features. For this purpose, alterations and datasets are composed and static detection is applied using the experimental toolkit Packing Box. Results of alterations are shown, in terms of information gain of features and accuracy of detection, on open source static packing detectors. Finally, their significant effects are highlighted and their effectiveness is evaluated.

Keywords: executable packing · packer detection · static analysis · adversarial examples · experimental toolkit

1 Introduction

Executable packing involves transformations that modify a binary file that preserve its semantics while having its layout changed at rest. Especially when compressing or encrypting parts of a binary, packing is useful for evading detection by static techniques like signature matching as used in an antivirus. Static techniques are those that only use information available at rest, that is, without executing the binary. On the contrary of dynamic analysis that leverages execution of the target binary to collect information at runtime, static analysis offers the major advantage of being much faster. The main challenge thus remains

F. Maggi et al. (Eds.): DIMVA 2024, LNCS 14828, pp. 174–183, 2024.
https://doi.org/10.1007/978-3-031-64171-8_9

performance, that is how to leverage as much information as possible, knowing that a static approach will capture less information than a dynamic one, while keeping processing time as low as possible, which is best achieved with a static technique. Therefore, approaches and methods in the field of static analysis continued to evolve since the early 2000's, increasingly focusing on machine learning, yielding various features sets. In this scope, a recent study on packing detection with machine learning by Bertrand Van Ouytsel *et al.* [3] highlighted a subset of *significant features* based on the features set of Biondi *et al.* [4].

On the one hand, many studies [14] addressed malware detection evasion, altering characteristics of malicious samples for targeting particular detection heuristics. In this setting, modifications are applied directly on samples and not on features. The underlying challenge is that feature modifications can usually not be mapped back to samples [14], which limits options for adversarial attacks. On the other hand, packing is widely used in malware and is sometimes even left as an open issue in malware analysis. According to statistics from Aghkhani *et al.* on a large-scale real-world dataset from 2017 and 2018, more than 60% of malicious samples were packed. As none of the currently malware-focused attacks specifically addressed the problem of packing yet, there is room for improvement in this particular field. While some alterations applicable to malware detection evasion apply to packing evasion too, they require further analysis and thorough experimentation on additional specific characteristics and dedicated detectors.

In this paper, focus is on attacking packing static detectors found in the wild, especially popular ones found on GitHub and commonly used by the community, and on finding sample alterations so that detection can be influenced and weak features can be identified. By purpose, the scope is restricted to targeting such detectors and not those based on machine learning as it will be the topic of another study. The objective is to answer the following research questions:

- Q1 Which *alterations* can be defined *to impact* common *significant features* that popular static packing detectors rely on ?
- Q2 Which *features* are *most vulnerable* to alterations on packed samples ?
- Q3 Which static packing *detectors* are *most vulnerable* to alterations ?

The contribution of this paper is to propose an extension of the Packing Box [9] that allows to define alterations and to show with some experiments based on five carefully crafted alterations to what extent features are impacted and how much accuracy of open source detectors can be decreased. It is shown that three out of these five alterations have a significant impact on some popular packing detectors and that weak features can be identified through this process.

The remainder of this paper first presents, in Sect. 2, background on packing and its related state-of-the-art static detection techniques. It also addresses some related works on adversarial attacks with some alterations that were already proposed. In Sect. 3, the Packing Box is introduced and improvements made for supporting alterations are presented, arguing on those created for targeting features used by in-scope detectors. Then in Sect. 4, the threat model of the adversarial setting is stated and some experiments are conducted based on datasets of packed samples, focusing attention on the accuracy of detectors to evaluate evasion through our alterations. Section 5 finally concludes on what could be achieved and what remains to be further developed and studied.

2 Background

This section presents important concepts for addressing static packing detection, including techniques based on heuristics and machine learning, but also adversarial attacks and their alterations and how it applies to executable packing.

Packing is defined as a set of transformations that tampers with the layout of a binary file at rest while preserving its working at runtime. In other words, the process of packing aims to refactor the syntax of the target binary without modifying its semantics. Such *transformations* range from simple compression or encryption to mutation of code or virtualization. It can also be slight modifications for anti-reverse engineering. It is used for legitimate purpose, e.g. for software license protection [13], as well as for malware, i.e. for evasion from detection [1]. It is worth being noted that this definition extends the traditional one commonly found in many studies that only consider compression and encryption.

Detection techniques range from static to dynamic analysis but can be a mix of both as well. Static techniques pertain to analysing a binary without running it, that is, only considering information at rest like structural features [7] or headers [16]. On the contrary, dynamic methods require execution for collecting information at runtime such as entropy changes [2] or write operations [18] in memory. As a consequence of the inevitable execution overhead and what can be inspected at rest or at runtime, static techniques always capture less information than dynamic ones. The greatest challenge is then to identify the best information that can be used for reliable detection. Many approaches have been addressed since the early 2000's including entropy-based methods [15], heuristics [11,16] or signature matching [12,17] and machine learning. Bertrand Van Ouytsel *et al.* [3] identified a set of *significant features* specifically for packing detection, based on the features set of Biondi *et al.* [4]. Their methodology relies on SHAP values, a metric that allows to evaluate the robustness of features.

Adversarial attacks are classified by their *threat model* that describes the attack surface and characterizes the attacker. *Attack surface* relates to the steps where an attack can take place. For machine learning, it corresponds to the steps of the learning pipeline. Poisoning typically involves the training step while evasion targets the test phase. For characterizing the attacker, Carlini *et al.* [5] proposes a comprehensive framework that defines three components. First, attacker's *goal* specifies if the attack is targeted or untargeted. When the attack is targeted, the attack is successful when the predicted class of the detector is a specific class of interest for the attacker. When untargeted, the predicted class can be any other than the original one. Second, the *capabilities* state what the attacker can do. In a realistic scenario, attacker's ability to modify examples is often limited and is typically defined by the space in which perturbations can be achieved. Third, attacker's *knowledge* is also a limiting factor. Two settings can be distinguished: white-box and black-box. In the white-box setting, the model is fully known while in the black-box setting, it is assumed that the attacker has no access to the model's inner workings so that only the output can be seen.

Two important spaces related to attacker's capabilities can be noted: the *feature* and *problem space*. Feature-space attacks are possible when a feature vector can be input directly to the model. The attacker can then apply a perturbation vector and analyze its effect on classification. In problem-space attacks, perturbations, called *alterations*, are applied directly to input samples. The major issue with feature-space attacks is that feature mappings are often not bijective, meaning that they do not capture the entire information from the problem space. In such cases, perturbation vectors from the feature space cannot be propagated back to the problem space [14]. Therefore, modified feature values could possibly not be mapped back to a realistic executable. For this reason, we consider only the problem space with *alterations* that are designed not to break executable samples' working. Moreover, as the focus is on common static packing detectors based on signatures or heuristics, there is no pipeline in which a feature space can be addressed for applying perturbations. Heuristics are treated as black-box models on which only the test phase is accessible.

Many studies addressed alterations of samples from the malware perspective, often focusing on a single characteristic of binaries like API imports [8,10,19] or appending bytes to the overlay [8,10,19], and targeting a specific type of malware detector. These characteristics that can be altered without breaking executables' working include headers [8,19], sections [8,19] or the signature [10,19].

In the next sections, focus is on the Portable Executable (PE) format of Windows, on its related static packing detectors integrated in the Packing Box and on a threat model relying on the problem space and evasion in a black-box setting. Section 3 addresses the material from the Packing Box required for the experiments of Sect. 4, including alterations targeting the set of significant features of Bertrand Van Ouytsel *et al.* [3]. Section 4 starts by describing the complete threat model as part of the experiments methodology.

3 Framework

This section presents the essentials of the Packing Box and how it was tuned to define alterations for our experiments. It explains functionalities required to further address our experiments of Sect. 4.

Packing Box [9] is the reference toolkit for managing experiments. It is an Ubuntu-based Docker image that contains a comprehensive toolset for controlling the machine learning pipeline from dataset generation to model training and testing. But more importantly, it integrates a series of detectors presented and evaluated on the basis of their performance by D'Hondt *et al.* [9], including Bintropy, Detect It Easy (DIE), Manalyze, PEiD, PEPack, PyPackerDetect, REMINDer and Retargetable Decompiler (RetDec). It aims to support multiple executable formats including Portable Executable (PE) from Windows. Given the wide variety of sources for PE samples and tools, the scope of this paper is limited to this format. Packing Box also provides the ability to manage experiments, datasets and models for enhancing repeatability and shareability of results with peer researchers. When creating an experiment, a dedicated workspace is set up for storing datasets and models but also for embedding resources such as features declarations and specific data such as lists of standard or packer section names. Packing Box defines items including packers, detectors, ML algorithms, features or *alterations* using *YAML*[1] *files*, which is a convenient and readable format for declaring aspects such as installation steps as well as for documentation for including comments or source and reference hyperlinks. In the scope of this study, we use experiments to manipulate and sort datasets and embed alterations definitions and scripts for evaluating the accuracy of detectors.

Alterations are applied in the problem-space, that is before feature computation, and aim to modify samples so that an impact can be observed at the test phase of the pipeline. Note that, as the scope of this paper is limited to black-box static packing detectors relying on signatures and heuristics, attack options are limited to the test phase anyway. Alterations are defined to target multiple kinds of features used in common heuristics, including section characteristics, e.g. name or flags, entropy and signatures, typically from the Entry Point.

A01 *rename packer sections* – This modifies sections with common packer or empty section names with standard names, targeting the Entry Point (EP) section first then other sections and choosing replacement names from the 6 most common standard names, including *.text*, *.data* or even *.bss*, then random standard names among a list composed from the PE format specification. It ensures that no section name is repeated. Targeted features: *number of standard sections* as common packer section names get replaced with standard ones.

A02 *move EP to new low-entropy code section* – This moves the EP to a new code section with a standard section name that plays the role of *trampoline* by defining a jump to the Original EP and it fills the section up to 64 bytes with a low-entropy pattern. Targeted features: whether the *EP is not in a standard section* as many packers use a code section with a non-standard name, *bytes after the EP* as then having a new series of bytes containing the trampoline instructions and a low-entropy pattern will not match signatures anymore.

A03 *fill sections with low-entropy pattern* – This fills sections with a pattern of 3 bytes when the virtual size, that is the size as expanded in memory at runtime, is higher than the size of section's data, that is the size on disk. This aims to lower entropy while not breaking the executable as sections normally gets expanded to their virtual size with zeros anyway. Targeted features: *byte entropy of the entire PE file* and *byte entropy of the code section*.

A04 *add low-entropy code section* – This adds a new code section with a size of 10% of the overall file size, ensuring it holds a low entropy by using a pattern of 3 bytes. Targeted features: *byte entropy of the entire PE file*

[1] Yet Another Markup Language.

as filling in the new section with 10% of the overall size with a low-entropy pattern necessarily impacts the overall entropy significantly.

A05 *add 20 common API imports* – This adds 20 common API imports to the Import Address Table (IAT), ensuring that there is no duplicate. Targeted features: *ratio of malicious API calls imported* as it will decrease as a result of adding common API imports.

The aforementioned alterations thus answer Q1, focusing on significant features from the current literature. The next section addresses the experimental setting and how these alterations were tested.

4 Experiments

This section first presents the setting and methodology used to conduct experiments aimed to answer Q2 and Q3. It then presents the results of the approach in terms of impact of alterations defined in Sect. 3 on the information gain of features and on the accuracy of in-scope detectors.

4.1 Methodology

The *environment* is set up by building Packing Box's Docker image and running its command-line interface. The reference dataset is `dataset-packed-pe`, available on GitHub, and is an alternative version of PackingData from Choi *et al.* [6] augmented with subsets of samples we packed with 5 new packers. It consists of subsets from 115 to 150 samples for each of the following packers: Alienyze, Amber, ASPack, BeroEXEPacker, Enigma VirtualBox, Eronana's Packer, Exe32pack, EXpressor, FSG, JDPack, MEW, Molebox, MPRESS, Neolite, NSPack, Packman, PECompact, PEtite, RLPack, Telock, Themida, UPX, WinUPack, Yoda's Crypter and Yoda's Protector.

The *threat model* is defined according to the framework of Carlini *et al.* [5]. In-scope detectors are those from the Packing Box presented in Sect. 3. Their attack surface is limited to the test phase, meaning that tools can be run and their output inspected; this scenario relates to evasion. Attacker's goal is targeted towards fooling the detector into concluding that an initially packed sample is not packed. Attacker's capabilities are limited to tampering with packed samples and feeding them to the detectors; the attacker's perspective is limited to the problem space by applying realistic perturbations on samples. Finally, attacker's knowledge is considered black-box, even though selected detectors are open source and their heuristics could therefore be reverse-engineered.

The *methodology* is divided into two parts, one for focusing on the impact of alterations on features and another on the impact on detectors, for the purpose of correlating results. Both involve three stages: first, a preparation of the baseline dataset (BL) and one dataset for each alteration (A0X), then second, application of alterations on these, and third, representation and analysis of results.

4.2 Impact of Alterations on Features

This experiment uses six datasets: BL for the reference dataset and A0X for this dataset altered with each alteration. Information Gain (IG) is measured using the Mutual Information (MI) between the feature values and the target label. As only packed samples are considered, this metric relies on the labels of the 25 different packers present in the baseline dataset (BL). Example in Fig. 1 shows values for feature *byte 0 after the EP* for 5 packers out of 25. On the left side, 4 distinct values are present in BL. On the right side, there is only 1 value left as A02 moves the EP to a new section starting with a trampoline instruction. This yields a high value of MI as depicted in Fig. 2.

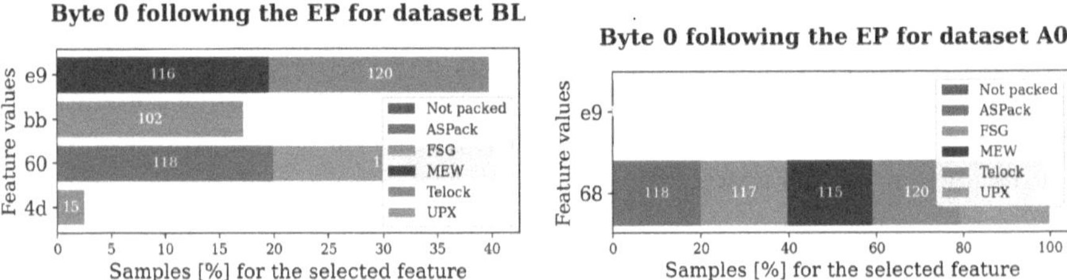

Fig. 1. Values of feature *byte 0 after EP* for BL and A02

In Fig. 2, values are normalized based on the overall highest value. It results in a heat map that highlights most significant effects of alterations.

Information Gain comparison with reference dataset BL

	A01	A02	A03	A04	A05
byte_[0-9]+_after_ep_mean	0.00	1.00	0.02	0.01	0.01
is_code_section_not_present	0.21	0.28	0.00	0.28	0.00
number_addresses_in_iat	0.00	0.00	0.03	0.00	0.42
entropy_code_section	0.15	0.10	0.01	0.03	0.00
is_ep_not_in_code_section	0.09	0.12	0.00	0.00	0.00
entropy	0.03	0.02	0.09	0.01	0.02
is_ep_not_in_standard_section	0.01	0.12	0.00	0.00	0.00
number_standard_sections	0.03	0.03	0.00	0.03	0.00
number_wx_sections	0.00	0.01	0.04	0.01	0.03
number_rwx_sections	0.00	0.00	0.04	0.00	0.03
number_func_imported_in_idt	0.00	0.00	0.00	0.00	0.04
ratio_malicious_api_imported	0.00	0.00	0.01	0.00	0.00

Normalized IG difference from reference dataset (scale 0 – 1)

Fig. 2. Effect of practical alterations on significant features

A01 mostly impacts features related to the code section since it renames packer sections to standard names starting with *.text*. However, the number of standard sections is not that impacted as many packers use standard names combined with only one or two packer-specific names, hence the resulting number of standard sections does not significantly vary. A02 mostly affects bytes after the EP as expected. As it creates a new section with trampoline instructions to the Original EP, it completely changes the 64 bytes after the EP. This alteration also affects features related to the absence of code section and those related to the EP as the new code section has a standard name. A03 slightly impacts *entropy* where binary samples have sections for which the raw size significantly differs from the virtual size. This is especially the case for packers like UPX that hold a first section of length zero that gets expanded in memory at runtime to a virtual size aimed for uncompressed data. As only a few packers work this way, the overall impact on Information Gain is not that significant. A04 mostly impacts the feature about the presence of a code section, as this alteration adds a new code section where some packed samples do not have one. As it aims to add 10% size of low-entropy bytes, it does not significantly impact entropy-related features, even though it can influence the overall entropy just enough to fool related heuristics. A05 mostly impacts the number of addresses of imported functions in the IAT as expected. This has no major impact on other features.

These results highlight a few sensitive features, answering Q2, among the most significant identified by Bertrand Van Ouytsel *et al.* [3]. These include *bytes after the EP* trivially exploited by moving the EP, features based on code section *.text* or entropy-based features exploited by injecting just enough low-entropy bytes sequence to bring entropy beneath heuristics' thresholds.

4.3 Impact of Alterations on Detectors

This experiment uses each per-packer subset from the reference dataset. Detection is applied for each detector on each subset and accuracy is aggregated with a weighted sum. Table 1 shows the results of mass-detection with unchanged results in gray, variations from zero to five percents in olive, variations of more than five percents in orange and significant variations yielding accuracy of less than two percents in red.

A01 is effective against Manalyze, which indicates that this detector relies solely on packer section names. The remaining 1.85% accuracy comes from NSPack samples that could not be altered without error. It also disturbs detection by DIE and PyPackerDetect to a lesser extent, meaning that these detectors solely rely on section names for some specific packers. A02 is highly effective against PEiD, indicating that this tool solely relies on signatures with many of them made to match from the EP. It is also effective against REMINDer which, on the contrary of PEiD, relies on a simple heuristic based on entropy of the section containing the EP. With this alteration, this entropy falls beneath the threshold of REMINDer as 64 low-entropy bytes are appended after the trampoline. This also shows significant effects on DIE, PEPack and RetDec. A03 breaks detection by Bintropy and significantly disturbs REMINDer as both rely

Table 1. Accuracy of detectors for BL and A0X

Detector	**BL**	A01	A02	A03	A04	A05
Bintropy	59.46%	59.76%	59.76%	**1.88%**	**0.01%**	56.83%
DIE	86.13%	77.52%	27.71%	84.66%	82.42%	83.50%
Manalyze	40.71%	**1.85%**	40.71%	44.44%	40.71%	40.71%
PEiD	79.69%	79.69%	**0.00%**	79.69%	77.78%	79.73%
PEPack	81.19%	81.19%	44.18%	81.09%	79.24%	81.19%
PyPackerDetect	95.03%	80.28%	95.87%	86.99%	95.03%	95.03%
REMINDer	42.07%	42.07%	**0.00%**	24.15%	42.07%	42.07%
RetDec	70.60%	70.60%	44.12%	70.57%	70.60%	66.99%

on entropy. In the case of Bintropy, even though many packers do not use a null-size section for expansion in memory for decompression, it is frequent that sections have differences between the raw and virtual size, causing enough drop in entropy to break Bintropy's heuristic. In the case of REMINDer, only the EP section matters, hence the lower efficiency. A04 has limited impact on the detectors except Bintropy. This shows that none of them rely on entropy except REMINDer which, this time, is not impacted as this alteration does not touch the EP section. The low-entropy pattern is enough to break Bintropy's heuristic, the remaining 0.01% coming from corrupted samples. A05 causes only slight perturbations, showing that detectors are not sensitive to API imports.

These results, answering Q3, confirm that signatures and heuristics relying on entropy and code sections are easy to fool. Moreover, heuristics relying on one feature like Manalyze (i.e.: common packer section names) can easily be evaded. Popular detector like DIE combining approaches based on signatures and heuristics can be attacked with a decent success rate. Given the results of PyPackerDetect, this requires further attention as it may use heuristics relying on features that are not covered here.

5 Conclusion

In this paper, the problematic of executable packing and performance of static detection was stated, addressing adversarial attacks for testing related significant features and their robustness. The framework relying on Packing Box was presented with its enhancement for supporting alterations of binary samples. Five of these alterations were defined, implemented and evaluated for testing against a chosen set of detectors (Q1). Furthermore, three out of these five were shown to have impact on significant features from the literature, including bytes after the Entry Point and features based on entropy and code sections (Q2), and to evade detection by popular open source packing detectors integrated into the Packing Box including Bintropy, DIE, PEiD and RetDec (Q3).

A future study will be conducted on machine learning based detectors with an extended set of alterations. The goal will be to study the robustness of features and algorithms to these alterations and to refine the set of significant features.

Acknowledgements. The authors are funded by the CyberExcellence project (RW, Convention 2110186). They want to express their gratitude to Romain Jennes who made a significant contribution through his master's thesis.

References

1. Anderson, H.S., Kharkar, A., Filar, B., Evans, D., Roth, P.: Learning to evade static PE machine learning malware models via reinforcement learning (2018)
2. Bat-Erdene, M., Park, H., Li, H., Lee, H., Choi, M.S.: Entropy analysis to classify unknown packing algorithms for malware detection. Int. J. Inf. Secur. **16**, 227–248 (2016)
3. Bertrand Van Ouytsel, C.H., Dam, K.H.T., Legay, A.: Analysis of machine learning approaches to packing detection. Comput. Secur. **136**, 103536 (2023)
4. Biondi, F., Enescu, M.A., Given-Wilson, T., Legay, A., Noureddine, L., Verma, V.: Effective, efficient, and robust packing detection and classification. Comput. Secur. **85**, 436–451 (2019)
5. Carlini, N., et al.: On evaluating adversarial robustness (2019)
6. Choi, M.J., Bang, J., Kim, J., Kim, H., Moon, Y.S., Díaz-Verdejo, J.: All-in-one framework for detection, unpacking, and verification for malware analysis. Secur. Commun. Netw. (2019)
7. Choi, Y.S., Kim, I.K., Oh, J.T., Ryou, J.C.: PE file header analysis-based packed PE file detection technique (PHAD). In: International Symposium on Computer Science and its Applications (2008)
8. Demetrio, L., Biggio, B., Roli, F.: Practical attacks on machine learning: a case study on adversarial windows malware. IEEE Secur. Priv. **20**, 77–85 (2022)
9. D'Hondt, A., Bertrand Van Ouytsel, C.H., Legay, A.: Experimental toolkit for manipulating executable packing. In: International Conference on Risks and Security of Internet and Systems (2023)
10. Fang, Z., Wang, J., Li, B., Wu, S., Zhou, Y., Huang, H.: Evading anti-malware engines with deep reinforcement learning. IEEE (2019)
11. Han, S., Lee, K., Lee, S.: Packed PE file detection for malware forensics. In: International Conference on Computational Science and Its Applications (2009)
12. Kancherla, K., Donahue, J., Mukkamala, S.: Packer identification using Byte plot and Markov plot. J. Comput. Virol. Hacking Tech. **12**, 101–111 (2015)
13. Khan, M., Akram, M., Riaz, N.: A comparative analysis of software protection schemes. Int. Arab J. Inf. Technol. (2014)
14. Ling, X., et al.: Adversarial attacks against windows PE malware detection: a survey of the state-of-the-art. Comput. Secur. **128**, 103134 (2021)
15. Lyda, R., Hamrock, J.: Using entropy analysis to find encrypted and packed malware. IEEE Secur. Priv. **5**, 40–45 (2007)
16. Pareek, H., Arora, R., Singh, A.: A heuristics-based static analysis approach for detecting packed PE binaries. Int. J. Secur. Appl. **7**, 257–268 (2013)
17. Shin, D., Im, C., Jeong, H., Kim, j., Won, D.: The new signature generation method based on an unpacking algorithm and procedure for a packer detection. In: International Journal of Advanced Science and Technology (2011)
18. Ugarte-Pedrero, X., Balzarotti, D., Santos, I., Bringas, P.G.: SoK: deep packer inspection: a longitudinal study of the complexity of run-time packers. In: IEEE Symposium on Security and Privacy (2015)
19. Wu, C., Shi, J., Yang, Y., Li, W.: Enhancing machine learning based malware detection model by reinforcement learning. In: International Conference on Communication and Network Security (2018)

Listening Between the Bits: Privacy Leaks in Audio Fingerprints

Moritz Pfister[1], Robert Michael[1], Max Boll[1], Cosima Körfer[1], Konrad Rieck[2,3], and Daniel Arp[2,3(✉)]

[1] Technische Universität Braunschweig, Braunschweig, Germany
[2] Berlin Institute for the Foundations of Learning and Data (BIFOLD), Berlin, Germany
d.arp@tu-berlin.de
[3] Technische Universität Berlin, Berlin, Germany

Abstract. Audio content recognition is an emerging technology that forms the basis for mobile services, such as automatic song recognition, second-screen synchronization, and broadcast monitoring. The technology utilizes *audio fingerprints*, short patterns that are extracted from audio recordings of a smartphone and enable the identification of specific content. These fingerprints are generally considered privacy-friendly, as they contain minimal information of the original signal. As a result, mobile applications have emerged in the past few years that silently monitor user habits by collecting such audio fingerprints in the background. In this paper, we systematically examine whether audio fingerprints leak sensitive information from the recording environment and potentially violate the privacy of smartphone users. To this end, we analyze three popular audio recognition solutions and develop attacks to infer sensitive information from their fingerprints. To the best of our knowledge, we are the first to show that the identification of speakers and words in the fingerprints is possible. Based on our analysis, we conclude that current audio fingerprints do not sufficiently protect privacy and should be used with great caution.

Keywords: audio content recognition · audio fingerprints · privacy · machine learning · mobile security

1 Introduction

Audio content recognition (ACR) is a recent technology for identifying content in audio recordings. In contrast to the direct analysis of audio signals, ACR relies on so-called *audio fingerprints* [37,40,47]. These small patterns are extracted from an audio signal and characterize its content in compressed form. Audio fingerprints can be efficiently transmitted and matched against large databases, enabling the identification of different types of audio content. While ACR has

M. Pfister and R. Michael—Authors contributed equally to this work.

F. Maggi et al. (Eds.): DIMVA 2024, LNCS 14828, pp. 184–204, 2024.
https://doi.org/10.1007/978-3-031-64171-8_10

been initially developed for recognizing music [31,48], it is also used for tracking the behavior of users via their smartphones. For instance, mobile frameworks, such as Zapr [53] and Alphonso [25], monitor the acoustic environment of the smartphone in regular intervals to track the user's broadcast media consumption. These frameworks are not mere proof-of-concepts, but have been discovered in popular applications with millions of downloads [30,44].

A premise underlying this use of ACR is that audio fingerprints are privacy-friendly. That is, by condensing a recording into patterns of a few bytes, most of the information of the input signal is lost. Fingerprints are thus considered as "hashes" that can only be matched against other fingerprints but provide no further clues on the signal. Some providers of ACR thus even claim the collected data to be non-personal and boast about their privacy-friendly services [52].

Technically, fingerprinting algorithms compress an audio signal by identifying prominent frequency peaks and aggregating them in a fixed-size format. This compression destroys any continuous signal that could directly leak information. However, when the fingerprints are generated in a personal environment, the extracted peaks may capture nearby voices and sounds. While the resulting fingerprints do not match any known patterns in the databases of ACR services, it is unclear whether they contain sensitive information about speakers present in the environment that can be potentially misused by adversaries.

To the best of our knowledge, this is the first paper to investigate the privacy risks of audio fingerprinting systematically. In particular, we analyze three popular ACR solutions and their underlying fingerprint generation. Based on this analysis, we develop three attacks for inferring sensitive information from the raw bits of the fingerprints. These attacks build on machine learning to model temporal relations between frequency peaks and uncover different types of information. They enable us to (a) differentiate between speech and music signals, (b) identify the voices of individual speakers, and (c) detect a fixed set of spoken words within the fingerprints. In our evaluation, these attacks allow identifying 40 speakers with up to 90% accuracy and 10 words (digits) with up to 82% accuracy, demonstrating that the generated audio fingerprints still contain sensitive information.

While our attacks cannot undo the strong compression of audio fingerprints, they serve as a proof of concept for the reconstruction of personal information. We conclude that current ACR techniques do not adequately protect privacy and must not be used for sensitive tasks, such as monitoring user habits.
In summary, we make the following contributions:

- *Analysis of ACR solutions.* We provide the first analysis of currently available ACR solutions and the underlying fingerprinting mechanisms.
- *New attacks on audio fingerprints.* We propose three attacks that infer sensitive information from audio fingerprints of popular ACR solutions.
- *Code release.* To foster further research in this direction, we publish the source code and data used in our experiments[1].

[1] https://github.com/acr-privacy/attacks.

The remainder of this paper is organized as follows: We provide an overview of related work in Sect. 2 and ACR solutions including the underlying fingerprint generation in Sect. 3. Our attacks are introduced in Sect. 4 and empirically evaluated in Sect. 5. We discuss countermeasures and limitations in Sect. 6. Finally, Sect. 7 concludes the paper.

2 Related Work

There exists various research related to this paper, which can be broadly divided into three different groups. We briefly discuss them in the following.

The first group of papers revolves around audio content recognition. In particular, early works already allow a reliable, noise-robust re-identification of audio content [15,47]. These are the basis for further improvements like increasing the robustness towards noise [20,33,39], and dealing with tempo and frequency shifts [40]. Although ACR algorithms are widely used, only few works exist that assess the potential privacy and security implications. On the security side, several works discuss evasion attacks challenging the robustness of audio fingerprinting [37,45]. On the privacy side, Knospe et al. [21] propose a privacy-enhanced fingerprinting algorithm that increases the entropy of extracted patterns to hinder attacks. Similarly, Celosi et al. [5] observe that Apple's digital voice assistant Siri is prone to dictionary attacks, allowing an adversary to identify pre-defined voice commands. Another large body of research demonstrates unintentional data leaks, e.g., methods for inferring sensitive information from encrypted network and telephone communication [26,36,50], inferring the TV content by analyzing the flickering lights emitted by TVs [51], or de-pseudomizing households by analysing pseudomized consumption data transmitted by smart meters [18]. More recently, various works have found further sources for data leakage in mobile and smart home environments [19,35]. Finally, Schlegel et al. [38] explore a mobile malware that stealthily extracts sensitive information through audio analysis.

Finally, there is a large amount of literature dealing with mobile device tracking. We discuss that various companies (e.g., [44,53]) developed techniques to profile the media consumption habits of smartphone users with audio fingerprints. Arp et al. [2] and Mavroudis et al. [29] reveal efforts where companies use ultrasonic beacons for similar purposes. The examination of potential privacy risks of tracking smartphone users via Bluetooth [14,23] demonstrates the possibility of cross-app tracking via fingerprints derived from BLE packets. Other works focus on privacy issues of cross-device tracking in detail [4,54] and mobile device tracking in general [6,35].

3 Audio Content Recognition

We start with an overview of the audio fingerprinting solutions considered in this paper and how they are used in practice (Sect. 3.1). Furthermore, we describe the technical principles, as these are essential to understand the attacks presented later in this paper (Sect. 3.2).

3.1 ACR Solutions

For our analysis, we consider three popular solutions for audio content recognition that focus on different application scenarios.

Quad-Based Fingerprinting (QBF). The first approach we investigate is an academic method proposed by Sonnleitner and Widmer [40]. We refer to this method as *quad-based audio fingerprinting (QBF).* We select this approach, as it is an extended variant of the classic method by Wang [47], which forms the foundation for many ACR solutions—including the popular Shazam app. Since the publication by Wang [47] lacks implementation details, we use QBF as a replacement, where we have concrete knowledge of its internal parameters and other essential information that allows us to fully re-implement the method.

Zapr. As the second approach, we consider the ACR solution from the company Zapr [22]. Zapr used to be a popular commercial ACR solution that was active until mid 2022. The company focused exclusively on media consumption monitoring and targeted advertising [53]. According to the company's website, their solution was present on more than 200 million mobile devices, backed by several large global media companies [53]. Although the company closed down, its technology can still serve as a prototypical example of how audio fingerprints are used for user tracking.

Zapr offered a software development kit (SDK) for media consumption monitoring to their customers. Their SDK included a dedicated recording service that kept running in the background and enabled apps to continuously monitor the acoustic environment—even if the apps were not active. To achieve this, the respective apps notified their users and requested permissions to access the location and the microphone of the device. Like companies with similar business models [24,44], Zapr claimed that it is impossible to reverse-engineer these audio fingerprints and gain insights into the recorded content [52]. Consequently, the privacy of users depended entirely on whether and what information are recoverable from the fingerprints submitted to Zapr.

ACRCloud. Finally, we consider the commercial ACR solution of the company ACRCloud. Among others, their customers include Deezer [8], MusixMatch [31], and Huawei. Together, these companies reach a global customer base with several millions of users, making ACRCloud one of the major solutions on the market. Besides music recognition for MusixMatch and Deezer, ACRCloud also focuses on custom content recognition, for example, in the segments of broadcast monitoring and advertising [1]. While apps like MusixMatch and Deezer require their users to start the audio recording explicitly, it is unknown whether this technology is also embedded in apps with a business model similar to Zapr. Moreover, due to the closed-source nature of ACRCloud, technical details of the fingerprinting algorithm are not publicly available. We address this problem when deriving our attacks in Sect. 4 using techniques for code instrumentation.

Further Approaches. Although there exist further approaches for employing ACR in the wild, we focus on the presented solutions, as they are prototypical for how audio fingerprints are used in practice, including foreground (ACRCloud) and background recording (Zapr).

3.2 Audio Fingerprinting

Before we introduce our attacks against ACR solutions, let us first briefly review the basic requirements and concepts of audio fingerprinting. The general pipeline that all ACR methods share can be broadly divided into three steps: *(a) pre-processing*, *(b) fingerprint extraction*, and *(c) fingerprint matching*. All solutions share various requirements they need to fulfill. We first discuss these requirements, and then explain the different steps of the generation pipeline. Here, we mainly focus on the method by Wang [47], because it has inspired several other recent approaches (e.g., [22,40]).

Requirements. In order to enable successful fingerprinting in different application scenarios, ACR solutions need to realize a trade-off between four requirements, which influence the characteristics of the underlying algorithms.

- *Reliable identification.* An ACR solution needs to reliably identify audio content solely based on a short recording. Moreover, it should produce only very few false-positive results during practical operation.
- *Noise robustness.* The solution needs to cope with strong noise. For instance, it should allow detecting media content even in locations with high background noise, such as in bars or on concerts.
- *High compression.* To enable efficient transfer and matching of fingerprints, they should only contain the necessary information for re-identifying a signal and compress the underlying recording significantly.
- *Computational efficiency.* Finally, the fingerprints need to be generated on devices with limited computational and memory resources, such as smartphones. Thus, the generation process needs to be very efficient.

Pre-processing. As the first step, the audio signal is prepared for extracting a fingerprint. Since not all frequency components are necessary for audio content recognition, the original signal is usually downsampled before a fingerprint is created (e.g., [15,47]). Audio recordings on smartphones are commonly generated as a stereo signal sampled with a frequency of 44.1 kHz. Hence, ACR solutions first merge the two channels into a mono track and then filter high frequencies using a simple low-pass filter.

Fingerprint Extraction. To access relevant features of an audio signal, most ACR methods map the underlying time series to the frequency domain and compute the spectrum of the signal. That is, they represent the signal as a mixture of frequency components using standard algorithms, such as the short-time Fourier transform (STFT). This frequency spectrum allows analyzing the changes of the

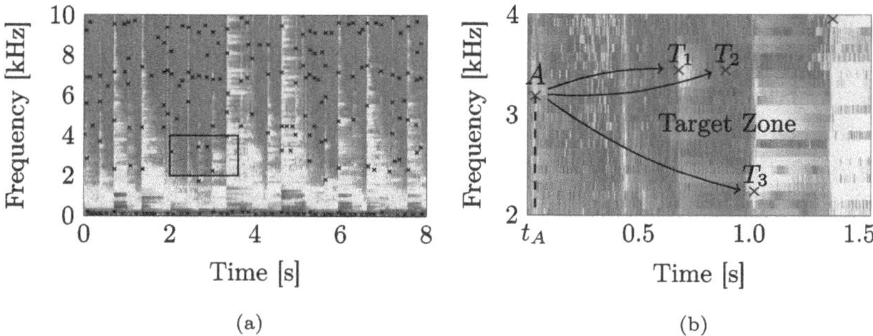

Fig. 1. (a) Spectrogram of an eight-second-long audio signal. Prominent frequency peaks are indicated by black markers. (b) Derivation of subfingerprints by combining the anchor point A (left) with peaks in the respective target zone.

signal's frequency components over time. As an example, Fig. 1a depicts the spectrogram of an audio signal that exhibits prominent frequency changes.

While most audio fingerprinting methods rely on a spectral analysis, the derived spectra usually differ in their time and frequency resolution. However, despite these differences, the algorithms proceed similarly and divide the frequency range in various non-overlapping frequency bands, also called *frequency bins*.

After calculating the spectrum of a signal, audio fingerprints are derived from prominent frequency components. The exact method used to create these fingerprints may differ, but they all have in common that they analyse the spectrum for characteristic peaks and their temporal relation. The extracted peaks form the basis for assembling the audio fingerprint. To give a concrete example of the fingerprint generation process, let us consider the approaches by Wang [47] and Sonnleitner et al. [40]. These methods first identify peaks in the spectrum of the signal. Afterward, they select so-called *anchor points* from these peaks and assign a *target zone* following each anchor point.

The target zone is a small area of the spectrum in a certain temporal distance to the corresponding anchor point A. Frequency peaks T_j in this zone are called target points (see Fig. 1b) and capture short temporal relations between frequencies in the signal. For each anchor point A, Wang then constructs multiple *subfingerprints* by combining A with a selected number of peaks T_j in the target zone. In particular, each resulting subfingerprint is a tuple $f_i = (b_A, b_T, \Delta t)$, where b_A denotes the frequency bin of the anchor point, b_T the frequency bin of the considered target point and $\Delta t = t_T - t_A$ their temporal distance.

In the method of Wang, a subfingerprint f_i comprises 30 bits, encoding each frequency bin and the time difference with 10 bits each. In contrast, Sonnleitner et al. extract 32-byte-long subfingerprints. Similarly, other parameters, such as the target zone width, differ depending on the approach. To give an intuition, Sonnleitner et al. [40] use each extracted peak once as an anchor point and consider a 0.8 s long window as target zone, which is 0.9 s away from the current anchor point. Despite these configuration details, most audio fingerprinting methods rely on characterizing the signal through subfingerprints that capture frequency pairs and their temporal distance.

Note that the extracted subfingerprints alone are not sufficient to identify a signal, as similar frequency combinations may occur in different audio content. Therefore, multiple subfingerprints are required together with the time coordinate of their respective anchor point t_A to allow for a reliable identification. The final audio fingerprint of a signal s is typically described as a sequence $f(s) = ((f_1, t_{A_1}), \ldots, (f_N, t_{A_N}))$. There exist fingerprinting methods where the number of extracted subfingerprints N is fixed (e.g., [15]), while for other approaches N depends on the audio signal to be fingerprinted (e.g., [40,47]). As we discuss in Sect. 4, our attack method partially leverages this structural information.

On the server side, the generated fingerprints are usually stored in a database along with some metadata of the original audio signal, such as a track name. Finally, the fingerprints in the database can be used to match incoming fingerprints derived from an audio signal recorded by a client app.

Fingerprint Matching. Once the database has been constructed successfully, one can use it to re-identify audio content solely based on small snippets of the audio signal. In particular, a client application records a short audio snippet of only a few seconds, and extracts a fingerprint from it as described before. Then, the app transmits the fingerprint to the server. On the server side, the fingerprint can be matched against the fingerprints stored in the database.

4 Attacks Against Audio Fingerprints

We now introduce our attacks against ACR. To this end, we first define the threat model underlying our attacks (Sect. 4.1), then discuss possible attack scenarios (Sect. 4.2), and finally present the details of our attack method (Sect. 4.3).

4.1 Threat Model

In our threat model, we assume that an attacker has access to audio fingerprints from an ACR solution. The attacker could be directly involved in the service that provides the solution or a third party that has access to the matched fingerprints. Numerous ways in which third parties could access these fingerprints are conceivable. These include data trading, service outsourcing, a false sense of privacy, law enforcement, or unintentionally through a data breach.

Furthermore, we assume that at least some of the collected audio fingerprints contain sensitive information the attacker is interested in. The attacker's goal is to reconstruct this information from the attacked fingerprints, such as speaker identities, spoken words, or further attributes of the acoustic environment at the time of recording. This implies that not only the privacy of the device owner is threatened, but even the privacy of all persons present during the fingerprinting process. As a result, our analysis is based on the following assumptions regarding the attacker's knowledge and capabilities:

- *Audio fingerprints.* The attacker has access to audio fingerprints from the device of a victim, for example, when they are transferred to the ACR service. However, the attacker neither needs to know the original signal nor a reverse mapping that allows her to reconstruct it. We refer to the collected audio fingerprints as *attacked (audio) fingerprints.*
- *Black-box access.* The attacker has at least black-box access to the fingerprinting algorithm that produces the attacked audio fingerprints. This capability enables them to generate new fingerprints at their will, for instance, by instrumenting the code of a mobile app.
- *Auxiliary data.* The attacker has access to auxiliary data related to the sensitive information of interest. Here, publicly available data can already be sufficient to run successful attacks. For instance, the attacker might use audio or video content published on social media platforms to target specific persons of interest in the attacked fingerprints.

We argue that such an attacker is realistic. First of all, there exist various solutions that generate audio fingerprints and send them to external servers [44,53]. Malicious employees or dubious companies might actively analyze the collected data to derive sensitive information from these audio fingerprints. Thus, users cannot always trust companies to handle their data responsibly. However, even if we assume that all companies are trustworthy, history has repeatedly shown that this does not eliminate the risk of data leaks to malicious actors [9,43]. For instance, Deezer, a company that uses ACRCloud for content recognition (see Appendix A), has been reported as a recent victim of data theft [9].

Secondly, it is a realistic assumption that the attacker has at least black-box access to the fingerprinting algorithm. In this paper, we demonstrate that a determined attacker does not need to know the exact implementation of the ACR algorithm in order to conduct the attack. Instead, they can instrument code in mobile apps, and generate fingerprints from custom data. More dedicated attackers could also use common reverse engineering methods to gain insights on the internals of specific methods, and use them to improve possible attacks. In the next section, we discuss in more detail how an attacker can generate fingerprints despite having only black-box access to the targeted ACR solution.

Finally, there are several public datasets for audio recognition that can be used as auxiliary data by an attacker. These datasets can be combined with attack-specific data, such as audio samples of a specific speaker, to provide the necessary foundation for the attack. The adversary might collect this data, for instance, from social media platforms, such as YouTube or Instagram, where many users upload personal media containing usable audio samples.

4.2 Attack Scenarios

The privacy impact of our attacks depends on the type of leaked information. Therefore, we define three attack scenarios, each focusing on a different information type. In these scenarios, we assume that the attacker has collected matching auxiliary data and now aims at inferring the corresponding information from captured audio fingerprints.

- *Distinguishing speech.* In the first scenario, the adversary wants to distinguish fingerprints derived from speech from those containing other signals. Since audio content recognition is often used to identify music tracks, we focus on distinguishing fingerprints derived from human speech and music, respectively. As auxiliary data, the attacker has access to a collection of corresponding audio recordings.
- *Identifying speakers.* In the second scenario, the attacker is interested in identifying the speaker of a voice signal that is captured by an audio fingerprint. This identification allows them to associate the recorded signal with a specific person, thus potentially compromising their privacy. As auxiliary data, we assume a dataset containing audio recordings of the targeted speaker and other voices.
- *Recognizing words.* Finally, the attacker tries to reconstruct parts of the spoken content from the fingerprints. If successful, the attacker can retrieve different sensitive information. We restrict our attack to identifying the English digits zero to nine, which already poses a privacy risk when, for example, a credit card or phone number is spelled. As auxiliary data, we assume that a dataset with spoken digits is available to the attacker.

Although we examine each of these attack scenarios individually, they can be chained together to generate a more powerful attack. Further, it is possible to add other recognition tasks, such as identifying ambient sounds or other spoken words, to strengthen our attack. We thus consider the three scenarios as prototypical for a group of recognition tasks that are applicable to audio fingerprints.

4.3 Attack Method

As the captured fingerprints contain only a fraction of the information of the original signal, the attacker needs to carefully engineer an attack method to reconstruct the information they are interested in. This method comprises three steps, which we discuss in the following.

Fingerprint Generation. As the first step, the attacker needs to generate a representative set of fingerprints based on the auxiliary data (see Sect. 4.1). This set serves as training data and enables machine-learning methods to infer a recognition model from the contained frequency peaks and their temporal relation. This fingerprint generation, however, depends on whether the adversary has white-box or black-box access to the underlying algorithms.

From the three considered ACR solutions, we only have full access to the QBF method and thus simply implement the fingerprinting algorithm described in the original publication [40]. We only make slight adaption to the algorithm's parameters, such that it can handle short audio snippets of less than a second, which the original configuration cannot. Unfortunately, we cannot simply implement the audio fingerprinting algorithms of the other two solutions, as they are closed-source software.

To tackle this problem, we reverse engineer the relevant parts of the SDKs of the two solutions. We first perform a static analysis of popular apps[2,3] containing the SDKs. This is done to identify the respective methods for generating audio fingerprints along with the used parameters. Using the determined parameters, we use the analysis toolkit Frida [34] to dynamically instrument the apps' code and generate fingerprints within the apps, enabling us to obtain the exact same fingerprints as sent to the companies' backends. That is, we provide the methods prepared audio buffers and intercept the returned audio fingerprint data.

For Zapr, we find that it implements five different fingerprinting methods. As the manual effort to reverse engineer all of them is too high, we decide to select two of them for further examination. In the following, we refer to them as Zapr Alg1 and Zapr Alg2. As the first algorithm, we select Zapr Alg1, since its reverse-engineered interface exhibits similarities to the only publicly available description provided by the company [22]. Moreover, we select Zapr Alg2, which appears to be the default algorithm at the time of our analysis.

Model Architecture. Once representative fingerprints have been generated, and the corresponding byte sequences are available, we can finally use machine learning for inferring recognition models. Due to the sequential nature of our dataset we base our models on the transformer architecture [46]. Since we want to solve a classification task, we only utilize the encoder part of the transformer architecture, which has shown promising results in other problem domains (e.g., [11,13]).

In detail, the model receives fingerprints as input. Each input fingerprint f is a sequence of subfingerprints $f_1, \ldots, f_k$, where each subfingerprint is l bits long. The sequence length k and the number of bits l are specific to the concrete ACR solution or classification task. Fingerprints containing fewer than k subfingerprints are padded with all-zero subfingerprints until they reach the sequence length of k. First, each subfingerprint is mapped to a d-dimensional embedding vector, which also encodes information about the subfingerprint's position within the sequence. The embedded subfingerprints are then processed by successive encoder blocks that allow capturing the context within a sequence of audio fingerprints. After the encoder blocks, the sequential output is condensed using the attention-based pooling mechanism *SeqPool* [16], and then mapped to an n-dimensional vector. The resulting vector contains the logits for the corresponding classification task, where n is the number of classes (e.g., the number of words or speakers).

Model Training. We peform a hyperparameter selection and train a distinct model for each of the attack scenarios and each ACR solution. All models follow the architecture described above. However, we vary the embedding dimension d,

[2] sha1: `3c0770204a5d769c1a22a4acb7f9d6a4dd12e55c`.

[3] sha1: `5de8eb4098d2e35a2c3951a169bf9e19a680e2d4`.

the number of encoder blocks, and the number of attention heads. Throughout the experiments, we use a fixed number of bits l for most of the considered ACR solutions. Only for Zapr Alg2, we vary this parameter, as the exact length of the subfingerprints is unknown for this solution. An overview of the selected hyperparameters can be found in Table 2 in the Appendix.

We train each model for 300 epochs using the weighted Adam optimizer, combined with a learning rate schedule based on cosine decay [27,28]. To avoid overfitting of our model, we apply three common regularization methods. In particular, we combine dropout [41], stochastic depth regularization [17], and label smoothing [42] during training.

Note that we make our code and the used datasets publicly available[4] to allow other researchers to reproduce the results, and access details on our feature extraction and model selection procedure.

5 Evaluation

We proceed to investigate the impact of the proposed attacks on the privacy of smartphone users. Following the three attack scenarios, we empirically evaluate whether an adversary can identify human speech, particular speakers, or spoken words in the content of the collected audio fingerprints.

5.1 Speech vs. Music

To run a successful attack, an attacker first needs to be able to filter out potentially interesting audio fingerprints that contain human speech. That is, the attacker wants to solve a binary classification problem, in which the learning model receives an audio fingerprint as input and predicts one of two possible classes.

Dataset. For this experiment, we combine audio samples of human speech from the popular LibriSpeech dataset [32] with music examples taken from the free music archive (FMA) dataset [10]. In particular, we randomly select the same number of samples from each of the datasets and split them into audio recordings with a duration of 3 s each. This is done to ensure that all audio samples in our dataset have a fixed length, so that the machine learning model cannot implicitly use the length of a sample as a shortcut for its decision [3]. We then employ the audio fingerprinting algorithms of the selected ACR solutions on these recordings. As all algorithms expect audio signals sampled at 8,000 Hz, we resample the files in the dataset accordingly. Our final dataset comprises a total of 61,624 audio recordings and respective fingerprints.

[4] https://github.com/acr-privacy/attacks.

Experimental Setup. To avoid overfitting the neural network, we split the data into separate training, validation, and test partitions. That is, we use 35,512 samples for training, 10,112 samples for validation, and 16,000 samples for testing. Note that we ensure that no audio recording used for obtaining the samples in one partition is used in another. We then determine the best setup for our attack on the validation data, where we vary the hyperparameters (see Sect. 4.3). We finally measure the performance of the best setup on the test set.

Results. Table 1 shows the results of this experiment. The attack achieves the highest success rates using the fingerprints of ACRCloud, with an accuracy of 98%. The results for QBF and Zapr Alg1 are only slightly worse, yielding reliable detection rates of 97% and 95%, respectively. While the attack is less effective against Zapr Alg2, we can yet classify 67% of the samples correctly—which still exceeds random guessing.

Table 1. Overview of experimental results. For each setting, the table shows the Accuracy (Acc), Precision (P), Recall (R), and F1-Score (F1). The last row shows the results of a random guess by the attacker (RND).

	Speech				Speaker				Word			
	Acc	P	R	F1	Acc	P	R	F1	Acc	P	R	F1
QBF	.97	.97	.97	.97	**.90**	**.90**	**.90**	**.91**	.70	.70	.70	.70
ACR	**.98**	**.98**	**.98**	**.98**	**.90**	**.90**	**.90**	**.91**	**.82**	**.82**	**.82**	**.82**
ZA1	.95	.95	.95	.95	.64	.64	.64	.64	.58	.58	.58	.59
ZA2	.68	.68	.68	.68	.05	.03	.05	.03	.40	.40	.40	.40
RND	.50	.50	.50	.50	.03	.03	.03	.03	.10	.10	.10	.10

Overall, our results show that audio fingerprints contain enough information to reliably distinguish between music and speech signals. This result is not unexpected, as the attack scenario is close to the original purpose of ACR solutions. Nevertheless, our attack can serve as a preparatory step for further analysis and inference of other relevant information.

5.2 Speaker Identification

For the next experiment, we assume that the attacker was able to obtain fingerprints that solely contain human speech. They now want to identify the speakers that were recorded by the mobile device. To mimic this scenario, we design an experiment with n speakers. That is, the attacker tries to distinguish between n distinct classes from the audio fingerprints. In this setting, the final layer consists of n output neurons, where each neuron corresponds to one of the speakers.

Dataset. To evaluate the attack in this setting, we use the LibriSpeech dataset [32]. We randomly select audio recordings of 2×40 distinct speakers from the LibriSpeech dataset. Like in the previous experiments, we resample the audio files at 8,000 Hz and split them into chunks of three seconds, which are then used to generate the audio fingerprints. The resulting dataset contains 33,920 samples, with at least 400 instances per speaker.

Experimental Setup. We first split the LibriSpeech data into a calibration set and an evaluation set, each comprising 40 *different* speakers, and containing 16,720 and 17,200 audio samples, respectively. This split ensures that we do not overfit to peculiarities of some of the speakers. We then perform experiments on the calibration set to select hyperparameters according to Sect. 4.3. Finally, we select the best attack setup and run the final experiment on the evaluation set.

Results. For QBF and ACRCloud, we receive a high accuracy of up to 90% (see Table 1). The attack success rate outperforms random guessing by large extent: As we deal with a 40-class problem, a random prediction only yields 2.5%. Only for Zapr Alg2, we obtain results close to random guessing. In contrast, however, we find that Zapr Alg1 is also prone to this attack, with an accuracy of 64%, proving that the attacked fingerprints still leak sensitive information.

In summary, we show with this experiment that audio fingerprints reveal enough information to identify individual speakers, which can be sensitive information in many cases. The attack fails on one algorithm only, i.e., Zapr Alg2. While at first glance this result suggests a higher robustness of Zapr Alg2, it should be interpreted with great caution due to our limited knowledge of the fingerprint structure. An attacker with greater knowledge of the internals of the fingerprinting algorithm might be able to also run a successful attack against these fingerprints as well.

5.3 Word Identification

In the third attack scenario, we also assume that the adversary has obtained audio fingerprints that contain human speech. However, unlike the previous experiment, the attacker is now interested in the spoken content stored in these fingerprints. To simulate this situation, we design an experiment where each audio fingerprint contains one of n spoken words. Here, we choose the English digits from zero to nine as the basis for these words, such that $n = 10$. The final layer of the neural network contains 10 neurons, where each output neuron represents one of the digits. While more complex attack scenarios are conceivable, even the identification of a limited set of words like digits poses a privacy problem, for example, when a PIN or credit card number is spelled out loud.

Dataset. For this experiment, we use the *Speech Commands* dataset [49] and extract all audio of spoken digits. The dataset already contains a prepared split into training and testing partitions. Therefore, we use this partitioning instead of generating our own. However, we prepare the training data by removing duplicates that would simplify the attack. Also, we balance the number of audio snippets per digit. Our final dataset includes a total of 36,462 audio samples, where 32,630 are used for training and 3,832 for testing.

Experimental Setup. We follow the same experimental procedure as described in Sect. 5.1, and select the setup that yields the best results on the validation data. The selected parameters are described in the Appendix (see Table 2). Finally, we evaluate its performance on the test set. In this case, however, the test set is not entirely balanced. Some digits occur slightly less frequently than others. To ensure a fair comparison, we macro-average the results to ensure that each digit is weighted equally.

Results. Table 1 shows the results of this experiment. In contrast to the identification of speakers, all considered solutions are vulnerable to this attack, although not to the same extent. In particular, we obtain the best results for ACRCloud with an accuracy of 82%, followed by QBF and Zapr Alg1 with 70% and 58%, respectively. To our surprise, the attack is also successful against Zapr Alg2 in 40% of the cases, where the identification of individual speakers failed. While the algorithm shows the highest robustness, the success of the attacker is still significantly higher than random guessing, which only yields 10% accuracy in this setting. Although our experiment builds on a simplified setup with a limited number of words, it is clear that spoken content can be identified in audio fingerprints, raising doubts about their privacy-friendliness. We suspect that similar attacks can be conducted with personal names, street addresses, and other sensitive terms that would compromise the privacy of smartphone users.

6 Discussion

Our experimental results in the previous sections show that audio fingerprints can indeed leak sensitive user information that might be misused by malicious parties. In the following, we sketch three possible countermeasures of increasing technical complexity to respond to this threat. Furthermore, we elaborate on limitations and possible biases in our experimental design.

Countermeasures. The first and simplest countermeasure is to notify users properly if the device is recording audio in the background. For instance, recent versions of mobile operating systems display an indicator on the device signaling an ongoing recording. While this approach should lower the risk that fingerprint generation takes place without the user's knowledge, it might be overlooked easily. An additional problem arises from the fact that most Android devices still run with Android 10 or lower and thus do not notify the users when the microphone is used in the background. Consequently, we argue that audio content recognition should only be performed at the user's request.

Secondly, cryptographic hashing can help to protect the sensitive data stored in the fingerprints. In particular, applying a cryptographic hash on the subfingerprints still allows matching them, while hindering the dissection and statistical analysis of the contained frequency peaks and temporal differences. However, to rule out dictionary attacks, the entropy of the fingerprint distribution has to be high enough. Further, a keyed hash function with a periodic key rotation should be used (see [21]). While the former remains questionable, the latter is ruled out by the business models in the ACR context.

Finally, we note the similarities between the technique of private set intersection (PSI) and the matching subfingerprints against a database. PSI enables two parties to generate an intersection of private sets without revealing the elements not in common. Several PSI approaches consider scenarios where a small set is intersected with a large set (e.g., [7,12]), thus reflecting the matching process in ACR solutions. Utilizing PSI, the service provider only learns of subfingerprints present in her database, thus reducing the information available to the attacker. This results in a reduced attack effectiveness, as matching specific samples needs preparation in advance. PSI, however, incures a notable overhead. While this may be acceptable for applications with sporadic ACR usage, such as music recognition, this renders continuous tracking infeasible. As with cryptographic hashing, if the fingerprint distribution is of insufficient entropy, creating a dictionary and matching it against the incoming fingerprints remains possible.

Limitations. Although we have designed our experiments with great care, we could not show the leakage of sensitive data in all examined cases. Unfortunately, this does not automatically imply a proper protection of sensitive data by the ACR solutions in these cases. Instead, we faced several limitations that might have affected our experimental outcome, hindering us to extract the sensitive information reliably.

First of all, our experimental setting is not free from bias, as we perform our experiments under controlled conditions on mostly balanced datasets. However, as we intend to demonstrate possible privacy threats of ACR solutions in this paper, we argue that our setup is still appropriate for this purpose. Still, further research is required to better assess and grasp the exact impact of our findings and develop effective defenses against this privacy threat. On one hand, we have not systematically assessed the robustness of the attacks in the presence of background noise, which could potentially even serve as a simple defense mechanism. Overall, more advanced attacks against ACR fingerprints along with corresponding defenses should be explored in the future.

Secondly, we find that the examined ACR solutions are not vulnerable to the same extent. Most notably, we could not show that distinguishing between multiple speakers is possible when generating fingerprints with the Zapr Alg2 algorithm. However, we note that we have not been able to fully reverse engineer and understand the internals of this particular solution. Hence, even attacks on this algorithm might still be possible, given an adversary with enough background knowledge of the underlying algorithms.

Finally, in this paper, we consider the application of ACR in mobile apps only. However, as these solutions are part of a broad ecosystem that is difficult for externals to grasp, they are deployed in many different application scenarios as well [1,15,40,47,53]. A complete survey of this ecosystem is thus out of scope. Nonetheless, we think that our gained insights generalize to other ACR solutions, as we have selected two popular examples that are widespread and offer services and SDKs not limited to a specific app or use case. Still, further research is required to assess the security of other ACR methods as well.

7 Conclusion

In this work, we provide insights into the ACR ecosystem and demonstrate that the employed audio fingerprints reveal private information. The evaluation of our attacks gives evidence that audio fingerprints leak sensitive information and should not be used as a privacy protection mechanism.

Within the conducted experiments the proposed attacks identify sensitive information, achieving high accuracy values that are significantly better than random guessing. In particular, we can show that it is possible to identify speakers and spoken words in the audio fingerprints of three major ACR solutions using machine learning techniques.

These results raise serious privacy concerns, especially if this technology runs in the background without the users' knowledge. We conclude that these approaches are not privacy-friendly and caution should be exercised, especially when using them for background monitoring.

Acknowledgements. This work was funded by the German Federal Ministry of Education and Research (BMBF) under the grants BIFOLD24B and 16KIS1142K.

Disclosure of Interests. The authors have no competing interests to declare that are relevant to the content of this article.

A Details on ACR Solutions

In this section, we provide information that we obtained through reverse engineering of the two commercial ACR solutions Zapr and ACRCloud.

Analysis Setup

We start by describing our experimental setup to reverse engineer the apps and discuss our findings of both solutions afterward.

Mobile Apps. For ACRCloud, we base our analysis on the Deezer app (version 6.1.14.99.) and verify that our insights also remain valid for more recent versions of the SDK (version 6.2.13.151). Similarly, we use the Android smartphone application *ABP Live TV News* (version 9.9.7) for Zapr.

Dynamic Analysis. Both solutions encapsulate the implementations of the ACR algorithms in a shared library, which is provided as a native binary object and accessed by the Android apps through the Android apps through the Java Native Interface (JNI). We treat the fingerprinting algorithms inside the shared object as a black box and observe its return values. To this end, we use the dynamic instrumentation toolkit *Frida*, which allows us to run the fingerprinting algorithms on controlled input signals, and extract the resulting audio fingerprints. A static analysis shows that all algorithms expect the input signal to be sampled at a frequency of 8,000 Hz with an audio bit depth of 16 bit. Providing the ACR implementations with properly preprocessed audio samples yields the required audio fingerprints, which can then be utilized for further analysis.

To learn more about the underlying structure of the fingerprints, we perform controlled experiments using specifically crafted audio signals from which we derive audio fingerprints. For instance, we use audio signals that contain only one particular frequency or even pure silence.

Table 2. Overview of final model parameters. The table shows the number of bits of the subfingerprints (l), sequence length (k), the embedding size (d), the number of encoder blocks ($encoders$), and the number of heads per encoder ($heads$) for each setting.

Experiment	Fingerprint	l	k	d	$encoders$	$heads$
Speech vs. Music	QBF	19	177	128	4	4
	ACRCloud	64	96	128	4	4
	Zapr Alg1	32	80	128	4	4
	Zapr Alg2	8	804	128	4	4
Speaker Identification	QBF	19	118	128	4	4
	ACRCloud	64	95	128	4	4
	Zapr Alg1	32	80	128	4	4
	Zapr Alg2	16	314	128	4	4
Word Identification	QBF	19	48	128	4	4
	ACRCloud	64	24	128	4	4
	Zapr Alg1	32	20	128	4	4
	Zapr Alg2	32	92	128	4	4

Fingerprint Structures

We find that the fingerprint structures do not only widely differ between ACRCloud and Zapr, but even between the two Zapr algorithms we selected for our analysis. In the following, we provide more details on our findings.

ACRCloud. For ACRCloud, we find that the generated audio fingerprints vary in length, although all audio snippets are three seconds long. In particular, the length of the generated fingerprints for our dataset varies between 344 and 752 bytes, with a median at 544 bytes. Each fingerprint consists of multiple subfingerprints (see Sect. 3.2) with a length of 8 bytes. The first two bytes of each subfingerprint encode the frequency of an identified peak. Here, ACRCloud seemingly segments the frequency band, which has a maximum frequency of 4,000 Hz, into 1024 distinct bins of equal size, leading to a frequency resolution of $f_{res} = \frac{4000\text{ Hz}}{1024} \approx 3.906$ Hz. The third and fourth byte of the subfingerprints encode the time offset Δt with a granularity of roughly 20 ms. For the last four bytes, we are unable to derive clear explanations. But we notice that the information stored in these bytes depend on the frequency bytes but not on the time offset.

Zapr. For Zapr Alg1, none of the observed fingerprints exceeds 340 bytes in length, which suggests a maximum length for the fingerprints. Additionally, each fingerprint's length is divisible by 4 bytes, indicating that they are composed of multiple subfingerprints, each 4 bytes long. The only exception we find is for silent signals, for which the algorithm does not output any fingerprints. The first two bytes of a fingerprint encode the time offset with a precision of 2 s. The last byte encodes frequency information, systematically partitioning the 4 kHz-band into 256 distinct frequency bins. The purpose of the third byte remains unclear. Unfortunately, for Zapr Alg2, we have not been able to derive information about its structure.

References

1. ACRCloud: ACRCloud: automatic content recognition services for doers (2022). https://www.acrcloud.com/. Accessed 22 Apr 2024
2. Arp, D., Quiring, E., Wressnegger, C., Rieck, K.: Privacy threats through ultrasonic side channels on mobile devices. In: Proceedings of IEEE European Symposium on Security and Privacy (EuroS&P) (2017)
3. Arp, D., et al.: Dos and don'ts of machine learning in computer security. In: Proceedings of USENIX Security Symposium (2022)
4. Brookman, J., Rouge, P., Alva, A., Yeung, C.: Cross-device tracking: measurement and disclosures. Proc. Priv. Enhancing Technol. (PETS) **2017**(2) (2017)
5. Celosia, G., Cunche, M.: Discontinued privacy: personal data leaks in apple Bluetooth-low-energy continuity protocols. Proc. Priv. Enhancing Technol. (PETS) **2020**(1) (2020)
6. Chatterjee, R., et al.: The spyware used in intimate partner violence. In: Proceedings of IEEE Symposium on Security and Privacy (S&P) (2018)
7. Chen, H., Laine, K., Rindal, P.: Fast private set intersection from homomorphic encryption. In: Proceedings of ACM Conference on Computer and Communications Security (CCS) (2017)
8. Deezer: Deezer | listen to music | online music streaming platform (2022). https://www.deezer.com. Accessed 22 Apr 2024

9. Deezer: Third party data breach – deezer support (2022). https://support.deezer.com/hc/en-gb/articles/7726141292317-Third-Party-Data-Breach. Accessed 22 Apr 2024
10. Defferrard, M., Benzi, K., Vandergheynst, P., Bresson, X.: FMA: a dataset for music analysis. In: Proceedings of the International Society for Music Information Retrieval Conference (ISMIR) (2017)
11. Devlin, J., Chang, M.W., Lee, K., Toutanova, K.: BERT: pre-training of deep bidirectional transformers for language understanding. In: Proceedings of the Conference of the North American Chapter of the Association for Computational Linguistics: Human Language Technologies (NAACL-HLT) (2019)
12. Dong, C., Chen, L., Wen, Z.: When private set intersection meets big data: an efficient and scalable protocol. In: Proceedings of ACM Conference on Computer and Communications Security (CCS) (2013)
13. Dosovitskiy, A., et al.: An image is worth 16×16 words: transformers for image recognition at scale. In: Proceedings of International Conference on Learning Representations (ICLR) (2019)
14. Faragher, R., Harle, R.: Location fingerprinting with Bluetooth low energy beacons. IEEE J. Sel. Areas Commun. **33**(11), 2418–2428 (2015)
15. Haitsma, J., Kalker, T.: A highly robust audio fingerprinting system. In: Proceedings of the International Society for Music Information Retrieval Conference (ISMIR) (2002)
16. Hassani, A., Walton, S., Shah, N., Abuduweili, A., Li, J., Shi, H.: Escaping the big data paradigm with compact transformers. CoRR abs/2104.05704 (2021)
17. Huang, G., Sun, Yu., Liu, Z., Sedra, D., Weinberger, K.Q.: Deep networks with stochastic depth. In: Leibe, B., Matas, J., Sebe, N., Welling, M. (eds.) ECCV 2016. LNCS, vol. 9908, pp. 646–661. Springer, Cham (2016). https://doi.org/10.1007/978-3-319-46493-0_39
18. Jawurek, M., Johns, M., Rieck, K.: Smart metering de-pseudonymization. In: Proceedings of Annual Computer Security Applications Conference (ACSAC) (2011)
19. Kennedy, S., Li, H., Wang, C., Liu, H., Wang, B., Sun, W.: I can hear your Alexa: voice command fingerprinting on smart home speakers. In: Proceedings of IEEE Conference on Communications and Network Security (CNS) (2019)
20. Kim, H.G., Cho, H.S., Kim, J.Y.: Robust audio fingerprinting using peak-pair-based hash of non-repeating foreground audio in a real environment. Cluster Comput. **19**(1) (2016)
21. Knospe, H.: Privacy-enhanced perceptual hashing of audio data. In: International Conference on Security and Cryptography (SECRYPT) (2013)
22. Konjeti, S., Potty, H., Kashyap, D.: Zapr audio fingerprinting (2017). https://www.music-ir.org/mirex/abstracts/2017/KP1.pdf
23. Korolova, A., Sharma, V.: Cross-app tracking via nearby Bluetooth low energy devices. In: Proceedings of ACM Conference on Data and Applications Security and Privacy (CODASPY) (2018)
24. LG Ads Solutions: Alphonso ACR technology and consumer choice - lg ads[3] (2018). https://tinyurl.com/yp9t2dmz. Accessed 22 Apr 2024
25. LG Ads Solutions: Automatic content recognition (2021). https://alphonso.tv/. Accessed 22 Apr 2024
26. Liberatore, M., Levine, B.N.: Inferring the source of encrypted http connections. In: Proceedings of ACM Conference on Computer and Communications Security (CCS) (2006)

27. Loshchilov, I., Hutter, F.: SGDR: stochastic gradient descent with warm restarts. In: Proceedings of International Conference on Learning Representations (ICLR) (2017)
28. Loshchilov, I., Hutter, F.: Decoupled weight decay regularization. In: Proceedings of International Conference on Learning Representations (ICLR) (2019)
29. Mavroudis, V., Hao, S., Fratantonio, Y., Maggi, F., Kruegel, C., Vigna, G.: On the privacy and security of the ultrasound ecosystem. Proc. Priv. Enhancing Technol. (PETS) **2017**(2) (2017)
30. Media, B.: Hotstar, newsdog and other Indian apps are spying on your phone's mic (2018). https://beebom.com/hotstar-newsdog-apps-spying-phone-mic/. Accessed 22 Apr 2024
31. Musixmatch S.p.A.: Musixmatch website (2022). https://www.musixmatch.com. Accessed 22 Apr 2024
32. Panayotov, V., Chen, G., Povey, D., Khudanpur, S.: LibriSpeech: an ASR corpus based on public domain audio books. In: IEEE International Conference on Acoustics, Speech and Signal Processing (ICASSP) (2015)
33. Park, M., Kim, H.R., Yang, S.H.: Frequency-temporal filtering for a robust audio fingerprinting scheme in real-noise environments. ETRI J. **28**(4) (2006)
34. Ravnås, O.A.V.: Frida • a world-class dynamic instrumentation framework (2022). https://frida.re/. Accessed 22 Apr 2024
35. Reardon, J., Feal, Á., Wijesekera, P., On, A.E.B., Vallina-Rodriguez, N., Egelman, S.: 50 ways to leak your data: an exploration of apps' circumvention of the Android permissions system. In: Proceedings of USENIX Security Symposium (2019)
36. Rimmer, V., Preuveneers, D., Juárez, M., van Goethem, T., Joosen, W.: Automated website fingerprinting through deep learning. In: Proceedings of Network and Distributed System Security Symposium (NDSS) (2018)
37. Saadatpanah, P., Shafahi, A., Goldstein, T.: Adversarial attacks on copyright detection systems. In: Proceedings of International Conference on Machine Learning (ICML) (2020)
38. Schlegel, R., Zhang, K., Zhou, X., Intwala, M., Kapadia, A., Wang, X.: Soundcomber: a stealthy and context-aware sound trojan for smartphones. In: Proceedings of Network and Distributed System Security Symposium (NDSS) (2011)
39. Son, W., Cho, H.T., Yoon, K.: Sub-fingerprint masking for a robust audio fingerprinting system in a real-noise environment for portable consumer devices. In: Digest of Technical Papers International Conference on Consumer Electronics (ICCE) (2010)
40. Sonnleitner, R., Widmer, G.: Robust quad-based audio fingerprinting. IEEE ACM Trans. Audio Speech Lang. Process. **24**(3) (2016)
41. Srivastava, N., Hinton, G., Krizhevsky, A., Sutskever, I., Salakhutdinov, R.: Dropout: a simple way to prevent neural networks from overfitting. J. Mach. Learn. Res. (JMLR) **15**(1), 1929–1958 (2014)
42. Szegedy, C., Vanhoucke, V., Ioffe, S., Shlens, J., Wojna, Z.: Rethinking the inception architecture for computer vision. In: Proceedings of IEEE Conference on Computer Vision and Pattern Recognition (CVPR) (2016)
43. The New York Times Company: All 3 billion yahoo accounts were affected by 2013 attack (2017). https://www.nytimes.com/2017/10/03/technology/yahoo-hack-3-billion-users.html. Accessed 22 Apr 2024
44. The New York Times Company: That game on your phone may be tracking what you're watching on TV (2017). https://www.nytimes.com/2017/12/28/business/media/alphonso-app-tracking.html. Accessed 22 Apr 2024

45. Thiemert, S., Nürnberger, S., Steinebach, M., Zmudzinski, S.: Security of robust audio hashes. In: IEEE International Workshop on Information Forensics and Security (WIFS) (2009)
46. Vaswani, A., et al.: Attention is all you need. In: Advances in Neural Information Processing Systems (NIPS) (2017)
47. Wang, A.: An industrial strength audio search algorithm. In: Proceedings of the International Society for Music Information Retrieval Conference (ISMIR) (2003)
48. Wang, A.: The shazam music recognition service. Commun. ACM **49**(8) (2006)
49. Warden, P.: Speech commands: a dataset for limited-vocabulary speech recognition. CoRR abs/1804.03209 (2018). http://arxiv.org/abs/1804.03209
50. White, A.M., Matthews, A.R., Snow, K.Z., Monrose, F.: Phonotactic reconstruction of encrypted VoIP conversations: Hookt on Fon-iks. In: Proceedings of IEEE Symposium on Security and Privacy (S&P) (2011)
51. Xu, Y., Frahm, J., Monrose, F.: Watching the watchers: automatically inferring TV content from outdoor light effusions. In: Proceedings of ACM Conference on Computer and Communications Security (CCS) (2014)
52. Zapr Media Labs: Privacy | Zapr Media Labs (Zapr discontinued its service in mid 2022. Thus, we can only provide a link to the snapshot of the website) (2022). https://tinyurl.com/rneknwyb. Accessed 22 Apr 2024
53. Zapr Media Labs: Zapr | TV analytics, integrated advertising, real-time surveys[3] (2022). https://tinyurl.com/2vhr6vmu. Accessed 22 Apr 2024
54. Zimmeck, S., Li, J.S., Kim, H., Bellovin, S.M., Jebara, T.: A privacy analysis of cross-device tracking. In: Proceedings of USENIX Security Symposium (2017)

Mobile and Web Application Security

Bringing UFUs Back into the Air with FUEL: A Framework for Evaluating the Effectiveness of Unrestricted File Upload Vulnerability Scanners

Sebastian Neef(✉) and Maath Oudeh

Security in Telecommunications, Technische Universität Berlin, Berlin, Germany
{neef,maath.oudeh}@tu-berlin.de

Abstract. Unrestricted file upload (UFU) is a class of web security vulnerabilities that can have a severe impact on web applications if uploaded files are not sufficiently validated or securely handled. A review of related work shows an increased interest in finding new methods to discover such vulnerabilities. However, each publication evaluates its new vulnerability scanner against a different set of artificial or real-world applications available at the time of writing. Thus, we identify the need for a comprehensive testing framework to allow a reproducible comparison between existing and future UFU vulnerability scanners.

Our contributions include the File Upload Exploitation Lab (FUEL), which models 15 distinct UFU vulnerabilities in isolated scenarios to enable a reproducible evaluation of UFU scanners' capabilities. The results of evaluating four black-box UFU scanners against FUEL show that no scanner manages to identify all UFU vulnerabilities, leaving real-world websites at risk of compromise due to false negatives. Our work aims to solve this problem by extending an existing UFU scanner with multiple new detection and exploitation techniques, which we call *Fuxploider-NG*, to increase its accuracy from ∼50% to over 90%, thereby surpassing the capabilities of existing UFU scanners and showcasing the importance of FUEL as a UFU vulnerability evaluation framework. To foster open science and future work in this area, we open-source FUEL and Fuxploider-NG.

Keywords: Web security · Vulnerability analysis · Unrestricted File Upload Vulnerabilities · Vulnerability scanner · FUEL · PHP · RCE · XSS

1 Introduction

The World Wide Web (WWW) is constantly evolving, offering its users the possibility to contribute content since the release of the web 2.0 [16]. According to [6], recent statistics count over 5.3 billion internet users, with a large quantity being social media users. Websites, including social media platforms, allow users to contribute content in various ways, e.g., by uploading files such as documents,

F. Maggi et al. (Eds.): DIMVA 2024, LNCS 14828, pp. 207–226, 2024.
https://doi.org/10.1007/978-3-031-64171-8_11

images, videos, or other types. Sharing files through web applications has become integral for businesses, social networks, personal websites, or other domains that rely on web applications.

Despite its 30th anniversary, the Hypertext Preprocessor (PHP) programming language remains the most popular for web development, with a share of over 75% based on statistics from [44], powering websites such as facebook.com or wikipedia.org [45]. Modern web servers such as Apache httpd or NGINX provide extensions to execute PHP code, further aiding its popularity [11,37]. For example, the PHP-based content management system WordPress claims to have powered over 40% of the web in 2021 [48].

PHP supports processing uploaded files with special variables, functions, and configuration options [39,42]. However, it is crucial to handle uploaded files carefully and securely, or unrestricted file upload (UFU) vulnerabilities may arise [29]. Especially in PHP-developed web applications, failure to properly validate uploaded files may result in full system compromise (e.g., remote code execution) or other high-impact attacks (e.g., cross-site scripting).

File upload vulnerabilities are prevalent to this day. Public vulnerability databases, such as [25], list over 1,000 public exploits in web applications related to UFU. Similarly, the Common Vulnerability Enumeration (CVE) database contains over 3,300 entries matching the keyword "file upload" [23]. Furthermore, WPScan documents several thousand security issues, including many authenticated and unauthenticated UFU vulnerabilities, in WordPress and its third-party plugins or themes [1].

The academic community has shown great interest in researching UFU vulnerability detection methods and contributing automated vulnerability scanners, as detailed in Sect. 2.4. Additionally, commercial (e.g., BurpSuite [32]) and open-source (e.g., Fuxploider [17] or ZAP [43]) scanners have emerged and are being actively developed to help identify UFU vulnerabilities.

However, none of the academic publications have used a standardized method for evaluating their new approaches but have used artificial or real-world PHP-based web applications available at the time of writing. Our results will show that the lack of a standardized evaluation environment has led to shortcomings in all UFU vulnerability scanners' detection capabilities. Consequently, users who trust the output of the currently available UFU scanners might assume their web application to be secure against UFU vulnerabilities, although the missing detection capabilities and false-negatives might allow attackers to still compromise the application.

Motivated by these discoveries, we aim to extend the research body in the UFU vulnerability detection domain with the following contributions:

- We present FUEL as an extensible and modular testing framework to facilitate reproducible and comprehensive evaluations of UFU scanners.
- We evaluate four academic, open-source, and commercial UFU scanners against a set of 15 unique UFU scenarios implemented in FUEL that reveals many shortcomings in these scanners.
- We implement *Fuxploider-NG*, which consolidates and surpasses the capabilities of the evaluated vulnerability scanners into a single tool.

2 Background and Related Work

Although the reader is assumed to be familiar with general concepts related to web technologies, e.g., the Hyper Text Transfer Protocol (HTTP), Hypertext Markup Language (HTML), or common web vulnerabilities, e.g., the Open Worldwide Web Application Project (OWASP) Top 10 [28], this section provides an overview of UFU related knowledge and related work to follow the remainder of this work.

2.1 PHP

The Hypertext Preprocessor (PHP) programming language was specifically designed for web development in 1994 [40] and has continuously evolved to reach version PHP 8.3 at the time of writing. We assume one reason for its high adoption is its simplicity, as HTML and PHP code can be mingled to create basic, dynamic web applications, e.g., <h1><?php echo $_GET['name'];?></h1>. To execute PHP code, a PHP file must be passed to the PHP interpreter, which is performed by modern web servers such as Apache, NGINX, or others [11,37]. Thus, the barrier to entry for developing feature-rich websites using PHP is low, and vulnerabilities, including UFU, might arise from simple programming mistakes [28].

2.2 File Upload Process

Understanding the fundamental steps in a file upload process is crucial for discovering UFU vulnerabilities. While the client-side handling of file uploads, e.g., letting the user choose a file and initiating the file transfer, is part of the HTML (e.g., [19,49]) and HTTP (e.g., [22,24]) specifications implemented in the browser, the server-side handling can differ based web application's implementation and web server's configuration. Figure 1 provides a simplified depiction of the basic components and interactions of a web-based file upload process.

File Selection. First, the web server sends the user a website containing an HTML form with an input field of type "file", as shown in Listing 1.1. The form's attributes inform the browser how to handle the file upload, e.g., the HTTP request method in `method`, the web server's receiving endpoint in `action`, and the form data's encoding in `enctype`. With the help of additional HTML properties or JavaScript, the user's file selection can be restricted and validated in the browser. However, these client-side restrictions can be circumvented, e.g., by intercepting and modifying the resulting HTTP request.

File Transfer. After the user has chosen a file to upload and submitted the form, the browser initiates the file transfer. RFC 7578 specifies the content type used by file uploads to be `multipart/form-data` [22]. Figure 2 shows the HTTP request issued by the browser to the server. In order to include

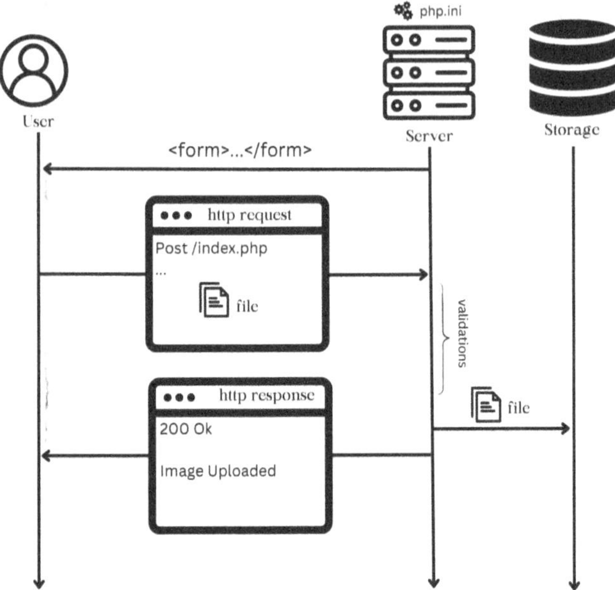

Fig. 1. A simplified schema of the file upload and retrieval steps.

```
<form enctype="multipart/form-data" method="POST" action="/
    index.php">
    <input type="file" name="fileToUpload"/>
    <input type="submit" value="Upload" name="submit"/>
</form>
```

Listing 1.1. A basic HTML file upload form.

the file's data in the request, the `multipart/form-data` encoding transforms all form fields into blocks separated by a randomly generated boundary and a `Content-Disposition` line. Additionally, the request contains information about the file, e.g., the name, content, and MIME type (cf. RFC 2045ff. [12]).

File Processing. PHP processes uploaded files by default and offers additional configuration options to restrict incoming file uploads [38], e.g., setting a maximum file size. Generally, uploaded files are assigned a random name and stored temporarily in a specifically configured directory. The PHP application accesses the files and their metadata through a 2D associative array named `$_FILES` [39,42]. The first dimension's keys are the file-fields' names, e.g., `fileToUpload`, and the second dimension provides a file's properties as shown in Listing 1.2. While the key `name` represents the file's name on the client's file system, `tmp_name` is the random file name generated by PHP.

```
POST /index.php HTTP/1.1
Host: localhost
Content-Length: 1337
Content-Type: multipart/form-data; boundary=----GCIl1OYbVERS6VWy
Connection: close

------GCIl1OYbVERS6VWy
Content-Disposition: form-data; name="fileToUpload"; filename="image.png"
Content-Type: image/png

[file content]
------GCIl1OYbVERS6VWy
Content-Disposition: form-data; name="submit"

Upload
------GCIl1OYbVERS6VWy--
```

Fig. 2. An example `multipart/form-data` encoded file upload HTTP request with the UFU-relevant manipulation points highlighted.

The attacker controls the `name` and `type` properties as well as the file's content. Therefore, web applications must correctly validate these to avoid UFU vulnerabilities [29]. As the next section clarifies, the example provided in Listing 1.2 is not secure. After validation, the web application may eventually wish to store the file by moving it from its temporary location to a permanent location, e.g., using `move_uploaded_file`.

2.3 Unrestricted File Upload

Although the actual risks and consequences of an insecure file upload process are dependent on several factors, e.g., its implementation, the web server's or storage layer's configuration, UFU vulnerabilities root in insufficient validation and typically result in code execution, cross-site scripting, or other high consequence attacks [29].

Improper Validation. The failure to validate attacker-controlled file properties may result in the upload of dangerous files [29,31].

```
$fname = $_FILES["fileToUpload"]["name"];
$ftype = $_FILES["fileToUpload"]["type"];
$fsize = $_FILES["fileToUpload"]["size"];
$ftmp  = $_FILES["fileToUpload"]["tmp_name"];
if($fsize < 100*1024 && $ftype = "image/png") {
    move_uploaded_file($ftmp, "/uploads/" . $fname);
}
```

Listing 1.2. A basic example of insufficient file property validation in PHP

Failure to correctly validate the file name may allow the attacker to create files in arbitrary storage locations, thus overwriting existing files or injecting HTML markup into the web application. The code in Listing 1.2 performs no file name validation whatsoever.

Incorrect file type validation might allow the attacker to trick the application into assuming a permissible file was uploaded, although the content is malicious. For example, the validation in Listing 1.2 assumes a `image/png` type but does not ensure that the type matches the file content. Attackers may use polyglot files [30], which are combinations of two or more file types into a single file, to bypass insufficient file type or file content checks.

Thus, secure web applications must verify that the file content matches the file's type to avoid confusion. However, examining the content is a complex task that depends on the file type or size, as large files may require more processing resources and time. Furthermore, attackers might exploit vulnerabilities in the parsers, libraries, or other components used for validation, e.g., ImageTragick [13].

Finally, failing to verify a file's size may result in large files being uploaded repeatedly, thus eventually exhausting the storage space. Depending on the web server's storage configuration, this may lead to consequent HTTP requests or uploads to fail, rendering the website unresponsive.

Vulnerabilities. Although the vulnerabilities resulting from insecure file uploads can be manifold [29], our evaluation focuses on code execution and cross-site scripting, similar to related work (cf. Sect. 2.4), as these have a high impact.

In the context of PHP web applications, several methods exist to achieve *Code Execution (CE)* against the web application through UFU. For example, one method is uploading a file containing PHP code and instructing the PHP interpreter to execute it, e.g., uploading `exploit.php` and subsequently accessing it. Other methods include changing the file's content to exploit outdated or vulnerable components that process the uploaded file, such as ImageTragick [13], or uploading configuration files that change the interpreter's or web server's behavior, e.g., `.htaccess` files [36].

In contrast to code execution, *Cross-Site Scripting (XSS)* attacks target the website's users. Insufficient validation might allow an attacker to upload `.html`,`.htm`, or other files containing XSS payloads in a way that the web server serves these with a `text/html` content type. Thus, a victim's browser rendering such a file will interpret the HTML markup and execute JavaScript payloads. Another conceivable attack is uploading files with an XSS payload in the file name, which is then insecurely emitted by the web application in its HTML context.

2.4 Related Work

File upload vulnerabilities are not a new research topic. First mentions of UFU issues can be traced back to 2009 when Barth et al. discussed content sniffing attacks [2] and 2011 when Barua et al. evaluated server-side content sniffing detection [3] against PhpBB2, Prestashop, and Serendipity.

Dahse et al. used six different PHP projects, including osCommerce, to evaluate their method to detect second-order vulnerabilities, including UFU, in 2014 [5]. In 2016, Riadi et al. analyzed *osCommerce* for UFU vulnerabilities [34], while Pooj et al. compiled 16 different file upload vulnerability cases. In 2017, De Meo et al. [7] used formal methods to detect UFU in DVWA.

Two years later, Huang et al. [14] developed *UChecker* to detect UFU vulnerabilities automatically and evaluated it on over 9,000 WordPress plugins. In 2021, they again used over 9,000 WordPress plugins to evaluate their new scanner *UFuzzer* [15]. Lee et al. proposed the vulnerability scanner *FUSE* dedicated to finding UFU vulnerabilities in 2020 and evaluated it on a variety of 33 real-world PHP applications from different sources [21].

In 2022, Yenduri et al. discussed vulnerabilities in PHP and their prevention, including UFU vulnerabilities, and implemented two custom PHP web applications for their evaluation [50]. The same year, Wichmann et al. [46] evaluated *FileUploadChecker* against WordPress, Drupal, XWiki, and OpenCMs. The most recent publication is by Chen et al. proposing *URadar*'s adaptive dynamic testing approach, which they evaluate on 18 PHP web applications from over three different sources [4].

Over time, several intentionally vulnerable web applications were developed, which also encompass UFU vulnerabilities: *WackoPicko* simulated a real-world application and was published by Doupé et al. in 2010 for a study about black-box vulnerability scanners [10]. In 2013, the *Damn Vulnerable Web Application* (DVWA) started development to become a PHP-based vulnerability training platform featuring multiple difficulty levels [47]. In 2014, the *JuiceShop* launched as a deliberately vulnerable online shop covering a plethora of vulnerability classes [18]. In contrast to DVWA or WackoPicko, the JuiceShop is developed in NodeJS, which leads to disparities in how uploaded files become dangerous. All three vulnerable web applications focus on a broad spectrum of vulnerability classes, thus having a limited amount of UFU scenarios, which we assume to be one reason for their seldom reference in related work.

Our literature review revealed that no standardized framework exists to evaluate UFU vulnerability scanners. Instead, each study chose its own set of artificial or real-world PHP applications, available at the time of writing, to evaluate their novel methodology or scanner. While this enables the discovery of previously unknown vulnerabilities in existing software, it renders a just comparison of the capabilities and improvements of vulnerability scanners infeasible. As our evaluation in Sect. 4 will show, this leads to several areas for improvement in the capabilities of the examined UFU vulnerability scanners.

3 FUEL

We identified the need for a *File Upload Exploitation Lab* (FUEL) that unifies a diverse set of 15 UFU-specific scenarios into a single test framework, enabling a comprehensive and reproducible comparison of UFU scanners.

3.1 Design

A framework for the evaluation of UFU vulnerabilities and scanners should follow several design goals to make it adoptable in future work:

- **Simplicity:** A simple setup is crucial to reduce friction for researchers to make it a go-to choice for their evaluation.
- **Extensiveness:** A diverse set of file upload vulnerability scenarios is necessary to evaluate the capabilities of vulnerability scanners.
- **Extensibility:** A modular system that allows future work to extend the framework with new UFU scenarios.

FUEL achieves the first and third design goals with the use of containerization. Docker [9] allows starting isolated containers agnostic to the operating system, thus facilitating the deployment and use of FUEL. Further, its *compose* extension simplifies the orchestration of multiple containers with a single configuration file; therefore, deploying FUEL becomes a single command in its project directory: `docker compose up`.

Moreover, FUEL implements each UFU scenario in a separate container, each with its individual build instructions, as depicted in Fig. 3. Hence, each scenario can have an individual configuration of the web server and PHP environment and distinct source code for the vulnerable application. All currently implemented scenarios in FUEL are based on PHP 8.3 and Apache httpd 2.4, with the web application being available from the host at port $10,000+N$ or the internal network 172.22.0.N/16. Consequently, FUEL is trivial to extend with additional scenarios while offering file-system and process isolation of each web application to avoid interference, thus achieving the second goal.

Although this work focuses on PHP-based UFU vulnerabilities, in its core, FUEL is language agnostic as indicated by the dotted lines in Fig. 3. Thus, can be extended with scenarios targeting other programming languages, such as NodeJS. Future work can easily extend FUEL with little engineering efforts by adding new container definitions to the docker-compose.yml file, providing their build instructions in Dockerfiles and assigning unused ports.

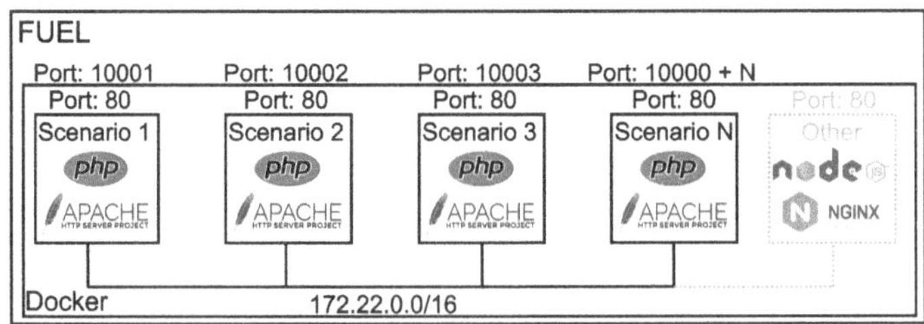

Fig. 3. Depiction of FUEL's design using multiple isolated containers.

3.2 UFU Scenarios

FUEL gives prominence to UFU vulnerabilities by implementing 15 distinct scenarios, which were distilled from related work, such as (†) Pooj et al. [31], (≀) OWASP [29] or (∗) Portswigger [33]. We selected scenarios that are applicable to PHP and primarily affect the web application (e.g., code execution) or its users (e.g. cross-site scripting). Scenarios targeting other programming languages or indirectly affecting other components or users, e.g. uploads of embedded or malicious files, were omitted. Each artificial scenario requires a specific exploitation or bypass technique to form a baseline for more complex vulnerabilities. These exploitation techniques can be found in many of the 1,175 exploits listed on a popular exploit database for the keyword *File Upload* [26]. Thus, these scenarios can approximate real-world vulnerabilities, whose exploitation might differ depending on the software's complexity or configuration.

All scenarios have in common that files are uploaded to `/uploads/`, a directory inside the web server's document root, to make the files accessible. While a brief description of each scenario follows, more detailed information is available in FUEL's repository (cf. Sect. 5.5).

- [†,∗]S1 *"Unrestricted"*: This scenario poses as a baseline since no validation is performed on the uploaded file, allowing the upload of any file, including `.php` files.
- [†]S2 *"Client-side"*: This scenario only implements client-side validation. By intercepting and manipulating the HTTP request, an attacker can circumvent the client-side validation to upload a PHP file.
- [†,≀,∗]S3 *"MIME-type"*: Only the uploaded file's MIME-type is validated server-side using the value provided in the `Content-Type` header. The intended bypass is uploading a malicious file and changing the header to a permissible value, e.g., uploading `exploit.php` with type `image/png`.
- [†,≀,∗]S4 *"Extension I"*: Checking for known dangerous file extensions, such as `.php`, is modeled in this scenario. Alternative extensions which are passed to the PHP interpreter, e.g., `.php5` or `.php8`, bypass this check.
- [†,≀,∗]S5 *"Extension II"*: Similar to the previous scenario, another bypass is changing the capitalization of individual file name's letters, e.g., `exploit.pHp`.
- [†,≀,∗]S6 *"Extension III"*: Misconfigured web servers might pass any file with `.php` in the file name, such as `exploit.php.png`, to the PHP interpreter, circumventing file extension checks.
- [†,≀,∗]S7 *"Extension IV"*: This scenario does not prevent the upload of configuration files, such as `.htaccess`, to change the web server's or PHP interpreter's behavior to execute malicious files.
- [≀,∗]S8 *"Extension V"*: Stripping prohibited extensions from file names is dangerous if not performed correctly, e.g., recursively. The intended bypass for this scenario is nesting the file extension, such as `.p.phphp` becoming `.php`.
- [≀,∗]S9 *"Magic number"*: Determining the file type from the first or last bytes of the file content is insufficient, as the file's content may contain executable

code. For example, GIF87a<?php phpinfo(); ?> uploaded as file.php might be detected as a image/gif file due to the correct file header.

≀,*S10 *"File content"*: Similar to the previous scenario, valid image files may contain PHP code in their metadata. Thus, checking the dimensions of an image file renamed to file.php succeeds, but so does the execution by the PHP interpreter when it encounters the PHP code.

*S11 *"Directory configuration"*: This scenario disables the PHP interpreter for the uploads/ folder. However, path traversal techniques may allow uploading in another location, such as /, where the PHP interpreter is not disabled.

†,≀,*S12 *"Special characters"*: Null bytes or other special characters may not be handled correctly by the application, allowing it to bypass validation checks. The intended bypass is placing a Null byte in the file name, e.g., file.php%00.png.

†S13 *"XSS"*: If file names are displayed to the user without escaping special characters, they may be used for XSS attacks, although the application processes no file content. This scenario models the upload of, e.g., <img src=x onerror=alert(1)>.png, which will trigger a JavaScript alert.

*S14 *"Race condition"*: External programs, such as antivirus scanners, might delete or block access to uploaded files if found to be malicious. Such an analysis usually incurs a delay, which might give attackers enough time to access and execute the file. The intended exploit is uploading a file and accessing it before the web server returns the HTTP response after deleting it.

*S15 *"PUT method"*: Web applications may use JavaScript or other technologies to dynamically perform HTTP requests to a backend server, including changes to the HTTP request method. This scenario models a situation in which an HTTP PUT request is used to create and update files on the server.

4 Evaluation

While the related work presented in Sect. 2.4 showed that UFU vulnerabilities were extensively studied over the recent years, continuously improving detecting techniques and discovering previously unknown vulnerabilities, the following question motivated the upcoming evaluation:

- RQ1: How capable are available UFU scanners against the file upload scenarios described in related work that we implemented in FUEL?

4.1 UFU Scanners

For the evaluation, we review (non-)academic work for UFU scanners and identify several that do not require access to the web application's source code, e.g., do not perform static analysis.

To our surprise, *FUSE* [20,21] was the only publicly available academic, black-box UFU scanner. For other publications implementing black-box vulnerability scanners, including *UChecker* [14], *UFuzzer* [15], *FileUploadChecker* [46] or *URadar* [4], we were unable to obtain source code. However, at the time of writing, the latter already had a GitHub repository[1] without content.

In contrast, we discovered two open-source and one commercial scanner to include in our evaluation. *Fuxploider* [17] is an open-source scanner specializing in flaws in file upload forms. The second open-source scanner is *ZAP* [43] in combination with the *FileUpload* plugin [35]. The commercial vulnerability scanner is *BurpSuite Professional* [32] with the *UploadScanner* plugin [27].

In general, all black-box UFU scanners follow a similar detection methodology. First, a new variation of a potentially malicious file or HTTP request is created and sent to the server. After the server responds, the scanner tries to access the uploaded file to identify successful XSS or CE exploits.

4.2 Methodology

As all scenarios in FUEL are isolated, we ran one UFU scanner against each scenario consecutively. Afterward, FUEL was reset to its initial state, and the next UFU scanner was tested. A timeout of 300 s was set for each scenario, after which the scanning process was interrupted to continue with the next scenario. Upon successful or forced termination, the scanner's output was analyzed for three aspects:

1. Was the scanner able to use the intended file upload bypass (iFUB)?
2. Was the scanner able to gain and report code execution (CE)?
3. Was the scanner able to gain and report cross-site scripting (XSS)?

Each vulnerability scanner requires its own specific configuration. *FUSE* and *Fuxploider* are command-line scanners while *ZAP* and *BurpSuite* provide a graphical interface. For example, the command-line scanners required a configuration file or a set of command-line arguments, e.g., specifying the upload endpoint and a detection criterion. In contrast, the graphical scanners required a full HTTP request and a detection criterion. We did two separate passes for the graphical scanners to initialize them with two different HTTP upload requests: A small image file (`fuel.png`) and a text file (`fuel.txt`) to avoid biases in the provided file format. The command-line scanners did not require any file or full HTTP request to run.

In general, we aimed to keep default settings, configure as little as possible, and provide as little manual assistance as needed to keep the results unbiased and reproducible. Naturally, we publish all details, such as configuration files, execution details, output logs, and software versions, to reproduce our results as part of our commitment to open science in our repository (cf. Sect. 5.5).

The evaluation was conducted on a single system with 64 GB of RAM, an 8-core/16-thread CPU (AMD Ryzen 7 4800H), and a PCIe NVMe SSD drive.

[1] https://github.com/Cycle183/URadar.

Table 1. Results from running each UFU scanner against FUEL's scenarios.

Scanner	**FUSE**			**Fuxploider**			**ZAP**			**BurpSuite**		
Type	iFUB	CE	XSS	iFUB	CE	XSS	iFUB	CE	XSS	iFUB	CE	XSS
Total	8	7	11	8	8	-	9	8	11	9	8	12
S1	✓	✓	✓	✓	✓	-	✓	✓	✓	✓	✓	✓
S2	✓	✓	✓	✓	✓	-	✓	✓	✓	✓	✓	✓
S3	✓	✓	✓	✓	✓	-	✓[1]	✓[1]	✓[1]	✓	✓	✓[1]
S4	✓	✓	✓	✓	✓	-	✓	✓	✓	✓	✓	✓
S5	✓	✓	✓	✓	✓	-	✓	✓	✓	×	×	✓
S6	×	×	✓	✓	✓	-	✓	✓	✓	✓	✓	✓
S7	✓	×	✓	✓	✓	-	×	×	✓	×	×	✓
S8	×	×	✓	×	×	-	×	×	✓	×	×	✓
S9	✓	✓	✓	×	×	-	✓	✓	×	✓	✓	✓
S10	✓	✓	✓	×	×	-	×	×	×	✓	✓	✓
S11	×	×	✓	×	×	-	×	×	✓	×	×	✓
S12	×	×	×	✓	✓	-	✓[1]	✓[1]	✓[1]	✓	✓	×
S13	×	-	×	-	-	-	✓	-	✓	✓	-	✓
S14	T	T	T	×	×	-	T	T	T	T	T	T
S15	×	×	×	×	×	-	×	×	×	×	×	×

✓: Found, ✓[1]: Only for `fuel.png`, ×: Not found, T: Timeout after 300 s

4.3 Results

All scanners required less than 300 s to finish their analysis, except in scenario 14. Table 1 shows the results of each UFU vulnerability scanner executed against FUEL's scenarios. Scenario 13 did not feature a way to achieve code execution, as indicated by -. Similarly, *Fuxploider* does not implement XSS detection techniques, leaving the column unmarked.

Cross-Site Scripting. The data shows that each scanner discovered more XSS than CE vulnerabilities. FUEL's current scenarios emphasize UFU vulnerabilities with a high impact, such as code execution, therefore not actively preventing the upload of files containing XSS payloads. Nonetheless, we include XSS for completeness in Table 1.

(Un)exploited Scenarios. Notably, all scanners successfully discovered the vulnerabilities in scenarios 1–4 if initialized with an image file. Otherwise, ZAP's FileUpload plugin failed to identify any vulnerability in scenario 3, and BurpSuite's UploadScanner did not recognize the XSS. Nonetheless, from scenario 5 on, the detection capabilities began to differ. No scanner achieved a successful exploit against the file upload mechanism to gain code execution in scenarios 8,

11, 14, and 15, which require file extension manipulation, path traversal, race condition, and request manipulation capabilities.

Scanner Capabilities. FUSE implements UFU exploitation techniques for 8 out of 15 scenarios, detecting a total of 7 CE and 11 XSS vulnerabilities. It is the only scanner not to exploit CE in scenario 6, while struggling with scenarios 8 and 10 through 15. Surprisingly, it detects the iFUB in scenario 7 but does not exploit the resulting CE.

Although *Fuxploider* lacks XSS capabilities, it always gains CE after identifying a UFU, which it achieves in 8 different scenarios. It did not identify the iFUB that requires file header or content manipulation, race condition, or path traversal capabilities. However, it is the only scanner to successfully exploit the CE in scenario 7.

The UFU vulnerability scanning plugin *FileUpload* for ZAP detects 9 of 15 iFUBs and exploits 8 out of 14 possible CE vulnerabilities but fails to identify the vulnerabilities in scenarios 7, 8, 10, and 11. However, it only achieves these comparatively good results when initialized with an image file. If it is initialized with a text file, it fails to find the vulnerabilities in scenarios 3 and 12.

While BurpSuite Professional's *UploadScanner* plugin discovers the most XSS vulnerabilities and 9 out of 15 iFUBs, it is the only scanner not to change the capitalization of a file's extension (scenario 5). Moreover, it fails to bypass the checks in scenarios 7, 8, and 11, which require uploading configuration files, nesting file extensions, or performing path traversal. Similar to the FileUpload plugin, the UploadScanner's results change if initialized with a non-image file, although it only affects the XSS detection in scenario 3.

4.4 Evaluating Our Fuxploider-NG

With the help of FUEL, we have identified several shortcomings in existing UFU scanners, as no scanner has the capabilities necessary to exploit all 15 scenarios. To demonstrate FUEL's effectiveness in improving existing work, we develop *Fuxploider-NG* by extending Fuxploider with additional UFU exploitation and XSS detection capabilities. In particular, we implemented placing PHP code after magic bytes or inside an image's metadata and using nested file extensions, path traversal, or race condition techniques. Further, we extended the existing UFU detection rule set and file samples, and implemented XSS detection based on matching a response's content against the set of predefined XSS payloads.

Table 2 shows the results from testing *Fuxploider-NG* with its improvements and updated configuration against FUEL. The results indicate that it successfully discovers all iFUBs, CE, and XSS vulnerabilities in scenarios 1 through 14. Only the last scenario could not be improved as it would require major changes to *Fuxploider*'s code base (e.g. how it issues HTTP requests) and is therefore considered future work.

Table 2. Our improved *Fuxploider-NG* scanner successfully exploits scenarios 1–14.

	S1	S2	S3	S4	S5	S6	S7	S8	S9	S10	S11	S12	S13	S14	S15
iFUB	✓	✓	✓	✓	✓	✓	✓	✓	✓	✓	✓	✓	✓	✓	×
CE	✓	✓	✓	✓	✓	✓	✓	✓	✓	✓	✓	✓	-	✓	×
XSS	✓	✓	✓	✓	✓	✓	✓	✓	✓	✓	✓	✓	✓	✓	×

5 Discussion

This section answers the research question RQ1 by interpreting the data from Table 1 and discussing the evaluated scanners' shortcomings. Then, it discusses the results from Fuxploider-NG and the importance of our FUEL framework.

5.1 Shortcomings of UFU Vulnerability Scanners

Albeit all scanners exploited the first two UFU scenarios, the results show that the UFU detection capabilities are inhomogeneous, as no scanner managed to exploit all of FUEL's scenarios. Exploitation of the first two scenarios by all scanners is unsurprising since no server-side validation is performed, and any malicious file can be uploaded.

To our surprise, ZAP's and BurpSuite's plugins have difficulties finding the vulnerabilities in scenario 3 if initialized with a text file but find them when initialized with an image file. One reason for ZAP's difference is assumed to be the missing manipulation of the MIME-type in the HTTP request.

Nonetheless, all scanners know to check for alternative, executable PHP extensions (scenario 4) and, except for Burp's plugin, change the extension's capitalization (scenario 5). Reuse of scenario 4's exploit in scenario 5 is apparent but impossible due to scenario 5's web server configuration. Although file extension bypasses appear well studied and implemented, FUSE is the only scanner not to exploit scenario 6, which requires an infix `.php` extension.

On the contrary, Fuxploider is the only scanner to exploit scenario 7 successfully by uploading an `.htaccess` configuration file to gain code execution. Although FUSE detects the possibility of uploading such files, it does not actively exploit it. Scenario 8 removes the string `.php` from the file's name to model an insecure attempt to mitigate dangerous extensions. According to the results, either no scanner has implemented such file name variations, or the exploit validation fails by trying to access the original file name but not the modified one.

Scenario 9 required a file with a valid image file header (e.g., `GIF87a` followed by PHP code) to bypass the upload filter and gain CE. Except for Fuxploider, all scanners successfully performed this iFUB. A more advanced version of this exploit is required for scenario 10, as the file's content must be valid image data, and the malicious PHP code must be placed in the image's metadata. This check tripped ZAP's plugin, so only FUSE and Burp's UploadScanner successfully gained CE.

No scanner successfully uses exploitation methods that rely on path traversal by manipulating the file name, as required for scenario 11. During this evaluation, all scanners knew that uploaded files would be stored in `/uploads/`, but this might not be the case for other web applications; thus, rediscovering the uploaded file becomes a difficult challenge. Even more so if the new file path leaves the document root or the accessible directory. Thus, such an exploitation method might have little chance of success but should be implemented anyhow, as demonstrated in Fuxploider-NG.

Except for FUSE, all scanners use special characters to gain code execution in scenario 12. Interestingly, ZAP's plugin only successfully exploits this scenario if initialized with an image file. We assume that while the plugin inserts special characters into the file name, it only manipulates the provided file name from the initialized HTTP request and thus does not change `fuel.txt` to `fuel.txt%00.png`, which would exploit this scenario. In FUSE's case, its authors might have assumed this bypass to be obsolete, as FUSE was published in 2020 and PHP handles Null bytes correctly since version 5.3 (released in 2011 [41]). Nonetheless, we included this scenario in FUEL and Fuxploider-NG for completeness and to match the capabilities of BurpSuite's plugin.

Scenario 13 does not process or store the file, except for displaying its name without escaping, making it susceptible to XSS. While this is not directly a UFU, it is related and may become a severe security issue, even if all other file properties are correctly validated. Still, only 50% of the scanners are capable of detecting this issue, while FUSE and Fuxploider lack the capabilities.

Scenario 14's validation implements a processing delay of 0.5 s and deletes the file before returning the HTTP response, which confuses all scanners. One possible explanation is strictly following the pattern of first uploading, then verifying, and waiting for a successful HTTP response to the upload request before attempting to access the file. In Fuxploider-NG, we implemented additional access attempts right after the upload without waiting for the HTTP response and thus successfully exploited this scenario.

Although we expected the scanners not to discover the iFUB in scenario 15, we still decided to include it for completeness since it was mentioned in related work. An explanation for the results is that the PUT upload requests are sent using JavaScript, and no scanner features functionality to successfully extract the request parameters or does not implement mutations of the HTTP method and request URL. Therefore, a longer timeout would not have changed the results.

5.2 Comparison to Fuxploider-NG

Due to the various shortcomings, it is not possible to determine the best-performing UFU scanner, as the capabilities differ greatly. Even if the unexploited scenarios 8,11,14, and 15 are not considered, each scanner fails at exploiting at least two scenarios. Consequently, one would need to use at least two scanners to evaluate a web application for a set cover of the exploited UFU scenarios. The three possible minimal combinations to cover scenarios 1–7, 9–10, and 12–13 are FUSE and ZAP's plugin initialized with an image, FUSE and BurpSuite's plugin, Fuxploider and BurpSuite's plugin.

As BurpSuite is a commercial tool with a subscription model costing several hundred Euros per year, FUSE and ZAP are the only free-to-use combination. However, in that case, the user must be aware that the ZAP plugin's accuracy depends on the initialization of the HTTP request, which is not documented in its repository. Furthermore, the need to use at least two scanners increases the friction due to the configurations' complexity and multiplies the testing duration, as most tests will be performed twice.

With Fuxploider-NG we contribute a single scanner that covers 14 out of 15 scenarios currently present in FUEL and successfully exploits them, thus reducing the complexity of testing a web application for UFU vulnerabilities.

5.3 Importance of FUEL

Using FUEL, we have shown that the capabilities of academic, open-source, and commercial vulnerability scanners are limited. Each vulnerability scanner has a different set of UFU exploitation techniques, thus leaving room for improvement to detect the 15 types of UFU vulnerabilities. This result is quite surprising, as related work has discussed these vulnerability scenarios since 2016, reaffirming the need for a standardized evaluation framework such as FUEL.

Consequently, using any of these scanners against real-world applications to assess their security entails the risk of false negatives, thus leaving its users with a false sense of security and web applications with a potential, undiscovered entry point for malicious actors.

A collection of tests to ensure new code does not break existing functionality is a core concept of test-driven development (TDD) [8]. Parallels can be drawn between a test suite of TDD and the set of UFU scenarios in FUEL, thus making FUEL an important framework in ensuring future work in the UFU domain covers all previously documented vulnerable scenarios in its goal to discover novel techniques.

Furthermore, FUEL's modularity embraces adding new scenarios, allowing the modeling of other UFU vulnerabilities. Finally, FUEL facilitates a fair, reliable, and reproducible way to evaluate improvements against related work.

5.4 Limitations and Future Work

For now, FUEL's scenarios solely focus on UFU vulnerabilities in PHP. While many websites use PHP, other languages are suitable for web development and potentially vulnerable to UFU, too, although the exploitation methods might differ. For example, not invoking an interpreter for each request might not directly execute malicious code inside a file. Since related work and the examined UFU scanners primarily focus on PHP, it naturally followed to implement FUEL's scenarios in PHP. However, future work can implement scenarios in other languages, as FUEL is language agnostic.

Each of the 15 scenarios was deliberately kept simple by modeling exactly one validation check. While this approach allows to efficiently evaluate UFU scanners, real-world applications are likely to perform multiple checks. However, FUEL is not limited to simple scenarios, as future work could implement scenarios based on real-world applications or combinations of existing scenarios, to evaluate the vulnerability scanners' performance against real-world UFU vulnerabilities.

Although we noticed variations in the scanners' performance, e.g., the number of requests or execution time, an analysis was considered future work for the following reasons. First, some scanners did not report these metrics or showed inconsistent runtime behavior. Second, it raises new questions, such as the trade-off between a faster scanner with less accuracy and a slower but more thorough and accurate scanning technique. Furthermore, some scanners terminated after the first finding, while others continued to examine the application.

Finally, it could be interesting to extend Fuxploider-NG to also detect more complex UFU vulnerabilities, e.g. combinations of the scenarios, which should be achievable with some engineering effort.

5.5 Open Science

It is unfortunate that several UFU scanners from related work were unavailable for this evaluation, thus limiting it to only four scanners. As the authors share strong beliefs in open and reproducible science, we will open-source FUEL[2], Fuxploider-NG[3], and all results related to our evaluation[4] on GitHub. Notably, we encourage and welcome contributions to FUEL, especially new scenarios, to make it a comprehensive evaluation suite for UFUs.

6 Conclusion

Our review of related work revealed a lack of a standardized evaluation methodology for UFU vulnerability scanners, which could be one possible reason for the divergence in detection capabilities of the evaluated academic, open source, and commercial vulnerability scanners. Thus, our first contribution is a novel framework, which we call FUEL, consisting of 15 distinct file upload scenarios distilled from related work.

The framework offers a standardized environment to facilitate a comprehensive and reproducible evaluation of existing vulnerability scanners, new UFU detection methods, or new vulnerable file-handling scenarios. Furthermore, FUEL's design goals were simplicity, extensibility, and reproducibility to make this framework suitable for future work in the unrestricted file upload vulnerability domain.

---

[2] https://github.com/FUEL-Project/FUEL-FileUploadExploitationLab.

[3] https://github.com/FUEL-Project/fuxploider-NG.

[4] https://github.com/FUEL-Project/FUEL-Evaluation.

With the help of FUEL, our results show that none of the evaluated black-box UFU scanners is capable of detecting all previously documented UFU scenarios. Each scanner has individual shortcomings, requiring users to test a web application with at least two scanners for a comprehensive analysis or leaving them with a false sense of security.

We propose *Fuxploider-NG*, which consolidates all UFU vulnerability detection methods into a single program, thus closing the capability gap and reducing the risk of false negatives when testing real-world applications for UFU vulnerabilities.

Finally, by open-sourcing FUEL and Fuxploider-NG, we aim to improve existing vulnerability scanners and lay the groundwork for more research to increase the number of detected UFU vulnerabilities in web applications.

References

1. AUTOMATTIC: Wordpress vulnerability statistics | WPScan. https://wpscan.com/statistics/. Accessed 31 Jan 2024
2. Barth, A., Caballero, J., Song, D.: Secure content sniffing for web browsers, or how to stop papers from reviewing themselves. In: 2009 30th IEEE Symposium on Security and Privacy, pp. 360–371. IEEE (2009)
3. Barua, A., Shahriar, H., Zulkernine, M.: Server side detection of content sniffing attacks. In: 2011 IEEE 22nd International Symposium on Software Reliability Engineering, pp. 20–29. IEEE (2011)
4. Chen, Y., Li, Y., Pan, Z., Lu, Y., Chen, J., Ji, S.: URadar: discovering unrestricted file upload vulnerabilities via adaptive dynamic testing. IEEE Trans. Inf. Forensics Secur. **19**, 1251–1266 (2024). https://doi.org/10.1109/TIFS.2023.3335885
5. Dahse, J., Holz, T.: Static detection of {Second-Order} vulnerabilities in web applications. In: 23rd USENIX Security Symposium (USENIX Security 2014), pp. 989–1003 (2014)
6. DataReportal, Meltware, Social, W.A.: Number of internet and social media users worldwide as of October 2023 (in billions). https://www.statista.com/statistics/617136/digital-population-worldwide/. Accessed 09 Jan 2024
7. De Meo, F., Viganò, L.: A formal approach to exploiting multi-stage attacks based on file-system vulnerabilities of web applications. In: Bodden, E., Payer, M., Athanasopoulos, E. (eds.) ESSoS 2017. LNCS, vol. 10379, pp. 196–212. Springer, Cham (2017). https://doi.org/10.1007/978-3-319-62105-0_13
8. Desai, C., Janzen, D., Savage, K.: A survey of evidence for test-driven development in academia. SIGCSE Bull. **40**(2), 97–101 (2008). https://doi.org/10.1145/1383602.1383644
9. Docker Inc.: Docker: Accelerated container application development. https://www.docker.com/. Accessed 01 Feb 2024
10. Doupé, A., Cova, M., Vigna, G.: Why Johnny can't Pentest: an analysis of black-box web vulnerability scanners. In: Kreibich, C., Jahnke, M. (eds.) DIMVA 2010. LNCS, vol. 6201, pp. 111–131. Springer, Heidelberg (2010). https://doi.org/10.1007/978-3-642-14215-4_7
11. F5 Inc.: PHP FastCGI example | NGINX. https://www.nginx.com/resources/wiki/start/topics/examples/phpfcgi/. Accessed 31 Jan 2024

12. Freed, N., Borenstein, N.S.: Multipurpose internet mail extensions (mime) part one: format of internet message bodies. RFC 2045, RFC Editor, November 1996
13. Greenbergand, Y., Ermishkin, N.: Imagetragick. https://imagetragick.com/. Accessed 07 Feb 2024
14. Huang, J., Li, Y., Zhang, J., Dai, R.: UChecker: automatically detecting PHP-based unrestricted file upload vulnerabilities. In: 2019 49th Annual IEEE/IFIP International Conference on Dependable Systems and Networks (DSN), pp. 581–592. IEEE (2019)
15. Huang, J., Zhang, J., Liu, J., Li, C., Dai, R.: UFuzzer: lightweight detection of PHP-based unrestricted file upload vulnerabilities via static-fuzzing co-analysis. In: Proceedings of the 24th International Symposium on Research in Attacks, Intrusions and Defenses, pp. 78–90 (2021)
16. Jacksi, K., Abass, S.M.: Development history of the world wide web. Int. J. Sci. Technol. Res. **8**(9), 75–79 (2019)
17. Jarry, V.: almandin/fuxploider: file upload vulnerability scanner and exploitation tool. https://github.com/almandin/fuxploider. Accessed 02 Feb 2024
18. Kimminich, B.: OWASP juice shop. https://owasp.org/www-project-juice-shop/. Accessed 31 Jan 2024
19. Kruisselbrink, M., Shmyroff, V.: File API. https://www.w3.org/TR/file-upload/. Accessed 07 Feb 2024
20. Lee, S.: WSP-LAB/FUSE: A penetration testing tool for finding file upload bugs. https://github.com/WSP-LAB/FUSE. Accessed 02 Feb 2024
21. Lee, T., Wi, S., Lee, S., Son, S.: Fuse: finding file upload bugs via penetration testing. In: NDSS (2020)
22. Masinter, L.: Returning values from forms: multipart/form-data. RFC 7578, RFC Editor, July 2015
23. MITRE: CVE - search results. https://cve.mitre.org/cgi-bin/cvekey.cgi?keyword=file+upload. Accessed 29 Jan 2024
24. Nebel, E., Masinter, L.: Form-based file upload in html. RFC 1867, RFC Editor, November 1995
25. OffSec Services Limited: Exploit database - exploits for penetration testers, researchers, and ethical hackers. https://www.exploit-db.com/?type=webapps. Accessed 29 Jan 2024
26. OffSec Services Limited: Exploit database - exploits for penetration testers, researchers, and ethical hackers. https://www.exploit-db.com/. Accessed 26 Apr 2024
27. Ospelt, T.: Upload scanner - PortSwigger. https://portswigger.net/bappstore/b2244cbb6953442cb3c82fa0a0d908fa. Accessed 02 Feb 2024
28. OWASP Foundation: OWASP Top Ten. https://owasp.org/www-project-top-ten/, Accessed 09 Jan 2024
29. OWASP Foundation: Unrestricted File Upload. https://owasp.org/www-community/vulnerabilities/Unrestricted_File_Upload. Accessed 01 Feb 2024
30. Polydet: Polydet/polyglot database: Database of polyglot files. By polyglot, we mean files readable in multiple formats. https://github.com/Polydet/polyglot-database. Accessed 31 Jan 2024
31. Pooj, K., Patil, S.: Understanding file upload security for web applications. Int. J. Eng. Trends Technol. **42**(7), 342–347 (2016)
32. Portswigger Ltd.: Burp Suite Professional. https://portswigger.net/burp/pro. Accessed 29 Jan 2024
33. Portswigger Ltd.: File uploads | Web Security Academy. https://portswigger.net/web-security/file-upload. Accessed 1 Feb 2024

34. Riadi, I., Aristianto, E.I.: An analysis of vulnerability web against attack unrestricted image file upload. Comput. Eng. Appl. J. **5**(1), 19–28 (2016)
35. Sasan Labs: Sasanlabs/owasp-zap-fileupload-addon: OWASP ZAP add-on for finding vulnerabilities in file upload functionality. https://github.com/SasanLabs/owasp-zap-fileupload-addon. Accessed 02 Feb 2024
36. The Apache Software Foundation: Apache HTTP server tutorial: .htaccess files. https://httpd.apache.org/docs/2.4/howto/htaccess.html. Accessed 20 Feb 2024
37. The Apache Software Foundation: Apache HTTP modules. https://httpd.apache.org/modules/. Accessed 31 Jan 2024
38. The PHP Group: PHP: Description of core php.ini directives. https://www.php.net/manual/en/ini.core.php#ini.sect.file-uploads. Accessed 31 Jan 2024
39. The PHP Group: PHP: $_FILES. https://www.php.net/manual/en/reserved.variables.files.php. Accessed 31 Jan 2024
40. The PHP Group: PHP: History of PHP. https://www.php.net/manual/en/history.php.php. Accessed 31 Jan 2024
41. The PHP Group: PHP: Null bytes related issues. https://www.php.net/manual/en/security.filesystem.nullbytes.php. Accessed 04 Feb 2024
42. The PHP Group: PHP: POST method uploads. https://www.php.net/manual/en/features.file-upload.post-method.php. Accessed 31 Jan 2024
43. The Software Security Project: ZAP. https://www.zaproxy.org/. Accessed 02 Feb 2024
44. W3Techs Web Technology Surveys: Historical trends in the usage statistics of server-side programming languages for websites, January 2024. https://w3techs.com/technologies/history_overview/programming_language. Accessed 31 Jan 2024
45. W3Techs Web Technology Surveys: Usage statistics and market share of PHP for websites, January 2024. https://w3techs.com/technologies/details/pl-php. Accessed 31 Jan 2024
46. Wichmann, P., Groddeck, A., Federrath, H.: Fileuploadchecker: detecting and sanitizing malicious file uploads in web applications at the request level. In: Proceedings of the 17th International Conference on Availability, Reliability and Security, pp. 1–10 (2022)
47. Wood, R.: digininja/DVWA: Damn vulnerable web application (DVWA). https://github.com/digininja/DVWA. Accessed 31 Jan 2024
48. Wordpress.org: Wordpress and the journey to 40% of the web. https://wordpress.org/40-percent-of-web/. Accessed 31 Jan 2024
49. World Wide Web Constortium: input type=file - file upload control - HTML5. https://www.w3.org/TR/2012/WD-html-markup-20120329/input.file.html. Accessed 07 Feb 2024
50. Yenduri, R., Al-khassaweneh, M.: PHP: vulnerabilities and solutions. In: 2022 2nd International Mobile, Intelligent, and Ubiquitous Computing Conference (MIUCC), pp. 391–396. IEEE (2022)

SandPuppy: Deep-State Fuzzing Guided by Automatic Detection of State-Representative Variables

Vivin Paliath(✉), Erik Trickel, Tiffany Bao, Ruoyu Wang, Adam Doupé, and Yan Shoshitaishvili

Arizona State University, Tempe, USA
{vivin,etrickel,tbao,fishw,doupe,yans}@asu.edu

Abstract. Current state-of-the-art automated fuzzing approaches cannot explore deep program-states without human assistance. Recently, IJON allowed humans to provide code-annotations on the target to expose program state to the fuzzer. However, this requires a human to read, understand, and annotate the program source, which limits scalability and applicability.

In this paper we introduce SANDPUPPY, a technique that automatically identifies potential state-representative variables and applies IJON-style instrumentation to expose corresponding state to the fuzzer. To identify these variables and their semantics, SANDPUPPY collects runtime variable-value traces from an initial fuzzing run and analyzes them along with the program source to instrument the program and expose internal state to the fuzzer. This process repeats and identifies additional variables, allowing the exploration of deeper states.

We evaluated SANDPUPPY against synthetic and real-world targets representing various fuzzing-challenges. The results show that SANDPUPPY can automatically solve problems that purely coverage-based approaches cannot solve without assistance from human experts. For example, SANDPUPPY can automatically play and solve levels of *Super Mario Bros*, maze programs, and identify complex, deep states in real-world targets such as LIBTPMS, resulting from combinations of previously identified states. Evaluating SANDPUPPY on real-world targets such as JSONCPP, LIBTPMS, PCAPPLUSPLUS, and READELF demonstrates that SANDPUPPY generally improves coverage compared to AFL, AFL++, LAFINTEL, REDQUEEN, SGFUZZ, and FERRY. In all, SANDPUPPY identified ten unknown vulnerabilities in PCAPPLUSPLUS and one unknown vulnerability in DMG2IMG.

1 Introduction

In recent years, fuzz testing, a dynamic-analysis technique that attempts to trigger bugs with pseudo-random input, demonstrated its effectiveness by uncovering a large number of vulnerabilities in diverse software. Hence, fuzzing received significant interest in both academia [3,4,10,32,35,39,45] and industry [1,7,22,46].

Modern fuzzing research has predominantly focused on maximizing *code coverage* — such as basic block coverage and branch/edge coverage — based on the

F. Maggi et al. (Eds.): DIMVA 2024, LNCS 14828, pp. 227–250, 2024.
https://doi.org/10.1007/978-3-031-64171-8_12

notion that covering more code triggers more vulnerabilities. This is usually done by satisfying previously-failing program branch conditions, whether by solving them using symbolic techniques [39,45], identifying specific parts of program input that should be stochastically mutated to increase the chance of satisfying conditions [4,10], or modifying the conditions themselves [32]. As a result of these interventions, code coverage is increased and additional bugs are found.

Interestingly, increased code coverage is not always correlated with more discovered vulnerabilities: Shoshitaishvili et al. showed that fuzzers could use human-assisted input mutation to find more vulnerabilities—*while triggering strictly less code*—than standard (automated) code-coverage guided fuzzers [38]. One explanation is that not all code coverage is equal [41], and coverage alone does not imply good exploration of a program's *state space.* IJON furthered this concept using manual source-code annotations to extend traditional code coverage feedback with *state-representative variables* [3]. The resulting *state-coverage guided fuzzer* showed promise in—with human help—triggering program states that standard code-coverage guided fuzzers cannot reach.

The next clear step in exploring the impact of state coverage on analysis outcomes is *automatic* reasoning about state-representative variables. Recent work in this area has made progress toward detecting state-representative variables [5,50], but either use them purely to increase code coverage (rather than *state* coverage) [50] or is only applicable to specific types [5].

In this paper, we propose SANDPUPPY, the first technique to automate the process of identifying and using state-representative variables in fuzzing to increase state coverage, and thus detecting *state-dependent* bugs.

In a reductive sense, every variable used in a program represents some sort of state. However, this does not necessarily imply that variables are equally representative to explore program state-space. Annotating too many variables overwhelms the fuzzer, as it explores unproductive execution paths due to the generation of useless inputs [3]. Therefore, a core challenge for SANDPUPPY then, is to select a minimal subset of state-representing variables from the set of all variables used by a program.

Our solution is inspired by recent observations that a significant number of program states are encoded in a subset of *integer* variables in a program [3,5,50]. Building on this, SANDPUPPY identifies candidate integer-variables in a target program and then uses simple, yet novel heuristics, on dynamic runtime value-traces aggregated across multiple executions, to gain a coarse understanding of their semantics. For such variables, SANDPUPPY uses one of several annotation methods (depending on the identified variable semantics) to add these variables to a fuzzer's coverage feedback. SANDPUPPY enables state-coverage guided fuzzing without human effort and without overwhelming the fuzzer with spurious feedback.

Contributions. In summary, our contributions are as follows:

- We design an approach for detecting state-representative variables in a program and for exposing associated state to guide fuzzing towards deep, complex states.
- We implement this approach into a research prototype called SANDPUPPY.

- We evaluate SANDPUPPY against custom targets representing specific fuzzing-challenges as well as real-world programs. SANDPUPPY not only identifies interesting state-representative variables, but the exposed state allows it to explore a deeper and more diverse set of program states and solve certain challenges that purely coverage-based fuzzing techniques cannot. SANDPUPPY also identifies existing and novel vulnerabilities and improves coverage, compared to existing tools.

To support open science, we will open source our research artifacts, including data sets, seeds, and the SANDPUPPY prototype.

Ethics. We reported all identified vulnerabilities (11 in PCAPPLUSPLUS and 1 in DMG2IMG). The PCAPPLUSPLUS developers agreed with us and have fixed all vulnerabilities. The DMG2IMG developers did not respond to our disclosure, and we are continuing to contact them.

2 Background

Fuzzers aim to generate inputs exploring novel program states. Intuitively, the more states explored, the greater the chance to find vulnerabilities. However, lacking a way to directly measure *state*, most fuzzers tend use *code coverage* as a proxy. Coverage-based fuzzers, such as AFL, use a code-coverage metric which approximates the program's state, to identify inputs triggering novel program states. AFL places instrumentation at the edges of the program's control flow graph (CFG) in order to track edge-to-edge transitions. The instrumentation identifies transitions using tuples of identifiers (id_s, id_d), where id_s is the source-block identifier and id_d is the destination-block identifier.

AFL approximates the execution state of a program with a *bitmap.* Roughly speaking, when the program executes a transition, AFL hashes the tuple (id_s, id_d) to an index in the bitmap and increments the counter at that index. AFL also uses a *global bitmap* to store the history of all bitmaps for the current fuzzing session. Monitoring for changes in the global bitmaps allows AFL to determine whether the input used for the current test is "interesting" (i.e., if it triggered previously-unseen program transitions and resulted in new coverage).

Directly measuring program state-space coverage during fuzzing is difficult, and modern coverage-guided fuzzers use code coverage as an approximation of program state-space coverage. However purely coverage-guided fuzzers cannot effectively explore the state space of a program [3], as *code coverage does not always correlate to program-state coverage.* IJON [3] addresses these issues by augmenting the coverage bitmap with values of human-annotated state-representative variables to *increase the correlation between bitmap coverage and state space coverage*: For example, consider a maze program where each valid input character permits a move direction, as shown in Fig. 1. Although the user has many possible positions and paths, a coverage-guided fuzzer that only considers code or branch coverage will struggle to explore them. IJON exposes x and y to the fuzzer, enabling the fuzzer to recognize inputs leading to new locations in the maze as interesting, essentially creating a *player-location-based coverage bitmap* (the bottom of Fig. 1). This results in a better exploration of the state space and a solution to the maze.

IJON requires human annotation of state-representative variables, which does not scale. In this paper, we explore techniques to *automatically* infer and use state-representative variables to improve fuzzing efficacy.

3 Overview

We propose SANDPUPPY, a technique that automatically identifies and instruments state-representative variables, thus exposing program state to the fuzzer.

Figure 2 provides an overview of SANDPUPPY's fuzzing pipeline. Given buildable program source, SANDPUPPY carries out the following steps iteratively, identifying and selecting new state-representative variables in each iteration:

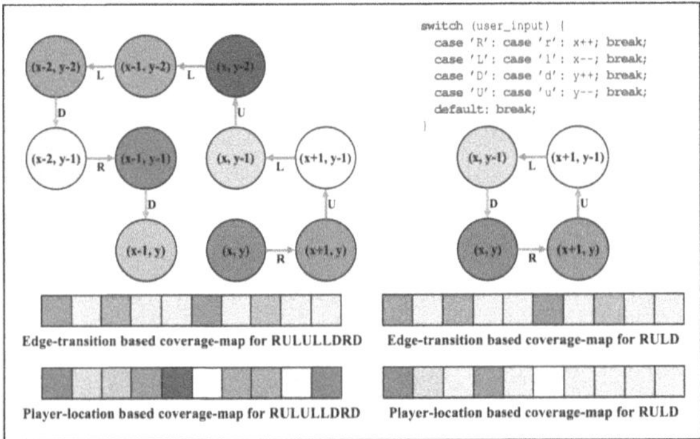

Fig. 1. Comparison of edge-transition based and player-location based coverage-maps for a maze program. Edge-transition cannot identify the distinct states resulting from different player locations whereas player-location can.

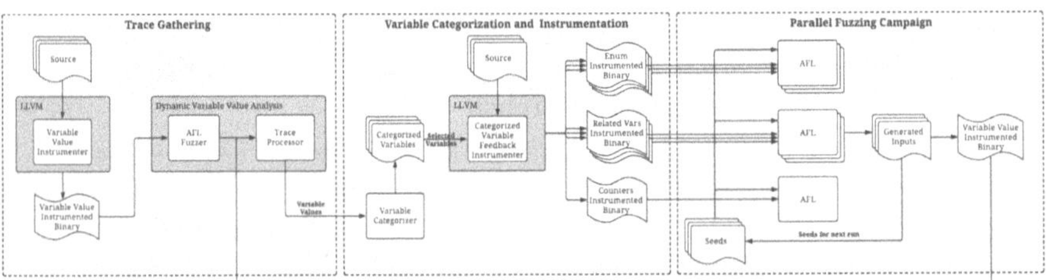

Fig. 2. SandPuppy's architecture is comprised of three phases: a trace-gathering phase that gathers variable value traces from an instrumented binary, a categorization and instrumentation phase that categorizes variables based on value traces and performs feedback instrumentation using selected variables, and a parallel-fuzzing phase where the instrumented binaries are fuzzed together. Inputs generated by the parallel-fuzzing campaign are used to generate additional traces and as seeds for subsequent runs.

Collect Variable-Value Traces: SANDPUPPY instruments the target program to output integer-variable value traces. It then fuzzes this instrumented binary using a standard fuzzer for a fixed amount of time, using its accumulated set of test-cases (seeds). SANDPUPPY uses the trace results to select a tractable subset of state-representative variables, as described in Sect. 4.

Identify State-Representative Variables: SANDPUPPY analyzes the collected traces to identify potential state-representative variables and their semantic categorization (Sect. 5) using a set of heuristics (defined in Sect. 6 and applied in Sect. 7). Traces from inputs that result in crashes or errors are excluded to minimize noise by ignoring invalid states.

Fuzz: SANDPUPPY generates instrumented binaries that collect feedback from corresponding state-representative variables, as detailed in Sect. 8. It then runs a parallel-fuzzing session using this set of binaries. Note that generated inputs are periodically synchronized between all fuzzing instances. While it is possible to generate a single binary for all identified variables, this significantly increases the likelihood of collisions with a single bitmap.

Repeat: The inputs generated by the parallel-fuzzing session are used as seeds for the next run's variable-value trace collection and fuzzing.

4 Trace Collection

SANDPUPPY classifies variables by observing their use by the program. It uses a trace collecting step to record the complete history of values that each integer variable receives throughout the instrumented program's execution.

SANDPUPPY needs a variety of traces. Consider a variable counting up to a number provided by input: if it is always zero, we have a single trace of the variable, which is indistinguishable from a trace of a constant variable set to zero and never modified.

For the first iteration of SANDPUPPY, we leverage an initial fuzzing phase to generate traces on multiple inputs, mitigating the influence of invalid inputs on later variable classification by discarding crashing and erroring traces. On subsequent iterations, we directly use inputs identified previously to generate traces. In this manner we can detect, categorize, and instrument any newly-detected variables. We also filter out the traces of any variable whose category has not changed since the last run, assuming that resources would be better spent on tracing and classifying still-changing variables.

For the variable v, we define the *value trace* T as a sequence of tuples (m, ν) where m is the source line-number where v was modified and ν is the current value of v. We also define $\mathbf{T}$ to be a sample of all traces collected for that variable from our fuzzing session; i.e., assuming the program was run against n inputs $\mathbf{T} = (T_1, T_2, \ldots, T_n)$.

```
void process_input(const char* input, state* state) {
  parsed_messages pmsgs = parse_input(input);
  for (int i = 0; i < pmsgs.num_messages; i++) {
    msg_type mtype = process_msg(pmsgs.messages[i], state);
    char* mtype_str = get_message_type(msg_str);
    if (strcmp(mtype_str, "type_1") == 0) {
        mtype = TYPE_1;
        process_type_1_msg(msg_str, state);
    } else if (strcmp(mtype_str, "type_2") == 0) {
        mtype = TYPE_2;
        process_type_2_msg(msg_str, state);
        ...
```

Listing 1.1. Enum variables exposing known state changes.

```
unsigned int processed_messages = 0;
void process_message(message* msg) {
  for (int i = 0; i < get_num_elements(msg); i++) {
    process_element(msg, i);
  }
  processed_messages++;
}
```

Listing 1.2. Counter variables exposing memory-related state.

5 Semantic Categories

SANDPUPPY instruments *state-representative* variables to approximate program states at runtime. Different variables contribute differently to this approximation. Here, we describe categories of state-representative variables and our intuition of how they represent the application's state space. As we are not attempting to devise a general and exhaustive taxonomy of "state-representative variables", we focus on specific categories that see near-ubiquitous use across a wide variety of programs.

Enums. *Enumerated types* represent a fixed set of named values. While internal representation is arbitrary, they typically compile to integer values. Enums are used to enhance readability and leverage the type system to prevent variables from being set to invalid values. They can also be correlated with input, especially those involving structured formats.

Format processing code often feature variables encoding state associated with user input, which are attractive instrumentation targets. Listing 1.1 shows a message-processing function, where different message combinations can affect program state in unique ways. Coverage-based fuzzers identify values of `mtype`, but may not recognize combinations of messages as interesting states as they trigger already-seen control-flow edges. Conversely, fuzzers that can reason about the value of `mtype` directly may be able to trigger additional program states. Similar code can be found in a wide variety of real-world software, such as FFMPEG, JSONCPP, LIBJPEG, and LIBPCAP.

Counters. Counters are variables that count up or down to some value. We assume that counters are monotonically increasing or decreasing, and that they may do so by arbitrary (and varying) amounts. Counters can either be *static*:

counting to a fixed value in every program execution, or *dynamic*: counting to different values in every program execution. Counters may also correlate with input as programs use them to count input elements or to sum input-element sizes. By guiding fuzzers towards maximum counter values, states revealing boundary and edge-case conditions can be explored.

Listing 1.2 shows an example of counter usage. By exposing the counter values `i` and `processed_messages` to the fuzzer and favoring executions that maximize their values, the fuzzer will explore program states involving both a larger number of messages and a larger number of elements per message.

Related Variables. Programs often feature *related variables* whose *combined* values represents a unique state. For example, variables x and y used to represent a player's location in a maze are related variables. By identifying related variables and exposing their combined state, fuzzers gain insight into a program's state space.

6 Semantic Category Heuristics

We use several heuristics on traces collected in Sect. 4 to classify variables as state-representative and sort them into categories that are described in Sect. 5.

Non-Semantic Categories. We start by defining heuristics to identify constants and booleans; two categories of variables that are not state-representative. Constants do not expose any useful state, as they do not change values. While booleans can expose useful state, coverage-based fuzzing is usually sufficient to explore these states as booleans often dictate control flow.

First, we define the observed set of values of the variable v as $\mathcal{V} = \bigcup_{T \in \mathbf{T}} \{\nu \mid (m, \nu) \in T\}$. We will also define the *number of unique values* $h_{nuniq} = |\mathcal{V}|$. If $h_{nuniq} = 1$, then it is highly probable that the variable is a constant. However, if $h_{nuniq} = 2$ and $\mathcal{V} = \{0, 1\}$, then it is likely a boolean. There is one surprising case of constants: a variable initialized to zero and then set to a value later. To handle this case SANDPUPPY categorizes variables with $\mathcal{V} = \{0, n\}$ where $n \neq 1$ as constants.

6.1 Semantic Categories

Enums. We define a heuristic for identifying enums that take on values that are either directly present in the input or are mapped to certain input values. Additionally, we also assume that an arbitrary number of such values exist in the input. This means that, for enums, the number of times it is modified may correlate with the input size. Let us define the function $isz : 2^T \rightarrow \mathbb{Z}_{\geq 0}$ that returns the size of the input used to generate the trace T. We then have $\mathbf{S} = (isz(T_i) \mid T_i \in \mathbf{T})$ and $\mathbf{M} = (|T_i| \mid T_i \in \mathbf{T})$.

We can now calculate the Pearson correlation coefficient for our sample, called the *times-modified to input-size correlation*, as $h_{tmisc} =$

$\frac{\sum_{i=1}^{|\mathbf{T}|}(s_i-\bar{s})(m_i-\bar{m})}{\sqrt{\sum_{i=1}^{|\mathbf{T}|}(s_i-\bar{s})^2}\sqrt{\sum_{i=1}^{|\mathbf{T}|}(m_i-\bar{m})^2}}$. The closer h_{tmisc} is to 1, the more likely that the variable is set based on values from the input. Note that this measure has some caveats, particularly if the discrete input elements are arbitrarily sized. Consider the example in Listing 1.1. With arbitrarily-sized messages, the program can process 3 messages of size n each, or a single message of size $3n$. In the first case the enum variable is modified three times, whereas in the second case it is modified only once. However, in both cases the input size is $3n$.

Consider another example demonstrated in Listing 1.1: The enum variable `mtype` is set based on the value of the string `mtype_str`. Also, `mtype` is modified on three lines; one for each message type. This means that if the number of lines where the variable is modified equals the number of unique values it can take, it is probably part of a construct such as the one in Listing 1.1. Therefore, we define the *number of modified lines* as $h_{nmod} = \left|\bigcup_{T\in\mathbf{T}} \{m \mid (m,\nu) \in T\}\right|$.

We also consider the case where values are assigned without being part of a structure similar to Listing 1.1. Here, the variable is only modified on one line.

Counters. We created several heuristics to identify if a variable is a counter. Our first intuition is to classify variables correlated with input size, as counters. This allows us to detect variables being used as such without detecting if they actually "count" (which other heuristics will do). If a strong correlation exists between a variable's maximum value in a trace and the input size for that trace, it is categorized as a counter. We first define $\mathbf{S} = (isz(T_i) \mid T_i \in \mathbf{T})$ and $\boldsymbol{\mathcal{V}}^{max} = (\nu_i^{max} \mid T_i \in \mathbf{T} \wedge \nu_i^{max} = \max(\{\nu_i \mid (m_i, \nu_i) \in T_i\}))$. Using this we can calculate the Pearson correlation coefficient for our sample, which we call the *maximum-value to input-size correlation*, as $h_{mvisc} = \frac{\sum_{i=1}^{|\mathbf{T}|}(s_i-\bar{s})(\nu_i^{max}-\bar{\nu}^{max})}{\sqrt{\sum_{i=1}^{|\mathbf{T}|}(s_i-\bar{s})^2}\sqrt{\sum_{i=1}^{|\mathbf{T}|}(\nu_i^{max}-\bar{\nu}^{max})^2}}$. Using h_{mvisc} as a measure of input-size correlation—the closer the value is to 1, the more likely it is that the variable holds values correlated with input size.

Next, we develop heuristics to detect counters and to distinguish between static and dynamic counters. We first define *maximum-values variance* as $h_{mvv} = \frac{\sum_{i=1}^{|\mathbf{T}|}(\nu_i^{max}-\bar{\nu}^{max})^2}{|\mathbf{T}|}$. This heuristic is the variance of the observed maximum values of the variable. As static counters count over the same set of values every time, the corresponding value of h_{mvv} will be 0, while dynamic counters will have a non-zero value.

The second heuristic, the *average value-set-cardinality ratio*, is defined as $h_{avscr} = \frac{\sum_{T\in\mathbf{T}} \frac{|\{\nu \mid (m,\nu)\in T\}|}{|\mathcal{V}|}}{|\mathbf{T}|}$. This heuristic is an average of the ratio of the number of unique values in a trace, and the number of unique values across *all* traces for a particular variable. If h_{avscr} is 1, then the variable takes on the exact same set of values in every trace. If this variable is a counter, then it must be static, and we can ignore it as it does not expose useful state.

The next set of heuristics for identifying counters requires pre-processing of variable traces. The intent is to identify segments of traces that could be *counter segments*, meaning that the segment reflects a variable counting to some value. This is important because counters are frequently used multiple times in a single run, and so a trace can contain multiple such segments. It also ensures that each segment contains at least two values and that there are no repeating values in a segment. Runs can occur if a piece of code with a counter is called repeatedly, where the counter loops only a single time. But it can also happen in the case where a variable is being repeatedly set to the same value. Due to this ambiguity, we collapse such runs into a single value.

The first heuristic using counter segments is called *directional consistency*: $h_c = \frac{|\sum_{T \in \mathbf{T}} \sum_{S \in counter\_segments(T)} dir(S)|}{\sum_{T \in \mathbf{T}} |counter\_segments(T)|}$ where $dir(S) = 1$ if $s_2 > s_1$ and -1 otherwise. The purpose of the directional consistency heuristic is to identify counter variables that are incremented or decremented consistently. For counters, h_c will be close to 1.

The final heuristic using counter segments is called *loop-sequence proportion*: $h_{lsp} = \frac{\sum_{T \in \mathbf{T}} \sum_{S \in counter\_segments(T) \wedge |S| > 2} |S|}{\sum_{T \in \mathbf{T}} |T|}$. This heuristic measures what proportion of a variable's traces are part of a loop sequence. For counter variables, this value should be close to 1. We deliberately exclude counter segments of length 2 or less, as segments of the form $[a, b]$ where $a > b$ or $a < b$ are common, meaning that even non-counter variables have a h_{lsp} close 1. However, it is far less likely for such variables to have segments of size 3 or higher where they are consistently incremented or decremented.

7 Identification of Semantic Categories

Once a set of variable-value traces are discovered, SANDPUPPY uses heuristics defined in Sect. 5 to categorize each variable into a semantic category. Note that the heuristics have thresholds associated with them, which we define here, and Sect. 9 describes the thresholds set during the evaluation. Variables deemed state-representative are instrumented to provide feedback during fuzzing (described in Sect. 8).

Enums and Counters. To determine an enum, SANDPUPPY checks if h_{tmisc} is greater than or equal to a threshold τ_{tmisc}. If it is modified on multiple lines, SANDPUPPY checks if the number of lines is the same as the number of values it receives.

SANDPUPPY identifies variables correlated with input size by checking if the variable's h_{mvisc} feature is greater than or equal to the threshold τ_{mvisc}.

For counters, we check to see if a variable's h_{lsp} and h_c are greater than or equal to the threshold τ_{lsp} and τ_c. To determine if a counter is static or not, SANDPUPPY checks if h_{mm} is 0 (i.e., it always counts to or from the same value) and h_{avscr} is 1 (i.e., it counts over the same set of values every time).

Related Variables. SANDPUPPY identifies related variables starting with all variables from the specified function and file that are not constants, booleans, or unclassified variables, to include those already been categorized as enums or counters.

SANDPUPPY then creates a candidate list of related variables for each variable. First, SANDPUPPY chooses one of the lines where the variable is modified.

Second, it scans up to δ_{max} lines before and after to identify other modified variables. SANDPUPPY does not do this for every line that modifies the variable because the intuition is that related variables are modified in close proximity to each other. Lastly, SANDPUPPY prunes each variable's candidate list to include only those variables that it considers related, by ensuring that (1) both variables are modified on the same number of lines and (2) their trace-length distributions are identical.

SANDPUPPY's pruning strategy does exclude variable pairs that have identical trace-length distributions. For example, SANDPUPPY will not identify related variables when one variable is modified on one line, but the second is modified on mutually-exclusive conditions. Here, for any trace generated from the same input, both variables be modified the same number of times, even if overall, the second variable is modified on more lines than the first. This makes feedback instrumentation simpler by avoiding reasoning about the context in which variables are modified.

8 Feedback Instrumentation

Having mapped variables to semantic categories, SANDPUPPY instruments state-representative variables to generate several variants of the program that expose corresponding feedback information.

Our implementation is based on the feedback mechanisms proposed in IJON [3]. Instrumentation uses LLVM passes based on the FUZZFACTORY framework [31]. In all cases, we use the FUZZFACTORY max reducer function to aggregate feedback across multiple inputs. The max reducer function marks an input as interesting if in the corresponding bitmap, a new location is set, or if a location has a higher value.

Permutation Feedback. We perform this instrumentation on enum variables to identify inputs resulting in unique permutations of values, potentially allowing exploration of associated complex-states. We generate a variant with this feedback for each identified enum variable. This instrumentation is similar to IJON's state-change log.

Maximization Feedback. This instrumentation is similar to the `IJON_MAX(x)` primitive [3]. For dynamic counters and variables correlated with input size, we employ instrumentation to guide fuzzing towards inputs that maximize their values. We generate a single variant with this feedback for all identified counters.

Related-Variable Maximization Feedback (RVMF). We perform this instrumentation on every related-variable pair; it is similar to the `IJON_MAX(slot, x)` primitive [3]. It maximizes the value of one variable with respect to another. A real-world example of such code can be seen in FFMPEG's DTS-HD demuxer. We generate two instrumented variants for each identified pair of related variables, as we do not know which variable is the dependent one.

Related-Variable Combination Feedback (RVCF). This instrumentation is performed on every related-variable pair, and is similar to `IJON_SET(hash_ints(x, y))` annotation in IJON [3].

9 Evaluation

We perform evaluations to understand SANDPUPPY's ability to automatically add state-representative variable feedback to fuzzing, and the impact of this to find bugs.

Reasoning About States. We first develop understanding of SANDPUPPY's core contribution by evaluating its variable semantic category identification in Sect. 9.1 and exploring the value of this identification in the actual exploration of complex states in software in Sect. 9.2.

Fuzzing for Vulnerabilities. We then perform a comparative evaluation of SANDPUPPY against state-of-the-art fuzzers, both in terms of achieved code coverage in Sect. 9.3 and triggered vulnerabilities in Sect. 9.4.

Understanding Variable Impact. For a deeper understanding of individual categories identified by SANDPUPPY, we perform an ablative evaluation in Sect. 9.5.

We performed initial fuzzing for trace-gathering and classification on a Linux Mint 20.2 machine with an Intel Core i9-8950HK CPU and 32 GB RAM. Parallel fuzzing was performed on a 40-node Kubernetes cluster with each node having two Intel(R) Xeon(R) CPU E5-2670 v2 processors. Each fuzzing instance receives identical resources during the parallel-fuzzing session.

We set initial fuzzing time to one hour, and found that the instrumentation slows the fuzzer down significantly, making it equivalent to around 5–10 min of fuzzing with a regular AFL instrumented binary. The results will demonstrate however, that even with this performance hit, SANDPUPPY can identify and categorize state-representative variables.

Through manual experimentation and observation on LIBTPMS and the test-bed described in Sect. 9.1, we derived the following threshold values: $\tau_{mvisc} = 0.6$, $\tau_{lsp} = 0.7$, $\tau_c = 0.8$, $\tau_{tmisc} = 0.5$, and $\delta_{max} = 6$. We note that these values may be program dependent and could potentially vary between programs. However, in our experiments these threshold values appeared consistent across multiple programs.

9.1 Identification of Semantic Categories

There is a fundamental trade-off between extraneous instrumented state representative variables (too many variants slow down fuzzing) and insufficient state representative variables (missing exposing state to fuzzing). As our goal is to improve fuzzing efficacy, we evaluated SANDPUPPY against a custom test-bed and against real-world programs to better understand its state reasoning. We

Table 1. Sample of classified variables with labels from various real-world programs. Variable names include the file and function name. The last column specifies whether the variable was identified as state-representative or not.

Target	Qualified Variable Name	Label	State Representative
libpng	`pngrutil.c:png_check_chunk_name:i:3140`	`static_counter`	N
	`pngrutil.c:png_check_chunk_name:c:3147`	`enum_from_input`	Y
	`pngrutil.c:png_read_filter_row_up:i:3956`	`dynamic_counter`	Y
dmg2img	`dmg2img.c:percentage:i:77`	`static_counter`	N
	`base64.c:decode_base64_char:c:49`	`enum_from_input`	Y
	`dmg2img.c:read_kolyblk:k.XMLLength:148`	`dynamic_counter`	Y
smbc	`PPU.cpp:PPU::writeDMAEh:i:501`	`static_counter`	N
	`Main.cpp:mainLoop:world_pos:309`	`related_variable`	Y
	`Main.cpp:mainLoop:pos_y:311`	`related_variable`	Y

present some results here, and detailed results for our synthetic test bed in the to-be-released extended version.

We evaluated SANDPUPPY's automatic variable categorization on the following programs: LIBPNG (1.6.37), LIBTPMS (0.8.3), JSONCPP (1.9.5), DMG2IMG (commit `a3e4134`), and SUPER MARIO BROS. In this experiment, we manually created ground-truth of the semantic category of the variables. We discovered minor discrepancies limited to dynamic counters being classified as enums and some enum variables that were not identified. Two factors cause this: (1) lack of a sufficiently-representative trace-distribution for a variable and (2) ambiguity inherent in our approach for detecting enums, which depends on correlation between input size and how often the variable is modified. Our analysis, a sample which is in Table 1, shows largely accurate identification of enums, static and dynamic counters, and related variables.

We also compared the state-representative variables SANDPUPPY identified to those *manually identified* by IJON [3] and found every variable manually identified by human-experts, including `command.index` in LIBTPMS and Mario's screen position in Super Mario[1]. SANDPUPPY identified the same variables as IJON, and with the appropriate semantic category.

9.2 Exploring Complex States

We evaluated SANDPUPPY's ability to explore complex state typically not accessible to purely coverage-based approaches; games and protocol parsers present a significant challenge as certain states can only be reached through a particular input sequence. We limited SANDPUPPY to a single run as initial trace-collection was sufficient to correctly identify and categorize variables that expose state information.

Super Mario Bros. We used the game version modified by IJON authors [3]: Changes include reading input from `stdin` and speedrunning changes. SAND-

[1] Note that LIBPNG is not included in the released data: https://github.com/RUB-SysSec/ijon-data.

Table 2. SANDPUPPY, AFL, and IJON playing Super Mario Bros (3 runs each). We show how often the level was solved and the median distance traveled. For SANDPUPPY and AFL, we include the median time (hh:mm) to find the solution.

SandPuppy				AFL			Ijon	
World	Solved	Time	% Distance	Solved	Time	% Distance	Solved	% Distance
1-1	✓	27:58	100%	✗	168:04	69%	✓	100%
1-2	✓	30:03	100%	✗	96:15	68%	✓	100%
1-3	✓	41:10	100%	✗	58:04	53%	✓	100%
1-4	✓	04:47	100%	✗	32:47	87%	✓	100%
2-1	✗	46:55	94%	✗	136:42	65%	✗	94%
2-2	✓	04:13	100%	✗	61:37	74%	✓	100%
2-3	✓	05:15	100%	✗	40:25	81%	✓	100%
2-4	✓	05:38	100%	2/3	125:36	100%	✓	100%
3-1	✓	52:19	100%	✗	10:00	29%	✓	100%
3-2	✓	09:42	100%	✗	54:29	66%	✓	100%
3-3	✓	12:02	100%	✗	90:18	94%	✓	100%
3-4	✓	07:09	100%	✗	26:21	74%	✓	100%
4-1	✓	09:51	100%	✗	72:03	95%	✓	100%
4-2	✓	03:54	118%	✗	18:18	57%	✓	118%
4-3	✓	13:44	100%	✗	26:31	56%	✓	100%
4-4	✗	04:13	83%	✗	128:20	82%	✗	82%
5-1	✓	12:05	100%	✗	192:01	45%	✓	100%
5-2	✓	18:01	100%	✗	179:52	82%	✓	100%
5-3	✓	08:42	100%	✗	91:03	63%	✓	100%
5-4	✓	05:16	100%	1/3	78:34	100%	✓	100%
6-1	✓	10:34	100%	✗	02:28	47%	✓	100%
6-2	✓	136:20	100%	✗	11:17	37%	✗	82%
6-3	✓	09:08	100%	✗	32:03	85%	✓	100%
6-4	✓	04:03	100%	1/3	63:22	100%	✓	100%
7-1	✓	22:34	100%	✗	196:51	68%	✓	100%
7-2	✓	10:17	100%	✗	178:09	82%	✓	100%
7-3	✓	12:43	100%	✗	19:40	81%	✓	100%
7-4	✓	08:55	100%	1/3	28:21	100%	✓	100%
8-1	1/3	164:29	100%	✗	21:16	22%	2/3	100%
8-2	1/3	199:26	100%	✗	47:05	60%	2/3	100%
8-3	✓	10:47	100%	✗	19:12	72%	✓	100%
8-4	✓	21:10	100%	✗	09:25	47%	✗	–

PUPPY identified a number of related variables, including those describing Mario's position in the world. We fuzzed SUPER MARIO BROS with AFL and SANDPUPPY for up to 200 h using identical seeds, recording Mario's progress. Results in Table 2 include how often the level was solved, the median time until the best solution was found (hh:mm), and the median percentage of distance covered.

SANDPUPPY solved all worlds except two and found the warp zones in worlds 1-2 and 4-2. World 2-1 could not be fully solved as Mario cannot use the spring due to the speedrunning code-changes. In world 4-4, Mario repeatedly gets stuck in a dead end. SANDPUPPY solved worlds 6-2 and 8-4 that IJON could not.

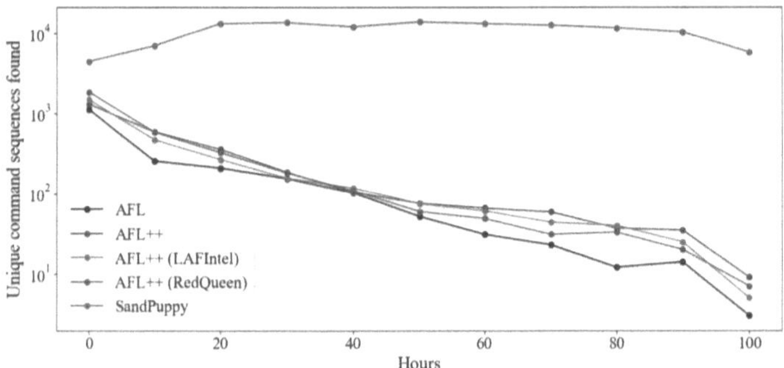

Fig. 3. Unique command-sequences with parameter size found over time in LIBTPMS. SANDPUPPY consistently finds more unique-sequences throughout the session compared to other fuzzers, which "stall" as fuzzing progresses. **The y-axis is in log scale.**

Table 3. Number of distinct command subsequences and total unique sequences found by all fuzzers in LIBTPMS, without accounting for command parameter-size and with. SANDPUPPY outperformed all other fuzzers.

Sub-seq. Length	AFL	AFL++	LAFIntel	RedQueen	SandPuppy
1	96/737	94/741	105/876	97/783	**111/115,751**
2	301/978	328/1,050	392/1,259	348/1,077	**671/116,613**
3	413/691	589/932	677/1,098	671/989	**879/2,659**
4	461/633	865/1,101	907/1,172	1,075/1,324	**1,186/1,792**
Unique Sequences	1,324/1,991	2,137/2,840	1,882/2,769	2,706/3,379	**2,998/119,278**

While most solutions were found by targets using RVMF from player coordinates, some were using RVCF from the same variables. In a few cases, targets using RVMF from Mario's world and screen x coordinates also found solutions. As the latter is constant, this normally causes Mario to deadend. However syncing inputs with other fuzzing instances appears to have a synergistic effect, allowing one with a simple but incomplete strategy to improve on a partial solution from another instance. We believe this allowed SANDPUPPY to solve levels 6-2 and 8-4 when IJON could not.

IJON also examined the exploration of complex state in maze programs, and we discovered that SANDPUPPY was able to successfully solve all mazes, including one of our own creation. We do not present the details due to space constraints.

State Exploration Evaluation. We evaluated SANDPUPPY's ability to explore the state-space of stateful parsers on the LIBTPMS library. This is a real-world software emulator for a Trusted Platform Module (TPM), and was also used in IJON [3]. The protocol defines various command-types and associated payloads, which affect processor-state in different ways.

To evaluate the system on generating complex messages over a significant time-frame, we fuzzed LIBTPMS for over 100 h with SANDPUPPY, AFL, AFL++, REDQUEEN, and LAFINTEL, using diverse seeds and equal compute.

Table 4. Number of variables detected after each trace-generation step of each run.

Target \ Run	Variables detected					
	Initial	1	2	3	4	5
jsoncpp	90	107	107	107	108	108
libtpms	554	872	912	912	912	945
PcapPlusPlus	384	495	499	500	506	506
readelf	314	595	626	636	637	640

We evaluate state-space exploration by tracking the following: total number of unique command-sequences, counts of unique command-sequences of different lengths, and counts of unique command sub-sequences of lengths 1, 2, 3, and 4. We also tracked command parameter-sizes, which helps distinguish between different varieties of the same command. SANDPUPPY was able to automatically identify and correctly categorize state-variables holding this information.

The results show that SANDPUPPY outperforms other fuzzers after a single run. SANDPUPPY found more unique sub-sequences overall, and more unique sequences of various lengths as well; especially when incorporating parameter size (Table 3). Based on sequences over time (Fig. 3), SANDPUPPY finds them at a higher and relatively-constant rate compared to other fuzzers.

9.3 Coverage Evaluation

We used JSONCPP, LIBTPMS, PCAPPLUSPLUS (commit a817631), and READELF (BINUTILS 2.32), and evaluated SANDPUPPY against AFL, AFL++, LAFINTEL, REDQUEEN, SGFUZZ and FERRY.

We ran SANDPUPPY for 6 iterations (runs), 4 h each, for a total fuzzing-time of 24 h. As the number of SANDPUPPY targets vary with each run, a one-to-one comparison based on equal compute is difficult. We therefore calculate the rounded-up average of the number of targets fuzzed in each SANDPUPPY run, and allocate that many parallel-fuzzing sessions for the other fuzzers, which we then fuzz for 24 h. The SGFUZZ and FERRY experimental setup was similar as described by the respective authors, and were run for 24 h as well.

Instrumented variables are recognized only if traces execute their code path. Table 4 shows that SANDPUPPY detects an increasing number of variables in the trace-generation phase, after each fuzzing run. This allows it to potentially explore additional complex states and also suggests that coverage increases over each run. To quantify, we calculated the percentage increase over baseline coverage, which is derived from the initial seeds provided to all fuzzers. We also calculated the percentage of uniquely covered blocks with respect to baseline coverage for each fuzzer, to quantify uniqueness of coverage (see Table 5).

The results (Table 5) show that SANDPUPPY superseded or matched the performance of the other fuzzers in raw coverage improvement on JSONCPP, LIBTPMS, and READELF. For PCAPPLUSPLUS, SANDPUPPY's achieved 85.70% coverage improvement compared to REDQUEEN's 86.23%. SANDPUPPY provides significant unique coverage on LIBTPMS, PCAPPLUSPLUS, and READELF.

A few cases stood out when inspecting unique basic blocks that were covered by SANDPUPPY. In LIBTPMS, SANDPUPPY triggered a path that detects and

Table 5. Improvement over baseline coverage. The value in parentheses is the percentage of unique coverage relative to the baseline. For FERRY comparison is done only using basic-block counts as it uses a symbolic approach and does not have initial seeds.

Fuzzer / Target	Coverage Improvement (%)							Blocks Covered	
	AFL	**AFL++**	**LAFIntel**	**RedQueen**	**SGFuzz**	**SandPuppy**	**Total**	**Ferry**	**SandPuppy**
jsoncpp	45.78 (0.00)	46.09 (0.00)	45.98 (0.00)	45.76 (0.00)	42.63 (0.33)	**46.65** (0.00)	47.10	623	**1314**
libtpms	96.78 (0.00)	111.77 (0.76)	112.26 (0.49)	111.06 (0.49)	0.33 (0.00)	**113.78** (5.67)	120.11	47	**3923**
pcapplusplus	28.37 (0.00)	34.23 (0.00)	35.97 (0.00)	**86.23** (10.67)	69.25 (0.00)	85.70 (10.40)	97.31	251	4910
readelf	235.58 (1.37)	188.52 (1.49)	174.74 (3.8)	178.11 (14.25)	n/a	**261.48** (9.6)	292.32	342	**8473**

Table 6. Bugs discovered by each fuzzer in PCAPPLUSPLUS. SANDPUPPY discovered 10 out of 11 bugs (2 uniquely). LAFINTEL, AFL++, REDQUEEN, and AFL discovered 9, 6, 4, and 3 bugs respectively. The last column specifies whether the vulnerable block was covered by other fuzzers.

Bug	AFL	AFL++	LAFIntel	RedQueen	SandPuppy	Covered by others
Heap overflow #1	✗	✗	✓	✗	✓	✓
Heap overflow #2	✓	✓	✓	✓	✓	✓
Heap overflow #3	✗	✓	✓	✓	✗	✓
Heap overflow #4	✓	✓	✓	✓	✓	✓
Heap overflow #5	✗	✓	✓	✓	✓	✓
Heap overflow #6	✗	✓	✓	✗	✓	✓
Heap overflow #7	✗	✗	✗	✗	✓	✗
Heap overflow #8	✓	✓	✓	✗	✓	✓
Heap overflow #9	✗	✗	✓	✗	✓	✓
Stack overflow #1	✗	✗	✓	✗	✓	✓
Stack overflow #2	✗	✗	✗	✗	✓	✓

mitigates dictionary attacks against the TPM processor by five successive authorization failures. In the case of PCAPPLUSPLUS, SANDPUPPY triggered parsing of DHCP requests, which rely on a specific combination of source and destination UDP ports. These findings show that in addition to improving coverage in general, SANDPUPPY can identify and cover multiple deep and complex states.

We did notice changes in variable classification from run to run and verified that re-classifications were accurate based on variable usage in the source code. Misclassifications have the potential to generate useless inputs, but otherwise do not appear to degrade performance. We also observed that some misclassified variables were accurately re-classified as non state-representative, which reduced the generation of useless inputs in subsequent runs.

Compared to SGFUZZ, SANDPUPPY exceeds coverage performance. The first reason is SGFUZZ's fragile regex-based static analysis; when used on BINUTILS source to instrument READELF, SGFUZZ crashed due to issues parsing enum definitions containing comments with braces. When used against LIBTPMS, SGFUZZ could identify certain enum definitions. It was not able to detect or instrument *any* variables, as their values are set using integer constants instead of enum constants.

SGFuzz also identifies false-positives that are not actually enums and the subsequent instrumentation causes a compilation error due to a type-mismatch. While an ignore-list solves this issue, SGFuzz does not distinguish between identically-named variables based on usage context.

To fairly evaluate Ferry, we used individual block counts as Ferry uses a symbolic approach that does not allow providing initial input seeds. SandPuppy significantly outperforms Ferry. When running against jsoncpp, PcapPlusPlus, and readelf, Ferry terminated early. In the case of libtpms, Ferry ran out of memory when given 64 GB.

9.4 Vulnerability Detection

We evaluated SandPuppy's ability to find vulnerabilities on two real-world targets: dmg2img and PcapPlusPlus.

We use **dmg2img** as it was fuzzed in academic research by WEIZZ [16] and Ijon [3]. SandPuppy identified multiple state-representative variables, including `kolyblk.XMLLength`, which was manually identified by Ijon authors. SandPuppy discovered 7 bugs; 4 were identified by WEIZZ and Ijon, and 2 are known issues (Bug 778819 and CVE-2021-32614). The last (unknown) bug is a one-byte read heap-buffer overflow in `convert_char4`, which we reported.

PcapPlusPlus is a library for capturing and manipulating network packets, featuring state-representative variables and stateful logic. Here we evaluated SandPuppy against AFL, AFL++, RedQueen, and LAFIntel, with equivalent fuzzing time and compute. The results are shown in Table 6: for a total of 11 previously unknown bugs (which we reported); including 2 stack-overflows and 9 heap-overeads. SandPuppy discovered 10 out of 11 bugs, out of which 2 were not found by other fuzzers. LAFIntel, AFL++, RedQueen, and AFL discovered 9, 6, 4, and 3 bugs respectively.

Of the two unique bugs, Heap Overflow #7 was triggered the "traditional" way: SandPuppy's state awareness triggered additional code. The buggy code containing Stack Overflow #2 was executed by *all* fuzzers, but SandPuppy triggered it with the correct *state* by identifying and instrumenting a computed length-variable.

Table 7. Number of distinct command subsequences and total unique sequences with parameter size found for each feedback-evaluation run in LIBTPMS. Preceding minus signs describe runs where corresponding feedback was excluded. Perm by itself outperforms all others and excluding it leads to a dramatic drop in performance.

Sub-seq. Length	RVCF	Perm	Max	RVMF	-RVCF	-Perm	-Max	-RVMF	Random	Regular
1	222	**41,492**	638	260	15,824	623	16,864	18,062	379	12,998
2	277	**41,754**	756	335	16,381	930	17,230	18,552	497	13,547
3	218	700	166	262	**1,237**	454	820	1,064	298	978
4	69	669	38	104	**702**	113	601	654	271	377
Unique Sequences	425	**43,073**	768	559	17,177	1,013	17,852	19,189	1,170	13,964

Table 8. Coverage (% of basic blocks hit) for each feedback-evaluation run in LIBPNG. Preceding minus signs describe runs where corresponding feedback was excluded. Excluding Perm increases performance vs. all other runs. RVMF performs best overall.

	RVCF	Perm	Max	RVMF	-RVCF	-Perm	-Max	-RVMF	Random	Regular
Coverage	32.10%	31.80%	31.76%	33.60%	32.99%	**34.47%**	31.97%	32.15%	31.87%	32.13%

9.5 Feedback Instrumentation Evaluation

The results so far demonstrate that SANDPUPPY identifies state-representative variables providing feedback sufficient enough to improve coverage, vulnerability identification, and state-exploration. However not *all* instrumented variables and associated feedback necessarily expose *useful* state enhancing these metrics. There is also a risk of fuzzing slowdown due to generating a large number of useless inputs [3].

We designed an experiment to measure the effect of each feedback instrumentation on coverage and state-exploration. We fuzzed LIBTPMS and LIBPNG for 24 h with a single feedback-type enabled/excluded, and with all feedback enabled on a random subset of variables. We evaluated these against a regular SANDPUPPY run.

Results in Table 7 show that permutation improves state exploration the most; excluding it reduces state exploration. In contrast, excluding other feedback types (RVCF, maximization, and RVMF) increases state diversity.

This trend holds when evaluating individual feedback: Permutation feedback by itself is almost four times better, while the remaining are not better than random. This suggests that, based on this particular diversity-metric, other feedback types do not expose useful state and generate slow inputs.

LIBPNG results are in Table 8. Here, *excluding* permutation feedback increases coverage. Variables instrumented include the chunk name and the chunk-header length. While the feedback it to recognize different combinations of chunk types and associated lengths, it results in a large number of inputs.

In contrast, RVMF improves coverage both by itself and when included with maximization feedback and RVCF. In either case coverage is better than when fuzzing with all feedback types, as some instrumented variable-pairs hold information related to image or chunk dimensions and sizes. These variables are also involved in a number of sanity or boundary checks and conditional parsing-logic.

Maximizing these variables with respect to each other is likely to exercise these paths, resulting in greater coverage. This is more likely when the fuzzing campaign is not overwhelmed by many uninteresting inputs, which is observed when we disable the permutation feedback.

Both cases demonstrate that fuzzing with all feedback types generally improves coverage and state exploration, the usefulness depends on the target and evaluation metrics. Better performance in future fuzzing runs could be achieved by analyzing the contributions of different feedback types, and including only those that enhance coverage.

10 Discussion, Limitations, and Future Work

We propose an approach that guides fuzzing towards deep states using feedback from automatically-identified state-representative variables. SANDPUPPY is used in an end-to-end, dynamic-feedback approach where test cases discovered after one full run are used for trace-generation and variable classification, and as initial seeds in a subsequent run. As coverage generally increases, new test cases have a chance to exercise new code paths, leading to the discovery of additional state-representative variables, thus allowing SANDPUPPY to explore deeper states within a program.

But all instrumented targets are not useful (Sect. 9.5). An evaluation step could attempt to optimize how each instrumented target improves coverage and their usage.

In large programs, SANDPUPPY can identify a large number of state-representative variables, resulting in a correspondingly large number of targets that need to be built and fuzzed. FUZZFACTORY can support up to four coverage maps at a time; this might allow exposing feedback from multiple variables through a single binary.

During coverage evaluation, we ran SANDPUPPY for six iterations of 4 h each to compare against 24 h runs of other fuzzers. However it is possible to run SANDPUPPY for any number of iterations of any duration. Furthermore, it is also possible to run SANDPUPPY for as long as needed, and terminate on other conditions.

As our intent was not to solve the variable semantic-categorization problem in general, we did not focus on a detailed analysis of false positives and negatives. The lack of a baseline and ground-truth also hampered this approach. While the results demonstrate that SANDPUPPY has practical benefits and improves code coverage, state coverage, and vulnerability discovery, a detailed analysis of the impacts of false positive and negatives on performance would be valuable.

11 Related Work

Core Fuzzing. Seed selection and scheduling enhancements have increased AFL performance in different situations [8,9,12,36,43,44]. Target information can be used to improve seed-mutation strategies [2,4,6,19,24,26,30,34]. Taint tracking improved efficiency by focusing mutation only on interesting input [10,17,27,28, 35].

Tweaking Coverage. ANGORA, COLLAFL, KAFL, and INSTRIM enhance coverage-tracking and feedback by improving recognition of novel coverage [10, 18,20,37]. ANGORA and COLLAFL reduce coverage-map collisions, thus helping AFL uniquely identify more program states. KAFL uses hardware-assisted techniques to improve coverage information when fuzzing kernels, while INSTRIM uses CFG-aware instrumentation to improve execution-path differentiation.

Input-State Correspondence. LAFINTEL [22] solves the magic-value comparison issue by splitting compare instructions into multiple smaller ones. STEELIX [25] uses lightweight analysis for comparison progress and identifying locations of magic bytes.

Concolic [11,13,21,39,45,49] and symbolic-execution [29,40,47,48] approaches can handle magic bytes or deeply-nested branches, but struggle to explore complex/deep states relying on a specific sequence of state updates. Symbolic execution can help solve some of these problems, but scalability is a concern [8] due to state explosion.

REDQUEEN [4] is an alternative to symbolic/concolic execution and taint tracking, as it derives input-state correspondence. While SANDPUPPY also relies on internal program-state, we expose it to influence feedback instead of correlating with input.

Domain-Specific State Reasoning. Domain-specific fuzzing [14,23,31,33,42] has solved certain deep-state fuzzing problems. PERFFUZZ [23] and SLOWFUZZ guide the fuzzer towards inputs to expose algorithmic-complexity vulnerabilities, while MEMFUZZ [14] augments coverage information by accounting for memory accesses.

General Program State Reasoning. IJON uses manual annotations [3] for deep-state fuzzing. While IJON overcomes a number of deep-state-fuzzing obstacles and surpasses prior fully-automated approaches, it requires a human to recognize state-representative variables that can improve coverage. SANDPUPPY builds on IJON by automating detection and instrumentation of state-representative variables.

INVSCOV [15] explores program state-space to improve coverage and detect vulnerabilities using runtime dynamic-analysis of program variables, similar to SANDPUPPY. But the focus is on identifying invariants and augmenting coverage feedback with this information, to partition program state-space and guide the fuzzer towards states violating these invariants, potentially triggering vulnerabilities.

SGFUZZ [5] explores state-spaces using regex-based static analysis to identify enums in protocol parsers. SGFUZZ also maintains a global variable-ignore-list of unqualified variable names. SANDPUPPY focuses on additional variable categories making it applicable to more program types. We also identify variables based on their semantics rather than syntax, allowing us to detect those that behave like enums.

FERRY [50] also explored identification of state-representative variables, categorizing them based on certain characteristics (different from SANDPUPPY). Rather than exploring the triggering of other program *states*, FERRY exclusively uses information from these variables to explore dependent *branches*. Based on our evaluation results, we saw that SANDPUPPY significantly outperforms FERRY.

12 Conclusion

SANDPUPPY is a fully-automated source-code based fuzzing approach that explores deep-states in programs by identifying and labeling state-representative integer variables using runtime value-traces, and instrumenting them to provide appropriate feedback based on their semantics. SANDPUPPY effectively automates IJON, requires no knowledge or manipulation of program source, and results in performance that either matches, or in certain cases slightly outperforms IJON. We believe that SANDPUPPY represents the first step toward fully-automated deep-state fuzzing.

Acknowledgements. This research project has received funding from the following sources: Defense Advanced Research Projects Agency (DARPA) Contracts No. FA875019C0003 and N6600122C4026; the Advanced Research Projects Agency for Health (ARPA-H) Grant No. SP4701-23-C-0074; the Department of the Interior Grant No. D22AP00145-00; and National Science Foundation (NSF) Awards No. 1663651, 2146568, 2232915, and 2247954. Any opinions, findings and conclusions or recommendations expressed in this material are those of the author(s) and do not necessarily reflect the views of the Government, and no official endorsement should be inferred.

References

1. Libfuzzer - a library for coverage-guided fuzz testing. https://llvm.org/docs/LibFuzzer.html. Accessed 26 Oct 2021
2. Aschermann, C., Frassetto, T., Holz, T., Jauernig, P., Sadeghi, A.R., Teuchert, D.: Nautilus: fishing for deep bugs with grammars. In: NDSS (2019)
3. Aschermann, C., Schumilo, S., Abbasi, A., Holz, T.: IJON: exploring deep state spaces via fuzzing. In: 2020 IEEE Symposium on Security and Privacy (SP), pp. 1597–1612. IEEE (2020)
4. Aschermann, C., Schumilo, S., Blazytko, T., Gawlik, R., Holz, T.: *REDQUEEN*: fuzzing with input-to-state correspondence. In: NDSS, vol. 19, pp. 1–15 (2019)
5. Ba, J., Böhme, M., Mirzamomen, Z., Roychoudhury, A.: Stateful greybox fuzzing. In: 31st USENIX Security Symposium (USENIX Security 22), pp. 3255–3272 (2022)
6. Blazytko, T., et al.: *GRIMOIRE*: synthesizing structure while fuzzing. In: 28th USENIX Security Symposium (USENIX Security 19), pp. 1985–2002 (2019)
7. Böhme, M.: Aflfast. https://github.com/mboehme/aflfast. Accessed 26 Oct 2021
8. Böhme, M., Pham, V.T., Nguyen, M.D., Roychoudhury, A.: Directed greybox fuzzing. In: Proceedings of the 2017 ACM SIGSAC Conference on Computer and Communications Security, pp. 2329–2344 (2017)

9. Böhme, M., Pham, V.T., Roychoudhury, A.: Coverage-based greybox fuzzing as Markov chain. IEEE Trans. Softw. Eng. **45**(5), 489–506 (2017)
10. Chen, P., Chen, H.: Angora: efficient fuzzing by principled search. In: 2018 IEEE Symposium on Security and Privacy (SP), pp. 711–725. IEEE (2018)
11. Chen, P., Liu, J., Chen, H.: Matryoshka: fuzzing deeply nested branches. In: Proceedings of the 2019 ACM SIGSAC Conference on Computer and Communications Security, pp. 499–513 (2019)
12. Chen, Y., Ahmadi, M., Wang, B., Lu, L., et al.: $MEUZZ$: smart seed scheduling for hybrid fuzzing. In: 23rd International Symposium on Research in Attacks, Intrusions and Defenses (RAID 2020), pp. 77–92 (2020)
13. Chen, Y., et al.: SAVIOR: towards bug-driven hybrid testing. In: 2020 IEEE Symposium on Security and Privacy (SP), pp. 1580–1596. IEEE (2020)
14. Coppik, N., Schwahn, O., Suri, N.: MEMFUZZ: using memory accesses to guide fuzzing. In: 2019 12th IEEE Conference on Software Testing, Validation and Verification (ICST), pp. 48–58. IEEE (2019)
15. Fioraldi, A., D'Elia, D.C., Balzarotti, D.: The use of likely invariants as feedback for fuzzers. In: 30th USENIX Security Symposium (USENIX Security 21), pp. 2829–2846 (2021)
16. Fioraldi, A., D'Elia, D.C., Coppa, E.: WEIZZ: automatic grey-box fuzzing for structured binary formats. In: Proceedings of the 29th ACM SIGSOFT International Symposium on Software Testing and Analysis, pp. 1–13 (2020)
17. Gan, S., et al.: $GREYONE$: data flow sensitive fuzzing. In: 29th USENIX Security Symposium (USENIX Security 20), pp. 2577–2594 (2020)
18. Gan, S., et al.: CollAFL: path sensitive fuzzing. In: 2018 IEEE Symposium on Security and Privacy (SP), pp. 679–696. IEEE (2018)
19. Han, H., Oh, D., Cha, S.K.: CodeAlchemist: semantics-aware code generation to find vulnerabilities in javascript engines. In: NDSS (2019)
20. Hsu, C.C., Wu, C.Y., Hsiao, H.C., Huang, S.K.: Instrim: lightweight instrumentation for coverage-guided fuzzing. In: Symposium on Network and Distributed System Security (NDSS), Workshop on Binary Analysis Research (2018)
21. Huang, H., Yao, P., Wu, R., Shi, Q., Zhang, C.: Pangolin: incremental hybrid fuzzing with polyhedral path abstraction. In: 2020 IEEE Symposium on Security and Privacy (SP), pp. 1613–1627. IEEE (2020)
22. Lafintel: Lafintel, August 2016. https://lafintel.wordpress.com/
23. Lemieux, C., Padhye, R., Sen, K., Song, D.: PerfFuzz: automatically generating pathological inputs. In: Proceedings of the 27th ACM SIGSOFT International Symposium on Software Testing and Analysis, pp. 254–265 (2018)
24. Lemieux, C., Sen, K.: Fairfuzz: a targeted mutation strategy for increasing greybox fuzz testing coverage. In: Proceedings of the 33rd ACM/IEEE International Conference on Automated Software Engineering, pp. 475–485 (2018)
25. Li, Y., Chen, B., Chandramohan, M., Lin, S.W., Liu, Y., Tiu, A.: Steelix: program-state based binary fuzzing. In: Proceedings of the 2017 11th Joint Meeting on Foundations of Software Engineering, pp. 627–637 (2017)
26. Lyu, C., Ji, S., Li, Y., Zhou, J., Chen, J., Chen, J.: Smartseed: smart seed generation for efficient fuzzing. arXiv preprint arXiv:1807.02606 (2018)
27. Mathis, B., Gopinath, R., Mera, M., Kampmann, A., Höschele, M., Zeller, A.: Parser-directed fuzzing. In: Proceedings of the 40th ACM SIGPLAN Conference on Programming Language Design and Implementation, pp. 548–560 (2019)
28. Mathis, B., Gopinath, R., Zeller, A.: Learning input tokens for effective fuzzing. In: Proceedings of the 29th ACM SIGSOFT International Symposium on Software Testing and Analysis, pp. 27–37 (2020)

29. Ognawala, S., Hutzelmann, T., Psallida, E., Pretschner, A.: Improving function coverage with munch: a hybrid fuzzing and directed symbolic execution approach. In: Proceedings of the 33rd Annual ACM Symposium on Applied Computing, pp. 1475–1482 (2018)
30. Padhye, R., Lemieux, C., Sen, K., Papadakis, M., Le Traon, Y.: Validity fuzzing and parametric generators for effective random testing. In: 2019 IEEE/ACM 41st International Conference on Software Engineering: Companion Proceedings (ICSE-Companion), pp. 266–267. IEEE (2019)
31. Padhye, R., Lemieux, C., Sen, K., Simon, L., Vijayakumar, H.: Fuzzfactory: domain-specific fuzzing with waypoints. Proc. ACM Program. Lang. **3**(OOPSLA), 1–29 (2019)
32. Peng, H., Shoshitaishvili, Y., Payer, M.: T-Fuzz: fuzzing by program transformation. In: 2018 IEEE Symposium on Security and Privacy (SP), pp. 697–710. IEEE (2018)
33. Petsios, T., Zhao, J., Keromytis, A.D., Jana, S.: Slowfuzz: automated domain-independent detection of algorithmic complexity vulnerabilities. In: Proceedings of the 2017 ACM SIGSAC Conference on Computer and Communications Security, pp. 2155–2168 (2017)
34. Pham, V.T., Böhme, M., Santosa, A.E., Căciulescu, A.R., Roychoudhury, A.: Smart greybox fuzzing. IEEE Trans. Softw. Eng. **47**(9), 1980–1997 (2019)
35. Rawat, S., Jain, V., Kumar, A., Cojocar, L., Giuffrida, C., Bos, H.: Vuzzer: application-aware evolutionary fuzzing. In: NDSS, vol. 17, pp. 1–14 (2017)
36. Rebert, A., et al.: Optimizing seed selection for fuzzing. In: 23rd USENIX Security Symposium (USENIX Security 14), pp. 861–875 (2014)
37. Schumilo, S., Aschermann, C., Gawlik, R., Schinzel, S., Holz, T.: $kAFL$: hardware-assisted feedback fuzzing for OS kernels. In: 26th USENIX Security Symposium (USENIX Security 17), pp. 167–182 (2017)
38. Shoshitaishvili, Y., et al.: Rise of the HaCRS: augmenting autonomous cyber reasoning systems with human assistance. In: Proceedings of the 2017 ACM SIGSAC Conference on Computer and Communications Security, pp. 347–362 (2017)
39. Stephens, N., et al.: Driller: augmenting fuzzing through selective symbolic execution. In: NDSS, vol. 16, pp. 1–16 (2016)
40. Wang, M., et al.: SAFL: increasing and accelerating testing coverage with symbolic execution and guided fuzzing. In: Proceedings of the 40th International Conference on Software Engineering: Companion Proceeedings, pp. 61–64 (2018)
41. Wang, Y., et al.: Not all coverage measurements are equal: fuzzing by coverage accounting for input prioritization. In: Network and Distributed System Security Symposium (2020). https://doi.org/10.14722/ndss.2020.24422
42. Wang, Y., et al.: Not all coverage measurements are equal: fuzzing by coverage accounting for input prioritization. In: NDSS (2020)
43. Wang, Y., Wu, Z., Wei, Q., Wang, Q.: Neufuzz: efficient fuzzing with deep neural network. IEEE Access **7**, 36340–36352 (2019)
44. Woo, M., Cha, S.K., Gottlieb, S., Brumley, D.: Scheduling black-box mutational fuzzing. In: Proceedings of the 2013 ACM SIGSAC Conference on Computer & Communications Security, pp. 511–522 (2013)
45. Yun, I., Lee, S., Xu, M., Jang, Y., Kim, T.: $QSYM$: a practical concolic execution engine tailored for hybrid fuzzing. In: 27th $USENIX$ Security Symposium ($USENIX$ Security 18), pp. 745–761 (2018)
46. Zalewski, M.: American fuzzy lop (2.52b). https://lcamtuf.coredump.cx/afl/. Accessed 26 Oct 2021

47. Zhang, B., Ye, J., Feng, C., Tang, C.: S2F: discover hard-to-reach vulnerabilities by semi-symbolic fuzz testing. In: 2017 13th International Conference on Computational Intelligence and Security (CIS), pp. 548–552. IEEE (2017)
48. Zhang, L., Thing, V.L.: A hybrid symbolic execution assisted fuzzing method. In: TENCON 2017-2017 IEEE Region 10 Conference, pp. 822–825. IEEE (2017)
49. Zhao, L., Duan, Y., Yin, H., Xuan, J.: Send hardest problems my way: probabilistic path prioritization for hybrid fuzzing. In: NDSS (2019)
50. Zhou, S., et al.: Ferry: state-aware symbolic execution for exploring state-dependent program paths, pp. 4365–4382 (2022)

Extended Abstract - Tracking Manifests - Persistent Identifiers in Progressive Web Apps

Dolière Francis Somé[1,2](✉)

[1] Stanford University, Stanford, USA
doliere@stanford.edu
[2] CISPA Helmholtz Center for Information Security, Saarbrücken, Germany
doliere.some@cispa.de

Abstract. Progressive Web Applications (PWAs) are web apps that blur the lines of differences with native apps thanks to advanced web technologies at their core. On the one hand, service workers (SWs) bring offline experience to users, background syncing, out-of-bound web push notifications, and turn web apps into client-side proxies that can intercept HTTP requests, and generate responses to fulfill them. On the other hand, web application manifests provide the metadata that makes PWAs installable on the user's device, just like any other native app. In this work, we provide the first privacy assessment of the design and implementation of web manifests, demonstrating that they can be leveraged by adversaries to persist unique identifiers which can be reused to identify and track users. To make matters worse, these identifiers can outlive privacy and tracking protection mechanisms like clearing all browsing data, unregistering SWs, etc. making them similar to supercookies, the most invasive type of persistent identifiers. From an empirical study of 37.6M websites in the wild, we report that the adoption of PWAs is prevalent on the Web, suggesting potential harm to user's privacy. We reported our findings to browser vendors, discussing countermeasures that amount to implementing in the browser additional options for users to uninstall PWAs when they clear their browsing data. With this work, we also raise awareness of the persistence-by-design nature of mechanisms like web app manifests and think that they should be held to the same privacy standards as other client-side storage like the infamous cookies that are under extensive scrutiny by regulations like the GDPR.

Keywords: PWAs · Web Manifests · Persistence · Tracking · Privacy

1 Web Applications Manifests

Manifests are prevalent in many software technologies, including mobile applications or browser extensions. A web app manifest (WAM) [7] is a JSON file with various fields (members/properties) describing metadata about a web app, candidate to installation (integration) onto the user's home screen, just like other

F. Maggi et al. (Eds.): DIMVA 2024, LNCS 14828, pp. 251–260, 2024.
https://doi.org/10.1007/978-3-031-64171-8_13

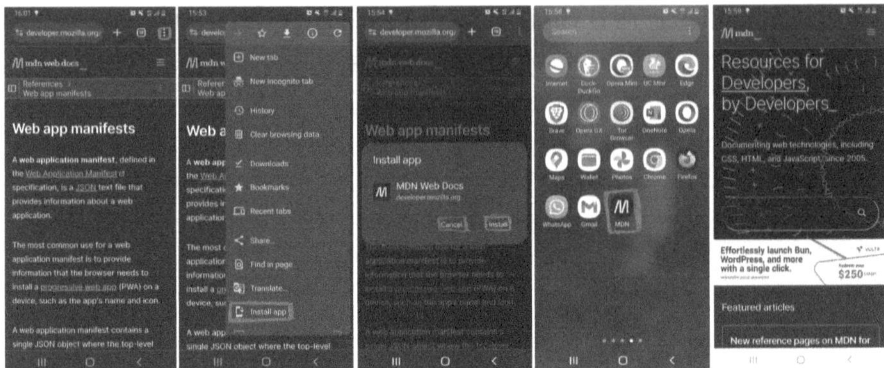

(a) Install Menu (b) Select Install (c) Installation (d) Launch App (e) Standalone

Fig. 1. Process of installing PWAs with Chrome Android. We take the example of https://developer.mozilla.org (MDN). At the end of the installation process, MDN is added to the user's device home screen just like other native apps. When launched, the app is rendered in a dedicated (standalone) window.

native apps. We want to stress that installing a web app [10] (See Fig. 1) is more evolved than simply creating a shortcut to a website on the user's device. Following is an excerpt of prominent members required or recommended to get the best and most seamless integration of a web app into the user's home screen (See Listing 1.3 for an example use of these members in a web app manifest).

- `name`: a directionality-capable (left-to-right by default) name of the app as displayed on the user's device. Related members like `dir` or `lang` explicitly set the direction and effective language of the name.
- `icons`: different URLs, types (e.g., png), and sizes (e.g., 192, 384, 512) of icons shown with the app on the user's device or when it is rendered.
- `start_url`: non-empty same-origin URL that the browser navigates when rendering the app. The browser may ignore or have the user alter it.
- `display`: the developer preferred display mode for a web app.
  - `fullscreen`: the app renders without any browser UI element, occupying the entire display area of the device.
  - `standalone`: the web app renders in a dedicated window, with minimal browser UI elements (no navigation or URL bars)
  - `minimal-ui`: similar to the `standalone` mode, but with navigation elements for going back or refreshing the page.
  - `browser`: default display mode, where the app is regularly rendered in a normal browser window, along with other websites.

  Browsers or the developer with the `display_override` member, (can) set a sequence of fallback modes when the preferred display is not supported.

To signal a web app manifest to a browser, one proceeds as follows:

```
<link rel="manifest" href="/manifest.json" />
```

Listing 1.1. Web app manifest inclusion in a web app HTML

This snippet can either be embedded as part of the response downloaded from the backend server or generated on the fly with JavaScript and added to the DOM (See Listing 1.3). The value of the `href` attribute can be a remote URL (as in Listing 1.1) or a Data URI [5] or inline (e.g., `data:`, `blob:`) manifest. The latter are only supported in Chromium browsers. When the manifest is loaded from a remote server, the `crossorigin` attribute (with values like *use-credentials* or *anonymous*) can be set to indicate whether the browser should download the manifest using user credentials, i.e. cookies, which are omitted by default.

1.1 Security Considerations

The same-origin policy (SOP) [8] does not impose restrictions on the origin of web app manifests. The *manifest-src* directive of the page Content-Security-Policy (CSP) [19] can be provided to impose restrictions. Cross-origin manifests responses must adhere to the cross-origin resource sharing (CORS) protocol [4], meaning that headers like *access-control-allow-origin* must be present. If credentials are included, then the *access-control-allow-credentials* header must also be present for the response to be processed by the browser. Installability — and therefore the origins of web app manifest and service workers — is available only to secure HTTPS (and localhost) websites.

1.2 Installation Requirements and Browsers Support

The installability of PWAs – from a browser window – is differently supported by all major browsers. Moreover, they disagree on the requirements that a web app should satisfy to be installable [6]. As shown in Table 1, Chromium browsers (e.g., Chrome, Edge) support installation on major desktop and mobile operating systems. A service worker [9] with a fetch event is recommended but no longer required for a web app to be installable. Other major browsers, Gecko (e.g., Firefox) and Webkit (e.g., Safari, Orion) only support installable PWAs on mobile, requiring both valid web app manifests and service workers.

Table 1. PWAs installation requirements and browsers support

	Desktop				**Mobile**		**Requirements**	
	Linux	*Windows*	*Mac OSX*	*ChromeOS*	*Android*	*iOS*	*Manifests*	*SWs*
Chromium	✓	✓	✓	✓	✓	✓	✓	✗
Gecko	✗	✗	✗	✗	✓	✓	✓	✓
Webkit	✗	✗	✗	✗	✗	✓	✓	✓

When the installation requirements are met, browsers provide various clues and options for the user to install the web app. Figure 1 demonstrates a typical installation process in a recent Chrome browser on Android. The process on other browsers is similar. At the end of the installation, the app is integrated into the user's device home screen, and when launched, the app will be typically rendered in a dedicated standalone window, just like other native apps. It is worth mentioning that Chrome apps have migrated to PWAs, and it is possible to bundle a PWA and distribute it via popular mobile marketplaces like Google Play and Apple App stores. `pwabuilder.com` provides such a service.

2 Persisting Unique Identifiers in Web App Manifests

The ability to persist unique identifiers in web app manifests is based on various observations. Unlike client-side storage like cookies, web app manifests are (i) not managed by the web app via dedicated APIs or headers, and (ii) so far browsers have not provided straightforward mechanisms for users to remove web apps manifests. Finally, by design, manifests must be persisted for the browser to integrate the installed web app into the user's home screen, make it searchable, and when the web app is launched, render it as expected by the developer.

2.1 Threats Models

We assume an adversary that aims at reidentifying users by leveraging PWAs components, in particular, web app manifests to persist unique identifiers on the user's device that will be later recovered. We distinguish 2 attackers. A (malicious) first-party is a website that the user visits deliberately or is tricked (via social engineering tactics) into visiting. The adversary can also be a (malicious) third-party script legitimately included by a benign first-party website or a web attacker that infected a benign but buggy website by successfully exploiting a cross-site scripting (XSS) [13] vulnerability for instance.

The attack consists of generating a randomly unique and user-specific identifier, embedding it in a web app manifest, and recovering it on subsequent user visits. Table 2 shows the attack mediums and affected browsers. We recall that data URI (`data:`, `blob:`) or inline manifests are only supported by Chromium browsers. Given that manifests can be embedded on the fly in web apps' DOM, inline manifests present the additional advantage of requiring no server-side logic to host them, which can be attractive for a third-party attacker. Moreover, we found that even if a manifest was already embedded in the web app, a third-party malicious script could simply remove it, as shown in Listing 1.2 and inject their own, as long as it does not violate the website' policy, e.g., CSP.

```
Array.from(document.getElementsByTagName("link")).filter(e =>
    e.rel == "manifest").map(m => m.remove())
```

Listing 1.2. Remove and replace legitimate manifests with malicious ones

2.2 Demonstration with Inline Manifests in Chromium Browsers

Listing 1.3 demonstrates how we leveraged inline web app manifests to persist unique identifiers in a Chromium browser. In this example, we are using the Google search engine (https://www.google.com) as the target of the attack (in

Table 2. Vectors for deploying unique identifiers with web app manifests

	Data URI	Remote URL
Chromium (Chrome, Edge, Opera, Brave)	✓	✓
Gecko (Firefox, LibreWolf, Waterfox)	✗	✓
Webkit (Safari, Orion)	✗	✓

particular for the icons that we constructed just for demonstration purposes) but the example can be adapted to any other website.

```
1  let uuid = self.crypto.randomUUID();
2  let manifest = {
3      "start_url": location.origin + "/#" + uuid, // or "/?uuid
           =" + uuid
4      "display": "standalone",
5      "name": "Google Search Engine",
6      "icons": [{
7              "src": "https://www.dropbox.com/scl/fi/
                  cnbmzyeph9xm02glw0hxf/google_384.png?rlkey=
                  zl8a9pzpxtn2vvp2c2vcgoci0&raw=1",
8              "sizes": "384x384",
9              "type": "image/png"
10         } // + other sizes  128, 192, 256, 512
11     ]
12 }
13 let mlink = document.createElement("link")
14 mlink.rel = "manifest";
15 mlink.href = "data:application/json," + encodeURIComponent(
       JSON.stringify(manifest) )
16 document.head.appendChild(mlink);
```

Listing 1.3. Persisting unique identifiers in inline web app manifests. This code snippet is functional and can be readily used to reproduce the issue.

One starts by generating a unique identifier, with the `crypto` API (Line 1), then adds it to the `start_url` member of the web app manifest (Line 3). The intuition is to add the unique identifier as the hash (fragment) of the URL. This is particularly interesting for a third-party attacker, as this identifier does not leave the browser when a request is made to the backend server to fetch the content of `start_url`. Alternatively and in general, the identifier could be stored in other sections of the URL, for instance as a query parameter, or as part of the URL path, etc. We recall (See Sect. 1) that `start_url` is the URL that will be navigated when the user launches the installed web app from their home screen. The rest of the manifest is quite standard and specifies among others a standalone display mode and icons that will be used during the presentation of the app to the user. Finally, the manifest is string-encoded and embedded as a data-URI (inline) manifest as shown in Line 15. Then, it is appended to the

DOM's head (Line 16). Finally, the browser detects and processes the manifest and offers the user to install the web app as shown in Fig. 2a.

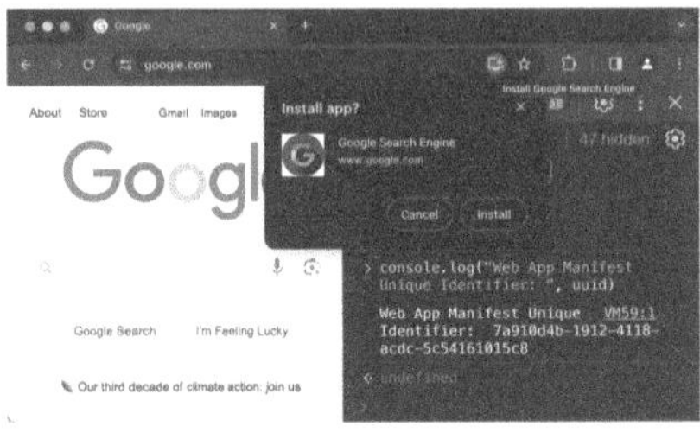

(a) Inline Web App Manifests

(b) Installed Web App Launch

Fig. 2. Unique identifiers in web app manifests are persisted by browsers

Each time that the user launches the installed app, the browser will render the value of the `start_url` member of the manifest. Though no UI elements are visible because of the standalone display mode chosen by the attacker, the unique identifier can be recovered from the fragment of the URL, by calling `location.hash.slice(1)` as shown in Fig. 2b. As one can see, this value matches the original unique identifier generated before installation (Fig. 2a). The attack goes unnoticed by the user because the URL is not displayed in the UI. Even when the user clears all their browsing data, the web app remains installed, the manifest persistent with the unique identifier. The attacker can detect an installed web app as shown in Listing 1.4 by checking the display mode (e.g., standalone or any other mode chosen by the attacker).

```
window.matchMedia('(display-mode:standalone)').matches ==true
```

Listing 1.4. Detecting an installed PWA by inspecting its display mode

3 Empirical Study

We performed an empirical study on the usage of PWAs in the wild in April 2023. The results are shown in Table 3 and concern valid web app manifests and service workers. The original dataset of 37,643,971 sites came from the CrUX [1] and Tranco [16] sites for the period 12/2021 to 11/2022.

Service workers and web app manifests are used in almost the same proportions. Nonetheless, on Chromium browsers where service workers are not required for installable PWAs, it is technically any website that can be turned installable by any third party as we have demonstrated in Sect. 2. In the wild, inline manifest usage is rather limited (4.4%). As for the rendering of installed apps, the `standalone` display mode stands out. In general, it appears that modes that require minimal browser UIs are the most popular (99%), which can be leveraged to hide unnoticeable unique identifiers in manifests.

Table 3. PWAs in the wild: SWs, manifests, and display modes.

	#Websites
Service Workers	852,404
Web App Manifests	756,428
Valid manifests	528,939
Inline (`data: blob:`) manifests	23,021 (4.4%)

(a) SWs and manifests usage

	#Websites
`standalone`	485,656
`minimal-ui`	20,272
`fullscreen`	17,746
`browser`	5,265

(b) Display modes in manifests

Overall, the prevalence of installable PWAs in the wild coupled with the attacks that we have demonstrated, could further lead to more websites exploiting web app manifests to persist unique identifiers in user's browsers. In the collected dataset, we found many occurrences of Urchin Tracking Module (UTM) parameters [11] in the `start_url` URL parameters, e.g., `utm_source` for 130,622 sites and `utm_medium` for 4,856 sites, but they did not include unique identifiers. URL fragments are less popular (found among 1,056 sites) and we did not find an active exploit of the attack we have demonstrated after manual inspection of all these fragments.

In our dataset, we found 18,153 (3.5%) websites with valid manifests that also deploy a CSP with restrictions on the web app manifest, i.e., either by including the `manifest-src` or its fallback `default-src` directive. Among these websites, 8,009 (44.1%) are unprotected from a third-party attacker perspective, as they allow inline or manifests from attacker-controllable origins (e.g., `https:`, `data:`). Overall, only 10,144 (1.9%) websites disallow third-party attackers.

4 Countermeasures

Content-Security-Policy: as we have shown, deploying unique identifiers is accessible to a malicious third party. An effective solution against this attacker is to deploy a Content-Security-Policy (CSP), by specifying the `manifest-src` directive with a list of allowed origins from which the manifest must be downloaded, e.g. `manifest-src 'self'`, to disallow malicious inline and remote URIs.

Clearing Browsing Data: browsers have built-in features and third-party technologies like browser extensions [2] that make it possible to manage users' browsing data, in particular clearing cookies and similar technologies, and even unregistering service workers. Figure 3a shows the current browsing data clearing options provided by Chrome desktop browsers. We think that users should be provided with an additional mechanism to uninstall web apps as suggested in Fig. 3b for Chromium browsers. Note that this is an option that we programmatically added from the browser console for demonstration purposes, to show how it would look like. Other browsers have similar privacy mechanisms that they could augment with the ability to uninstall web apps.

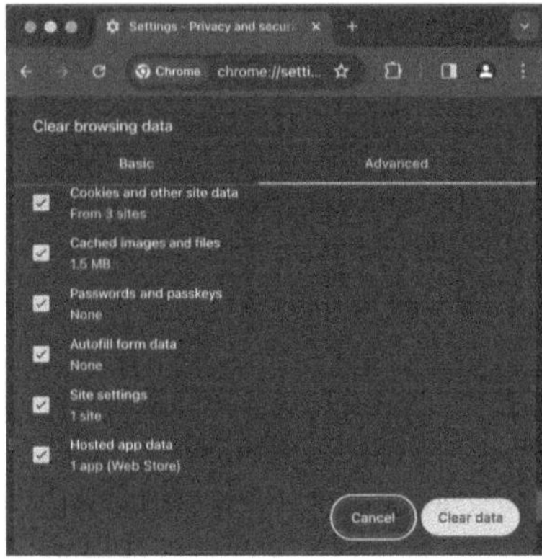

(a) Options to clear browsing data

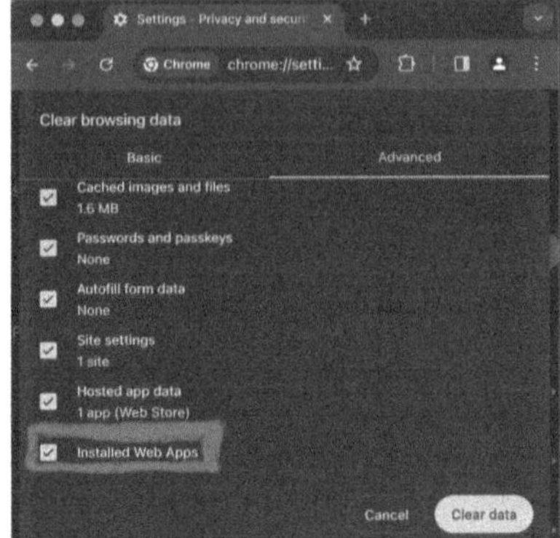

(b) Option to clear (uninstall) web apps

Fig. 3. Proposal for adding an option for the user to clear installed Web apps

5 Discussion

Responsible Disclosure: Data URI or inline manifests are only supported in Chromium browsers, the browser engine with the largest market share, that powers Chrome, Edge, Opera, Brave, etc. We reported to the Chromium team the ability to persist unique identifiers as demonstrated in Sect. 2. The vendor was able to reproduce the issue in all recent browsers and marked it as untriaged for further analysis. We also suggested the implementation of an additional option dedicated to the uninstallation of web apps. Naturally, such an option should also be reflected in other mechanisms like the `browsingData` API of extensions [2]. At the time of this writing, we are not aware of any time and whether such an option will be provided to users.

Differences with Native Apps: one may argue that it is expected that installed web apps permanently persist unique information in the user's device. We want to argue that the persistence of unique identifiers in web app manifests is an undesirable side effect and not a feature. Apart from their integration to the user's device, installed web apps do not enjoy special features compared to non-installed websites. With that said, our main motivation was to sketch for browser vendors to implement additional options for users to opt out of installed apps and therefore potential tracking, including the many (Google) web apps installed by default on Chrome desktop browsers e.g., Google Docs, Sheets, Slides, Drive, YouTube, and Gmail.

Service Workers and Other Persistent Mechanisms: Service workers are the programmatic component of PWAs, bringing among others, the ability to intercept and manipulate HTTP requests and responses, receive and process web push notifications, provide offline experience to users, or even perform background syncing. Because they are persisted by browsers, service workers themselves can be leveraged to persist unique identifiers. Unlike web manfests however, service workers are cleared when the user clears all their browsing data. The Cache is the other persistent mechanism that is extensively leveraged in PWAs, in particular by SWs to store and retrieve HTTPS responses for performance or offline purposes. Just like other client-side storage mechanisms, the ability for the user to flush the cache is already supported by browsers.

6 Related Work

Traditionally, the community distinguishes stateful and stateless tracking mechanisms [14,15,17,18]. Stateful are mechanisms that store unique identifiers on the client side to reidentify users and track them. Among these, cookies are undoubtedly the most known. A class of cookies, referred to as supercookies, are all the built-in or third-party (Adobe Flash, Java Applets) mechanisms abused for persistence and tracking, that were not intended as such by browsers in the first place [12]. The capabilities we have demonstrated with web app manifests can be categorized as supercookies. Indeed, from the perspective of the browser, web app manifests were not meant to persist unique identifiers. But more interestingly, the mechanisms provided by the browser to clear websites data, like clearing the cache or removing cookies do not affect installed web app manifests.

Links to Privacy Regulations: Tracking mechanisms like cookies have been under special scrutiny in particular with the advent of (inter)national privacy regulations like the General Data Protection Regulation (GDPR) [3]. With this study, we want to also raise awareness that mechanisms like service workers and web app manifests can be categorized as *similar technologies*. Though the web app does not use dedicated APIs to persist identifiers, these mechanisms are persistent by design and therefore should be held to the same standards as other client-side storage mechanisms like cookies.

7 Conclusion

In this work, we have shed a privacy light on the core components of progressive web applications. We have shown that web app manifests that comprise the metadata describing installed apps can be leveraged by malicious websites and third parties to persist unique identifiers that can be recovered to track users, bypassing privacy mechanisms where the user clears all of their browsing data. With the relaxation of requirements for installable PWAs, Chromium browsers even make it possible to turn any website into an installable app by generating on the fly an inline manifest, potentially embedding unique identifiers. We conclude that web app manifests, because they are persistent by design, can be categorized as supercookies, the most persistent and resilient type of identifier, all without relying on an external third-party mechanism like a plugin. We performed an empirical study and found that installable progressive web apps are prevalent in the wild, and reported our findings to vendors, urging for the implementation of additional options for users to uninstall web apps when they see fit, to conserve their privacy.

References

1. Chrome User Experience Report — CrUX. https://developer.chrome.com/docs/crux/
2. chrome.browsingData. https://developer.chrome.com/docs/extensions/reference/api/browsingData
3. Cookies, the GDPR, and the ePrivacy Directive. https://gdpr.eu/cookies/
4. CORS protocol. https://fetch.spec.whatwg.org/#cors-protocol
5. Data URLs. https://developer.mozilla.org/en-US/docs/Web/HTTP/Basics_of_HTTP/Data_URLs
6. Making PWAs installable - MDN. https://developer.mozilla.org/en-US/docs/Web/Progressive_web_apps/Guides/Making_PWAs_installable
7. Progressive Web Application — Add to Home screen — Manifest. https://developer.mozilla.org/en-US/docs/Web/Progressive_web_apps/Add_to_home_screen#manifest, https://w3c.github.io/manifest/
8. Same-Origin Policy. https://developer.mozilla.org/en-US/docs/Web/Security/Same-origin_policy
9. Service Workers 1 — W3C Candidate Recommendation. https://www.w3.org/TR/service-workers/
10. Technical features of PWAs. https://developer.mozilla.org/en-US/docs/Web/Progressive_web_apps/Guides/What_is_a_progressive_web_app#technical_features_of_pwas
11. UTM parameters. https://en.wikipedia.org/wiki/UTM_parameters
12. Kamkar, S.: Evercookie (Supercookie). https://en.wikipedia.org/wiki/Evercookie, https://github.com/samyk/evercookie
13. KirstenS: Cross Site Scripting (XSS). https://owasp.org/www-community/attacks/xss/
14. Laperdrix, P., Bielova, N., Baudry, B., Avoine, G.: Browser fingerprinting: a survey. ACM Trans. Web **14**(2) (2020). https://doi.org/10.1145/3386040
15. Mayer, J.R., Mitchell, J.C.: Third-party web tracking: policy and technology. In: IEEE Symposium on Security and Privacy, SP 2012, 21–23 May 2012, San Francisco, California, USA, pp. 413–427. IEEE Computer Society (2012). https://doi.org/10.1109/SP.2012.47
16. Pochat, V.L., van Goethem, T., Tajalizadehkhoob, S., Korczynski, M., Joosen, W.: Tranco: a research-oriented top sites ranking hardened against manipulation. In: 26th Annual Network and Distributed System Security Symposium, NDSS 2019, San Diego, California, USA, 24–27 February 2019. The Internet Society (2019). https://www.ndss-symposium.org/ndss-paper/tranco-a-research-oriented-top-sites-ranking-hardened-against-manipulation/
17. Solomos, K., Ilia, P., Karami, S., Nikiforakis, N., Polakis, J.: The dangers of human touch: fingerprinting browser extensions through user actions. In: 31st USENIX Security Symposium (USENIX Security 22), pp. 717–733. USENIX Association, Boston, MA, August 2022. https://www.usenix.org/conference/usenixsecurity22/presentation/solomos
18. Su, J., Kapravelos, A.: Automatic discovery of emerging browser fingerprinting techniques. In: Proceedings of the ACM Web Conference 2023, pp. 2178–2188. WWW '23, Association for Computing Machinery, New York, NY, USA (2023). https://doi.org/10.1145/3543507.3583333
19. West, M.: Content Security Policy Level 3. https://www.w3.org/TR/CSP3/

PayRide: Secure Transport e-Ticketing with Untrusted Smartphone Location

Marazzi Michele(✉), Patrick Jattke, Jason Zibung, and Kaveh Razavi

Computer Security Group, ETH Zurich, Zürich, Switzerland
mmarazzi@ethz.ch

Abstract. The smartphone location is the basis for a plethora of popular applications, such as traffic navigation, games, and geotagging. Since the user can manipulate the reported location, it is possible to compromise these applications with fake locations. These attacks generally have a limited impact, but this is changing with the increasing level of trust in the smartphone location. As a prominent example, recent transport e-ticketing applications perform *financial transactions* based on the assumption that the smartphone location represents that of the user. Unfortunately, this assumption leads to location-based attacks with direct financial implications. We present FreeRide, a real-world attack that allows a malicious user to ride public transports for free. Existing mitigations against FreeRide are either ineffective or impractical since they attempt to enforce the integrity of the smartphone location. Instead of enforcing location integrity, our proposed mitigation, PayRide, establishes the user's location using the position of the public transport. We have formally verified the PayRide protocol and evaluated its boundary conditions based on a range of possible accuracies reported by the smartphone and public transport.

Keywords: Smartphone · Location · GPS · Transport · Attack

1 Introduction

Virtually everyone with a smartphone today can report their geographic location to mobile applications. This has led to a viable market that companies have been quick to capitalize on. Examples range from navigation to online dating services. More recent ones include the mobile applications of public transport companies in several European countries that rely on the smartphone's location to provide automated e-ticketing [7,30]. These applications make the crucial assumption that the location reported by the smartphone is the same as the user's location. This assumption leads to practical location-spoofing attacks, which we demonstrate in this paper. To protect against these attacks, transport e-ticketing applications must preserve the link between the user and the smartphone despite inaccuracies and lack of trust in the location data provided by the smartphone. This paper presents PayRide, the first formally verified mitigation that preserves this link and protects such applications against location-spoofing attacks.

F. Maggi et al. (Eds.): DIMVA 2024, LNCS 14828, pp. 261–282, 2024.
https://doi.org/10.1007/978-3-031-64171-8_14

Smartphone Location Security. Mobile applications implicitly trust the provided interface for receiving location information. Thus, these applications are exposed to Iago attacks [2], where either software or hardware manipulates the location information before it reaches the implicitly trusted interface. Recent Sybil attacks on Google Maps, for example, show that it is possible to fake traffic jams by spoofing the location of multiple emulated devices [5].

A straightforward approach to mitigate such attacks is to provide the smartphone with a mechanism to obtain a *proof of location*, either via anchor nodes [12,15,22] or from other smartphones in the vicinity [8,19,33]. The application can use the information from the nodes to attest the location provided by the smartphone. These mitigations are not deployed due to the cost of deploying a new infrastructure and the possibility of peer collusions. Another set of mitigations instead provides *location integrity* by ensuring that the GPS sensor data is signed by a Trusted Execution Environment (TEE) [9,14,25]. The application can then verify the integrity of the signed location information. Although location integrity is potentially interesting to mitigate known (low-severity) attacks, it does not provide sufficient protection in all scenarios of interest.

FreeRide. Recent transport e-ticketing applications make financial transactions based on the assumption that the location of the user is the same as the location reported by the smartphone. As we show in this paper, this extension of trust has dire security implications. In the context of malware, previous work has discussed that a malicious application could redirect requests intended for the TEE on a target device to another malicious device [3,21]. We make a key observation that reversing this attack, i.e., using location information from another phone, enables location-spoofing attacks that bypass location integrity in the context of transport e-ticketing. We show a real-world instantiation of such attacks, which we call FreeRide, on the automated e-ticketing application of the Swiss public transport called SBB EasyRide. SBB EasyRide is designed to track the movement of a user based on the location provided by the smartphone and to charge the correct fare based on the distance traveled by different means of public transport. FreeRide enables free traveling across Switzerland while generating a valid ticket. The same applies to public transport of other European countries that rely on the exact same backend technology as SBB EasyRide [7].

PayRide. Given the increasing popularity of location-based e-ticketing applications and the severity of attacks such as FreeRide, developing a secure and deployable mitigation is of the utmost importance. Preserving the link between the user's actual location and the location reported by the smartphone is necessary for designing such a mitigation, which requires overcoming three main challenges. First, the location reported by the smartphone cannot be trusted. Second, the location information is inherently inaccurate, making it difficult to establish this link. Last, potentially unreliable Internet connections imply that this link cannot be preserved and verified in real-time.

We address these challenges in the design of PayRide, which is, to the best of our knowledge, the first solution to secure transport e-ticketing applications against location-spoofing attacks. To overcome the first challenge, PayRide relies on the transport location as a trusted anchor. This means that our second challenge must consider the location inaccuracies of both the smartphone user and the transport. We formalize and calculate the minimum distances that can be charged based on these inaccuracies. Using these calculations, PAYRIDE can then distinguish malicious from honest users without needing to trust the reported location. To address the last challenge, PAYRIDE relies on a TEE to enforce location sampling and to preserve the order of the recorded information for calculating ticket prices or issuing fines in retrospect. Our evaluation of PAYRIDE considers its boundary conditions based on the smartphone location accuracy, and our formal analysis proves its security.

Contributions. We make the following contributions:

1. We provide a security analysis of emerging transport e-ticketing applications, showing that existing mitigations cannot adequately protect against location-spoofing attacks.
2. To demonstrate the practical relevance of this threat, we build FREERIDE, a real-world attack against the Swiss public transportation mobile application, SBB EasyRide.
3. We design and evaluate PAYRIDE, a new mitigation against attacks such as FREERIDE, and we formally verify the security of its protocol.

Responsible Disclosure. We coordinated a responsible disclosure with SBB and were able to discuss FreeRide and its mitigation PayRide in a meeting with the company. SBB has assured us that effective measures have since been implemented to prevent the use of PAYRIDE. We would like to point out that using the public transport without a valid ticket in Switzerland has consequences under criminal law. During our experiments with FreeRide, we therefore always carried a valid ticket.

2 Background and Motivation

2.1 GPS on a Smartphone

A smartphone obtains its accurate location using a global navigation satellite system (GNSS). The acronym "GPS" refers to the GNSS used in the United States. Because of its widely accepted use, we will use GPS interchangeably with GNSS throughout this paper. GPS is based on multiple satellites orbiting around the Earth. Each GPS satellite continuously broadcasts a message to the Earth, which is then received by the device. The message contains a time and a position that allows the device to calculate its coordinates on Earth. Before the era of smartphones, GPS was exclusively used for its designed purpose: assisted navigation. GPS allows a user to pinpoint their location on a map in challenging environments such as deserts or oceans. Once smartphones became popular, applications like Google Maps made this service available to everyone. Further, the Internet connectivity of smartphones allows using the user's location in participatory sensing applications. For example, Google Maps links aggregated user locations to functional information such as car jams and crowded areas.

2.2 Current Attacks on the User's Location and Mitigations

Attacks on the user's location have focused on two scenarios: spoofing the GPS of the victim's device and poisoning participatory sensing databases [5,13,23,28, 31,32]. In the first case, the victim is the user relying on their device to navigate. An attacker spoofing the GPS signal can hijack the victim's movement. In the second case, the attack victims are applications that rely on the user's location for participatory sensing. Researchers have shown that an attacker can poison a database by faking many identities, known as a Sybil attack [5]. In the context of the user's location, this translates to confusing services, such as making Google Maps report a car jam on a completely empty street.

Mitigations. To mitigate these attacks, a substantial amount of research describes how to generate a user's "proof of location". The majority of proof-of-location systems require deploying new infrastructures such as secure access points that attest a user's position [12,15,22]. However, none of them has been deployed due to the high cost of new infrastructures. A second category proposes to use nearby "witnessing" peers to verify a user's location [19,33]. This raises privacy concerns and requires a certain number of trusted peers nearby.

Trusted Execution Environment. Previous work suggests moving the GPS sensors reading to a restricted Trusted Execution Environment (TEE) that would sign the data and guarantee its integrity [9,14,25]. Even with the highest privileges, a user cannot read or modify the memory areas of the TEE. The user will access the TEE functionalities via specific API calls, typically used for cryptographic operations [16]. Recent work shows potential security problems of using TEEs as a trusted anchor [3,21]. In particular, if malware is deployed on a victim machine, the user might have their TEE requests hijacked to a maliciously controlled TEE. As we will show in this paper, variations of these issues also affect transport e-ticketing applications.

2.3 Transport e-Ticketing

An emerging class of applications tracks the user's location to provide automatic ticketing for public transports [26]. While traveling, the user can show an application-generated ticket, which is deemed valid by ticket controllers. Then, when the user has reached their destination, these applications analyze the journey to determine the traveled distance, and the user is billed accordingly. This mechanism was first introduced in Switzerland [26], and more recently, extended to other countries such as Austria [34] and Germany [6]. It is currently being implemented for Belgian [29] and French [7] public transports.

We show for the first time that the divergence from the intended GPS purpose results in direct financial benefits to malicious users, who can perform practical attacks on transport e-ticketing. This raises the following research question that we address in this paper: *can we achieve secure transport e-ticketing by relying on the untrusted location provided by the smartphone?*

3 Threat Model

We assume a victim transport e-ticketing application that relies on the smartphone location to provide services to the malicious smartphone user. In particular, the victim application generates a public transport ticket and bills the user at the end of their journey. The fare is derived from the user's collected location data. We assume that the malicious user wants to gain a direct financial benefit by providing the application with locations that do not correspond to their real physical positions. The aim of the attacker is to evade the detection of standard ticket control techniques that catch classic fraudulent behavior. In other words, we consider classical fraudulent behavior that would apply to standard tickets out of scope. This includes, for example, using public transport without any ticket, hoping not to get controlled. The application backend can rely on a TEE deployed on the user's smartphone, which is considered secure. We consider the user's GPS signal to be untrusted, which means that a malicious user can spoof and control their reported GPS location. We assume the transport company knows the location of their transport vehicles in real-time and that a controller can be present on the transport to perform checks on the user's ticket. These are standard assumptions for companies employing e-ticketing.

4 Challenges Overview

Our first challenge is to demonstrate that the critical assumption that the smartphone location is the same as the user location can be exploited by a malicious user targeting e-ticketing applications.

Challenge (C1). Demonstrating location-based attacks on real-world transport e-ticketing applications.

To solve this challenge, we describe the architecture of transport e-ticketing applications and the attacks against them, which we refer to as FREERIDE (Sect. 5). Then, in Sect. 6, we perform real-world FREERIDE attacks on the Swiss public transport mobile application, *SBB Mobile*. Building these attacks required us to reverse-engineer parts of the application. We show the scalability of FREERIDE by implementing location spoofing as a service, which significantly increases the impact of the attack by making it accessible to non-technical users.

Our next challenge is understanding whether any existing solution is practical and can secure transport e-ticketing applications.

Challenge (C2). Anaylsis of the security of existing proof-of-location proposals and their applicability to transport e-ticketing applications.

We address this challenge in Sect. 7, where we present our security analysis of existing mitigations towards attacks such as FREERIDE. We show that the existing proposals cannot protect e-ticketing applications or are severely impractical. Therefore, we consider the end-to-end deployment of a mitigation and study the design requirements to make it practical and secure. Our last challenge is developing a mitigation based on the derived requirements.

Challenge (C3). Designing a secure mitigation against location-spoofing attacks for transport e-ticketing applications.

We address this challenge with the design of PAYRIDE, our proposed mitigation discussed in Sect. 8. PAYRIDE preserves the link between the location of the smartphone and the user without needing to trust the position reported by the user. Designing PAYRIDE requires answering the following fundamental questions: (i) How are GPS accuracies handled? (ii) How are communication delays addressed? (iii) Which impact do GPS accuracy and communication delays have on the application security? We provide a detailed analysis of PAYRIDE boundary conditions based on the smartphone location accuracy and formally prove the security guarantees of its protocol. Because multiple European states share the same backend technology provider [7], FREERIDE urges for the deployment of PAYRIDE.

5 Functioning and Security of e-Ticketing Applications

We now describe the functioning of tracking-based e-ticketing applications.

Honest Behavior. A user taking public transport starts the application and checks in to activate the tracking mode. From now on, their location is tracked as they travel, and a QR code representing a valid ticket is created. The ticket becomes immediately valid as the user can be controlled at any time during their journey. In case of a ticket inspection, the user shows the QR code, which is verified by the controller's application. After they reach their destination, the user stops the tracking mode and is billed according to the crossed geographical *zones* during their journey. Crossing more zones will result in a higher final ticket price. If the ticketing application does not detect that the user has traveled on public transport, the user is not billed [27].

Dishonest Behavior. The public transports covered by e-ticketing applications do not employ physical barriers for checking the tickets. Instead, passengers traveling without a ticket are fined once a controller checks them. This is the same for both e-ticketing and paper-based ticketing users. To avoid fines, users must activate the tracking functionality *before* accessing the train and keep it active throughout the whole journey, as the user can be checked at any time. Users that have started the tracking mode *after* the controller entered the transportation are fined. In the next section, we will provide more details about the functioning of e-ticketing applications and reverse engineer SBB EasyRide.

Ticket Applicability. Depending on the transport company and pricing model, the tracking ticket can be valid for multiple means of transportation. In the case of Swiss public transport, the same ticket is valid for all buses, trams, and trains. **For simplicity and without loss of generality, we consider trains as an example in the remainder of this paper.**

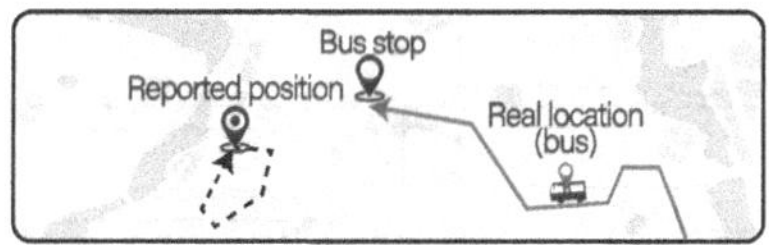

Fig. 1. FreeRide$_{\text{local}}$. A malicious user is on public transport. The application backend receives location updates corresponding to walking, providing the malicious user with a free but valid ticket.

5.1 Attacking Public Transport e-Ticketing

In our attack FREERIDE$_{\text{LOCAL}}$, a malicious user makes use of the public transport service without paying for a ticket or getting fined. To this end, they exploit the public transport e-ticketing application, performing a local (on-device) location-spoofing attack. During the attack, the malicious user wants the application backend to receive user locations that do not correspond to a train ride. The locations can be crafted to appear as realistic movements, representing low-fare pathways or simulating low GPS accuracy with high communication delays. In Fig. 1, we report an example of FREERIDE$_{\text{LOCAL}}$. In this scenario, the malicious user generates a walking pattern to bypass straightforward mitigations, such as invalidating a ticket corresponding to a user who is standing still. As we discuss in Sect. 7, this simple attack could be mitigated with GPS integrity supported by TEE [14,25]. However, slight variations bypass these mitigations, as we show next.

5.2 Bypassing Mitigations

We now discuss how an attacker can exploit the fact that *the location of the smartphone and the user are not necessarily the same* to bypass strong mitigations. Conceptually, this type of attack is similar to the *Cuckoo attack* originally commented in the context of malware [3,21]. In a Cuckoo attack, malware installed on the victim device redirects requests that are originally intended for the local TEE to an attacker-controlled device. In this scenario, the victim cannot verify that they are interacting with their trusted device and not a malicious one. The owner of the device is thus the victim of the attack. In the context of wireless systems, this might also be interpreted as a *wormhole attack* [10].

FreeRide$_{\text{remote}}$. We make a key observation that "reversing" the Cuckoo attack, i.e., using location information from another phone, enables location-spoofing attacks that bypass location integrity in the context of transport e-ticketing. We refer to this attack as FREERIDE$_{\text{REMOTE}}$. This attack type applies to any scheme that aims to obtain location integrity via a TEE or similar technologies. This is because the backend cannot verify that it is communicating with the smartphone that the malicious user is *physically traveling with* in place of a different one.

In FREERIDE$_{\text{REMOTE}}$ (Fig. 2), two or more phones are used; one that the malicious user carries and others at different locations. For simplicity, in this example, we consider only one extra phone at a different location. Whenever requested, an

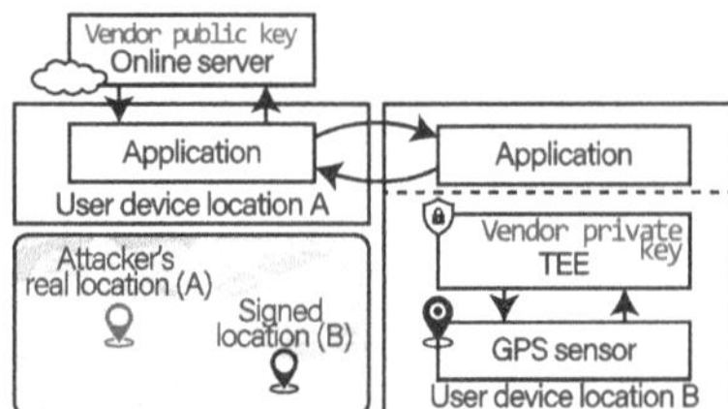

Fig. 2. FreeRide$_{\text{remote}}$ attack scheme. An online server requests the user's location. The malicious user forwards the request from the first to the second device. The second device is used to sign a different location with its TEE and sends it back.

honest user sends the position reported by the smartphone to the application. A malicious user, however, *replays* the decoy location provided by the second phone (placed at a different location) to the application. Consequently, without any possible way to check the link between the location of the user and the smartphone, the application believes that the location of the user is the same as the (decoy) location of the second phone. Variations of this scenario are possible. For example, the phone is *not* required to be completely still. It could be held by a different user who does not take public transport, or it could be kept on local trains that remain in the same billing zone to travel a longer distance for a reduced fare. Further, the second device might contain the user private key.

Impact on Mitigations. The impact of FreeRide$_{\text{REMOTE}}$ is severe as it bypasses proposed mitigations based on location integrity [9]. These mitigations propose that a secure device (e.g., TEE) would sign the user location as proof of its integrity. However, the phone with the decoy location forwards a signed location to the phone carried by the malicious user. As such, the application cannot detect the ongoing attack. Unfortunately, of the previously proposed mitigations, location integrity is the only practically deployable solution, as we discuss in Sect. 7. Our mitigation, discussed in Sect. 8, assumes that the user location is untrusted and will not rely on location integrity to provide security.

6 FreeRide

We now describe our attacks performed on the e-ticketing application of the Swiss national railway company, SBB Mobile. First, we explain how Android applications obtain the user location (Sect. 6.1). Then, we obtain important insights by reverse engineering the SBB application (Sect. 6.2), and based on them, we execute both FreeRide$_{\text{LOCAL}}$ and FreeRide$_{\text{REMOTE}}$ (Sect. 6.3). We conclude by demonstrating a Location-Spoofing-as-a-Service platform (Sect. 6.4). To the best of our knowledge, this is the first time in the literature that a location-based attack creates a *direct* financial advantage for the user.

6.1 Android Location API

To protect the users' privacy, applications that seek to obtain the smartphone location need to be granted specific permissions. The permissions are divided into (i) foreground, (ii) background, (iii) approximate, and (iv) precise location. The difference between foreground and background permissions is the allowed duration of the location access, while precise and approximate refers to the location accuracy. Lastly, the application will further specify requirements such as the location update interval and priority, both affecting power consumption.

Fused Location. The API provides a fused location, which is based on (i) power consumption, (ii) interface availability, and (iii) reported accuracy. GPS is preferred if network scanning is unavailable or if a higher precision is needed. Due to the nature of the transport e-ticketing applications, i.e., high precision and coverage, the location relies on the GPS sensor. Currently, to the best of our knowledge, there exists no TEE implementation of an integrity-protected GPS.

6.2 Application Reverse Engineering

We reverse engineer the SBB application (i.e., APK) using the dex-to-Java decompiler *jadx*. Using static analysis, we do not find any efforts to prevent the execution on a rooted device. To understand how the application works while the location data is used, we instrument it using the dynamic instrumentation framework Frida. EasyRide (the application tracking functionality) uses a rebranded version of the tracking technology by `fairtiq`, a service provider that generates automatic billing for various partners in Europe [7], including SBB. This is confirmed by the relevant EasyRide code being located in the `com.fairtiq.sdk` package. By dynamically instrumenting the application, we record its behavior during a normal trip and make the following four observations.

(O1) Tracking Events. The application sends various events to the online server. The events describe the beginning and end of a ticket's validity, the device location, and the identified user activity. The application page specifies that the activity data is used to send a reminder to the user to perform a checkout [27].

(O2) Location Provider. The application calls the fused location provider for the user position. To understand the type of request sent (see Sect. 6.1), we dynamically hook the necessary function call. We found that the location is requested with the "accurate" level and has an update interval of 2 s. With our mitigation (Sect. 8), we will describe how the update interval plays a major role in ensuring application security.

(O3) Location Processing. The raw data location is transmitted to the backend without any preprocessing. In particular, the information sent to the server contains the latitude and longitude of the user position.

(O4) Messages Batching. We found that data can be sent in batches and is flushed every 30 s. That means even though the location data is sampled frequently **(O2)**, it can be sent with large delays. As we will discuss (Sect. 8), reducing data delays plays an essential role in mitigating attacks.
Given our observations **(O1–O4)**, we conclude that only the location data is considered for the fare calculation. Next, we proceed with the design and evaluation of our FREERIDE attacks.

6.3 Attacks Design and Evaluation

The attacker's goal is to deceive the SBB application to pretend that they are standing still or walking, while in reality, they are using public transport. Based on our observations, we hypothesize that the ticket billing only depends on the GPS location. However, a limitation of the previous analysis is that we do not know how the data is processed in the backend. To better understand this, we exploit FREERIDE$_{\text{LOCAL}}$ by testing three different movement models of the spoofed location: (i) a fixed position (i.e., no movement), (ii) a simulated walk that does not cross stations, and (iii) a simulated walk that crosses stations.

FreeRide$_{\text{local}}$. In FREERIDE$_{\text{LOCAL}}$, the Android *location service* returns spoofed values to an API call in a completely transparent way for the victim application. To achieve this, we build an Android application that spoofs the user location and reproduces the three different types of movement. The application is designed as a mock location service, which is typically used to protect the user's privacy or to aid application development. The attack only requires changing the return value of the `isMock()` API, which we perform using the Smali Patcher tool. For our attack, we used a Pixel 5 running on Android 11, which we rooted using Magisk.

FreeRide$_{\text{remote}}$. Our second attack can bypass location integrity. As described before, we use two devices: one is static (the decoy), and the other one is traveling with the malicious user. Because there is currently no location integrity mechanism deployed in TEEs, we simulate it by keeping the decoy location fixed. For simplicity and without loss of generality, the decoy runs on an emulated Android. We provide more details about device emulation in Sect. 6.4. We generate a valid ticket on the decoy device, and we transmit it to the malicious user's phone. To perform the following attack evaluation, we use the latest unmodified version of the application installed from the Google Play Store.

Evaluation. We verified that in all three categories of spoofed movement (i.e., standing still, not crossing any stations, and crossing stations) with FREERIDE$_{\text{LOCAL}}$, as well as FREERIDE$_{\text{REMOTE}}$, the application always generates a ticket. The case in which the movement starts at a station and crosses a different one is the only occasion that results in the user being billed. This indicates that the attacker has complete control over the application by changing the location only. However, even if the application correctly generates a valid ticket, it could be that once a ticket controller checks it, some internal controls are applied to counteract fraud. To evaluate this, we performed several trips with different free tickets and were repeatedly checked by ticket controllers. To reduce our attack tests to the minimum necessary, we performed them on a limited distance that only crossed two zones and we always carried valid tickets as well. On all occasions, the free ticket was always considered to be valid by the controllers and never followed by any charges. Because the SBB application is used for all buses, trains, and trams, the attack allows a malicious user to travel for free across Switzerland.

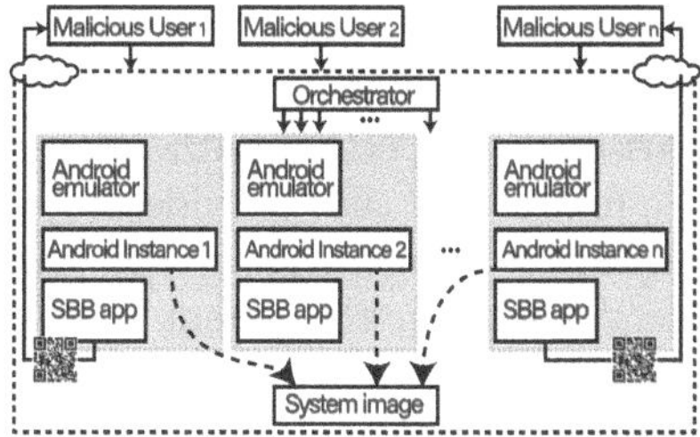

Fig. 3. LSaaS architecture. Malicious users connect to an online webpage and receive a valid and free ticket (QR code). Internally, the LSaaS orchestrator initializes the required copies of a virtualized Android environment. Each copy has a shared base system image that is prepared for the SBB application.

Table 1. Comparison of location-spoofing efforts. Of the practically-evaluated attacks, FREERIDE is the first that provides a malicious user a direct financial benefit and is executed considering signed GPS data. It is also the first attack that scales to non-technical users via LSaaS.

Effort	Direct financial benefit	Practically evaluated	On-device spoofing	GPS-signal spoofing	FreeRide$_{remote}$ attack	LSaaS
FreeRide	●	●	●	○	●	●
GPS attacks [32]	○	●	○	●	○	○
Sybil attacks [5]	○	●	●	○	○	○
Hobbyist/gaming [20]	○	●	●	○	○	○

6.4 LSaaS: Location-Spoofing-as-a-Service

We now demonstrate that it is practical for a malicious user to develop a *Location-Spoofing-as-a-Service* (LSaaS) platform, extending the attack to non-technical users. To this end, we designed an LSaaS platform for FREERIDE as depicted in Fig. 3. We used Android Studio, which is bundled with a device emulator based on QEMU. The emulator runs an entire device, including GPS and networking, and the GPS location can be changed programmatically by using `adb`. Upon starting their ride, a malicious client obtains a valid ticket from the web server. First, the user provides account credentials for the SBB application. Then, the internal orchestrator creates and starts an ad-hoc Android emulator instance. The instance boots a preconfigured system with the SBB application, logs the user in, starts a new journey, and returns the QR code of the valid ticket to the user. The position remains fixed throughout the journey, resulting in no charge to the malicious user. We confirmed that the SBB application cannot detect that the source of the GPS location is fictitious, as the location service reports `false` for the `isMock()` API.

Scaling Emulated Devices. The orchestrator (Fig. 3) clones emulated devices on demand. To allow the LSaaS to scale to a large number of emulated devices, (i) a common system image is shared across all devices and (ii) a resource-optimized device configuration is used with a single core and a low-resolution screen.

In conclusion, by implementing a Location-Spoofing-as-a-Service, we showed that it is simple to scale our attack to a larger number of non-expert users. Table 1 provides a comparison of FreeRide with other known location-spoofing attacks. FreeRide is the first attack that provides a direct financial benefit for the attacker while bypassing location integrity. The results of $\text{FreeRide}_{\text{LOCAL}}$, $\text{FreeRide}_{\text{REMOTE}}$ and LSaaS highlight the urgent need to develop and deploy a practical and secure mitigation against FreeRide and other attacks based on bypassing location integrity.

7 Mitigation Requirements

We analyze proposed mitigations and evaluate their protection against the two variants of FreeRide. While we present previous work, we derive requirements that will guide us to the design of our mitigation PayRide (Sect. 8).

7.1 Existing Mitigations

In 2009, it was predicted that location would become a key aspect of mobile applications [24]. Since then, many researchers have focused on ways to generate proof of location for devices. However, the vast majority of these solutions are not practical, and as we will show, the few practical ones do not protect against $\text{FreeRide}_{\text{REMOTE}}$. There are three categories by which previous work can be classified into: (i) mitigations requiring a new infrastructure, (ii) mitigations involving peer interactions, and (iii) mitigations relying on signed GPS data.

Mitigations Requiring a New Infrastructure. These mitigations are based on deploying new infrastructure and require an access point in each area to be mapped [12,15,22]. An access point provides a marker for verifying that the user is present at a particular location. However, there are multiple limitations to this approach: (i) the deployment of such infrastructure does not scale with the expansion of the world roads, (ii) without relying on GPS, the location accuracy may be strictly limited, and moreover, (iii) applications would still be vulnerable to $\text{FreeRide}_{\text{REMOTE}}$. For these reasons, we do not consider this type of mitigation to be a viable solution. We argue that deploying a new infrastructure is a significant obstacle to the real-world adaption of a mitigation. Hence, we derive the following requirement for our mitigation:

Requirement (R1). PayRide should be based on already existing and deployed infrastructure (i.e., GPS).

Mitigations Involving Peer Interactions. This group of mitigations is based on a paradigm named *community-verified locations* [19,33], in which the validation of a user's position is delegated to peers in their proximity. The approach (i) assumes that there are always multiple users at the same location, which might become difficult at low-traffic locations and during off-peak hours, (ii) requires that people are close enough to establish an ad-hoc network connection, (iii) may interrupt the user's connectivity as either Bluetooth or WiFi would need to be disconnected temporarily, (iv) requires a mechanism to identify malicious peers providing false location data, and (v) potentially compromises the users' privacy. Similar to the first class of mitigations, these proposals are impractical. As we consider the reliance on other users limiting and a potential security risk, we define the following requirement:

Requirement (R2). PAYRIDE should provide application security without relying on any other users.

Mitigations Relying on Signed GPS Data. Lastly, some work proposed signing the GPS location data using a TEE or any equivalent secure environment [9,14,25]. In these setups, device vendors embed a private key into the TEE memory during manufacturing. This key never leaves the device and is used to sign the GPS sensor values. After the online backend receives the signed user location, it verifies the validity of the signature using the device vendor's public key. To protect against replay attacks, the backend generates and transmits a random nonce to the device. This nonce must be included in the signed message and is incremented in each new position update.

Such a scheme cannot be used to protect against FREERIDE$_{\text{REMOTE}}$. First, the physical signal of the GPS can still be spoofed. Second, even if the source of the user location would be more secure than GPS, the TEE cannot validate the link between the user and device, which forms the basis of FREERIDE$_{\text{REMOTE}}$. Third, there is no way to prevent a malicious user from starting tracking directly on a different device. For the same reasons, the attack would also succeed if more complex schemes are employed, such as TEEs including the user's private key.

Mitigating FreeRide$_{\text{remote}}$ Attacks. Relying on FREERIDE$_{\text{REMOTE}}$, a malicious user appears to have a perfectly valid position outside or inside of the public transport network. Their position is valid, as it can be signed with a trusted key. This leads us to the third requirement for our mitigation:

Requirement (R3). PAYRIDE should be able to match the physical location of the user to the location reported by the smartphone. However, it should assume the reported location to be untrusted.

7.2 The Ticket Controller

We now define the requirements for the ticket controller. There are multiple aspects to consider: first, relying on the ticket controller to visually inspect a reported user location creates a serious privacy concern caused by exposing the user's location history.

Second, differentiating between an honest and a malicious user might not be straightforward for the ticket controller. A malicious activity might be difficult to manually identify due to imprecise location data, network connection delays, or substantial variations in the user's movement speed. Lastly, the ticket validation should not require complex interactions for the ticket controller. We conclude with our last requirement:

Table 2. Comparison of proposed mitigations. Proposals are either secure (✓) or vulnerable (✗) depending on the class of attack considered. Depending on their requirements, they can be practical (✓) or not (✗). No previous mitigation protects against FreeRide$_{\text{REMOTE}}$ attacks.

Approach	On-device spoofing	GPS-signal spoofing	FreeRide$_{\text{remote}}$ attacks	Practical
PayRide	✓	✓	✓	✓
New infrastructure [15]	✓	✓	✗	✗
Peer interactions [8,19,33]	✓	✓	✗	✗
GPS signing [9]	✓	✗	✗	✓

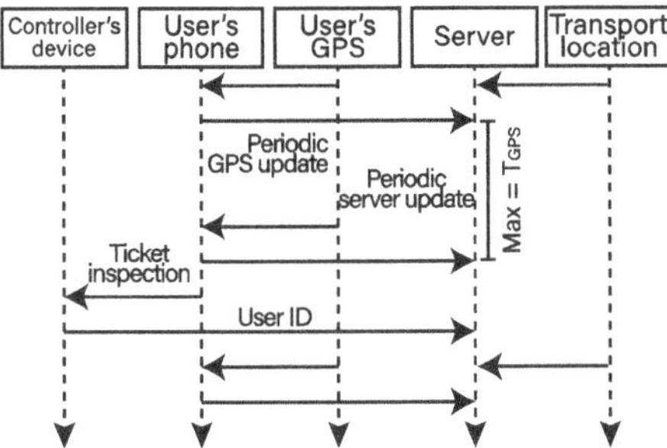

Fig. 4. PayRide protocol. The user periodically sends position updates to the application server. These updates must happen within a certain time limit. If a controller verifies a ticket, the user identifier is sent to the server. The server verifies that the user's location is within an acceptable range from the transport location.

Requirement (R4). PayRide should not rely on the ticket controller to check if the user is spoofing its location.

Given these four requirements, we now design our new mitigation PayRide. Table 2 summarizes the differences between the requirements in the design of PayRide and existing mitigations.

8 PayRide

We now explain the principles of PayRide (Sect. 8.1), formalize its constraints (Sect. 8.2) and discuss how to safely define them (Sect. 8.3). Finally, we describe its formal verification model (Sect. 8.4) and conclude by describing the extension PayRide$_{\text{TEE}}$.

8.1 Operating Principles

With PayRide, a user who spoofs their location in a way that would give them financial benefit is equivalent to a user who does not possess a valid ticket. Both users are fined when their ticket is checked at any point along their journey. In the case of tracking ticket, the fine will be issued by the server backend. Following the requirements **R1–R2**, we have designed PAYRIDE to rely exclusively on the reported GPS location. Instead of relying on location integrity, PAYRIDE security is based on regular position updates sent by the user. While PAYRIDE does not depend on trusting the sampled location (**R3**), TEE enables it to handle poor internet connectivity, as discussed in Sect. 8.5 with PAYRIDE$_{TEE}$.

PayRide Protocol and Ticket Validation. We summarize the PAYRIDE protocol in Fig. 4. The user periodically provides position updates to the server backend. These positions can correspond to the real user locations (GPS sensor) or carefully crafted pathways. If a maximum cumulative delay between updates (T_{GPS}) is not respected, the ticket is considered invalid. We relax this requirement in Sect. 8.5. As standard practice of the affected parties, the ticket can be controlled one or multiple times at any given point in time during the journey. The controller behavior remains the same, scanning the user's ticket and verifying that it was created before they entered the train (**R4** and Sect. 5). With PAYRIDE, the user's identifier is sent to the application backend. In the backend, PAYRIDE validates the user's location and considers that position for the fare calculation.

The user is fined if their reported location is not inside a defined margin of error from the transport location, considering the prior and next position, as reported by the user. A user moving faster than the train is considered malicious because users could misuse it to fake movements from the previous station to their current position during ticket inspections. In what follows, we formally describe the allowed error margin, the required timing for the location updates, and how they relate to PAYRIDE security.

8.2 Formalizing Restrictions

PAYRIDE should prevent two cases: (i) a user traveling for free (i.e., not crossing any stations) and (ii) a user paying less than they should (i.e., tracking movements across fewer zones). We consider fraudulent behaviors that apply to standard tickets (e.g., not having a ticket) to be out of scope.

Modeling the Margins of Error. To define a valid user position, we need to consider (i) the GPS margin of error and (ii) the update delay. The *GPS error margin* corresponds to the distance between the real physical location of the user and the one determined by the GPS (Fig. 5-❶-❸). We refer to the maximum allowed error as l_u for the user's position and as l_c for the transport position. The *update delay* represents the cumulative time between two consecutive location updates. This includes the time needed to calculate the GPS position and the time required for a network packet to travel to the backend server. Without loss of generality, we refer to the GPS update period (T_{GPS}) as the maximum time between position updates, measured at the time they arrive at the server (Fig. 5-❺). A longer delay results in an invalid ticket. We consider this maximum period T_{GPS} to be the same for user and transport location updates.

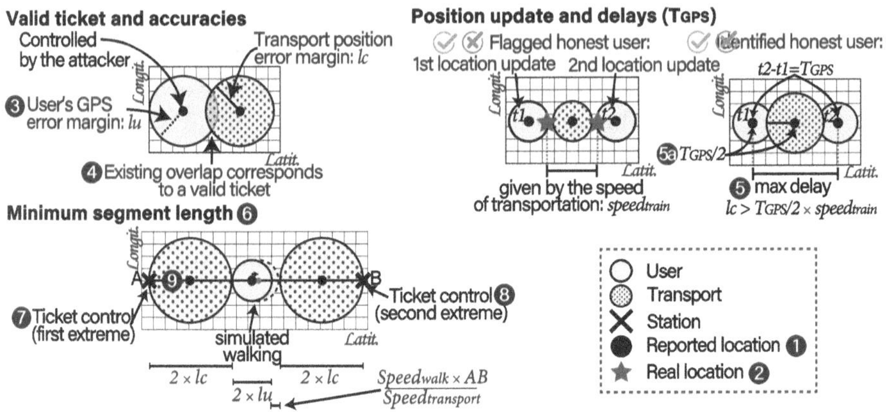

Fig. 5. Maximum delay and GPS accuracies. The user and train have a reported (❶) and a real (❷) position. The real position is within the radius l_u and l_c from the reported position (❸). The ticket is considered valid if there exists an intersection between the allowed error margin of the train and of the user's position (❹). The maximum accepted delay for a position update depends on the transportation speed and the train GPS error margin l_c (❺). The minimum segment length that can be secured (❻) depends on the allowed margins of error. For it to be secure, the user must never be able to generate a free-but-valid ticket, regardless of where they are controlled along the journey (❼, ❽).

These margins need to be fixed for PayRide to correctly distinguish between an honest and a dishonest user. To be noted, l_c and T_{GPS} are not independent: if the allowed l_c is very small, it is difficult for the reported train position to match the user's position, even if they are close. This effect is worsened by a slow update rate (T_{GPS}) on a moving vehicle. If the update is not frequent enough, and the train is moving, the *reported* user's position can differ substantially from the train. However, PayRide needs to correctly identify an honest user, and hence, it is essential to consider the relationship between the locations reported by the user and the train. Note that l_u limits the shortest trip that PayRide can protect. For example, if l_u is larger than a train ride, a malicious user can spoof a static location throughout the entire trip, which would be correct upon ticket control. We calculate in Sect. 8.3 the minimum distances that PayRide can protect.

GPS Update and l_c. The user's location needs to be updated frequently enough for an honest user to be correctly recognized during the ticket control. Because the train moves, the user's location is likely stale when compared in the backend with the train position. The distance between the user's real position (represented by the train position) and the stale position becomes smaller with a higher sampling rate. Given a certain T_{GPS}, we now describe the l_c boundaries to ensure honest users are always correctly identified.

For our analysis, we assume the worst-case scenario where the train position reaches the backend between two user location updates (Fig. 5-5a). In other words, we consider that the user's location is stale by $\frac{T_{GPS}}{2}$. This is the worst condition, as PAYRIDE considers the prior and next position reported by the user. Given the train movement of $speed_{train}$, the total distance is equal to $speed_{train} \times \frac{T_{GPS}}{2}$. Therefore, given the maximum speed and that both the user and train can report an inaccurate location, we require l_c to be higher than the following threshold to correctly identify an honest user:

$$l_c > \frac{T_{GPS}}{2} \times speed_{train} \tag{1}$$

8.3 Attack Based on the Error Margins

A malicious user can exploit location errors to falsely appear static (i.e., performing *distance fraud*). This allows them to travel for free if not controlled, while holding a valid ticket in case they are. We now analyze the link between location errors (l_c and l_u) and the shortest journey that PAYRIDE can secure.

Definitions. For a malicious user to possess a valid ticket, their reported position must always match the train reported position within the defined margins. This has to hold for the entire trip, as otherwise, there would exist a position along the route where the malicious user would be fined if checked. This is shown in the example in Fig. 5-6, where a controller is checking tickets at the beginning of the trip (7) or at the end (8) on a train moving from station A to station B (length $\overline{AB}$). We remind the reader that a user who moves faster than a defined walking speed ($speed_w$) is billed as traveling.

Shortest Secured Journey. We now describe the attack that a malicious user can perform by exploiting l_c and l_u. After that, we derive the shortest journey of length $\overline{AB}$ such that malicious users cannot obtain a free but valid ticket.

As discussed, the malicious user aims to spoof their location such that (i) any ticket control along the journey $\overline{AB}$ is valid and (ii) their movement is never faster than walking speed, as otherwise, it would result in a fare. Intuitively, as shown in Fig. 5, the user can spoof their location to be in between A and B. This ensures that the error margin, $2 \times l_u$, covers as much as possible of their journey, increasing their chances of a matching position with the train. If the journey $\overline{AB}$ is shorter than $2 \times l_u$, a ticket where the user is standing still is always valid. We now extend the calculation by also considering the train location margin.

The first position where the user can be checked is at the beginning of the journey (7). The real position of the train is A, but the reported position can be, in the worst case for the mitigation, distant by l_c (9). This distance enhances the possible matching area of a malicious user. The last position where the user can be checked is at the end of the journey (8) with the same possible error of l_c. Because the segment is contiguous, it follows that if the user has a valid position where both of these two extremes report a valid match, any position along the journey will be considered valid. Any non-linear connection between A

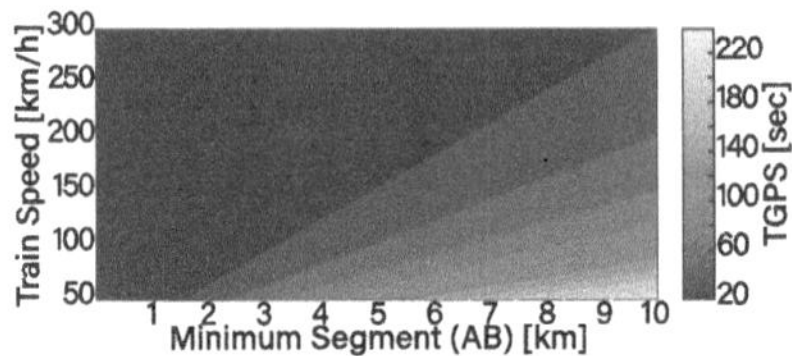

Fig. 6. Location update period. Relationship between ($\overline{AB}$), the train speed, and valid values for T_{GPS} (Eqs. (1) and (3)). The faster the train, the more frequent location updates are required.

and B can be reduced to a straight connection, which represents the worst-case for the mitigation. Therefore, the train margin of error adds $2 \times l_c$ matching coverage on the part of the segment *before* the user and adds $2 \times l_c$ matching coverage on the part of the segment *after* the user, for a total of $4 \times l_c$.

As a last step, we must consider that the user can generate a spoofed location that moves. The malicious user can further increase the covered length that matches with the train by slowly moving their position. The maximum coverage is obtained if they move at the maximum speed ($speed_w$) for the entire duration of the trip ($\frac{\overline{AB}}{speed_{train}}$). Summing all the contributions, we obtain the following maximum segment length that allows a malicious user to travel for free ($\overline{AB_{free}}$):

$$\overline{AB_{free}} = speed_w \times \frac{\overline{AB_{free}}}{speed_{train}} + 2 \times l_u + 4 \times l_c \approx 2 \times l_u + 4 \times l_c \tag{2}$$

It follows that the condition on the segment length to achieve security is:

$$\overline{AB} > 2 \times l_u + 4 \times l_c \tag{3}$$

Example of Margins of Error. We now provide example values for a train ride. We assume a high-speed transportation of up to 250 km/h with a minimum distance between stations of 5 km and a maximum walking speed of 5 km/h. As an example, we consider location accuracies of $l_u = l_c = 500$ m which satisfy Eq. (2). Given Eq. (1), this results in a T_{GPS} of roughly 14 s. Figure 6 shows T_{GPS} for an $\overline{AB}$ in the range between 500 m and 10 km, and for a maximum train speed between 50 km/h and 300 km/h.

8.4 Tamarin Verification

We use Tamarin [17] to formally verify the correctness and security of PayRide. Tamarin is a symbolic formal verification tool widely used for protocol verification [1]. To this end, we define a formal Tamarin model of PayRide, based on the PayRide protocol and the described margins of error. In the model, two positions that are considered the same represent two locations that fall inside the allowed margins of error, which we defined before. We model the slowest possible (i.e., *worst*) GPS update period T_{GPS}. The user can behave in a malicious way (dishonest) by sending a *fake* location or behave correctly (honest) by sending correct location updates. The controller *might* check the user's ticket at any time. There is no restriction imposed on the user's behavior, meaning that we allow users to send a mix of honest and dishonest position updates. Lastly, we assume that the communication between the controller's device and the server is secure and that the train position is trusted.

Given the model, we prove that if the user is checked during a ride (i) a malicious user is always fined (i.e., no false negatives) while (ii) an honest user is never fined (i.e., no false positives). The formal model is simple (170 lines of code) and does not require implementing new cryptographic primitives on top of an already secured communication channel. The tool takes less than a minute to formally verify the model properties.

8.5 Handling Unreliable Internet Connections

So far, PAYRIDE requires Internet connectivity to secure applications. A user with a loss of connectivity for more than T_{GPS} will incur a fine. This is an emerging condition due to Eq. (1). Yet, it might be difficult to meet this in practice in areas with frequent loss of Internet connectivity. We now discuss how PAYRIDE can be extended to use TEEs to remove this requirement.

We assume a TEE to be present, as they are available on most modern computing devices. As described earlier (Sect. 7.1), TEEs cannot be used to sign the integrity of the GPS signal to provide security. However, we consider the TEE by itself to be secure. For this reason, we use TEE not to protect integrity but to protect the **order** and **sampling period** of the positions that the client must send to the backend. We refer to our novel design as PAYRIDE$_{\text{TEE}}$.

TEE to Enforce Timing. Previous work [9,14,25] employ TEEs to achieve integrity and preserve order of the sampled data. However, we make the observation that ordering alone does not entail *timing*. If a specific period for the location updates is not enforced, the application backend can receive a valid but outdated user location. For example, a transport e-ticketing application could receive trusted location updates from a walking user while the user is currently on a train.

PayRide$_{\text{TEE}}$. All the described principles of PAYRIDE and our formalizations remain the same. However, when using a TEE, T_{GPS} represents the location *sampling* period. By using a secure interrupt, the TEE samples the location respecting T_{GPS} and signs it while including a counter to preserve the order. In the communication with the backend, a nonce is included to prevent replay attacks. In case of Internet connectivity loss, the signed updates are stored, and once the connectivity is restored, sent to the backend. To avoid attacks based on *stopping* location updates, the check-out operation is performed after the TEE has sent *all* position updates to the backend. In case the smartphone is powered off while the user is still checked in, the user is considered traveling without a valid ticket.

9 Related Work

We summarize research that focuses on (i) spoofing the GPS signal, (ii) poisoning participatory sensing applications, and (iii) spoofing the user's location.

Spoofing GPS Signals. For many years, efforts have been made to demonstrate that GPS is not secure [13,23,31]. In these attacks, differently from FreeRide, the device owners are the victims of GPS spoofing. Typical attacks include jamming the GPS signal [28] or hijacking guided directions [32]. The main targets of GPS spoofing attacks have been unmanned aerial vehicles (UAVs), but they could also target people using driving assistants on the streets or the open sea.

Poisoning Participatory Sensing Applications. Recently, researchers showed that participatory sensing applications are vulnerable to poisoning attacks [5], which is in line with previous work [11,25]. Authors were able to artificially create areas detected as car traffic jams in Google Maps, making the application believe that a street is busy. Sybil attacks are a well-known threat for participatory applications and peer-to-peer systems [4,18], in which malicious users influence the system functioning by using multiple identities.

Spoofing User Locations. User-location spoofing has been used by hobbyists to unlock applications or features that are based on certain locations. For example, games like Pokémon GO rely on the user's location to provide events to players and were subject to location spoofing attacks in the past.

10 Conclusions

We show that the assumption that the location of the smartphone matches that of the user has fundamental security implications for transport e-ticketing. We show-case this using a real-world attack, called FreeRide, that enables a malicious user to travel with a valid *free* ticket across Switzerland. Our analysis of existing mitigations shows that they are insufficient to protect against attacks such as FreeRide. To fill this gap, we designed PayRide, the first secure mitigation for transport e-ticketing applications. To make PayRide practical, we had to overcome a number of challenges, including relying on untrusted smartphone locations, GPS inaccuracies, and sporadic connectivity. We formally verified the PayRide protocol and evaluated its boundary conditions based on the smartphone location accuracy.

Acknowledgments. We thank our anonymous reviewers for their valuable feedback and Ralf Sasse for his help with Tamarin. This work was supported by the Swiss National Science Foundation under NCCR Automation, grant agreement 51NF40 180545, and the Swiss State Secretariat for Education, Research and Innovation under contract number MB22.00057 (ERC-StG PROMISE).

Disclosure of Interests. The authors have no competing interests to declare that are relevant to the content of this article.

References

1. Basin, D., Cremers, C., Dreier, J., Sasse, R.: Symbolically analyzing security protocols using TAMARIN. ACM SIGLOG News **4**(4), 19–30 (2017)
2. Checkoway, S., Shacham, H.: Iago attacks: why the system call API is a bad untrusted RPC interface. ACM SIGARCH Comput. Archit. News **41**(1), 253–264 (2013)
3. De Oliveira Nunes, I., Ding, X., Tsudik, G.: On the root of trust identification problem. In: ACM IPSN 2021 (2021)
4. Douceur, J.R.: The Sybil attack. In: Druschel, P., Kaashoek, F., Rowstron, A. (eds.) IPTPS 2002. LNCS, vol. 2429, pp. 251–260. Springer, Heidelberg (2002). https://doi.org/10.1007/3-540-45748-8_24
5. Eryonucu, C., Papadimitratos, P.: Sybil-based attacks on Google maps or how to forge the image of city life. In: ACM WiSec 2022 (2022)
6. FAIRTIQ Ltd.: Area of Validity. https://fairtiq.com/en/passengers/area-of-validity
7. FAIRTIQ Ltd.: Partnership with FAIRTIQ. https://fairtiq.com/en/partner-with-fairtiq/public-transport-agencies
8. Gambs, S., Killijian, M.O., Roy, M., Traoré, M.: PROPS: a privacy-preserving location proof system. In: IEEE SRDS 2014 (2014)
9. Hu, H., Chen, Q., Xu, J., Choi, B.: Assuring spatio-temporal integrity on mobile devices with minimum location disclosure. IEEE Trans. Mob. Comput. **16**(11), 3000–3013 (2017). https://doi.org/10.1109/TMC.2017.2683492
10. Hu, Y.C., Perrig, A., Johnson, D.B.: Wormhole attacks in wireless networks. IEEE J. Sel. Areas Commun. **24**(2), 370–380 (2006)
11. Huang, K.L., Kanhere, S.S., Hu, W.: Are you contributing trustworthy data? The case for a reputation system in participatory sensing. In: ACM MSWiM 2010 (2010)
12. Javali, C., Revadigar, G., Rasmussen, K.B., Hu, W., Jha, S.: I am alice, i was in wonderland: secure location proof generation and verification protocol. In: IEEE LCN 2016 (2016). https://doi.org/10.1109/LCN.2016.126
13. Larcom, J.A., Liu, H.: Modeling and characterization of GPS spoofing. In: IEEE HST 2013 (2013)
14. Liu, H., Saroiu, S., Wolman, A., Raj, H.: Software abstractions for trusted sensors. In: ACM HotMobile 2010 (2010). https://doi.org/10.1145/2307636.2307670
15. Luo, W., Hengartner, U.: Proving your location without giving up your privacy. In: ACM HotMobile 2010 (2010). https://doi.org/10.1145/1734583.1734586
16. McGillion, B., Dettenborn, T., Nyman, T., Asokan, N.: Open-TEE: an open virtual trusted execution environment. In: IEEE TrustCom 2015, vol. 1 (2015)
17. Meier, S., Schmidt, B., Cremers, C., Basin, D.: The TAMARIN prover for the symbolic analysis of security protocols. In: Sharygina, N., Veith, H. (eds.) CAV 2013. LNCS, vol. 8044, pp. 696–701. Springer, Heidelberg (2013). https://doi.org/10.1007/978-3-642-39799-8_48
18. Newsome, J., Shi, E., Song, D., Perrig, A.: The sybil attack in sensor networks: analysis & defenses. In: IEEE IPSN 2004 (2004)
19. Nosouhi, M.R., Sood, K., Yu, S., Grobler, M., Zhang, J.: PASPORT: a secure and private location proof generation and verification framework. IEEE TCSS **7**(2), 293–307 (2020)
20. Paay, J., Kjeldskov, J., Internicola, D., Thomasen, M.: Motivations and practices for cheating in Pokémon GO. In: ACM MobileHCI 2018 (2018)

21. Parno, B., McCune, J.M., Perrig, A.: Bootstrapping Trust in Modern Computers. Springer, New York (2011). https://doi.org/10.1007/978-1-4614-1460-5
22. Pham, A., Huguenin, K., Bilogrevic, I., Dacosta, I., Hubaux, J.P.: SecureRun: cheat-proof and private summaries for location-based activities. IEEE Trans. Mob. Comput. **15**(8), 2109–2123 (2015). https://doi.org/10.1109/TMC.2015.2483498
23. Psiaki, M.L., Humphreys, T.E., Stauffer, B.: Attackers can spoof navigation signals without our knowledge. Here's how to fight back GPS lies. IEEE Spectr. **53**(8), 26–53 (2016)
24. Saroiu, S., Wolman, A.: Enabling new mobile applications with location proofs. In: ACM HotMobile 2009 (2009). https://doi.org/10.1145/1514411.1514414
25. Saroiu, S., Wolman, A.: I am a sensor, and i approve this message. In: ACM HotMobile 2010 (2010). https://doi.org/10.1145/1734583.1734593
26. SBB: EasyRide - the ticket that does things your way—SBB. https://www.sbb.ch/en/timetable/mobile-apps/sbb-mobile/easyride.html
27. SBB: Help with EasyRide
28. Setiadji, M.Y.B., Aji, B.P., Amiruddin, A.: Deceiving smart lock trusted place in Android smartphones with location spoofing. In: IEEE ICOIACT 2020 (2020)
29. SNCB: Seamless Ticketing. https://www.belgiantrain.be/en/about-sncb/en-route-vers-mieux/innovation/seamless-ticketing
30. Swiss Federal Railways: The SBB online portal for trains and public transport—SBB. https://www.sbb.ch/en
31. Tippenhauer, N.O., Pöpper, C., Rasmussen, K.B., Capkun, S.: On the requirements for successful GPS spoofing attacks. In: ACM CCS 2011 (2011)
32. Zeng, K.C., Shu, Y., Liu, S., Dou, Y., Yang, Y.: A practical GPS location spoofing attack in road navigation scenario. In: ACM HotMobile 2017 (2017)
33. Zhu, Z., Cao, G.: Toward privacy preserving and collusion resistance in a location proof updating system. IEEE Trans. Mob. Comput. **12**(1), 51–64 (2011)
34. ÖBB Group: SimplyGo! https://www.oebb.at/en/tickets-kundenkarten/online-mobile-ticketing/oebb-app/simplygo

Knocking on Admin's Door: Protecting Critical Web Applications with Deception

Billy Tsouvalas(✉) and Nick Nikiforakis

Stony Brook University, Stony Brook, NY, USA
{vtsouvalas,nick}@cs.stonybrook.edu

Abstract. In this paper, we introduce PageKnocker, a deception-based supplementary authentication mechanism, aimed primarily at protecting the public-facing authentication endpoints of critical web applications. PageKnocker is inspired by the network security concept of port knocking, and uses the requests to the web application as a means of authentication for the login page. Specifically, the authentication is successful (i.e. the user gets to access the login page of a web application), if the user's request sequence matches their personal predefined request sequence. In this manner, PageKnocker offers web application administrators the comparative advantage of knowing the nature of a visitor even before that visitor sends the first set of credentials. Alongside PageKnocker, we introduce two deceptive login environments, one overtly deceptive and one clandestine, towards which we direct any unauthenticated user attempting to reach the real login page. To evaluate the security and usability of page knocks, we deploy PageKnocker-protected honeypots in the wild and perform a separate user study, showing that PageKnocker can resist tens of thousands of brute-forcing bots, while remaining usable and intuitive.

1 Introduction

Following the increasing popularity of interactive web services over the last two decades, reliance on web applications for everyday tasks has become common place. Whether it be email services, e-commerce, online learning, news outlets, or social media, web applications have permeated almost every aspect of the online experience. As a result, web applications have become primary targets for malicious actors, whose objectives span a wide array, ranging from violations in confidentiality and data privacy to larger scale operational disruptions.

The main and first line of defense of web applications is credential-based authentication. In order to accommodate users, web applications provide login pages, where users are asked to provide their usernames and passwords. These public-facing login endpoints are of critical importance since they are the entry points to sensitive and private parts of the web application. Apart from implementation-based security vulnerabilities [1], login pages as public-facing endpoints are inherently unprotected, since anyone can access the login page and attempt to bypass the authentication. Particularly for the authentication

F. Maggi et al. (Eds.): DIMVA 2024, LNCS 14828, pp. 283–306, 2024.
https://doi.org/10.1007/978-3-031-64171-8_15

endpoints intended for a web application's administrator, there is no reason for them to be accessible (and therefore targetable) by all visitors of that web application.

With the weaknesses of credential-based authentication being well-documented (including users choosing low-quality passwords and reusing them across different websites [2–4]), alternative and/or supplementary authentication schemes have been proposed and adopted. Although such supplementary authentication techniques tend to improve certain usability and security aspects of web authentication, they have not managed to sufficiently and effectively address the issues surrounding passwords [5–8]. Thus, credential-based authentication still remains the primary, and oftentimes, sole means of authentication for the majority of web applications.

This combination of easily-accessible authentication endpoints *and* weak or reused passwords has been abused extensively by credential stuffing bots, with estimates as high as one in five of all authentication requests coming from malicious automated systems [9]. In credential stuffing attacks, attackers obtain username, password pairs from third-party sources (most commonly products of breaches [10]) and employ bots to test these credentials by injecting them in the respective fields of a web application's login page. Given the elevated privileges of administrators, it follows that administrator accounts are very often primary targets in such campaings [11], with examples such as the Uber 2016 data breach that lead to over 57M personal details of Uber users being leaked after a credential stuffing attack targeted the company's administrators and developers [12].

Recognizing the unprotected nature of public-facing login endpoints, and considering the extent of bot-driven credential stuffing attacks against administrator accounts, in this paper, we introduce PageKnocker, a novel deception-based supplementary authentication mechanism on top of passwords, which protects public-facing authentication endpoints. The authentication scheme of PageKnocker is based on request sequences, meaning that a user is granted access to the login page by making a predefined user-specific sequence of requests to the web application. For example, in order to access the https://example.com/admin authentication endpoint, PageKnocker enforces that a user must first request https://example.com/about followed by https://example.com/categories/ and https://example.com/news, all of which are preexisting pages on the protected web application, which are turned into page-knocking sequences.

PageKnocker is inspired by the widely known network security concept of port knocking [13]. By employing PageKnocker, we capitalize on the familiarity of administrators with the web application - specifically with its User Interface (UI) - and we manage to effectively protect public-facing login pages, while allowing for a usable system that requires no special software or hardware. Furthermore, PageKnocker works as a gateway to deception, by introducing a deceptive layer to the authentication, enabling effective active defenses and attacker behavior analysis, all while protecting exposed endpoints of the web application. This paper makes the following original contributions:

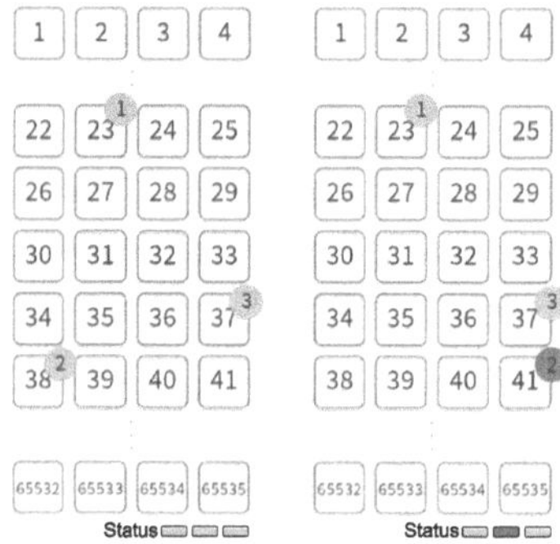

Fig. 1. Successful (left) and unsuccessful (right) port knocking completion on arbitrary port knocking sequence $\{23, 38, 37\}$. Numbered square represent the system ports, while numbers on the upper-right edges of the squares represent the order of the knocks.

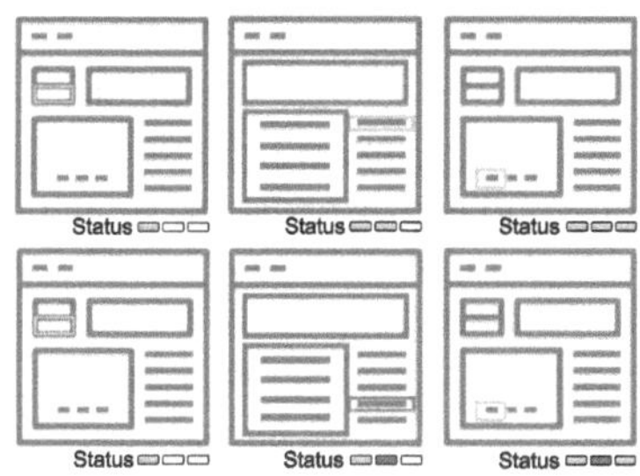

Fig. 2. Successful (top) and unsuccessful (bottom) page knock completion on abstract webpage example. Green contours represent correct knocks, while red ones represent incorrect ones. (Color figure online)

- We introduce PageKnocker, a novel application-agnostic deception-based authentication mechanism based on request sequences, aimed at limiting access to web-application authentication endpoints.
- We implement PageKnocker using two different deceptive techniques related to what attackers see when trying to illicitly access a protected endpoint.
- We evaluate PageKnocker using both honeypots in the wild (collecting over a million requests from over 21,000 distinct IP addresses) as well as through a user study with security experts as participants.

To inspire additional research in the area of deception-based defenses, the source code along with a video demo of PageKnocker are available at https://pageknocker.github.io/.

2 Background

PageKnocker draws inspiration from the network-security concept of "port knocking" [13]. Port knocking is employed as an additional authentication method, to allow only authorized clients to reach a specific network port of a remote server. Namely, port knocking allows access to a client only if it has sent packets to a secret predefined sequence of ports of the server.

Taking the SSH daemon as an example, TCP port 22 is by default closed to all connecting clients on a server employing the port-knocking mechanism. If a client sends a sequence of packets to other ports on that server in the correct order, that client is considered authorized and will now be able to connect to the SSH daemon listening on port 22 (that port remains closed for everyone else except that specific client). Figure 1 shows a high-level visual representation of the port-knocking concept in action. For an arbitrary port knocking sequence consisting of the ordered set of ports $\{23, 38, 37\}$, a client is authenticated if they send packets to port 23, then to port 38, and finally to port 37. The left part of Fig. 1 represents a successful port knocking attempt, while the right part depicts a failed authentication attempt. Once a client successfully completes the port-knocking authentication, it can now establish a connection to that network service and proceed with traditional password-based/key-based authentication.

In the system described in this paper, we employ the same logic where users are allowed to reach the public-facing login page of the web application, only if they have previously visited the correct URLs of the web application in the correct order. We underline that, while PAGEKNOCKER is inspired by the concept of port knocking, the two are completely different in design and implementation, primarily due to being active at different layers of the OSI model (application layer vs. transport layer).

3 Design

PAGEKNOCKER is a supplementary authentication mechanism on top of passwords that protects the public-facing login page of critical web applications. It operates by evaluating a sequence of requests to the web application against predefined user-specific sequences of requests. The user chooses their own page knocking sequence, in the same manner that they would choose their passwords, and they are allowed to reach the login page of the web application, only if they recreate that sequence.

3.1 Page Knocks

We define a page knock as a sequence of GET requests to the web application. These requests can be arbitrary but they work best when they are on contiguous (i.e. sequentially reachable) pages, so that users can conveniently navigate through them. To authenticate, the user needs to make a specific series of requests to the web application in a particular order. If this request sequence matches the user's predefined page knocking sequence, they are allowed to proceed (e.g. be allowed to access the web application's login form). If web clients try to visit the PAGEKNOCKER-protected authentication endpoint *before* finishing their page knock, PAGEKNOCKER can either deny them access, or engage them in cyber deception.

In the context of page knocks, matching a user's request sequence to a predefined page knock refers to finding a subset of the user's requests, which includes the predefined knocks in the correct order. For example, provided a predefined page knocking sequence: (A_2, A_5, A_{17}) where A_i represent pages of the web application, for $i = 1, ..., n$, with n being the total number of available pre-authentication pages of the web application, $(A_1, A_3, A_2, A_5, A_{17}, A_2)$ is a correct match, while (A_2, A_1, A_5, A_{17}) is not. Putting page-contiguity aside, the possible combinations of a page knocking sequence of length N is n^N.

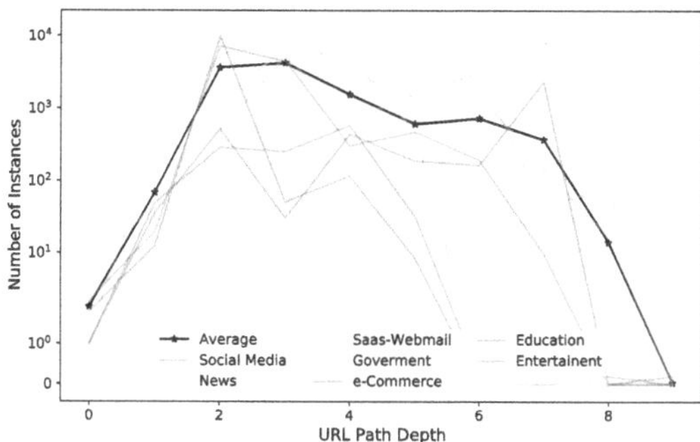

Fig. 3. URL Path Depth for Different Website Categories

In this manner, a PAGEKNOCKER-protected web application is able to distinguish between legitimate users and attackers even before they reach the web application's login form. Thus, PAGEKNOCKER allows us to adjust our defenses based on a priori information and effectively protect the main authentication of the web application. Figure 2 shows a visual representation of how users interact with a PAGEKNOCKER-protected web application in the context of a three-page knock. The length and specific pages of a knocking sequence can be selected in any manner of way, e.g., capturing the actions that a user already performs when they visit a webpage before they attempt to authenticate.

Since the strength of a page-knocking sequence is tied to the number of contiguous, pre-authentication pages that are available to users, we evaluate the number of such pages on different types of popular web applications. Using the Tranco dataset [14], we collect all the pages of the most popular websites which make their sitemap available and categorize them based on their target industry. Since page knocks are made up of contiguous pages, we divide the collected pages by path depth, i.e. *example.com/foo* has a depth of one, *example.com/foo/bar* has a depth of two, etc. We present the number of pages per path depth of the website categories along with their average in Fig. 3. We observe that on average, websites contain upwards of 1,000 webpages for a path depth between two and four. This demonstrates that there is a significant number of webpages that may be employed as parts of a page knock sequence. Moreover, considering the pages of other path depths as well, we confirm the substantial number of possible page knocking sequence combinations for all website categories.

3.2 Applicability and Deception Strategies

PAGEKNOCKER's primary objective is to protect the public-facing login page of web applications by introducing an additional layer of authentication to the login endpoint that is invisible to web clients. Protecting the login page is of critical importance, since the vast majority of login pages are public-facing and accessible by anyone, leaving them entirely unprotected against brute force and credential stuffing attacks. Table 4 of the Appendix lists the predefined login page paths for administrators and backend users for the top ten Content Management System (CMS) based on their market share. We observe that for the vast majority of CMS applications (over 70% of the market share), the publicly available predefined login page paths allow anyone to reach a web application's sensitive login form by merely requesting a well-known path.

PageKnocker allows web applications to differentiate between legitimate users who are attempting to login vs. attackers who have stolen credentials or are about to start brute-forcing their way into the application. Given an incomplete pageknock, web applications can deny access to the login form, request more information in the form of Multi-Factor Authentication (MFA) and the solving of CAPTCHAs, or engage users in deception. In this paper, we focus on the last approach by implementing support for completely hiding a login form (i.e. giving users an HTTP 404 error) or allowing them to reach a login form but always denying them access, even if attackers provide correct credentials. The former strategy can be used to confuse attackers about the nature of the web application (in relation to Table 4 of the Appendix), whereas the latter keeps them engaged in order to study their authentication attempts and protect other systems from attacks.

3.3 Threat Model

PageKnocker can protect all publicly-accessible authentication endpoints from attackers launching brute-force attempts or operating with stolen credentials. While all authentication endpoints are in scope for PageKnocker, it is best suited to protect endpoints which are intended to be used by a small number of administrators and are the ones primarily targeted by web bots [11].

Even though PageKnocker is situated at the server side and therefore completely invisible to benign and malicious clients, we operate under the assumption that an attacker may have knowledge of the existence of page knocks, but no knowledge of the specific sequences chosen by users.

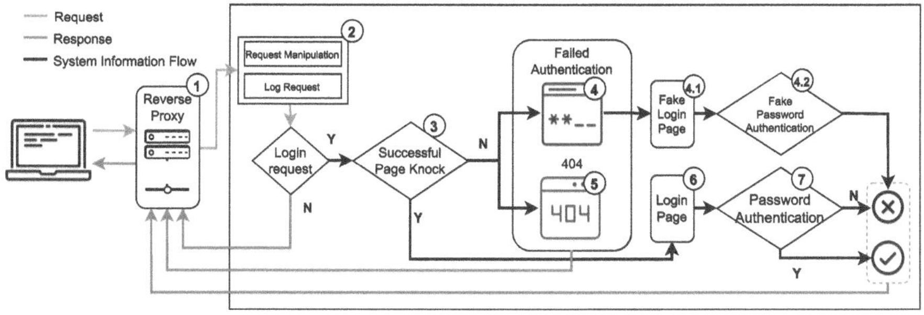

Fig. 4. **PageKnocker** architecture with request and response flows

3.4 Security Properties

Here, we briefly describe a number of non-obvious security properties of PageKnocker that differentiate it from existing authentication countermeasures.

- **No special software/hardware necessary.** PageKnocker does not require any special software or hardware. The organic clicks on a web application's UI are converted into page knocks and used to protect authentication endpoints.
- **Resistant to shoulder surfing.** A user's page knocks on a PageKnocker-protected web application are indistinguishable from regular browsing. As such, PageKnocker increases the complexity of a shoulder-surfing attack, which now necessitates both the correct page knocks, as well as the credentials.
- **Invisible to MITM.** Unlike traditional port knocking, page knocking happens at the application layer and inherits all the confidentiality and integrity guarantees of TLS.
- **Gateway into deception.** PageKnocker can be employed alongside other technologies and, even before attackers send their first authentication attempt, transparently engage them in deception, i.e., route their traffic to different servers, serve them content containing honeytokens, etc.
- **Disincentivizing credential reuse.** Since PageKnocker turns a web application's own UI into a means of authentication, it makes it difficult for users to reuse credentials across unrelated websites (e.g. users cannot use the same page-knocking sequence for a blog and an online bank, even if they choose the same password).
- **Compatible with existing authentication mechanisms.** Along with protecting the login page of a web application, PageKnocker can be used in parallel with existing supplementary authentication mechanisms, such as MFA and Single Sign-On (SSO), thus enhancing the application's overall security stance.

4 Implementation

In terms of PageKnocker's implementation, our objective is for the system to be lightweight, application-agnostic, and capable of resisting real attackers, without being cumbersome for benign users.

4.1 Overview

Below, we outline the main functionality of the proposed framework, along with the main components that address these functions.

Containerization. PageKnocker employs Docker containers for every component of its architecture, allowing us to easily interchange parts of the system, without disrupting the remaining components.

Server-Side Request Manipulation. Given that our system needs be able to monitor and manipulate requests and responses to the web application, we employ mitmproxy [15]. We place the `mitmproxy` in front of the web application, allowing us to intercept, inspect, and manage the HTTP requests to the application server, and manipulate the HTTP responses that the web application sends to the user.

Reverse Proxy. We employ an Nginx reverse proxy, to manage the TLS certificate of the web application's domain, handle TLS termination, and forward the traffic to the `mitmproxy` and web application docker containers. This reverse proxy is placed between the client and the dockerized containers of the web application.

4.2 Request-Response Flow

Figure 4 shows PageKnocker's architecture, along with the request and response flows. Below, we elaborate on each step of these flows.

User Request - Reverse Proxy ①. A client request to the web application is first delivered to the Nginx reverse proxy. The reverse proxy handles the TLS termination and forwards the request's URI to the host of the web application. It is the reverse proxy that manages the domain redirection and manages the TLS certificates.

Server-side Request Manipulation ②. After the TLS termination of the Nginx reverse proxy, the request reaches `mitmproxy`. Regarding the received requests, we use `mitmproxy` in two ways. First, we log the incoming request, along with the IP address of the client, the cookie for our web application, and the request method, path, and timestamp. We then evaluate whether it is a login request by comparing the request's endpoint with the known authentication path of the protected web application (popular endpoints shown in Table 4 of Appendix). This is a configuration option that must be provided the first time PAGEKNOCKER is deployed to protect a web application. Non-login requests are ignored by PAGEKNOCKER, whereas login ones are processed according to the next steps. We note that the entirety of the pageknock-authentication procedure takes place in the mitmproxy module, before a request reaches the web application.

PageKnocker and Login Environments ③. Having established that a visitor is requesting access to the web application's login page, we evaluate whether they have completed their page knocking sequence. PAGEKNOCKER consults the logs that it collects for each request made to the web application, identifying each visitor by existing web-application cookies or by IP address, depending on the deployment configuration. We then associate to each user the requests that they make to the web application sorted by their timestamps. Thus, we have time-ordered request sequences for all visitors to the web application, and we can evaluate if a user has managed to successfully complete the predefined page knock. We only authenticate users whose page knocks were matched against their predefined knocking sequence within a configurable window of time (set to 10 min for our prototype). This means that a correct page knock is only valid for a short period of time, and needs to be repeated after this time expires.

Dealing with Page Knock Violations. Given that PAGEKNOCKER's authentication scheme provides information on the nature of the user before they reach the login page, we are able to adjust the response of the system. In this context, we develop two different deception-based login environments, which are triggered by a failed page knock authentication; a Failed Authentication (FA), and an HTTP 404 environment. A user that has failed the page knocking authentication, is redirected to one of these two environments, depending on the deployment configuration, upon requesting to access the actual login page.

Failed Authentication ④. The Failed Authentication (FA) login environment looks exactly like the real login page of the web application. The deceptive element of this environment lies in the functionality of the main authentication. In particular, we circumvent the credential (username/password) authentication of the web application using mitmproxy, and we deny access to the user, whatever credentials they may provide; even correct ones. Specifically, this is achieved by manipulating the web application response, when the predefined page knocking sequence has not been matched. An attacker that has failed to complete the predefined page knocks, is directed to a page that is a clone of the authentic login page (4.1), mimicking the wrong-credentials error message of the protected web application, regardless of which credentials the attacker uses (4.2).

Attackers who are unaware of PAGEKNOCKER will eventually conclude that the stolen credentials were not valid (even if they really were) against the protected web application and stop attacking. Sophisticated attackers who are certain about the validity of their credentials may eventually infer the presence of a PageKnocker-like defense system leaving them with unclear next steps. An attacker who keeps engaging with a PageKnocker-protected web application risks getting studied, fingerprinted, and potentially engaged in additional deception.

Page Not Found (HTTP 404) ⑤. When attackers who have not completed the predefined page knocks request the login page, PageKnocker responds with a HTTP 404 page. The goal of this deceptive strategy of PageKnocker is to confuse attackers regarding the nature of the attacked web application. Bots that engage in fingerprinting to identify the nature of a web application (so that they can attack the right authentication endpoint) may conclude that the fingerprinting was faulty and move to another target. We report on the behavior of bots when encountering a PageKnocker-protected web application in Sect. 5. As before, sophisticated attackers attacking a specific target may again conclude that the web application is protected by PageKnocker and is therefore engaging them in deception. Attackers who choose to keep attacking the target, do so knowing that all information that they have discovered regarding that web application may be part of a deception strategy aimed at studying them and keeping them engaged.

Login Page ⑥. The FA and 404 login environments are triggered by a request to the login page from a user who has not completed the predefined page knocks. A user that has completed the predefined page knocks is authenticated, as per the page knocks, and is therefore allowed to access the true login page and the true password-authentication component of the protected web application ⑦, without any further interventions/content modifications by PageKnocker.

5 Security Evaluation

In this section, we report on a three-month long experiment where we deployed PageKnocker-protected web applications on the public web, to understand how bots interact with them. In the next section (Sect. 6) we present a separate user study simulating both sophisticated attackers with knowledge of PageKnocker's presence, as well as characterizing the usability of our system.

To evaluate our approach against bot attacks, we deployed 12 WordPress honeypots secured by PageKnocker using different page knock pages and knock lengths. Half of the honeypots engaged attackers in failed-authentication (FA) deception (all credentials attempted by non PageKnocker-authenticated users are always wrong) and the other half presented unauthenticated attackers with our 404 (Page Not Found) approach.

For the six FA deployments, we specify two with a page knocking sequence of a single request (which we will refer to as "short" page-knock length), two with a sequence of three requests ("medium"), and two with a sequence of five requests ("long"). We establish page-knocking sequences of the same lengths for the six 404 honeypots. Lastly, we note that every predefined page knocking request sequence was unique among all of the aforementioned deployed honeypots.

Table 1. Requests received in the wild (2 honeypots per page-knock length)

Page Knock Length	Total	GET [%]	POST [%]
Short	138,919	17.19	82.15
	387,749	59.15	40.71
Medium	133,886	19.00	80.61
	128,826	42.32	57.28
Long	129,963	39.50	60.09
	128,443	19.78	79.79
Overall	1,047,786	32.82	66.77

(a) Requests to the six FA honeypots

Page Knock Length	Total	GET [%]	POST [%]
Short	21,387	94.07	4.52
	21,823	96.54	1.99
Medium	27,712	97.37	1.46
	17,028	94.52	2.55
Long	22,390	93.84	4.75
	17,918	94.90	2.14
Overall	128,258	95.21	2.90

(b) Requests to the six 404 honeypots

5.1 Experimental Setup

All 12 versions of the page-knocking-adapted web applications were deployed on the public web for a period of 90 days on newly registered domain names. We associate each request sent to PAGEKNOCKER with the IP address of the web client. The page knocking authentication is therefore carried out by identifying the page knocks that each individual IP address makes to the web application. To ensure that our honeypots are comparable to real-world deployed web applications we used Let's Encrypt [16] to obtain valid TLS certificates for all of them. An additional advantage of using TLS certificates is that, in addition to IP-address-scanning web bots, the announcement of our certificates on Certificate Transparency (CT) [17] is also likely to attract CT bots [18].

5.2 Results

In the span of three months, our deployed honeypots received 1,176,044 requests from 21,935 unique IP addresses, with 1,047,786 (89.09%) and 128,258 (10.91%) requests made to the FA and the 404 honeypots, respectively. Specifically towards authentication endpoints, the deployed honeypots received a total of 790,541 requests (67.22 % of all requests), with 761,909 (96.38%) requests towards the FA honeypots, while only 28,632 (3.62%) directed towards 404 honeypots. Regarding the effective security of PAGEKNOCKER, out of the total of 21,935 unique IP addresses, we observe only two cases of correct page-knock guesses (0.0091%).

Overall, across all deployed honeypots, we observed large discrepancies in the number of received requests between FA and 404 configurations. In order to examine these divergences, we evaluate the received requests with respect to their GET and POST distributions.

PageKnocker Authentication Robustness. Regarding the two correctly guessed page knocking sequences, in both cases, the traffic was directed towards a honeypot with a predefined page knock sequence of a single request (the easiest possible scenario). These originated from two different IP addresses, namely 34.76.104.x and 34.140.32.x which sent requests to just one of our 12 honeypots

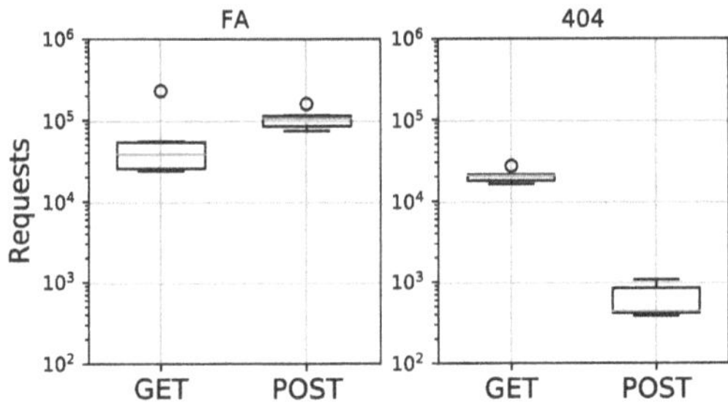

Fig. 5. Distribution of GET and POST requests in FA and 404 honeypots in the wild

exhibiting clear web-scanning behavior. They both initially made a request for a `robots.txt` file, and subsequently proceeded to consecutively request front end pages of the web application. Both were eventually redirected by the web application to the login form, when attempting to access privileged content. The fact that they managed to clear the single request page knock was due to pure chance, since they managed to knock on the correct path before landing on the login form.

Given the rarity of this event and the fact that a single request page knock represents the least secure scenario that we evaluate, we consider these cases to be statistically insignificant (two cases out of 21,935 unique IPs, which is 0.0091%). We can therefore conclude that PageKnocker successfully protected the deployed web applications from almost 800K illicit accesses.

GET and POST Requests. Tables 1a and 1b show the GET and POST requests to the FA and 404 honeypots respectively. On both tables, we present the requests sent to each honeypot, along with the predefined page knocking sequence length associated with each deployed honeypot. By observing the distribution of GET and POST requests, it becomes apparent that the POST requests are significantly fewer for the 404 honeypots. This is due to the fact that the FA login environment allows the unauthenticated user to access and interact with the login page where bots can keep trying credentials (submitted via POST requests) and keep getting wrong-credentials errors. Contrastingly, in our 404 honeypots, PageKnocker makes the login page unavailable so that the vast majority of bots stop their scanning attempts much earlier. Figure 5 shows the overall distribution of GET and POST requests across all honeypots. In accordance with the measurements of Tables 1a and 1b, we observe similar distributions for GET requests, but the accumulative distribution of requests further reveals that the discrepancy for the POST requests, approximates two orders of magnitude. We note that for clients who get a 404 error on the authentication endpoint, one would expect that the number of POST requests be near zero, since there is no page to receive the POST requests. Interestingly, we observed that that is not the case, revealing faulty application logic for a small fraction of the malicious bots that attacked our honeypots and kept on sending POST requests, even

Table 2. Common bot appearances across FA and 404 honeypots

Appearances in Different Honeypots	IPs	Only FA[%]	Only 404 [%]	Both[%]
2–4	4,582	35.49	10.30	54.21
5–6	1,696	55.66	0.06	44.28
7–9	817	0	0	100
10–12	1,496	0	0	100

after having knowledge that the login page was not there. The tables breaking down the distribution of GET/POST requests just towards the authentication endpoint are available in Tables 5a and 5b (in the Appendix).

Common Bots Across Honeypots. Since all 12 honeypots were in operation at the same time, we can identify the specific bots that targeted more than one honeypot, with a particular focus on the ones that targeted multiple honeypots across deception types. In this way, we can compare the strategies of specific bots when encountering our failed-authentication (FA) deception honeypots vs. the 404 ones.

In Table 2, we present the number of IP addresses that make requests to multiple honeypots, along with the distribution of appearances towards only FA, only 404, or both types of honeypots. As an example, we observe that there are 1,696 bots (characterized by their unique IP addresses) which make requests towards 5 to 6 independent honeypots, with 55.66 % of them directing traffic towards only FA honeypots, and 0.06 % directing traffic only to 404 honeypots. Since we have six FA and six 404 deployed honeypots, bots making requests to more than six honeypots are, by definition, reaching both types of honeypots. This is also the reason behind the selection of presented ranges of honeypots.

Overall, from Table 2, we observe that there are significantly fewer bots that make requests only towards 404 honeypots, compared to the ones making requests only towards FA, or towards both honeypot types. This means that bot interaction with FA or both types of honeypots can potentially result in a larger accumulation of bot traffic around those honeypots. We suspect that bots visiting FA honeypots do not register any abnormal behavior on the part of the web application (covert deception), while bots that visit the 404 honeypots recognize the misbehavior of the login page (overt deception) and, in some manner, flag the web application as non-conforming for future reference. It is therefore likely that FA honeypots are added to databases of potentially vulnerable websites that can be attacked again in the future, either by the same attacker or by other attackers through data exchanging. Contrastingly, the non-conforming behavior of PAGEKNOCKER's HTTP 404 honeypots may be precluding them from lists of potential targets and thereby attracting overall fewer unique attackers, compared to the FA honeypots. Focusing on the bots (as identified by their IP addresses) that appear in both types of honeypots, we present in Fig. 6 a scatter plot depicting the GET and POST requests made from bots that visit all deployed honeypots.

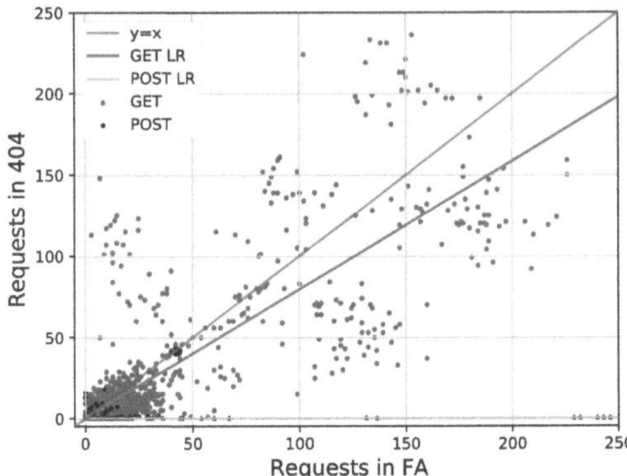

Fig. 6. GET and POST requests of common IPs in FA and 404 honeypots in the wild

The vertical axis represents the requests made to 404 honeypots, whereas the horizontal axis represents the requests towards a failed-authentication (FA) honeypot. A data point above $y = x$ indicates that there are more requests to a 404 than a FA honeypot, while the opposite is true for data points below that line. First, we observe the large accumulation of GET requests at the region of fewer than 50 requests (for both the FA and the 404), whereas the POST requests are concentrated along the horizontal axis. To extract insights from this data, we provide the Linear Regressions (LRs) for both the GET and POST requests. The LR curves demonstrate the following:

- Since the slope of the GET-requests LR curve is relatively close to that of the $y = x$, we can conclude that the GET requests between the FA and the 404 honeypots do not present a significant difference. This is to be expected from our prior experimentation, given that GET requests essentially represent reconnaissance or discovery attempts by bots, which we do not interfere with our PageKnocker mechanism.
- On the other hand, the POST-requests LR curve is practically parallel to the horizontal axis, meaning that the overwhelming majority of POST requests is directed towards FA honeypots. This matches our previous results (see Fig. 5), where we showed that the 404 deployments receive fewer POST requests than the FA ones, to the difference of two orders of magnitude. Thus, we observe that this result holds when considering common clients that appear in both FA and 404 honeypots.

We note that in Fig. 6 we also include nine IPs that demonstrate very large numbers of POST requests, which fall outside the confines of the presented plot. This was a conscious choice for ease of viewing, and we present these data points in Table 6 (Appendix). Taken together, these observations show that, for the vast majority of bots, PageKnocker triggered different paths of their attack logic. That is, given the deceptive "wrong-credentials" message, bots keep sending POST requests with additional credentials. The same bots, however, when encountering our 404-deception honeypots stopped attacking our honeypots. System administrators can therefore choose between different deception modes of PageKnocker depending on whether they want bots to go away as fast as possible (404 deception) or keep them engaged in order to study them, without any danger of successfully authenticating to the targeted web application (failed-authentication deception).

Time-Based Analysis of Requests. In this section, we evaluate the behavior of bots in terms of their successive requests towards our PageKnocker honeypots. As we established in our prior analysis of GET and POST requests, bots employ GET requests as part of their reconnaissance and proceed with POST requests when attempting to brute-force the application's authentication endpoints. Since PageKnocker disrupts this sequence of events, we examine how the bots react to our mechanism in terms of what requests they make. In Fig. 7, we present the distribution of request methods (GET, POST) for the first 1,000 bot requests towards our FA and 404 honeypots. The choice of the first 1,000 requests stems from the fact that clients making at least 1,000 requests approximate the totality of our dataset, as we show in Fig. 8 (Appendix).

In Fig. 7, we observe the average percentage of GET/POST requests to PageKnocker honeypots evaluated over the first 1,000 time-ordered requests. The horizontal axis represents the time-ordered requests, which are made by all bots interacting with the respective honeypots, meaning that a value of 200 on the horizontal axis represents the 200th request that a bot makes. Thus, we observe that for FA honeypots, bots begin their interaction with slighlty more GET requests, and after the 100th request, on average, the POST requests have clearly surpassed the number GET requests. This indicates that the bots have transitioned from reconnaissance to actual attack. This POST-majority pattern continues for the better part of the first 1,000 requests, and only presents a relatively small disparity, as we approach the last requests.

Regarding the 404 honeypots, we notice diametrically opposite results. The GET requests represent the majority, being well above 80% of the total requests for the first 200 requests. Furthermore, we notice that after initially attempting relatively few POST requests, after the 400th request, both GET and POST requests drop to zero. This reveals that the majority of bots are directed away from the application, with just a few bots with faulty application logic attempting to send POST requests despite not finding the actual authentication endpoints.

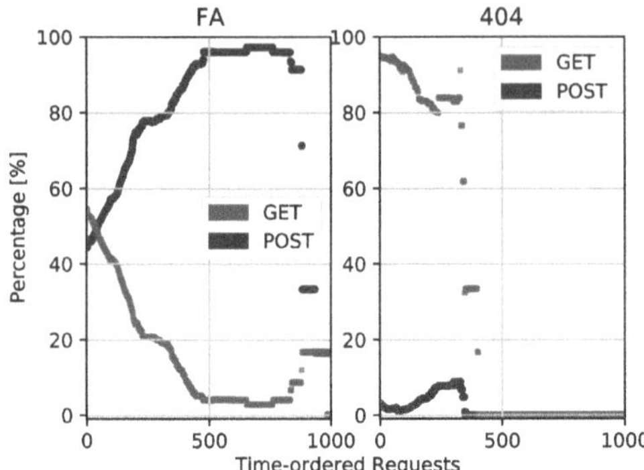

Fig. 7. Distribution of request methods in time for FA and 404 honeypots in the wild

6 User Study

In Sect. 5 we evaluated PAGEKNOCKER's ability to protect web applications against non-targeted attacks by malicious bots. In this section, we report on a study involving real users to understand i) whether they can defeat PAGEKNOCKER when they are told about its existence (simulating attackers), and ii) their ability to choose, recall, and use page knocks when interacting with a PAGEKNOCKER-protected web application (simulating web application administrators).

6.1 Setup of Study

In order to best approximate the roles for both scenarios (attackers and administrators), we recruited 16 users from our computer science department (graduate students and faculty) who work in the area of cyber security and are knowledgeable in the subject of attacks and authentication. In contrast with the 22K unique bots that attacked our PAGEKNOCKER-protected honeypots in Sect. 5, in this user study we consciously trade attack volume in favor of attacker sophistication.

The user study setting comprises of two parts addressing the aforementioned objectives. For the first part, the participants are introduced to the PAGEKNOCKER mechanism and its functionality. Subsequently, the participants are asked to carry out the two following tasks:

1. Each participant is directed towards a deployed web application that uses a predefined page knocking sequence of a *single* request. Their task is to figure out the single request, which is necessary to match the predefined sequence, and eventually log in to the web application. The participants have 10 min to complete this task.
2. The second task is the same as the first, with the exception that the predefined page knocking sequence contains *three* requests.

We note that for both tasks, the deployments are exactly the same as the in-the-wild deployed honeypots, and the participant pool was equally split between FA and 404 login environment implementations. The participants are also provided with correct login credentials, which lead to a webpage informing them of a successful task completion.

For the second part of the user study, having knowledge of PAGEKNOCKER's functionality and first-hand experience of the implementation, the participants are asked to create their own page knocking sequence. Once they submit their page knocks, they are asked three times in the two subsequent weeks after their submission, to re-visit the web application, and perform their page knock. We advised the participants to act like legitimate users of the PAGEKNOCKER-protected web applications and consider security as well as usability aspects in their selection of personal page knocks, based on their theoretical understanding and empirical experience with the framework.

6.2 Ethical Considerations

After the review of our user-study protocol by our IRB, we were informed that there was no requirement for approval or exemption, since we only collected the paths visited in our deployed web applications which contain no private, user-specific information. The participants were asked to volunteer, in the context of a security seminar, and each of them was entered in two raffles for two $50 gift cards, contingent upon their participation in the respective part of the user study.

6.3 Results

Regarding the first part of the user study, out of the 16 participants, only a single one managed to break the PAGEKNOCKER-authentication for the single-request instance, while none were able to break the three request sequence instance. This clearly demonstrates that, even though attackers may be able to brute-force their way through extremely short page knocking sequences, correctly guessing longer sequences is far from trivial.

In Table 3, we present the results of the second part of our user study. As a user identifier, we use an arbitrary index number, and for each user, we provide four metrics: the completion results of their personally submitted page knocking sequence upon their three returns in the two subsequent weeks, and the respective lengths of the freely-chosen, page knocking sequences. Two of the participants (Users #10 and #11) did not return for the page knock recall part of the study, while another two (Users #8 and #9) returned only once. We note that for the latter case (Users #8 and #9), their returns, although not completed fully, were successful, indicating that the users correctly remembered and reproduced their page knocks at least once. Out of the 12 participants that completed the return study in its entirety, 11 (91.67%) of them consistently completed their page knocks correctly for all three return days. The one participant who did not (User #1) never managed to complete their page knocks correctly.

Table 3. User Study Usability Results

User	Day 1	Day 2	Day 3	Knock Length
1	✗	✗	✗	2
2	✓	✓	✓	3
3	✓	✓	✓	3
4	✓	✓	✓	3
5	✓	✓	✓	3
6	✓	✓	✓	3
7	✓	✓	✓	3
8	✓	-	-	3
9	✓	-	-	3
10	-	-	-	3
11	-	-	-	3
12	✓	✓	✓	4
13	✓	✓	✓	4
14	✓	✓	✓	4
15	✓	✓	✓	4
16	✓	✓	✓	5

Regarding the choice of page knocks by the participants, we observe the majority opted for page knocking sequences of length three and four. All but one of the participants who participated in the second part of the study, managed to consistently reproduce their page knocks correctly. Given that the overwhelming majority of users managed to correctly remember and reproduce their page knocks, and considering that all users submitted longer page knocks, we attribute the failure of User #1 to potentially initial confusion upon submission of the personalized sequence. We note that this user still managed to reproduce half of their sequence in all return days.

From the two parts of the user study, we make the following observations:

- First, PAGEKNOCKER is resistant against expert users that have knowledge of its existence and its functionality. We experimentally confirm that longer page knocks are more secure, even if these are just 3 link clicks on a web application's UI.
- Observing the choices of page knocking sequence lengths, the vast majority of users recognize the necessary sequence length threshold to be between three and four requests.
- Lastly, we showed that PAGEKNOCKER exhibits high usability, since the participants remembered their knocks consistently.

7 Limitations

Given the ever-increasing sophistication of cyber attacks, PAGEKNOCKER is not a silver bullet against all possible authentication attacks. That is, while PAGEKNOCKER can protect web applications against credential brute-forcing, guessed/reused passwords, and traditional phishing, it cannot defend against the MITM phishing kits where attackers can observe all traffic from a user's browser and replay it (i.e. clicks, typed credentials, MFA codes, etc.) to the victim web application [19]. Like with any real-world defense mechanism, PAGEKNOCKER would have to be layered alongside other countermeasures (i.e., anomaly detection systems), to defend users against these types of attacks.

An attacker that has knowledge of the presence and functionality of PAGEKNOCKER, could attempt a brute-forcing strategy to find the correct page-knocking sequence. Such a strategy would consist of performing arbitrary page knocks and subsequently visiting the login page, to evaluate each page knocking sequence. Excluding the fact that such a brute force attack can only be carried out against the 404 deceptive configuration, where the attacker can evaluate the correctness of each attempted page knocking sequence based on the response of the web application (404, or login page), such back-and-forth actions would need to be carried out repeatedly, potentially hundreds of thousands of times. The repetitive nature of this attack (page knock paths to login page loop) constitutes an extremely distinct client behavior, which could easily be detected by the server.

Although adapting PAGEKNOCKER to protect existing web applications is a relatively uncomplicated procedure, as described in Sect. 3.2, the different in-page structures and a web application's content-update frequency, could have an impact on the efficacy of the system (i.e. a regularly updated blog or a news website with many of the webpage paths changing often). To counterbalance this, we note that PAGEKNOCKER can still protect the authentication of such web applications, using a smaller subset of the webpage's sitemap; i.e. the pages that do not change.

8 Related Work

Deception-Based Defenses. The field of deception-based defenses comprises of all security techniques aiming to protect digital elements through redirection, diversion, or the deflection of attacks, by deceiving the perpetrator. In this context, honeypots have been integrally linked with deceptive security, providing the aforementioned qualities, while also allowing further analysis of an attacker's behavior by monitoring and tracking intruders. Although the deceptive defenses predate them [20], honeypots as virtual entities attracting and engaging attackers have been on the forefront of network and web security for the better part of the last 20 years [20,21]. In order to tackle diverse security issues, the concept of honeypots has been extended to analogous elements, such as honeypot networks (also known as honeynets) [22], honeytokens [23], honeywords [24], and decoy documents and files [25–28], overall characterized as honey-x entities. Such honey-x approaches have been employed in different security applications, from intrusion and malware detection [25,29,30], to software and web application defenses [31–34]. Introducing believable and attractive decoy and bait elements in the system enhances the detection capability, while the different deployments and characteristics of trap-based architectures allow for ad hoc deceptive defenses addressing various attacks [26,30,35,36].

Deception and Web Applications. In the context of web applications and web servers, deception-based approaches have focused on augmenting the attack surface of the web application by introducing deceptive elements. This concept is founded on the idea of using deceptive elements as detection points, where any interaction with them is labelled suspicious. More specifically, strategies such as providing falsified system version information, tampering with *robots.txt* files, generating virtual honey documents, cookie scrambling, and Javascript obfuscation have been attempted [37,38]. Furthermore, approaches generating decoy files in lieu of tripwires deployed on real systems have proven effective for intruder detection, while achieving low rates of false positives [26,27].
On the other hand, deceptive techniques have also been employed in a parallel world context, where the attackers are covertly redirected to an isolated and appropriately curated clone of the original environment after detection [33,39, 40]. Thus, the attacker is under the impression that they are still attacking the real system, while the defender is able to observe the attacker's behavior, without the risk of lateral movements or exposure of critical resources. While PageKnocker's main objective is to protect the authentication endpoint, it also serves as a detection mechanism, and may be employed as a trigger for a parallel world setting.

Regarding the authentication itself, in order to alleviate the inadequacies of password-based web authentication, deception-based approaches, such as login rituals and web tripwires have been suggested as supplementary authentication mechanisms [41]. In the case of login rituals, users are considered "truly" authenticated by performing specific actions post credential-based authentication, e.g., interact with specific elements of the web application. Tripwires are elements that authenticated users must avoid to remain authenticated. PageKnocker also uses web-application elements as deception primitives but brings these to pre-authenticated users. In this way, we can protect web applications from tens of thousands of would-be attackers (Sect. 5) *before* they even reach the login page. Provided the improved defensive posture afforded by both overt and covert deceptive defenses [42], PageKnocker employs two deceptive environments; one manifestly deceptive and one clandestine.

9 Conclusion

In this paper we introduced PageKnocker, a deception-based supplementary web authentication mechanism on top of passwords that can protect public-facing authentication endpoints from attackers. Inspired by the concept of port knocking, PageKnocker checks the requests that a user makes towards the web application against a predefined user-specific sequence of requests. The user is authenticated and allowed to reach the login page only if the sequence of requests they make matches their predefined sequence. The objective of this supplementary authentication scheme is to protect the login page of web applications, while allowing the administrators to detect attackers *before* they reach the credential-based authentication. Utilizing this a priori knowledge on the nature of the visitor, we construct two different deceptive login environments, towards which we direct the unauthenticated attacker; the first directs the user to a fake login page, while the second directs them to an HTTP 404 error. We evaluate the security and usability of PageKnocker using both in-the-wild experiments with honeypots as well as via a user study. From these experiments, we conclude that PageKnocker manages to effectively protect the main authentication of the web application, while being user-friendly. Furthermore, we demonstrated that overt placement of deceptive elements leads to bot traffic disengaging quicker, while concealing such elements encourages further bot interaction.

Acknowledgements. We thank the anonymous reviewers for their helpful feedback. This work was supported by the Army Research Office (ARO) under grant W911NF-24-1-0051.

Data Availibility Statement. A video demonstration of PageKnocker along with source code are available at https://pageknocker.github.io/.

A Appendix

Table 4. Predefined login page paths for top 10 CMS web applications by market share [43]

CMS	Predefined Login Path	Market Share [%]
Wordpress	/wp-admin, /wp-login.php	63.6
Joomla	/administrator, /administrator/index.php	2.7
Drupal	?q=user/login, landing page	1.8
PrestaShop	/admin	1
Magento	/admin	0.9
TYPO3	/typo3	0.6
Craft	/admin/login	0.2
Ghost	/ghost, /ghost/#/signin	0.1
Contao	/contao, /contao/login	0.1
ProcessWire	/processwire	0.1

Table 5. Requests received to the login page

Page Knock Length	Total	GET [%]	POST [%]
1	131,596	10.33	89.67
1	182,182	11.41	88.59
3	133,201	16.21	83.79
3	94,108	17.93	82.07
5	95,857	16.85	83.15
5	124,965	15.31	84.69
Overall	761,909	14.67	85.33

(a) Requests to the login page of FA honeypots

Page Knock Length	Total	GET [%]	POST [%]
1	5,056	43.73	56.27
1	4,542	49.43	50.57
3	4,602	49.35	50.65
3	4,546	48.88	51.12
5	5,409	42.91	57.09
5	4,177	49.16	50.84
Overall	28,632	47.24	52.76

(b) Requests to the login page of 404 honeypots

Table 6. Common IPs appearing in both FA and 404 honeypots making large amounts of POST requests

IP	Requests	GET	POST	Country
20.83.147.x	168,604	48	168,556	United States
79.110.62.x	16,248	519	15,729	Switzerland
31.14.75.x	10,280	144	10,136	Czech Republic
81.161.229.x	9,648	126	9,522	United States
20.162.210.x	6,132	126	6,006	United States
89.187.163.x	5,040	72	4,968	Turkey
185.207.35.x	3,572	84	3,488	United States
89.187.163.x	2,689	72	2,617	Germany
20.166.249.x	1,927	84	1,843	Czech Republic

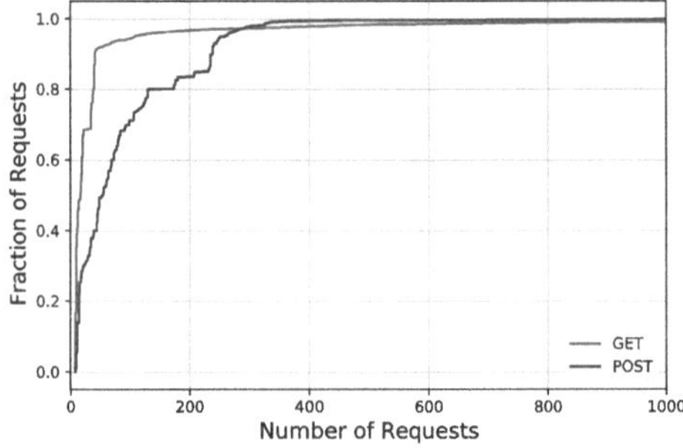

Fig. 8. CDF of requests per number of requests FA and 404 honeypots in the wild

References

1. Van Acker, S., Hausknecht, D., Sabelfeld, A.: Measuring login webpage security. In: Proceedings of the Symposium on Applied Computing (2017)
2. Das, A., Bonneau, J., Caesar, M., Borisov, N., Wang, X.F.: The tangled web of password reuse. In: NDSS, vol. 14, pp. 23–26 (2014)
3. Florencio, D., Herley, C.: A large-scale study of web password habits. In: Proceedings of the 16th International Conference on World Wide Web (2007)
4. Bonneau, J., Herley, C., Van Oorschot, P.C., Stajano, F.: The quest to replace passwords: a framework for comparative evaluation of web authentication schemes. In: 2012 IEEE Symposium on Security and Privacy (2012)
5. Reynolds, J., et al.: Empirical measurement of systemic {2FA} usability. In: 29th USENIX Security Symposium (2020)
6. Petsas, T., Tsirantonakis, G., Athanasopoulos, E., Ioannidis, S.: Two-factor authentication: is the world ready? Quantifying 2FA adoption. In: 8th European Workshop on System Security (2015)
7. Lin, X., Ilia, P., Solanki, S., Polakis, J.: Phish in sheep's clothing: exploring the authentication pitfalls of browser fingerprinting. In: 31st USENIX Security Symposium (2022)
8. Pearman, S., Zhang, S.A., Bauer, L., Christin, N., Cranor, L.F.: Why people (don't) use password managers effectively. In: 15th Symposium on Usable Privacy and Security (SOUPS 2019) (2019)

9. F5 labs 2023 identity threat report: The unpatchables. https://www.f5.com/labs/articles/threat-intelligence/2023-identity-threat-report-the-unpatchables
10. Troy Hunt - Have I Been pwned? (2018). https://haveibeenpwned.com/
11. Li, X., Azad, B.A., Rahmati, A., Nikiforakis, N.: Good bot, bad bot: characterizing automated browsing activity. In: 2021 IEEE Symposium on Security and Privacy (2021)
12. Ilascu, I.: Bleeping Computer: Uber Fined for Covering Up 2016 Data Breach (2018). https://www.bleepingcomputer.com/news/security/uber-fined-for-covering-up-2016-data-breach/
13. Degraaf, R., Aycock, J., Jacobson, M.: Improved port knocking with strong authentication. In: 21st Annual Computer Security Applications Conference (ACSAC'05)
14. Pochat, V.L., Van Goethem, T., Tajalizadehkhoob, S., Korczyński, M., Joosen, W.: Tranco: a research-oriented top sites ranking hardened against manipulation. arXiv preprint arXiv:1806.01156 (2018)
15. Cortesi, A., Hils, M., Kriechbaumer, T.: Mitmproxy: a free and open source interactive HTTPS proxy (2010). contributors
16. Let'sEncrypt (2023). https://letsencrypt.org/
17. Certificate Transparency (2023). https://certificate.transparency.dev/
18. Kondracki, B., So, J., Nikiforakis, N.: Uninvited guests: Analyzing the identity and behavior of certificate transparency bots. In: 31st USENIX Security Symposium (2022)
19. Kondracki, B., Azad, B.A., Starov, O., Nikiforakis, N.: Catching transparent phish: analyzing and detecting mitm phishing toolkits. In: Proceedings of the 2021 ACM SIGSAC Conference on Computer and Communications Security (2021)
20. Stoll, C.: Stalking the wily hacker. Commun. ACM (1988)
21. Provos, N., et al.: A virtual honeypot framework. In: USENIX Security Symposium (2004)
22. The Honeynet Project (2023). https://www.honeynet.org/
23. Bercovitch, M., Renford, M., Hasson, L., Shabtai, A., Rokach, L., Elovici, Y.: HoneyGen: an automated honeytokens generator. In: Proceedings of 2011 IEEE International Conference on Intelligence and Security Informatics (2011)
24. Juels, A., Rivest, R.L.: Honeywords: making password-cracking detectable. In: Proceedings of the 2013 ACM SIGSAC Conference on Computer & Communications Security (2013)
25. Bowen, B.M., Hershkop, S., Keromytis, A.D., Stolfo, S.J.: Baiting inside attackers using decoy documents. In: Chen, Y., Dimitriou, T.D., Zhou, J. (eds.) SecureComm 2009. LNICST, vol. 19, pp. 51–70. Springer, Heidelberg (2009). https://doi.org/10.1007/978-3-642-05284-2_4
26. Voris, J., Jermyn, J., Boggs, N., Stolfo, S.: Fox in the trap: thwarting masqueraders via automated decoy document deployment. In: Proceedings of the Eighth European Workshop on System Security (2015)
27. Ben Salem, M., Stolfo, S.J.: Decoy document deployment for effective masquerade attack detection. In: Holz, T., Bos, H. (eds.) DIMVA 2011. LNCS, vol. 6739, pp. 35–54. Springer, Heidelberg (2011). https://doi.org/10.1007/978-3-642-22424-9_3
28. Nikiforakis, N., Balduzzi, M., Van Acker, S., Joosen, W., Balzarotti, D.: Exposing the lack of privacy in file hosting services. In: 4th USENIX Workshop on Large-Scale Exploits and Emergent Threats (LEET 2011) (2011)
29. Vasilomanolakis, E., Srinivasa, S., Cordero, C.G., Mühlhäuser, M. Multi-stage attack detection and signature generation with ICS honeypots. In: NOMS 2016 IEEE/IFIP Network Operations and Management Symposium (2016)

30. Bowen, B.M., Prabhu, P., Kemerlis, V.P., Sidiroglou, S., Keromytis, A.D., Stolfo, S.J.: BotSwindler: tamper resistant injection of believable decoys in VM-based hosts for crimeware detection. In: Jha, S., Sommer, R., Kreibich, C. (eds.) RAID 2010. LNCS, vol. 6307, pp. 118–137. Springer, Heidelberg (2010). https://doi.org/10.1007/978-3-642-15512-3_7
31. Park, Y., Stolfo, S.J.: Software decoys for insider threat. In: Proceedings of the 7th ACM Symposium on Information, Computer and Communications Security (2012)
32. Araujo, F., Hamlen, K.W., Biedermann, S., Katzenbeisser, S.: From patches to honey-patches: lightweight attacker misdirection, deception, and disinformation. In: 2014 ACM SIGSAC conference on Computer and Communications Security (2014)
33. Sahin, M., Hebert, C., De Oliveira, A.S.: Lessons learned from sundew: a self defense environment for web applications. In: Proceedings of the 2020 Measurements, Attacks, and Defenses for the Web (MADWeb) Workshop in the Network and Distributed System Security Symposium (NDSS) (2020)
34. Sahin, M., Hebert, C., Cabrera Lozoya, R.: An approach to generate realistic HTTP parameters for application layer deception. In: Ateniese, G., Venturi, D. (eds.) Applied Cryptography and Network Security. ACNS 2022. LNCS, vol. 13269, pp. 337–355. Springer, Cham (2022). https://doi.org/10.1007/978-3-031-09234-3_17
35. Bowen, B.M., Kemerlis, V.P., Prabhu, P., Keromytis, A.D., Stolfo, S.J.: Automating the injection of believable decoys to detect snooping. In: Proceedings of the 3rd ACM Conference on Wireless Network Security (2010)
36. Reti, D., Fraunholz, D., Elzer, K., Schneider, D., Schotten, H.D.: Evaluating deception and moving target defense with network attack simulation. In: Proceedings of the 9th ACM Workshop on Moving Target Defense (2022)
37. Fraunholz, D., Reti, D., Duque Anton, S., Schotten, H.D.: Cloxy: a context-aware deception-as-a-service reverse proxy for web services. In: Proceedings of the 5th ACM workshop on Moving Target Defense (2018)
38. Fraunholz, D., Schotten, H.D.: Defending web servers with feints, distraction and obfuscation. In: 2018 International Conference on Computing, Networking and Communications (ICNC) (2018)
39. Anagnostakis, K.G., Sidiroglou, S., Akritidis, P., Xinidis, K., Markatos, E., Keromytis, A.D.: Detecting targeted attacks using shadow honeypots. In: 14th USENIX Security Symposium (2005)
40. Urias, V.E., Stout, W.M., Lin, H.W.: Gathering threat intelligence through computer network deception. In: 2016 IEEE Symposium on Technologies for Homeland Security (HST) (2016)
41. Barron, T., So, J., Nikiforakis, N.: Click this, not that: extending web authentication with deception. In: Proceedings of the 2021 ACM Asia Conference on Computer and Communications Security (2021)
42. Ferguson-Walter, K.J., Major, M.M., Johnson, C.K., Muhleman, D.H.: Examining the efficacy of decoy-based and psychological cyber deception. In: USENIX Security Symposium (2021)
43. W3Techs. Usage statistics and market share of WordPress (2023). https://w3techs.com/technologies/details/cm-wordpress

AI for Security

Approach for the Optimization of Machine Learning Models for Calculating Binary Function Similarity

Suguru Horimoto[1,2](✉), Keane Lucas[3], and Lujo Bauer[3]

[1] National Police Agency, Tokyo, Japan
[2] Visiting Researcher at CyLab Security and Privacy Institute of Carnegie Mellon University, Pittsburgh, USA
shorimot@andrew.cmu.edu
[3] Carnegie Mellon University, Pittsburgh, USA
{kjlucas,lbauer}@andrew.cmu.edu

Abstract. Binary function similarity comparison is essential in a variety of security fields, such as software vulnerability detection and malware analysis, because it enables engineers to accelerate otherwise time-consuming tasks. While various approaches for binary function similarity comparison have been proposed, in an experiment of previous work to fairly evaluate existing methods, a method combining graph neural network (GNN) and bag-of-words (BoW) exhibited the highest performance. In this method, each basic block (BB) in a function is embedded into a vector by BoW. As a result, the function vector is derived from sparse vectors. In this paper, we propose a method combining a GNN with fastText, instead of BoW. Furthermore, in order to optimize machine learning models for calculating binary function similarity, we apply early stopping based on mean reciprocal rank (MRR) to our machine learning training. Our method outperformed the previous method combining GNN and BoW by up to 2% in AUC, up to 9% in Recall@1 and up to 7% in MRR10 in a certain case. Additionally, through a function search case study in malware analysis, our method has been found to be applicable for finding distinctive functions present in LockBit Ransomware.

Keywords: Malware analysis · Graph learning · Similarity

1 Introduction

Binary function similarity comparison is essential in a variety of security fields, such as software vulnerability detection and malware analysis. For example, in software vulnerability detection, the similarity comparison helps engineers quickly find the same vulnerability in different binaries [7,18]. However, binaries compiled with different compilers and to different architectures make binary function similarity comparison very challenging. For instance, each function in a binary consists of a control flow graph (CFG) and basic blocks (BBs), where each BB consists of assembly instructions. When binaries are compiled from the same source code with different compilers, the structure of a CFG and assembly

F. Maggi et al. (Eds.): DIMVA 2024, LNCS 14828, pp. 309–329, 2024.
https://doi.org/10.1007/978-3-031-64171-8_16

instructions of BBs are drastically different from each of the binaries because the structure of a CFG and assembly instructions of BBs in the binaries rely on their respective compiler [9,13].

To tackle this issue, many researchers in security fields have proposed methods to compare binary function similarity with machine learning [13,18,25]. While numerous methods for calculating function similarity have been proposed, Marcelli et al. [17] conducted an evaluation to perform a fair comparison of these methods. Their evaluation found that Graph Neural Network (GNN) and Graph Matching Network (GMN) models [13] outperformed other methods in terms of binary function similarity comparison. In the GNN and GMN models Marcelli et al. [17] prepared, however, function vectors are generated using CFGs and Bag-of-Words (BoW)-based BB vectors. In the BoW approach, a BB vector is generated from the counts of opcode occurrences within the top 200 opcodes selected by Marcelli et al. [17]. As a result, the function vector is derived from sparse vectors. Besides, the effectiveness of the GNN and GMN models in malware analysis remains unclear because the search performance of the GNN and GMN models in malware analysis has not been evaluated by Marcelli et al. [17].

In this paper, we propose a new method for binary function similarity comparison by using *fastText* [4] (Sect. 3.1). fastText is an open-source natural language processing (NLP) tool developed by Facebook, well-suited for various NLP tasks, including word embeddings generation. One of the features of fastText is to split a word into subwords, which are parts of letters within a word, and learn embeddings for these subwords. This approach enhances the handling of out-of-vocabulary words and language variations, allowing for the consideration of fine-grained language details. In our research, we obtain embeddings of opcodes using fastText and use these embeddings to generate BB vectors. Even across different architectures, there are multiple opcodes that share partial common subwords. For instance, opcodes for copying data, such as "mov" or "movzx", contain the partially common subword "mov". Therefore, using fastText to generate embeddings of opcodes appears to intuitively be a more effective approach.

Furthermore, we apply early stopping [20] based on Mean Reciprocal Rank (MRR) in training. Early stopping is a methodology used to halt the learning process when evaluation metrics no longer exhibit continuous improvement during the training. MRR is an evaluation metric for ranking accuracy and is used for evaluating binary function comparison methods [17,27]. Early stopping based on MRR enables to enhance the search performance of our method (Sect. 3.3).

In the evaluation of search performance in malware analysis (Sect. 4.7), we conducted a search for the functions of LockBit Ransomware [5] obtained from Vx Underground [1] using our method. Our method has been found to be applicable for finding distinctive functions present in LockBit Ransomware.

In a nutshell, the contributions of this paper are as follows:

- We improve the performance of GNN and GMN models by leveraging fastText as a method of embedding BBs (Sect. 3.1). We published the source code for training the models, some trained multi-architecture models and the testing dataset for the multi-architecture models.[1]

[1] https://github.com/sgr-ht/mam-for-cbfs.

- We evaluate the effectiveness of early stopping based on MRR in training to enhance the search performance of the GNN and GMN models (Sect. 4).
- We evaluate a method combining the GNN model and fastText for its effectiveness in calculating the similarity of binary functions in malware analysis (Sect. 4.7).

2 Related Work

Creating *embeddings* is a popular method for binary function similarity comparison. The embeddings refer to vectors in a low-dimensional space where semantically similar inputs are associated with points that are close to each other, irrespective of the differences in their original representations [17]. Two methods commonly employed for creating embeddings are code embeddings and graph embeddings [17]. Code embeddings is an approach to create embeddings using NLP by considering assembly code as text. Graph embeddings is another approach to create embeddings using machine-learning models which compute embeddings using a CFG and features in each function.

2.1 Code Embeddings

SAFE [18] can convert a function into a vector using a self-attentive network [14], which comes from the seq2seq model, a commonly used NLP structure. In SAFE, assembly instructions in a function are embedded into vectors and then each vector of assembly instruction in a function is fed into the self-attentive network to embed a function vector.

2.2 Graph Embeddings

Gemini. The CFG and BBs of a function are first transformed into an annotated CFG, which is a graph containing manually selected features such as the number of assembly instructions and the number of string constants in Gemini [25]. After that, Gemini creates a function vector using structure2vec [6].

GNN and GMN Models. Li et al. [13] proposed GNN and GMN models, which can create a function vector from the CFG and BBs of a function and calculate the similarity between pairs of functions. GNN and GMN models achieve the best results in the experiments of prior work [17], which performs a comparison among existing approaches that are publicly available, such as Gemini and SAFE.

3 Proposed Method

Figure 1 gives an overview of the process of converting a function into a vector in order to compare binary function similarity. The process that we use is based

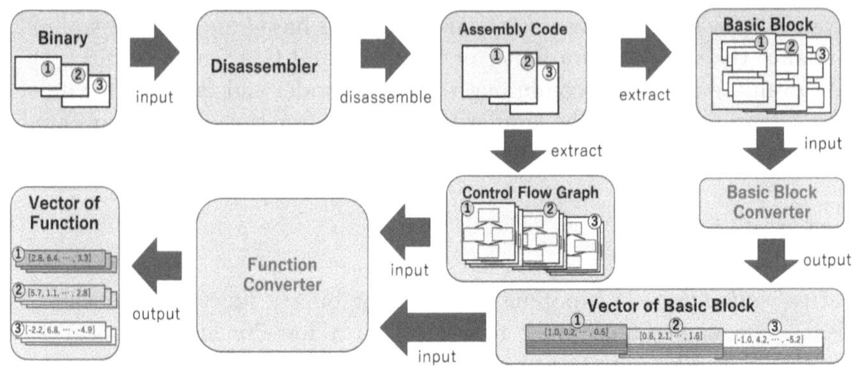

Fig. 1. The overview of the process of converting functions into vectors.

on Li et al. [13]. In this process, using fastText [4] as a BB converter and the training of a function converter with early stopping based on MRR are new. We first input binaries as a dataset to be disassembled. We then extract BBs and CFGs from the assembly code. After that, the BBs are fed into the BB converter, which converts a BB into a vector. Finally, the BB vectors and the CFGs are fed into the function converter, which embeds a function vector. Each function vector has the embedded meaning of the BBs and the CFG in each function. By repeating these processes to each function of each binary, it enables us to transform all functions of a dataset into vectors.

3.1 BB Converter

As the BB converter we use fastText, which is an NLP tool developed by Facebook used to generate word embeddings. Other work has shown that fastText outperforms word2vec [19], which is an NLP tool to generate word embeddings [4]. Our results presented here suggest that using fastText for vectorizing BBs is a better approach for calculating binary function similarity, particularly because of fastText's utilization of opcode embeddings to vectorize BBs. In terms of similar improvements in other applications, other work has shown that fastText can also be beneficial for vectorizing assembly instructions, which can then be used to vectorize a function [2,24]. Other work has also shown that fastText can lead to improvements in type inference [11].

In our research, we obtain embeddings of opcodes using fastText and use these embeddings to generate BB vectors. The advantage of fastText is to transform unseen opcodes not included in a training dataset into vectors. If there is a BB containing only unseen opcodes, i.e., opcodes not included in the training dataset for BoW and fastText, then BoW would be incapable of extracting any features. On the other hand, fastText is likely to extract at least some features owing to the benefits of subword embeddings. For example, in one of our test datasets, there is a BB containing only one "mov.s", which is an opcode of MIPS (e.g., mov.s dest, src), and "mov.s" is not included in the training dataset. BoW cannot extract any features of the BB, but fastText can extract a feature thanks to the subword embedding of "mov". In examining our dataset, we found at least 60 other BBs where fastText has this advantage for "mov.s", and many other BBs for other unseen opcodes.

The benefit of subword embedding leads to another advantage of fastText, which is to transform BBs with similar meanings into vectors that are spatially close to each other. In BoW, BBs with similar meanings do not necessarily become vectors that are spatially close. This is because BoW is based on the frequency of opcode occurrence, and does not take into account the semantic aspects of opcodes. On the other hand, in fastText, opcodes are vectorized considering the context information of the surrounding opcodes, and semantically similar opcodes tend to become more spatially close vectors. For example, comparing the similarity between a BB consisting of only "mov" and a BB consisting of only "movzx" using cosine similarity, fastText can capture semantics of the BBs better compared to BoW. "mov" and "movzx" are opcodes related to data movement. In our dataset, we found the similarity of fastText is 0.515 and the similarity of BoW is 0 and many BBs consist of only "mov" or "movzx". Therefore, these two advantages help inter-node GNN's classify better.

In order to train fastText, we first preprocess all assembly instructions in each function to reduce the vocabulary size. The reduction of the vocabulary size enables us to train the BB converter faster and to generate assembly instruction vectors which capture the meaning of the assembly instructions more accurately [18,29]. Based on Marcelli et al. [17], we extract opcodes from all assembly instructions in each function. For example, "mov" is extracted from the following instruction.

```
mov eax, 800
```

After preprocessing, opcodes and BBs are considered as words and sentences, respectively. This methodology regarding contents in BBs as words and BBs as sentences is commonly used to create BB vectors [10,28,29]. Then, opcodes in each BB are fed into fastText. In order to generate opcode vectors, each opcode is divided into subwords by N-gram. N-gram is a representation method of contiguous sequences of N characters in a word [4]. For example, when N equals to 3, the mov opcode is divided into as follows.

```
<mo, mov, ov>
```

The < and > are boundary symbols at the beginning and the end of a word [4]. fastText generates opcode vectors using the subword embeddings. After that, we create a BB vector by averaging of opcode vectors in the BB.

3.2 Function Converter

As the function converter, we use GNN and GMN models [13]. The difference between the GNN model and the GMN model lies in their approach to generating vectors for functions. In the GNN model, each function vector is generated based only on its respective BB vector and CFG. On the other hand, in the GMN model each generated function vector is influenced by the BB and CFG of its paired counterpart, which is another function from a pair of functions (Sect. 2.2). We use and tweak GNN and GMN models based on prior work [17], which provides detailed instructions on how to use and tweak them.

3.3 Early Stopping Based on MRR

In order to optimize GNN and GMN models for calculating function similarity, we leverage early stopping [20] based on MRR. Early stopping is a methodology used in training to halt the learning process when evaluation metrics, such as the area under curve (AUC) of the receiver operating characteristic (ROC) curve, no longer exhibit continuous improvement during the training. The primary objective of applying early stopping in training is to prevent overfitting. Overfitting is a problem in which a model is trained extremely to match trends in the training dataset, such that the model achieves high accuracy on the training dataset, but fails to generalize well to data outside the training dataset.

MRR is an evaluation metric for ranking accuracy and is used for evaluating binary function comparison methods [17,27]. The computation of MRR involves calculating the reciprocal of the highest rank position for the correct answer for each search query and then taking the average of these reciprocals. The value of MRR falls within the range of 0 to 1, with the value closer to 1 indicating superior ranking of search performance. For instance, MRR10 assesses search performance within the top 10 positions, and if, for a specific search query, the correct answer is ranked second, the MRR10 will be 0.5.

In Sect. 4, we show that applying early stopping based on MRR to training enhances the search performance of our method.

4 Evaluation

Our evaluation is based on the comparison performed by Marcelli et al. [17]. We trained and tested GNN and GMN models [13] as multi-architecture models and single-architecture models. Functions of ARM, x86 and MIPS are fed into multi-architecture models for training. The functions come from ELF format binaries. We trained and tested multi-architecture models on an Amazon EC2 P3 instance (p3.2xlarge), which is equipped with Ubuntu 20.04, Intel(R) Xeon(R) CPU E5-2686 v4 @ 2.30 GHz and 61 GB memory.

In regard to single-architecture models, we fed the models functions of x86 32-bit binaries in PE format. We trained and tested single-architecture models on a workstation equipped with Ubuntu 18.04 on Windows 10 Pro 64-bit with WSL, Intel(R) Core(TM) i7-7800X CPU @ 3.50 GHz and 64 GB memory.

For preparing datasets, we used Python programs which are open to the public and were provided by prior work [17]. Some of these Python programs were originally created to work with IDA Pro [8] 7.3, and we worked on rewriting them to work with IDA Pro 7.7. Thus, some of the dataset preprocessing, such as disassembling binaries and counting BBs in a function, depends on IDA Pro.

We implemented multi-architecture models and single-architecture models in Tensorflow 1.14 and conducted training based on Python programs publicly available via prior work [17].

4.1 Dataset

We prepared two datasets for multi-architecture models, and one dataset for single-architecture models.

Dataset-1. `Dataset-1` comprises seven open source projects: `ClamAV`, `Curl`, `Nmap`, `OpenSSL`, `Unrar`, `Z3` and `Zlib` [17]. These open source projects are compiled for ARM, x86, MIPS in 32- and 64-bit versions and are compiled by Clang and GCC with 4 different versions each and with 5 optimizations (O0, O1, O2, O3 and Os) [17]. 5,489 binaries of Dataset-1 are in ELF format and are used for the training and the evaluation of multi-architecture models. We obtained Dataset-1 preprocessed by Marcelli et al. [17]. Prior work removed duplicate functions based on their names and the hash values of their instructions [17].

In total, Dataset-1 comprises approximately 790k functions, each of which has more than four BBs. Dataset-1 is split into training, validation and testing datasets. The training dataset comprises approximately 260k functions from ClamAV, Curl, OpenSSL and Unrar. The validation dataset comprises approximately 10k functions from Zlib. The testing dataset comprises approximately 520k functions from Nmap and Z3.

Dataset-BINKIT. We obtained `Normal-dataset` from prior work [12] and selected some binaries of Normal-dataset as another testing dataset for the evaluation of multi-architecture models. We call this testing dataset Dataset-BINKIT. In order to investigate the impact of the number of software used to a testing dataset on performance, we minimized the differences between Dataset-1 and Dataset-BINKIT as much as possible, excluding the number of software. We selected 7,200 binaries in ELF format, which are compiled for ARM, x86, MIPS in 32- and 64-bit versions and are compiled by Clang and GCC with 2 different versions each and with 4 optimizations (O0, O1, O2 and O3). We preprocessed Dataset-BINKIT using prior work's code [17], which enables us to remove duplicate functions based on their names and the hash values of their opcodes. We also removed functions in Dataset-BINKIT which are the same as functions in Dataset-1 in terms of the hash values of their opcodes. In total, Dataset-BINKIT comprises approximately 510k functions which have more than four BBs. The selected binaries in Dataset-BINKIT are the following GNU software packages; `gsl`, `gss`, `gzip`, `hello`, `inetutils`, `libiconv`, `libidn`, `libmicrohttpd`, `libtasn1`, `libtool`, `libunistring`, `lightning`, `macchanger`, `nettle`, `osip`, `patch`, `plotutils`, `readline`, `recutils`, `sed`, `sharutils`, `spell`, `tar`, `texinfo`, `time`, `units`, wdiff, `which` and `xorriso`.

Dataset-Win32. We first collected six open source project source codes: `Curl`, `FFmpeg`, `OpenSSL`, `PuTTY`, `SQLite` and `Zlib`. Then, we compiled them for x86 (32-bit) and by Clang and GCC with 2 different versions each and with 5 optimizations (O0, O1, O2, O3 and Os). 480 binaries of Dataset-Win32 are in PE format and are used for the training and the evaluation of single-architecture models. We preprocessed Dataset-Win32 using prior work's code [17]. In total, Dataset-Win32 comprises approximately 260k functions which have more than four BBs. In the evaluation of single-architecture model, we performed 5-fold cross validation because Dataset-Win32 is smaller than the datasets for multi-architecture models. For convenience, we represent each separate cross validation fold as cv1, ..., cv5. In each cross validation fold, the training dataset comprises approximately 210k functions and the validation dataset comprises approximately 50k functions.

4.2 Machine Learning Models

For multi-architecture models, we evaluated the models using holdout validation to compare them with the GNN model [17]. When data imbalance occurs upon splitting datasets using the holdout validation, the data imbalance can introduce potential bias in the evaluation of a model [22]. Also, other work has shown that k-fold cross validation tends to give better results compared to holdout validation [26]. To mitigate the bias, we employed five different random seeds to conduct the evaluation and obtained the mean results of the all random seeds. Altering the random seeds leads to changes in the training of GNN and GMN models, such as model initialization and variations of function pairs used for training. We first employed random seed 11 because the GNN model [17] to which we compare our results appeared to be trained on random seed 11, according to prior work [17]. We also tried random seeds 12 through 15, with similar results.

In regard to the evaluation of single-architecture models, we evaluated the models using cv1, ..., cv5 and random seed 11. Before training the models, We modified the training process to ensure that the function pairs were not duplicated in each epoch. The rationale behind this is to avoid the duplication of function pairs due to the lower abundance of compiler versions and architectures compared to the training dataset for the multi-architecture models.

Baseline. In the evaluation of multi-architecture models, we prepared three baseline models. One of them is a GNN model which is created by Marcelli et al. [17] and is open to the public. The GNN model uses BB vectors created by BoW. The others are a GNN model and a GMN model which use BB vectors created by BoW and are trained by early stopping based on MRR. As a BB converter, BoW serves to convert BBs into BB vectors, which are generated from the counts of opcode occurrences within the top 200 opcodes in all BBs of the training dataset from Dataset-1. We selected the top 200 opcodes using prior work's code [17]. The other training configuration for the GNN model and the GMN model, such as the dimension of function vector (128-dimension), is the same as Marcelli et al. [17].

In regard to the evaluation of single-architecture models, we prepared a GNN model and a GMN model which uses BB vectors created by BoW as baseline models. BB vectors are generated from the counts of opcode occurrences within the top 200 opcodes in all BBs of the training dataset from cv1, ..., cv5. We selected the top 200 opcodes using prior work's code [17]. The top 200 opcodes vary depending on cv1, ..., cv5. The baseline models are trained by early stopping based on MRR. The other training configuration for the GNN model and the GMN model is the same as Marcelli et al. [17].

Our Proposed Model. In the evaluation of multi-architecture models, we prepared a GNN model and a GMN model which use BB vectors, which are the average of the embeddings of opcodes in each BB. The embeddings are created by a fastText [4] model. The fastText model is trained using BBs of the training dataset from Dataset-1. The training configuration for the fastText model includes skipgram, 200 iterations, minn 3, maxn 6, window size 5, minCount 1, and a BB vector dimension of 200. After training, the fastText model takes each of the BBs of Dataset-1 and Dataset-BINKIT as input and generates corresponding BB vectors for each of the BBs. Our proposed models are trained by early stopping based on MRR. The other training configuration is the same as Marcelli et al. [17].

In regard to the evaluation of single-architecture models, we prepared a GNN model and a GMN model which use BB vectors, which are the average of the embeddings of opcodes in each BB. The embeddings are created by a fastText model. The training configuration for the fastText model includes skipgram, 200 iterations, minn 3, maxn 3, window size 5, minCount 1, and a BB vector dimension of 128. The fastText model is trained using BBs of the training dataset from Dataset-Win32. In total, we created five fastText models from the training datasets in cv1, ..., cv5. The other training configuration is the same as the evaluation of multi-architecture models.

4.3 Measures of Performance

Based on Marcelli et al. [17], in order to evaluate the performance of multi-architecture models, we prepared four different tasks to evaluate: (1) XA: two functions of a pair come from different architectures and bitness, but the same compiler, compiler version, and optimization. (2) XC: two functions of a pair come from different compiler, compiler versions, and optimizations, but the same architecture and bitness. (3) XC+XB: two functions of a pair come from different compiler, compiler versions, optimizations, and bitness, but the same architecture. (4) XM: two functions of a pair come from randomly selected architectures, bitness, compiler, compiler versions, and optimizations. We also prepared two different tests based on Marcelli et al. [17]: (i) AUC, which is one of the metrics to evaluate classifiers. (ii) MRR10 and the recall at different K thresholds (Recall@K), which are common ranking metrics. Ranking metrics are useful to assess the search performance, particularly in applications requiring the exploration of candidate functions within a large database [17].

Pair Creation. We created positive pairs and negative pairs based on function names and tasks. In the case of training, we created them based on function names. For instance, when two functions have the same function name although the two functions are compiled by different configurations, such as optimization, the two functions are regarded as a positive pair. On the other hand, when two functions have different function names, the two functions are regarded as a negative pair regardless of the compilation settings. We created positive pairs and negative pairs using prior work's code [17]. Additionally, two functions of a positive pair for AUC test, the ranking test and early stopping come from binaries with the same name but compiled by different configurations.

Calculation of AUC. For the first test, we prepared two datasets of 50k positive pairs and 50k negative pairs for each task from Dataset-1 and Dataset-BINKIT using prior work's code [17]. As for Dataset-1, we first tried to use two datasets from Marcelli et al. [17] to replicate their experiment. However, we found duplicate pairs from the positive pairs from Dataset-1 and some positive pairs for XA, XC, and XC+XB from Dataset-1 partially do not satisfy the constraint of each task. We also fixed a bug in prior work's AUC code, which caused an error of approximately 0.3% in the AUC calculation on Dataset-1. We performed the first test in the evaluation of multi-architecture models.

Calculation of MRR10 and Recall@K. For the ranking test of the XM task, we randomly selected 20k positive pairs and 2,000k negative pairs, that is 100 negative pairs for each positive one.

In order to evaluate the performance of single-architecture models, we prepared the XM task. The XM task is evaluated according to two commonly used ranking metrics, MRR10 and Recall@K. For the ranking test of the XM task, we randomly selected 20k positive pairs and 6,000k negative pairs, that is 300 negative pairs for each positive one.

Configuration of Early Stopping Based on MRR10. We prepared four different patience values to stop training, early stopping patience (EP) 5, EP10, EP15 and EP20. For example, training on EP5 would be halted when the value of MRR10 does not improve for five consecutive epochs. In the evaluation of multi-architecture models, MRR10 in each epoch is calculated by 500 positive pairs and 50k negative pairs of the validation dataset from Dataset-1 for XM task. In regard to the evaluation of single-architecture models, MRR10 in each epoch is calculated by 500 positive pairs and 150k negative pairs of the validation dataset from Dataset-Win32 for XM task. The positive and negative pairs in early stopping and in the ranking test are not duplicated.

4.4 Results of Multi-architecture Models

From this point onward, we shall refer to the GNN model and the GMN model of our proposed models as GNN+fastText and GMN+fastText respectively. Like-

Table 1. AUC, Recall@1 and MRR10 comparison between our proposed models and the baseline models on Dataset-1. Each value is the average from random seed 11 to 15. Using fastText [4] improves the baseline performance in almost every case.

Model	XA	XC	XC+XB	XM	Recall@1(XM)	MRR10(XM)
GNN+BoW	0.98	0.82	0.82	0.88	**0.48**	0.55
GNN+fastText	0.98	**0.86**	**0.86**	**0.90**	**0.48**	**0.56**
GMN+BoW	0.98	0.79	0.78	0.85	0.43	0.51
GMN+fastText	**0.99**	0.83	0.83	0.88	0.44	0.53

wise, we shall refer to a GNN model created by Marcelli et al. [17] as the GNN+BoW epoch10 [17], and a GNN model and a GMN model which use BB vectors created by BoW and are trained by early stopping based on MRR as GNN+BoW and GMN+BoW.

Results of Training. Figure 2 shows the results of training with random seed 11. On EP15, GNN+fastText is better than the other models according to MRR10 although GNN+fastText necessitates a greater number of epochs in comparison to the other models. The MRR10 of GNN+fastText is 0.814 at epoch 75, whereas the MRR10 of GNN+BoW is 0.797 at epoch 24. Also, the MRR10 of GMN+fastText is 0.738 at epoch 20, whereas the MRR10 of GMN+BoW is 0.745 at epoch 12. For your reference, we calculated AUC in each epoch using 10k positive pairs and 10k negative pairs for the only XM task, which come from prior work [17]. Recall@1 is also calculated by the same pairs as the ones to calculate MRR10. In comparison to MRR10, AUC converges more quickly after 20 iterations rather than 60 in the result of GNN+fastText. The similar trend observed in training with random seed 11 was confirmed for the other random seeds as well. In the evaluation of Multi-Architecture Models, GNN+fastText and GNN+BoW are trained on EP15 for the evaluation. We confirmed that the result of GNN+BoW on EP20 with random seed 11 was slightly worse than on EP15 in Dataset-1, and GNN+BoW on EP20 with random seed 12 and 15 were slightly worse than on EP15 in Dataset-BINKIT. To mitigate the impact of overfitting, we evaluated GNN+fastText and GNN+BoW which are trained on EP15. On the other hand, GMN+fastText and GMN+BoW are trained on EP20.

Results of the Testing Dataset of Dataset-1. Table 1 and Table 2 show the AUC, Recall@1 and MRR10 of multi-architecture models on Dataset-1. Each value in Table 1 is the average from random seed 11 to 15, and each value in Table 2 is the result of random seed 11. Also, Table 2 shows the comparison between GNN+fastText, GNN+BoW and the GNN+BoW epoch10 [17]. In Table 1, the AUC values of 0.86, 0.86 and 0.90 for XC, XC+XB and XM respectively, the Recall@1 value of 0.48 and the MRR10 value of 0.56 show that GNN+fastText is better than the other models. In Table 2, GNN+fastText out-

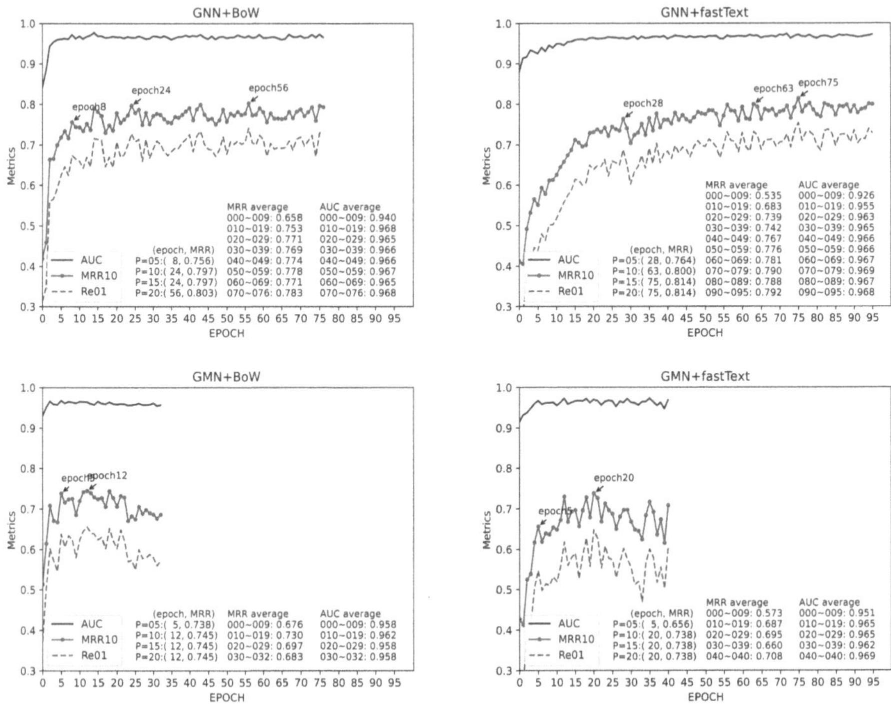

Fig. 2. The training results of GNN+fastText, GMN+fastText, GNN+BoW and GMN+BoW with random seed 11 on Dataset-1. Re01 denotes Recall@1 and P denotes early stopping patience (EP). Epoch indices start at zero. In the results of training on EP15, the MRR10 value of 0.814 shows that GNN+fastText is better than the other models.

performed the GNN+BoW epoch10 [17] by up to 3% in AUC, 5% in Recall@1 and 5% in MRR10.

Results of Dataset-BINKIT. Table 3 and Table 4 show the AUC, Recall@1 and MRR10 of multi-architecture models on Dataset-BINKIT. Each value in Table 3 is the average from random seed 11 to 15, and each value in Table 4 is the result of random seed 11. In a manner akin to Table 2, Table 4 shows the comparison between GNN+fastText, GNN+BoW and the GNN+BoW epoch10 [17]. In Table 3, the AUC values of 0.98, 0.95 and 0.96 for XA, XC+XB and XM respectively, the Recall@1 value of 0.64 and the MRR10 value of 0.72 show that GNN+fastText is better than the other models. In Table 4, GNN+fastText outperformed the GNN+BoW epoch10 [17] by up to 2% in AUC, 9% in Recall@1 and 7% in MRR10.

Table 2. AUC, Recall@1 and MRR10 comparison between our proposed model and the baseline models on Dataset-1. Each model is trained with random seed 11. Using fastText [4] improves the baseline performance in almost every case.

Model	XA	XC	XC+XB	XM	Recall@1(XM)	MRR10(XM)
GNN+BoW	**0.98**	0.81	0.81	0.88	**0.49**	**0.56**
GNN+fastText	**0.98**	**0.85**	**0.84**	**0.90**	0.48	**0.56**
GNN+BoW epoch10 [17]	0.97	0.82	0.82	0.88	0.43	0.51

Table 3. AUC, Recall@1 and MRR10 comparison between our proposed models and the baseline models on Dataset-BINKIT. Each value of AUC, MRR10 and Recall@1 is the average from random seed 11 to 15. Using fastText [4] improves the baseline performance in almost every case.

Model	XA	XC	XC+XB	XM	Recall@1(XM)	MRR10(XM)
GNN+BoW	0.97	0.95	0.94	0.95	0.62	0.70
GNN+fastText	**0.98**	0.95	**0.95**	**0.96**	**0.64**	**0.72**
GMN+BoW	0.97	0.95	0.93	0.94	0.59	0.68
GMN+fastText	0.97	**0.96**	0.94	0.95	0.58	0.67

4.5 Case Study of Vulnerability Search Using Multi-architecture Models

In order to validate the effectiveness of early stopping based on MRR in enhancing the search performance of multi-architecture models, we replicated the case study of vulnerability search, using a dataset from Marcelli et al. [17]. The dataset includes preprocessed functions from OpenSSL1.0.2d, which is compiled for x86, x64, ARM 32-bit and MIPS 32-bit architectures, and from two firmware images: Netgear R7000(ARM 32-bit) and TP-Link Deco M4(MIPS 32-bit). OpenSSL1.0.2d compiled for the four architectures contain ten vulnerable functions, Netgear R7000 contains four of them and TP-Link Deco M4 contains nine of them [17]. We searched for the four functions of Netgear R7000 and the nine functions of TP-Link Deco M4 using the ten functions of OpenSSL1.0.2d. We also calculated MRR10 from these search results using prior work's code [17].

Table 5 shows the MRR10 of GNN+fastText, GMN+fastText, GNN+BoW and GMN+BoW and the results of Marcelli et al. [17] on the case study. In this case study, we used GNN+fastText and GNN+BoW trained on EP15 with random seed 11 and GMN+fastText and GMN+BoW trained on EP20 with random seed 11. Also, GNN+BoW [17] and GMN+BoW [17] in Table 5 denote the results of Marcelli et al. [17]. In regard to Netgear R7000, the MRR10 values of 0.88, 1.00, 1.00 and 0.80 for x86, x64, ARM32 and MIPS32 respectively show that GMN+fastText is better than the other models in all architectures. As for TP-Link Deco M4, the MRR10 values of 0.79 and 0.75 for x86 and x64 respectively show that GMN+fastText is better than the other models in x86 and x64. Also, the MRR10 values of 0.78 and 0.78 for ARM32 and MIPS32

Table 4. AUC, Recall@1 and MRR10 comparison between our proposed model and the baseline models on Dataset-BINKIT. Each model is trained with random seed 11. Using fastText [4] improves the baseline performance.

Model	XA	XC	XC+XB	XM	Recall@1(XM)	MRR10(XM)
GNN+BoW	0.96	0.95	0.93	0.95	0.62	0.70
GNN+fastText	**0.98**	**0.96**	**0.95**	**0.96**	**0.66**	**0.74**
GNN+BoW epoch10 [17]	0.97	0.95	0.93	0.95	0.57	0.67

Table 5. MRR10 comparison between our proposed models and the baseline models on the case study of vulnerability search of Netgear R7000 and TP-Link Deco M4. (11) represents the result of random seed 11 and [17] denotes the result of prior work [17].

Model	Netgear R7000				TP-Link Deco M4			
	x86	x64	ARM32	MIPS32	x86	x64	ARM32	MIPS32
GNN+BoW (11)	0.60	0.58	0.75	0.58	0.33	0.53	0.50	0.40
GNN+fastText (11)	0.75	0.63	0.78	0.75	0.51	0.51	0.38	0.64
GMN+BoW (11)	0.67	0.50	0.56	0.38	0.56	0.67	**0.78**	**0.78**
GMN+fastText (11)	**0.88**	**1.00**	**1.00**	**0.80**	**0.79**	**0.75**	0.52	0.73
GNN+BoW [17]	0.33	0.32	0.56	0.30	0.49	0.56	0.36	0.61
GMN+BoW [17]	**0.88**	0.54	**1.00**	0.79	0.67	0.73	0.70	**0.78**

respectively show that GMN+BoW are better than the other models in ARM 32-bit and MIPS 32-bit.

4.6 Results of Single-Architecture Models

Figure 3 shows that the results of training with random seed 11 in cv1. On EP15, GNN+fastText is better than the other models according to MRR10 although GNN+fastText necessitates a greater number of epochs in comparison to the other models. The MRR10 of GNN+fastText is 0.819 at epoch 82, whereas the MRR10 of GNN+BoW is 0.813 at epoch 42. Also, the MRR10 of GMN+fastText is 0.778 at epoch 32, whereas the MRR10 of GMN+BoW is 0.715 at epoch 3. For your reference, we calculated AUC and Recall@1 in each epoch during training. Each AUC is calculated by 40k positive pairs and 40k negative pairs of the validation dataset from Dataset-Win32 for XM task. Each Recall@1 is calculated by the same pairs to calculate MRR10. In comparison to MRR10, AUC converges more quickly after 10 iterations rather than 40 in the result of GNN+fastText. The similar trend observed in training in cv1 was confirmed for cv2, ..., cv5 as well. To mitigate the impact of overfitting, we evaluated GNN+fastText and GNN+BoW on EP15 for the evaluation. This is because the result of GNN+BoW in cv2 and of GNN+fastText in cv3 on EP20 with random seed 11 were slightly worse than on EP15. On the other hand, GMN+fastText and GMN+BoW are trained on EP20.

Table 6 shows Recall@1 and MRR10 of GNN+fastText, GMN+fastText, GNN+BoW and GMN+BoW on Dataset-Win32. Each value in Table 6 is the

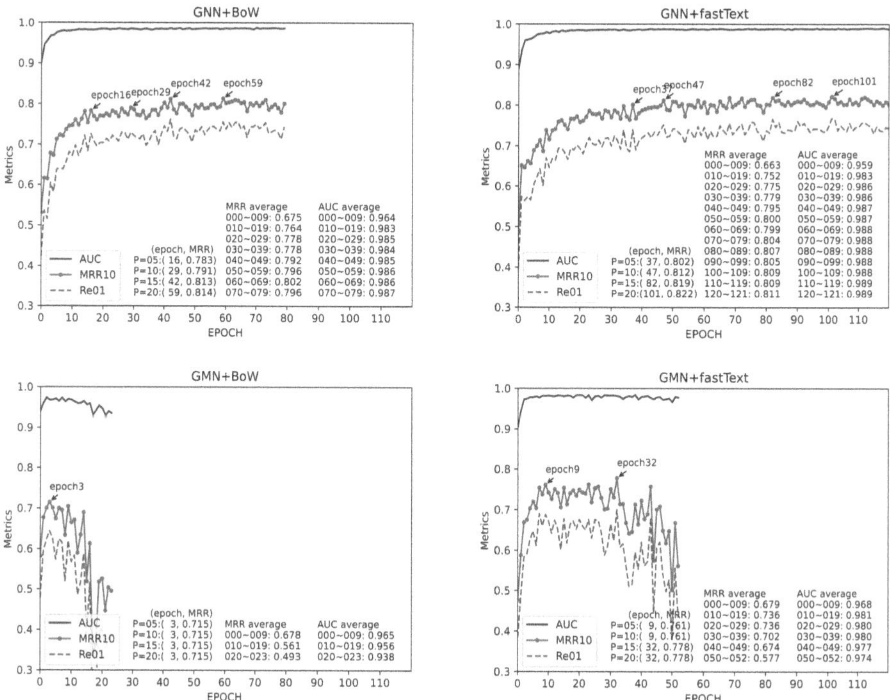

Fig. 3. The training results of GNN+fastText, GMN+fastText, GNN+BoW and GMN+BoW with random seed 11 on cv1. Re01 denotes Recall@1 and P denotes early stopping patience (EP). Epoch indices start at zero. In the results of training on EP15, the MRR10 value of 0.819 shows that GNN+fastText is better than the other models.

average from cv1 to cv5. In Table 6, the Recall@1 value of 0.73 and the MRR10 value of 0.79 show that GNN+fastText is better than the other models.

4.7 Case Study of Function Search Using GNN+fastText

In order to validate the search performance of our proposed model in malware analysis, using the samples of LockBit version 3.0 (LockBit 3.0) and a LockBit 3.0 builder obtained from Vx Underground [1] we conducted a search for the functions of LockBit 3.0 using the GNN+fastText, which is trained in cv1 with random seed 11 on EP15. LockBit is a variant of ransomware that was prominent in 2022, and a builder for LockBit 3.0 was leaked [5], making it easier to analyze. Analysis of LockBit 3.0 [23] revealed that there was a variant of LockBit 3.0 which has four distinct phases "Unpack Sections", "Reconstruct IAT", "Escalate Privilege", and "Ransom Main". The variant was also found to call corresponding functions for each of these phases within the `.itext` section [23]. For experimental purposes, using the LockBit 3.0 builder which is uploaded on

Table 6. MRR10 and Recall@1 comparison between our proposed models and the baseline models on Dataset-Win32. Each value is the average from cv1 to cv5.

Model	Recall@1(XM)	MRR10(XM)
GNN+BoW	0.72	0.78
GNN+fastText	**0.73**	**0.79**
GMN+BoW	0.64	0.71
GMN+fastText	0.66	0.74

Vx Underground [1], we created a variant (LockBit 3.0 encryptor), which was unpacked and included each phase excluding the Unpack Sections.

In light of this, using four functions each from the Reconstruct IAT phase (`Reconstruct IAT`, `create_heap_API_jmp_table`, `hash_api_address_resolve`, `gen_random`), the Escalate Privilege phase (`Escalate Privilege`, `Load Configuration`, `System Language Discovery`, `Set Icon`), and the Ransom Main phase (`Ransom Main`, `Drop Ransom Note`, `Set Wallpaper`, `Wipe Recycle Bin`), we conducted the case study of function search, which is based on the case study of vulnerability search by Marcelli et al. [17]. As for the twelve function names, `create_heap_API_jmp_table`, `hash_api_address_resolve`, `gen_random` were chosen based by prior work [21], and the rest were chosen based by prior work [23]. We prepared three targets for the case study, which are obtained from Vx Underground [1]. The SHA-256 hash values of the targets are as follows.

`Sample-58260:58260a6687486e39dc46461270b391280b7d59997d84b6639230d95e3bdfca23`

`Sample-87b76:87b76f35740262abb8da224b94779ff56eb6346318b4f9fb1988a59a72a4e6c9`

`Sample-a56b4:a56b41a6023f828cccaaef470874571d169fdb8f683a75edd430fbd31a2c3f6e`

In our function search, we confirmed that 12 out of 12 functions are detected in each target. In this case study, "detected" means that the result of a function search, using a specific function as a query for a given target, ranked first. We also observed that 31 out of 36 functions in the targets matched those of the LockBit 3.0 encryptor when comparing functions by hashopcode. The hashopcode, in this context, is a hash value of opcodes of a function. We computed the hashopcode of each function using the Python program from prior work [17]. In terms of hashopcode, `gen_random` in each target, `Reconstruct IAT` and `create_heap_API_jmp_table` in Sample-58260 are functions which did not match the corresponding ones of the LockBit 3.0 encryptor.

5 Discussion

5.1 Multi-architecture Models

GNN+fastText outperformed the GNN+BoW epoch10 [17] by up to 2% in AUC, 9% in Recall@1 and 7% in MRR10 in the evaluation of multi-architecture models with random seed 11 on Dataset-BINKIT. In regard to the results of AUC, GNN+fastText is higher than that of the other models in almost every case.

On the other hand, in the average results of random seed 11 to 15 in the ranking test (Table 1 and Table 3), a difference of up to 2% in MRR10 and up to 2% in Recall@1 was observed between GNN+fastText and GNN+BoW. From these results, it is considered that the superior performance of GNN+fastText and GNN+BoW over the GNN+BoW epoch10 [17] on Dataset-1 and Dataset-BINKIT can be attributed to the effectiveness of early stopping based on MRR.

Additionally, it is necessary to prepare multiple seeds for the evaluation because altering the random seeds will change the search performance. For example, in the results of GNN+fastText on Dataset-BINKIT (Table 3 and Table 4), there was a 2% difference in Recall@1 and MRR10 between the average results and those obtained with random seed11.

In terms of training time, a tendency was observed that GNN+fastText required more time for training compared to GNN+BoW (Sect. 4.4). This is attributed to the denser BB vectors created by fastText, which is believed to have contributed to the increased time requirement in comparison to those created by BoW. On the other hand, no significant difference was observed between GMN+fastText and GMN+BoW. The GMN models (GMN+fastText and GMN+BoW) conclude their training relatively quickly compared to GNN models (GNN+fastText and GNN+BoW). However, the GMN models exhibit more pronounced fluctuations in MRR10 during training, making it challenging to enhance the search performance of the GMN models.

When comparing the performance of the GNN models and the GMN models, it was observed that the GNN models outperformed the GMN models on Dataset-1 and Dataset-BINKIT (Sect. 4.4), whereas in the case study of vulnerability search, the GMN models yielded better results (Sect. 4.5). Notably, the case study involved searches for four or nine functions, while Dataset-1 and Dataset-BINKIT encompassed searches across 20k functions. Therefore, the results of Dataset-1 and Dataset-BINKIT are more reliable in terms of evaluating the search performance of each model compared to the case study. Furthermore, the function vectors generated by the GMN models exhibit slight variations depending on the pairs [13]. This results in an increased number of searches when searching for a specific function. For instance, when searching for one function similar to a particular function among 100 functions, the GMN models require creating 100 function pairs, necessitating 200 conversions of functions into vectors. In contrast, the GNN models only require 101 conversions of functions into vectors, making it nearly twice as fast in terms of function-to-vector conversion speed. In the realm of malware analysis, where new malware is being developed at a rate of 4.2 samples every second [3], the speed of analysis is important. Therefore, leveraging the GNN models in malware analysis are highly desirable. However, it cannot be discounted that with more meticulous hyperparameter tuning and the addition of training data, the performance of the GMN models could potentially surpass that of the GNN models.

We conducted a further experiment because we noticed that there were duplicate functions in terms of the hash values of their opcodes in Dataset-1 we obtained. Although prior work removed duplicate functions based on their names and the hash values of their instructions [17], we hypothesized that removing duplicate functions based on the hash values of their opcodes would be appropriate for machine learning models that use features of opcodes, not instructions. In the further experiment, we removed approximately 12% and 25% of functions from the training dataset and validation dataset, respectively, based on for their names and the hash values of their opcodes based on prior work's code [17]. Using the revised Dataset-1, we recreated the top 200 opcodes and a fastText model, and trained GNN+BoW and GNN+fastText with random seed 11 to 15. After the training, we tested the models on EP15 using the same testing datasets from Dataset-BINKIT in Table 3 and Table 4. As a result, we observed on average from random seed 11 to 15 that GNN+fastText outperformed GNN+BoW on the XM task for AUC and MRR10 while GNN+BoW outperformed GNN+fastText on the XC task for AUC. As for the comparison between GNN+BoW epoch10 [17] and GNN+fastText with random seed 11, GNN+fastText outperformed GNN+BoW epoch10 [17] by up to 1% in AUC, 5% in Recall@1 and 3% in MRR10.

5.2 Single-Architecture Models

In terms of AUC Recall@1 and MRR10, the single-architecture models exhibited a similar trend to the multi-architecture models. It is difficult to assess the effectiveness of early stopping based on MRR compared to the evaluation of multi-architecture models. However, because single-architecture models can detect functions of LockBit 3.0 accurately (as shown in the case study), they are considered to be sufficiently optimized for search performance.

5.3 Future Work

To further enhance the search performance in malware analysis using our proposed models, we plan on implementing data augmentation utilizing Malware Makeover [15,16]. In existing methods for calculating binary function similarity, function pairs based on function name are required to train the existing methods. However, the function name of malware typically has been stripped, making it exceedingly challenging to conduct training using functions of malware. Moreover, even if some functions have the same function name, there is no guarantee that they possess the same functionality. Therefore, we leverage Malware Makeover to enhance the search performance in malware analysis. Malware Makeover can create variants from malware, which retain the functionality of original malware. By conducting additional training using Malware Makeover, we aim to achieve further improvement in the search performance of our proposed models. We also plan to compare binary program similarity using a GNN model [13]. We hypothesize that the GNN model may perform well for calculating binary program similarity because the GNN model can convert a call-tree into a vector.

6 Conclusion

In this paper we proposed the method combining GNN and GMN models [13] and fastText [4] to transform functions in a binary into vectors for binary function similarity comparison. GNN+fastText outperformed the GNN+BoW epoch10 [17] by up to 2% in AUC, up to 9% in Recall@1 and up to 7% in MRR10 in a certain case when training multi-architecture models with random seed 11 (Sect. 4.4). In the case study of vulnerability search (Sect. 4.5), specific values of MRR10 in Table 5 show that GMN+fastText are better than the baseline models. Furthermore, our findings suggest that early stopping based on MRR enables to improve the search performance of GNN+fastText and GNN+BoW. In the case study of single-architecture models (Sect. 4.7), our work shows that GNN+fastText is effective to search for malicious functions of LockBit 3.0 because GNN+fastText was able to find 12 out of 12 functions which we prepared for the case study. In future work, to further enhance the search performance in malware analysis using our proposed models, we plan to implement data augmentation utilizing Malware Makeover [15,16] (Sect. 5.3).

Acknowledgments. This work was supported in part by the U.S.\ Army Research Office under MURI grant W911NF-21-1-0317; by Carnegie Mellon CyLab; and by National Police Agency of Japan.

References

1. Vx Underground. https://www.vx-underground.org/. Accessed 12 Dec 2023
2. Ahmad, I., Luo, L.: Unsupervised binary code translation with application to code clone detection and vulnerability discovery. In: Findings of the Association for Computational Linguistics: EMNLP 2023, pp. 14581–14592 (2023)
3. AV-TEST: SECURITY REPORT 2019/2020. https://www.av-test.org/fileadmin/pdf/security_report/AV-TEST_Security_Report_2019-2020.pdf. Accessed 10 Nov 2023
4. Bojanowski, P., Grave, E., Joulin, A., Mikolov, T.: Enriching word vectors with subword information. Trans. Assoc. Comput. Linguist. **5**, 135–146 (2017)
5. Cybersecurity and Infrastructure Security Agency: Understanding Ransomware Threat Actors: LockBit. https://www.cisa.gov/news-events/cybersecurity-advisories/aa23-165a/. Accessed 09 Nov 2023
6. Dai, H., Dai, B., Song, L.: Discriminative embeddings of latent variable models for structured data. In: Proceedings of the 33rd International Conference on Machine Learning, pp. 2702–2711. PMLR (2016)
7. Gao, J., Yang, X., Fu, Y., Jiang, Y., Sun, J.: VulSeeker: a semantic learning based vulnerability seeker for cross-platform binary. In: Proceedings of the 33rd ACM/IEEE International Conference on Automated Software Engineering, pp. 896–899 (2018)
8. Hex-Rays: IDA Pro. https://hex-rays.com/ida-pro/. Accessed 15 Oct 2023
9. Hu, Y., Wang, H., Zhang, Y., Li, B., Gu, D.: A semantics-based hybrid approach on binary code similarity comparison. IEEE Trans. Softw. Eng. **47**(6), 1241–1258 (2021)

10. Ito, N., Hashimoto, M., Otsuka, A.: Feature extraction methods for binary code similarity detection using neural machine translation models. IEEE Access **11**, 102796–102805 (2023)
11. Ivanov., V., Romanov., V., Succi., G.: Predicting type annotations for python using embeddings from graph neural networks. In: Proceedings of the 23rd International Conference on Enterprise Information Systems - Volume 1: ICEIS, pp. 548–556. SciTePress (2021)
12. Kim, D., Kim, E., Cha, S.K., Son, S., Kim, Y.: Revisiting binary code similarity analysis using interpretable feature engineering and lessons learned. IEEE Trans. Softw. Eng. **49**(4), 1661–1682 (2023). https://doi.org/10.1109/TSE.2022.3187689
13. Li, Y., Gu, C., Dullien, T., Vinyals, O., Kohli, P.: Graph matching networks for learning the similarity of graph structured objects. In: Proceedings of the 36th International Conference on Machine Learning, pp. 3835–3845. PMLR (2019)
14. Lin, Z., et al.: A structured self-attentive sentence embedding. arXiv preprint arXiv:1703.03130 (2017)
15. Lucas, K., Pai, S., Lin, W., Bauer, L., Reiter, M.K., Sharif, M.: Adversarial training for raw-binary malware classifiers. In: 32nd USENIX Security Symposium (USENIX Security 23), pp. 1163–1180 (2023)
16. Lucas, K., Sharif, M., Bauer, L., Reiter, M.K., Shintre, S.: Malware makeover: breaking ml-based static analysis by modifying executable bytes. In: Proceedings of the 2021 ACM Asia Conference on Computer and Communications Security, pp. 744–758 (2021)
17. Marcelli, A., Graziano, M., Ugarte-Pedrero, X., Fratantonio, Y., Mansouri, M., Balzarotti, D.: How machine learning is solving the binary function similarity problem. In: 31st USENIX Security Symposium (USENIX Security 22), pp. 2099–2116. USENIX Association, Boston, MA, August 2022. https://www.usenix.org/conference/usenixsecurity22/presentation/marcelli
18. Massarelli, L., Di Luna, G.A., Petroni, F., Baldoni, R., Querzoni, L.: SAFE: self-attentive function embeddings for binary similarity. In: Perdisci, R., Maurice, C., Giacinto, G., Almgren, M. (eds.) DIMVA 2019. LNCS, vol. 11543, pp. 309–329. Springer, Cham (2019). https://doi.org/10.1007/978-3-030-22038-9_15
19. Mikolov, T., Chen, K., Corrado, G., Dean, J.: Efficient estimation of word representations in vector space. arXiv preprint arXiv:1301.3781 (2013)
20. Shen, R., Gao, L., Ma, Y.A.: On optimal early stopping: Over-informative versus under-informative parametrization. arXiv preprint arXiv:2202.09885 (2022)
21. Tanaka, Y.: (2022). https://jp.security.ntt/tech_blog/102hz18. Accessed 21 Oct 2023
22. Tantithamthavorn, C., McIntosh, S., Hassan, A.E., Matsumoto, K.: An empirical comparison of model validation techniques for defect prediction models. IEEE Trans. Softw. Eng. **43**(1), 1–18 (2017)
23. TXOne Networks: LockBit 3.0 Analysis: How to Enhance Ransomware and Malware Protection. https://www.txone.com/blog/malware-analysis-lockbit-3-0/. Accessed 21 Oct 2023
24. Wang, M., Interrante-Grant, A., Whelan, R., Leek, T.: COBRA-GCN: contrastive learning to optimize binary representation analysis with graph convolutional networks. In: Cavallaro, L., Gruss, D., Pellegrino, G., Giacinto, G. (eds.) DIMVA 2022. LNCS, vol. 13358, pp. 53–74. Springer, Cham (2022). https://doi.org/10.1007/978-3-031-09484-2_4
25. Xu, X., Liu, C., Feng, Q., Yin, H., Song, L., Song, D.: Neural network-based graph embedding for cross-platform binary code similarity detection. In: Proceedings of

the 2017 ACM SIGSAC Conference on Computer and Communications Security, pp. 363–376 (2017)

26. Yadav, S., Shukla, S.: Analysis of k-fold cross-validation over hold-out validation on colossal datasets for quality classification. In: 2016 IEEE 6th International Conference on Advanced Computing (IACC), pp. 78–83 (2016)
27. Yang, S., et al.: Asteria-pro: enhancing deep learning-based binary code similarity detection by incorporating domain knowledge. ACM Trans. Softw. Eng. Methodol. **33**(1), 1–40 (2023)
28. Yu, Z., Cao, R., Tang, Q., Nie, S., Huang, J., Wu, S.: Order matters: Semantic-aware neural networks for binary code similarity detection. In: Proceedings of the AAAI Conference on Artificial Intelligence, vol. 34, no. 01, pp. 1145–1152 (2020)
29. Zuo, F., Li, X., Young, P., Luo, L., Zeng, Q., Zhang, Z.: Neural machine translation inspired binary code similarity comparison beyond function pairs. arXiv preprint arXiv:1808.04706 (2018)

Inferring Recovery Steps from Cyber Threat Intelligence Reports

Zsolt Levente Kucsván[1(✉)], Marco Caselli[3], Andreas Peter[1,2], and Andrea Continella[1]

[1] University of Twente, Enschede, The Netherlands
{z.kucsvan,a.continella}@utwente.nl
[2] Carl von Ossietzky Universität Oldenburg, Oldenburg, Germany
andreas.peter@uni-oldenburg.de
[3] Siemens AG, Munich, Germany
marco.caselli@siemens.com

Abstract. Within the constantly changing threat landscape, Security Operation Centers are overwhelmed by suspicious alerts, which require manual investigation. Nonetheless, given the impact and severity of modern threats, it is crucial to quickly mitigate and respond to potential incidents. Currently, security operators use predefined sets of actions from so-called playbooks to respond to incidents. However, these playbooks need to be manually created and updated for each threat, again increasing the workload of the operators. In this work, we research approaches to automate the inference of recovery steps by automatically identifying steps taken by threat actors within Cyber Threat Intelligence reports and translating these steps into recovery steps that can be defined in playbooks. Our insight is that by analyzing the text describing threats, we can effectively infer their corresponding recovery actions. To this end, we first design and implement a semantic approach based on traditional Natural Language Processing techniques, and we then study a generative approach based on recent Large Language Models (LLMs). Our experiments show that even if the LLMs were not designed to solve domain-specific problems, they outperform the precision of semantic approaches by up to 45%. We also evaluate factuality showing that LLMs tend to produce up to 90 factual errors over the entire dataset.

Keywords: CTI reports · Recovery steps · LLM · Semantic model

1 Introduction

The number of threats that organizations face nowadays increases annually [12], reaching volumes up to multiple thousands per day [2]. To defend themselves, organizations employ teams that are in charge of monitoring the infrastructure, identifying attacks (*alert evaluation*), and responding to incidents (*incident response*). These teams operate in so-called Security Operation Centers (SOCs).

F. Maggi et al. (Eds.): DIMVA 2024, LNCS 14828, pp. 330–349, 2024.
https://doi.org/10.1007/978-3-031-64171-8_17

Both alert evaluation and incident response rely on heavy and repetitive manual tasks, which eventually lead to burnout and alert fatigue of the human operators [35]. In recent years, several works proposed approaches to reduce the workload of the security operators [10,13,38] and automate different tasks within SOCs, such as alert triaging [5,11,14]. However, most of the literature only focuses on the alert evaluation phase, leaving automation in the incident response phase mostly unaddressed.

Security operators use playbooks to respond to incidents, which are manually-crafted sets of instructions containing *actions* and *artifacts*—e.g., in the instruction **delete malicious executable** the action is **delete** and the artifact **malicious executable**. When being under attack, these documents help security operators identify adequate countermeasures. Creating playbooks is often a complex task, most importantly for organization-specific playbooks, which are influenced by many factors [31]. Hence, the creation of the playbooks often results in a bottleneck in terms of manual work and time consumption, again contributing to the alert fatigue of the operators.

In this work, we research approaches to automate the generation of the key elements of playbooks, namely the actions and artifacts of the countermeasure steps to be deployed on the affected systems to recover from threats. We refer to these steps as *recovery steps*. Note, that while generating the recovery steps is a step towards automating the playbook creation process, constructing playbooks is out of scope in our work.

In our context, we consider recovery steps that can be potentially automated, such as deleting files, restoring backups, or blocking connections. Actions that are not related to affected systems from a technical perspective (i.e., legal actions, policy-related actions, or similar) are out of scope. In particular, we focus on analyzing the behaviour of threats and extracting the threat actions to infer recovery steps. We take the commonly used Cyber Threat Intelligence (CTI) reports as a reference, as they contain descriptions of attacks written by security operators. These reports are written in natural language and require NLP techniques to analyze them in an automated system.

To the best of our knowledge, there are no prior works that explore automated inference of recovery steps for incident response. However, several prior works aim to solve some underlying problems, such as extracting the threat actor's behaviour or artifacts. Legoy et al. [19] proposed a multi-label classifier to extract attack behaviour; Lim et al. [21] proposed a machine learning based framework to annotate security-related text, detecting malware actions and capabilities; Park et al. [25] designed a semantic NLP based framework to extract information from CTI reports and identify relations between these. While the previously mentioned works are based on traditional approaches, it is important to mention the recent advances of large language models (LLMs) and their promising capabilities to solve domain-specific problems despite the fact that they were trained on a general corpus of data. The industry already started to propose solutions for *incident response* that are based on the LLMs, such as Microsoft Security Copilot, which delivers a step-by-step guide on how to investigate and respond to incidents [18], or the Trend Micro's Trend Companion AI, which helps security operators with information to accelerate response time [23]. Unfortunately, these tools are not openly available, nor researched properly. This stresses the importance of this work of comparing LLMs and traditional approaches for extracting CTI knowledge to infer actionable recovery steps.

In this paper, we compare traditional NLP techniques that are based on semantic models to generative models that rely on advanced LLMs. First, motivated by the lack of available systems, we design and implement our own semantic-based pipeline (Sect. 3.1). Second, we study and experiment with different types of LLMs to assess how they perform against semantic approaches. In our assessment, we include both an open-source LLM (LLaMa 2 [37]) and multiple versions of a closed-source LLM (GPT 3 [8] and GPT 4 [4]).

For the assessment, due to the unavailability of an existing dataset, we create a new labeled dataset of recovery steps from 200 CTI reports, randomly chosen from the Trend Micro Encyclopedia [1]. This dataset is composed of sentences labeled with their corresponding recovery steps for each report.

Our results show that both the semantic and generative models have their own advantages and disadvantages. The semantic model is more consistent, based on factual rules that give a better explainability, while LLMs outperform it up to 45% in precision and 53% in recall. On the other hand, LLMs can produce up to 90 factual errors over the entire dataset, that are inferred recovery steps unconnected to the information given in the input.

The contributions of our work are the following:

- We design the first system, based on traditional NLP techniques to analyze CTI reports and infer threat recovery steps.
- We build the first labeled dataset of recovery steps, composed of labeled sentences from 200 CTI reports.
- We evaluate semantic and generative approaches, assessing and comparing the performances and capabilities of NLP-based models against multiple off-the-shelf LLMs. Based on our empirical evaluation, we provide insights and suggestions for future research in this field.

In the spirit of open science, we provide public access to the dataset we created and the source code of the implemented semantic model at https://github.com/utwente-scs/recovery-inference.

2 Background and Challenges

In this section, we describe how CTI reports can be used as input for our research and how can we leverage different NLP approaches to process threat descriptions. Additionally, we describe the challenges we need to overcome for designing a recovery steps inferring system.

2.1 Cyber Threat Intelligence Reports

CTI reports are documents with descriptions of threats, containing information such as attacker behavior, indicators of compromise (IOCs), or courses of action to mitigate the threat [30]. We consider the following simple CTI example:

> The malware arrives on a system via a USB device. It **encrypts** the **folder**: %User Profile%\Documents. It **drops** a **malicious file** to the desktop. After the execution, the malware deletes itself.

By isolating core actions in the example above, such as **encrypts folder** or **drops malicious file**, a human operator, understanding the behavior of the malware, takes appropriate course of action to recover from the attack. This is useful, especially in cases, when the available playbooks are too generic, with high-level instructions instead of concrete actions for specific threats. Extending generic playbooks with information extracted from CTI reports results in a more comprehensive and complete course of action to recover from the threat, but without an automated system, this requires additional manual effort by searching, reading, and understanding reports.

2.2 Traditional NLP Techniques

To process text written in natural language, we employ NLP techniques. Traditional NLP techniques that rely on capturing the meaning and relationships between words, through linguistic analysis are often used. Linguistic analysis in NLP offers a variety of tasks, such as *tokenization*, *lemmatization*, *part of speech (POS) tagging*, *dependency parsing*, etc. In the following, we give a short insight for each of these tasks, that will be used later on in this paper.

Tokenization is the process of splitting the text into multiple pieces. These pieces are usually words or parts of words. E.g., The malware deletes the file. can be tokenized into: [The], [malware], [deletes], [the], [file], [.].

Lemmatization is the process of transforming tokens into their dictionary base form to enhance text understanding. Due to the grammar, words might have different forms (e.g., delete, deleted, deletion) and by lemmatizing we make sure we identify all forms as the same word.

POS tagging is the process of labelling the words into their corresponding grammatical category, such as nouns or verbs.

Dependency parsing is the process of identifying connected tokens that create phrases. All the connections create a dependency tree that describes the grammatical structure of a sentence or text.

2.3 Generative LLMs

In the last couple of years, LLMs went through an impressive development, enabling them to address a wide range of general tasks. Especially they are very good at generating coherent, logical text based on an input prompt. Input prompts are usually instructions or questions fed into LLMs to initiate responses. Based on a prompt and the background knowledge of the acquired from the large corpora of training datasets, the LLMs provide a piece of text as output, which is statistically the most likely continuation of the initial prompt.

There are two main categories of generative LLMs: *open-source* LLMs that are open for the users, and *closed-source* LLMs that have their inner architecture hidden from the public. Another difference between these two types of LLMs is their training process. Open-source LLMs usually use public data, while closed-source LLMs mostly leverage proprietary datasets for the training.

2.4 Challenges

Extracting recovery steps involves multiple research domains with unsolved challenges, including Cyber Threat Intelligence and NLP-based Information Extraction. This section outlines these challenges and their relevance to our research.

Cyber Threat Intelligence. A research by Abu et al. [3] on Cyber Threat Intelligence emphasizes two main challenges relevant for our work: the large quantity of CTI data and their timeliness. Time to process the data is a crucial factor in incident response, to avoid as much damage as possible. Moreover, having a large quantity of CTI data implies an increased effort for security operators to search, analyze, and identify relevant information from the reports to respond to incidents. These challenges require automated systems to reduce manual effort for operators. Our research therefore focuses primarily on *automation*.

Information Extraction. There are several solutions proposed to extract different types of information from CTI reports such as [17,21] or [19]. One general challenge comes from the unstructured nature of these reports which often contain ambiguities that might be hard to interpret properly by the NLP system, leading to wrong parsing of the sentences, wrong POS tagging, or misidentified dependencies. These ambiguities might refer to the narrative style of the report (i.e., active or passive voice), but also to other noises such as special characters or words that are not common in natural language (e.g., IOCs). Another challenge is identifying and filtering relevant information for recovery steps. As an example, if the extracted information refers to a user executing a non-malicious activity such as copying a file, we need to distinguish it from the threat actions. The last challenge in this process is to link threat actions to the recovery steps which requires a deeper understanding of the threat behaviour. To sum up, the three challenges of inferring recovery steps from CTI reports are the following: *text ambiguity*, *information relevance*, and *recovery linking*.

3 Methodology and Implementation

To address the challenges described in Sect. 2.4 we take two approaches: semantic-based and generative. This section describes the methodology and implementation details of the two approaches.

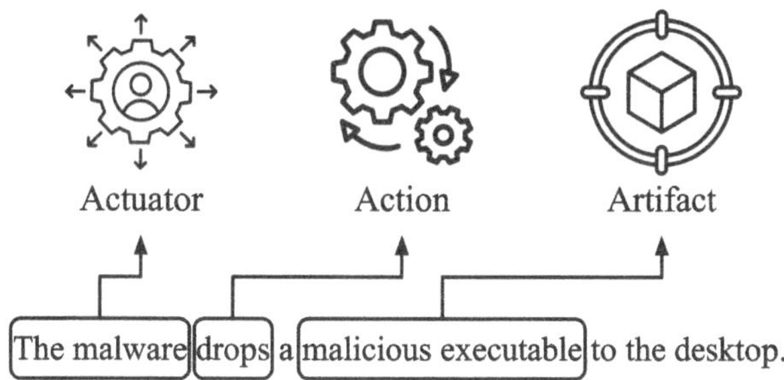

Fig. 1. Example of the identified components by the semantic model. When extracting the information from this sentence, we identify **The malware** as actuator, **drops** as action and **malicious executable** as artifact.

3.1 Semantic Approach

To address the *automation* challenge described in Sect. 2.4, we propose an automated system that infers recovery steps based on the description of an attack. To process the threat descriptions, first, we need to model the sentences, which gives a structured form that is more interpretable for the computers. One common approach used in similar works is identifying three main components: *action*, which is the main event described in the sentence, *actuator*, which is the entity performing the action, and *artifact*, which is the entity acted upon or affected by the action [30] (e.g., see Fig. 1).

Our system has two main components at its core: the *extractor* and the *mapper*. The extractor is responsible for extracting information from a given text based on our sentence model described above. The mapper searches the extracted actions in a predefined list of threat actions that have a mapping to a recovery action. For the actions that are found in this mapping, the inferred recovery action is given to the operator. However, some actions are actually threat actions, such as **to steal** in the context of stealing personal information, but they do not have a recovery action that can be deployed on the system. Instead, these threat actions require making legal actions, change of policies or raise awareness of the employees. In this research, we consider non-technical recovery actions out of scope.

The mapping is a tool that can be customized based on the preferences and needs of every SOC. While this mapping becomes another resource that needs maintenance over time as the policies of the organization change, maintaining it does not imply a significant increase in the manual effort of the operators. Giving the flexibility to customize the mapping is needed, because if we take as an example the artifacts created by a threat on a system, some organizations might want to delete the artifact completely, others might put it just in a quarantine to be able to analyze it later.

To implement our system we create pipelines where each component of the pipeline has a clearly defined task. The main components of our pipeline are the following: *text preprocessor* for cleaning the text, the extractor for identifying the

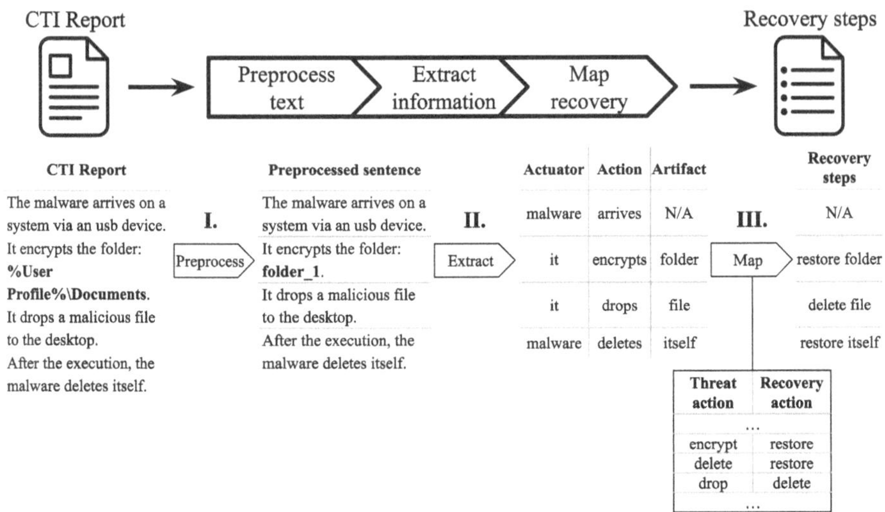

Fig. 2. Example of how the designed system works. As input the system gets a CTI report in an unstructured format which in the I. step is preprocessed to mask IOCs and split the input text into sentences, then in the II. step the actuator, action and artifact triplets are extracted from the sentences, and in the III. step the recovery steps are inferred from the triplets based on a mapping.

(actuator, action, artifact) triplets, and the mapper for linking the threat actions to recovery actions. The full pipeline (see Fig. 2) is implemented in Python, using different libraries that we describe in the remainder of this section.

Text Preprocessing. As described in Sect. 2.4, CTI reports often contain ambiguous words (e.g., file paths, IP addresses, or hashes) that are unusual in the generic English text. To partly address the *text ambiguity* challenge, we use the IOCSearcher [9] tool to mask the IOCs.

In the reports, paragraphs are delimited by new lines, but we observe that in some cases (e.g., bullet point enumeration) some information positioned in new lines is still connected logically to a previous paragraph. Using simple heuristics, we split the text into logically connected parts to facilitate proper sentence parsing at the next step.

As the last preprocessing step, we use the SpaCy library [16], to tokenize and parse the sentences. The motivation of using SpaCy is that this library provides a large set of advanced natural language processing functions for extracting information from text. Also, this tool is commonly used in research that is connected to NLP. For our work, we use the built-in `en_core_web_trf` model that allows us to process English text.

Triplet Extraction. This part of the pipeline is responsible of identifying the actuator, action, and artifact triplets, using the POS taggings and dependency tree got from processing the sentences. First, we search for the main action in a sentence. Then, using the dependency relations, we check for other connected actions to allow processing sentences, where instead of a single action we find a chain of actions (e.g., the malware **connects** to remote site and **exfiltrates** information). The last step is to iterate through the nouns and check for their dependency towards the tokens in the list of actions, which determines if the noun represents a subject (i.e., this is the actuator of the action), a direct object (i.e., this is the artifact of the action) or another dependency.

Recovery Action Mapping. Inferring recovery steps relies on a mapping between threat actions and recovery actions. The predefined list of threat actions has the role to address the *information relevance* challenge, while the mappings address the *recovery linking* problem described in Sect. 2.1. This mapping is stored in a JSON file that can be configured based on the needs of the user. The JSON file contains key-value pairs, where the keys represent the threat action while the values represent the corresponding recovery actions.

3.2 Generative Approach

In the semantic approach, we used a pipeline, where individual components solve a smaller sub-task in inferring recovery actions. Conversely, generative models can be used to directly request the desired output, e.g., using their question-and-answer capabilities. In this setup, we need to engineer an input prompt that instructs the LLM to infer recovery steps. Some good practices of AI-experts in prompt engineering is to be explicit about the task specification and requirement and give examples of desired interactions [40]. Based on these practices, we design our prompt, that has three components: a system prompt that instructs the LLMs on how to behave, an example prompt showing the desired interaction, and the actual prompt for the task. For the system prompt, we first define the role the LLM should assume (i.e., incident responder in a SOC), then we give the instructions for inferring the recovery steps with the clarification of what to do if there is no recovery step that can be inferred, and last, we indicate the desired output format (JSON). For the example prompt we include a short CTI report and provide the expected, correct recovery steps in JSON format. For the actual prompt, we always include the CTI report from which we would like the LLM to infer the recovery steps. The final prompt template we have created is included in Appendix A.

The interaction with the LLMs is different for the open-source and closed-source LLMs. While the open-source LLMs are available to download and run on a local infrastructure, the closed-source LLMs offer only an API service for the interaction.

4 Evaluation

We have implemented a semantic-based model that infers recovery steps by analyzing CTI reports. To evaluate this model, we compare it against different off-the-shelf generative LLMs to gain a deeper understanding of what are the advantages and disadvantages of each model. We start by creating a labeled dataset of recovery steps, described in Sect. 4.1. Next, we perform our experiments presented in Sect. 4.2. Then Sect. 4.3 describes the metrics used to measure the performance of the approaches. Finally, we evaluate the results of the experiments in Sect. 4.4.

4.1 Dataset

To evaluate our results we use CTI reports collected from the Trend Micro Threat Encyclopedia [1]. These reports describe several technical aspects of the malware such as arrival details, autostart technique, routines, modification it does in the system, information it gathers for exfiltration or propagation.

The reports are written in natural language, thus we create a manually labeled subset of the original reports. The dataset, manually labeled by us, consists of 200 randomly selected reports. The manual labels of one report contain the following:

- manually parsed sentences from the report;
- manually extracted actuator, action, artifact triplets for each sentence; and
- manually mapped recovery action and artifact pairs for each triplet that has a recovery countermeasure.

In this manually labeled dataset, we have a total of 1,603 sentences of which 818 are unique. Amongst the unique sentences, we identified 829 actuator, action, artifact triplets, from which 580 have recovery steps based on our evaluation.

4.2 Experimental Setup

For this research, we conduct an experiment of processing all the CTI reports from our labeled dataset presented in Sect. 4.1. This experiment is repeated on multiple models, that we detail in the remaining of this subsection.

Running the experiment on the semantic model does not require special hardware, hence we run them on regular workstation with the following characteristics: 11th Gen Intel Core i7-11800H processor running at 2.3 GHz, 16 GB of RAM, running Windows 11.

For the open-source LLM experiment, we choose the 13 billion parameter version of LLaMa 2. This requires a more powerful hardware than a regular workstation, so we run the experiments on a GPU cluster using an NVIDIA RTX 6000 GPU with 48 GB of memory.

For the closed-source LLM experiment we do not have the possibility to run the experiment locally. Instead, using the paid API service provided by OpenAI, we send our prompts to a specific version of GPT, the LLM remotely generates the answer and sends us back the result. We repeat the experiment on multiple versions of GPT, namely GPT 3.5 Turbo, GPT 4, and GPT 4 Turbo, which at the time of experiments point to the following models (in the same order): `gpt-3.5-turbo-1106`, `gpt-4-0613` and `gpt-4-0125-preview`.

Table 1. Overall results of average metrics and the total number of factual errors made over the entire dataset.

Model	Avg. precision (%)	Avg. recall (%)	Factual errors (count)
GPT 3.5 Turbo	65.11	76.42	22
GPT 4	52.81	**84.57**	24
GPT 4 Turbo	**78.02**	83.15	**16**
LLaMa	45.24	63.97	90
Semantic model	32.85	31.50	N/A

4.3 Metrics

The problem of inferring recovery steps does not fit the traditional definition of a classification task. While in a classification task the main task is to assign a label or category to a given input, in our case the emphasis is on identifying and representing entities and their relations within sentences. In NLP this is referred as a relation extraction task [7].

To evaluate this relation extraction task, we use the analogous metrics to those employed in classification tasks. These metrics are computed at sentence-level, considering the following:

- **True Positives (TP)**: the count of correctly inferred recovery steps based on the ground truth.
- **False Positives (FP)**: the count of incorrectly inferred recovery steps that are absent within the ground truth.
- **False Negatives (FN)**: the count of ground truth recovery steps that were not inferred by the system.

To evaluate the performance of the system at report-level, we aggregate the respective TP, FP, and FN counts across all sentences within a report and we compute the precision (P) and recall (R) by their classical definition.

$$P = \frac{TP}{TP + FP}, \qquad R = \frac{TP}{TP + FN}$$

We also measure the runtime of each model on a report-level. Furthermore, in the experiments with the LLMs we include two additional measurements regarding factuality and output format of these models. LLMs tend to infer recovery steps cannot be mapped to any existing sentence in the input report. We call these *factual errors*. To evaluate the factuality, we compute on a report level the number of factual errors. LLMs were designed to provide natural-language-like output, however they can be instructed to provide a more specific format. To evaluate our models, we count on the dataset-level for how many input reports LLMs provide a JSON formatted output as they were instructed. Note, that for evaluating the performance and factuality of the models, we include only the outputs that provided a JSON format.

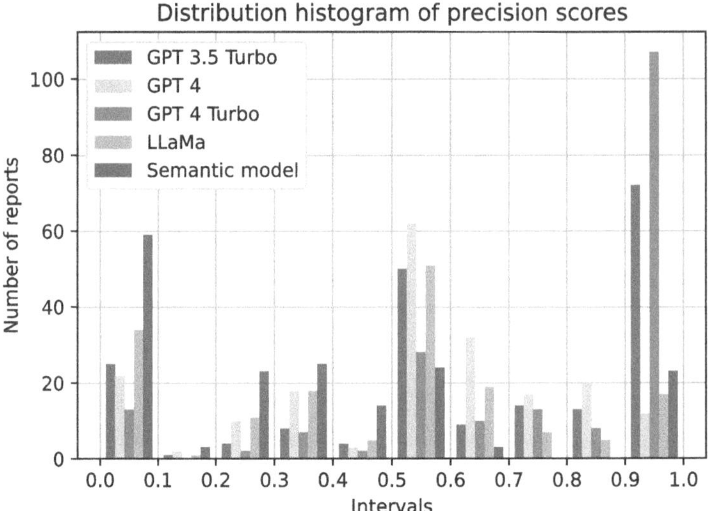

Fig. 3. Distribution histogram of precision scores for the different models. Vertical grid lines mark intervals, and the bars show the number of reports with precision values within each interval.

4.4 Results

Performance. Fig. 3 presents the distribution histogram of precision scores obtained for each model, while Fig. 4 shows the analogous histogram for recall scores. Additionally Table 1 presents the average precision and recall over the entire dataset. Note, that we removed the data points where the precision or recall were undefined (i.e., both TP and FP or FN counts were zero).

From the results, we can observe that overall the entire dataset, GPT 4 Turbo performs the best for precision, while GPT 4 outperforms it for recall. This means, that while GPT 4 Turbo produces fewer FPs (i.e., inferred steps that are not recovery steps), GPT 4 is better at inferring the correct recovery steps, leaving out fewer unidentified steps. Both in terms of precision and recall, the semantic model performs the worst compared to the LLMs.

Factuality. The semantic model has a limited set of possibilities to output depending on the actions included in the mapping. The output depends on the extracted information combined with this mapping, hence it is not possible to get an output that cannot be mapped to a word extracted from the input report.

One example would be the following factual error: **restore user folder**, which according to the LLM comes from the sentence **The executable encrypts the user folder.** At first glance, this recovery step suggested by the LLM makes sense with the provided explanation. However, upon closer investigation, we found that there was no mention of encrypting user folders in the

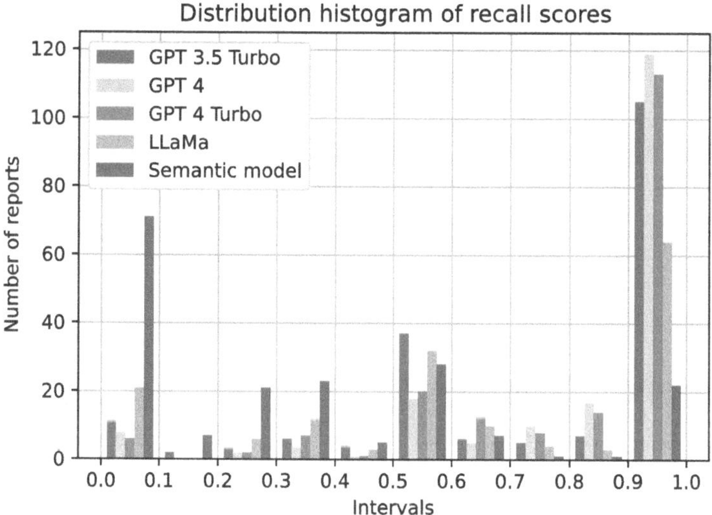

Fig. 4. Distribution histogram of recall scores for the different models. Vertical grid lines mark intervals, and the bars show the number of reports with recall values within each interval.

input CTI report in the first place. Table 1 shows the total number of factual errors in the dataset.

Output Format. Our semantic model always provides JSON formatted output by design. The generative LLMs were instructed to provide the output in JSON format, however, only the GPT models were able to comply to this request formulated in the input prompt. The LLaMa model provided for 170 reports the output in a JSON format and for the rest 30 reports it provided only natural-language-like answers.

Runtime. Table 2 shows a comparison of the runtime to process a single report from our dataset. We note that the semantic model outperforms the LLMs in terms of runtime, the average runtime of this model being almost ten times faster than the LLMs. While on average both LLaMa and GPT models perform similarly, we note that the minimum and maximum runtime is significantly higher for GPT.

5 Discussion

This section of the paper discusses the results presented in Sect. 4.4. First, we compare the two approaches based on the results and discuss key takeaways. Then, we present the limitations and future directions of our work.

Table 2. Runtime measurements of the different models. The table shows the minimum, average, and maximum processing time per report for each model. *For GPT, the times shown represent the average runtime of the three models.

Model	Time (s)		
	Min.	Avg.	Max.
Semantic model	0.43	2.64	28.51
LLaMa	2.91	20.67	104.44
GPT*	7.33	19.61	612.66

5.1 Semantic vs. Generative Models

While in terms of performance, the results shown in Sect. 4.4 clearly show that LLMs outperform the semantic approach, there are other aspects we can compare the approaches on. In the following, we make a theoretical comparison between the semantic and generative models and we formulate the key takeaways of our work.

Factuality. Semantic approaches rely on different rules and heuristics, thus in this case we cannot speak about their factuality. On the other hand, our results show, that factuality is a problem of generative LLMs that tend to produce factual errors.

Output Format. When implementing a semantic approach, the developer has always the choice to decide on the format of the output. Using LLMs, we have the option to provide the desired output format in the input prompt, but there is no guarantee that this will be strictly applied.

Predictability. The semantic model we implemented is deterministic, meaning that users can expect to always get the same answer for the same input. More than this, the mapping used to infer the recovery steps guarantees that for a certain threat action the inferred recovery action is always the same. Conversely, the LLMs tend to use multiple synonyms for the same recovery action (e.g., **delete**, **remove**, **clean**) instead of using one, deterministic form of the action, which might bring challenges in the further processing of the output.

Hardware Requirements. While we are able to run the semantic model on a personal computer, LLaMa requires a system with high-performance GPU. The GPT models are not available to run locally, but given that they have significantly more parameters in their architecture than LLaMa we can safely assume that these models have the highest requirements from our experiments.

Takeaways. The most important takeaways of this work are the following:

- LLMs reach higher performance in terms of precision and recall than semantic approaches, even if they were not specifically trained to solve the specific problem at hand.
- Semantic approaches focus on factual accuracy and consistent reasoning, in contrast to LLMs which tend to produce unpredictable or factually incorrect answers.
- A hybrid approach, combining semantic NLP with LLMs, has a potential to surpass the limitations of both approaches.

5.2 Limitations

Our results show that both approaches have several limitations. While the semantic model has the focus on understanding the meaning of the sentences, it lacks a broader understanding of the context. One example of this limitation is the sentence **It deletes itself after execution.** In this example **delete** seems to be a threat action that has **restore** as recovery, but focusing on a broader context, **itself** refers to the malware, which means in this case the system should not infer any recovery steps.

On the other hand, LLMs are better at taking into account this contextual information, but have their own limitations in explainability. LLMs are to some extent able to provide explanations to which sentence of the input is the recovery step connected to, but occasionally these explanations are factual errors. This leads to the main limitation of the LLMs, namely the factualness of their output: not every answer coming from the LLMs is based on facts, instead the answers can be sometimes just creative choosing of seemingly correct words.

5.3 Future Work

To address the limitations above, we propose as a future direction the combination of the two approaches. Leveraging the wide-range background knowledge of the LLMs together with the explainability of the semantic approach could allow for a high-performance system in which the decisions can be traced back to specific knowledge sources or rules. Such a system would avoid factual errors while having a consistent set of recovery actions to choose from.

Another possible direction for future work is to use the inferred recovery steps to generate actual playbooks tailored to a specific infrastructure that can be used in incident response. This would result in new challenges that need to be addressed such as ordering the recovery steps by their dependencies (e.g., first stop malicious process, then restore backup) and taking into consideration the infrastructure, the security tools, the policies of the organization.

6 Related Work

We can categorize the related work into three main parts: *CTI*, *knowledge extraction*, and *incident response playbook* related works. These three are the key components that our work relies on. CTI is the key source of information for our system, knowledge extraction is the main process we use to analyze the behaviour of the threats and playbooks represent the direction towards which our system is headed by inferring recovery steps. In the following, we present some of the most important works on these three fields that are related to our research.

6.1 Cyber Threat Intelligence

In 2018 Abu et al. [3] present in their paper an important challenge that we still face: the overload of CTI data, which requires to automation solutions for operators not to be overwhelmed with their work. Tounsi et al. [36] in their survey highlight another important issue, namely the quality of the available threat intelligence, especially describing the technical details such as potential attack vectors and TTPs. Besides the overload and quality, another challenge identified by Wagner et al. [39] is focused on sharing the intelligence, which raises issues of trust establishment but also interoperability of the data. A more recent study is presented by Schlette et al. [30] on the usage of CTI, analyzing different formats and identifying core concepts required for incident response.

The previously mentioned works provide valuable perspectives on recent advancements in the CTI domain. They shed light on the technical characteristics of CTI reports and the mechanisms by which these reports are shared. This highlights the potential of using CTI reports as input data for designing a recovery step inferring system.

6.2 Knowledge Extraction

Knowledge extraction is a common process in the field for gathering, analyzing, and leveraging information that helps us improve the defense of IT systems. There are lots of works proposing solutions for this processes. Two main categories of approaches are the traditional NLP techniques and data-driven models.

Traditional NLP Techniques. These models focus on capturing the meaning of data beyond the meaning of the individual words by identifying different entities, relationships, and attributes. There are several works that build upon traditional NLP techniques to extract information such as IOCs, attack patterns, or temporal information, that are useful for security operators.

One of the most influential papers is TTPDrills [17], which presents an approach that combines NLP with information retrieval techniques to extract threat action. Park et al. [25] describes a full-stack information extraction system, SecIE, that extracts entities, relationships, and even temporal information with high accuracy. Another work, iACE [20], presents a technique based on semantic models to extract IOCs from blog posts and articles crawled from the internet. Some works focus on creating knowledge graphs from processing CTI reports, such as Extractor [28], that build a complex, detailed knowledge graph including the temporal order of the attack.

Data-Driven Models. We can distinguish two subcategories of data-driven models: supervised and unsupervised models.

First, we present some of the related works that are based on supervised models. Lim et al. [21] were the first ones to introduce an ML-based annotator that targets to label CTI reports by identifying entities, relations, and attributes, while classifying sentences based on their relevance (i.e., if they describe malware actions and capabilities or not). Similar to this, Manikandan et al. [22] presents a conditional random fields approach combined with a convolution neural network-based system to predict malware token labels and classify sentences (with the same criteria as [21]). Some other works combine ML with traditional NLP techniques, such as CASIEs [29], that combines deep learning with linguistic features to extract information about cybersecurity event.

In the following, we present the related works based on unsupervised models, where we accentuate the focus on the novel generative LLMs that are a special type of data-driven models trained on large corpora of data to learn producing human-like text. LLMs are not designed for solving domain-specific problems, instead they act as a very knowledgeable human (i.e., they produce natural-language-like answers, just as a human knowing the answer to the question would). One recent work that looks into extracting information from CTI reports using the question-and-answer capabilities of the LLMs is presented in the paper of Siracusano et al. [32] describing a system for extracting entities and attack patterns. Several other papers look into the usage of these models on different areas of the security domain such as writing reports [27], generating secure code [15], investigating software vulnerabilities [26], or creating honeypots [33], but all these works leverage the text generation ability rather than investigating the domain-specific question and answer capabilities of the LLMs.

6.3 Incident Response Playbooks

Since playbooks play an important role in the incident response process, researchers actively look into this topic. In 2018 MITRE introduced their RECAST [6] framework for efficiently choosing prioritized courses of action. Later, Onwubiko et al. [24] proposes SOTER, which is a general incident management playbook, and together with this, the authors create an incident management lexicon and propose an incident classification and prioritization scheme. A more recent study by Stevens et al. [34] looks into the process of creating playbooks using standard playbook design frameworks and identifies that these playbooks often lack sufficient information to be used in an actual attacks. Probably the most up to date and most comprehensive study on openly available playbooks is presented by Schellte et al. [31]. In this study the authors categorize playbooks into community and organization-specific playbooks, while analyzing the key characteristics of community playbooks and identifying the influencing factors that play a role in creating organization-specific playbooks.

7 Conclusion

In this work we implemented a semantic approach to infer recovery steps from CTI reports, comparing it with off-the-shelf LLMs. To measure the performance of the approaches, we constructed a labeled dataset composed of malware CTI reports. Our experiments show that LLMs outperform the semantic approach by 45% in precision and 53% in recall. However, we also found that the semantic approach is more consistent, inferring fewer incorrect recovery steps than LLMs.

Acknowledgments. We would like to thank our reviewers for their valuable comments. This work has been supported by the SeReNity project, Grant No. cs.010, funded by Netherlands Organisation for Scientific Research (NWO). The findings reported herein are the sole responsibility of the authors.

A Prompt Used for the LLM Experiments

```
<s>[INST] <<SYS>>
You are an Incident Responder from a Security Operations Center.
Your task is to execute the following instructions:
1. analyze the given report;
2. extract the recovery actions to be deployed under the described
attack;
3. identify the object on which the recovery action should be
executed.
If there is no recovery action for a certain sentence, discard it
and don't include it in the output.

Output the answer in JSON using the following format:
[
  {
    "sentence": sentence,
    "recovery_action": recovery_action,
    "object": object
  }
]
<</SYS>>

The report of the attack is the following:

The malware adds a malicious executable. The executable encrypts
the user folder. This attack has high risk. [/INST]
[
  {
    "sentence": "The malware adds a malicious executable.",
    "recovery_action": "delete",
    "object": "executable"
```

```
  },
  {
    "sentence": "The executable encrypts the user folder.",
    "recovery_action": "restore",
    "object": "user folder"
  }
]
</s>

<s>[INST]
The report of the attack is the following: {content}
[/INST]
```

References

1. Threat Encyclopedia—trendmicro.com. https://www.trendmicro.com/vinfo/us/threat-encyclopedia/
2. 2023 State of Threat Detection (2023). https://www.vectra.ai/resources/research-reports/2023-state-of-threat-detection
3. Abu, M.S., Selamat, S.R., Ariffin, A., Yusof, R.: Cyber threat intelligence-issue and challenges. Indonesian J. Electr. Eng. Comput. Sci. **10**(1), 371–379 (2018)
4. Achiam, J., et al.: GPT-4 technical report. arXiv preprint arXiv:2303.08774 (2023)
5. Aminanto, M.E., Ban, T., Isawa, R., Takahashi, T., Inoue, D.: Threat alert prioritization using isolation forest and stacked auto encoder with day-forward-chaining analysis. IEEE Access **8**, 217977–217986 (2020)
6. Applebaum, A., Johnson, S., Limiero, M., Smith, M.: Playbook oriented cyber response. In: National Cyber Summit, NCS. IEEE (2018)
7. Bach, N., Badaskar, S.: A review of relation extraction. Literature review for language and statistics II **2**, 1–15 (2007)
8. Brown, T., et al.: Language models are few-shot learners. In: Advances in Neural Information Processing Systems, vol. 33, pp. 1877–1901 (2020)
9. Caballero, J., Gomez, G., Matic, S., Sánchez, G., Sebastián, S., Villacañas, A.: The rise of GoodFATR: a novel accuracy comparison methodology for indicator extraction tools. Future Gener. Comput. Syst. **144**, 74–89 (2023)
10. Chyssler, T., Burschka, S., Semling, M., Lingvall, T., Burbeck, K.: Alarm reduction and correlation in intrusion detection systems. In: Proceedings of the Conference on Detection of Intrusions and Malware & Vulnerability Assessment (DIMVA). LNI, GI (2004)
11. van Ede, T., et al.: DeepCASE: semi-supervised contextual analysis of security events. In: Proceedings of the IEEE Symposium on Security and Privacy (S&P) (2022)
12. Fleck, A.: Cybercrime expected to skyrocket in coming years (2022). https://www.statista.com/chart/28878/expected-cost-of-cybercrime-until-2027/
13. Ghribi, S., Makhlouf, A.M., Zarai, F., Guizani, M.: Fog-cloud distributed intrusion detection and cooperation. Trans. Emerg. Telecommun. Technol. **33**(8), e3835 (2022)

14. Hassan, W.U., et al.: NoDoze: combatting threat alert fatigue with automated provenance triage. In: Proceedings of the Symposium on Network and Distributed System Security (NDSS) (2019)
15. He, J., Vechev, M.T.: Large language models for code: security hardening and adversarial testing. In: Meng, W., Jensen, C.D., Cremers, C., Kirda, E. (eds.) Proceedings of the ACM Conference on Computer and Communications Security (CCS) (2023)
16. Honnibal, M., Montani, I.: spaCy 2: natural language understanding with Bloom embeddings, convolutional neural networks and incremental parsing (2017). To appear
17. Husari, G., Al-Shaer, E., Ahmed, M., Chu, B., Niu, X.: TTPDrill: automatic and accurate extraction of threat actions from unstructured text of CTI sources. In: Proceedings of the 33rd Annual Computer Security Applications Conference (2017)
18. Jakkal, V.: Introducing Microsoft security copilot: empowering defenders at the speed of AI (2023). https://blogs.microsoft.com/blog/2023/03/28/introducing-microsoft-security-copilot-empowering-defenders-at-the-speed-of-ai/
19. Legoy, V., Caselli, M., Seifert, C., Peter, A.: Automated retrieval of ATT&CK tactics and techniques for cyber threat reports. arXiv preprint arXiv:2004.14322 (2020)
20. Liao, X., Yuan, K., Wang, X., Li, Z., Xing, L., Beyah, R.: Acing the IOC game: toward automatic discovery and analysis of open-source cyber threat intelligence. In: Proceedings of the ACM Conference on Computer and Communications Security (CCS) (2016)
21. Lim, S.K., Muis, A.O., Lu, W., Ong, C.H.: MalwareTextDB: a database for annotated malware articles. In: Proceedings of the 55th Annual Meeting of the Association for Computational Linguistics (Volume 1: Long Papers) (2017)
22. Manikandan, R., Madgula, K., Saha, S.: TeamDL at SemEval-2018 task 8: cybersecurity text analysis using convolutional neural network and conditional random fields. In: Proceedings of The 12th International Workshop on Semantic Evaluation (2018)
23. Murphy, S.: Your new AI assistant: trend vision one^{TM}-companion (2023). https://www.trendmicro.com/en_my/research/23/f/companion-ai-assistant-trend-vision-one.html
24. Onwubiko, C., Ouazzane, K.: SOTER: a playbook for cybersecurity incident management. IEEE Trans. Eng. Manag. **69**(6), 3771–3791 (2022)
25. Park, Y., Lee, T.: Full-stack information extraction system for cybersecurity intelligence. In: Proceedings of the Conference on Empirical Methods in Natural Language Processing: Industry Track (2022)
26. Pearce, H., Tan, B., Ahmad, B., Karri, R., Dolan-Gavitt, B.: Examining zero-shot vulnerability repair with large language models. In: Proceedings of the IEEE Symposium on Security and Privacy (S&P) (2023)
27. Ranade, P., Piplai, A., Mittal, S., Joshi, A., Finin, T.: Generating fake cyber threat intelligence using transformer-based models. In: Proceedings of the International Joint Conference on Neural Networks, IJCNN (2021)
28. Satvat, K., Gjomemo, R., Venkatakrishnan, V.: Extractor: extracting attack behavior from threat reports. In: Proceedings of the IEEE European Symposium on Security and Privacy (EuroS&P) (2021)
29. Satyapanich, T., Ferraro, F., Finin, T.: CASIE: extracting cybersecurity event information from text. In: Proceedings of the AAAI Conference on Artificial Intelligence (2020)

30. Schlette, D., Caselli, M., Pernul, G.: A comparative study on cyber threat intelligence: the security incident response perspective. IEEE Commun. Surv. Tutor. **23**(4), 2525–2556 (2021)
31. Schlette, D., Empl, P., Caselli, M., Schreck, T., Pernul, G.: Do you play it by the books? A study on incident response playbooks and influencing factors. In: Proceedings of the IEEE Symposium on Security and Privacy (S&P). IEEE Computer Society (2024)
32. Siracusano, G., et al.: Time for aCTIon: automated analysis of cyber threat intelligence in the wild. CoRR abs/2307.10214 (2023)
33. Sladić, M., Valeros, V., Catania, C., Garcia, S.: LLM in the shell: generative honeypots. arXiv preprint arXiv:2309.00155 (2023)
34. Stevens, R., et al.: How ready is your ready? Assessing the usability of incident response playbook frameworks. In: Proceedings of the CHI Conference on Human Factors in Computing Systems (2022)
35. Sundaramurthy, S.C., et al.: A human capital model for mitigating security analyst burnout. In: Proceedings of the Eleventh Symposium on Usable Privacy and Security (SOUPS) (2015)
36. Tounsi, W., Rais, H.: A survey on technical threat intelligence in the age of sophisticated cyber attacks. Comput. Secur. **72**, 212–233 (2018)
37. Touvron, H., et al.: Llama 2: open foundation and fine-tuned chat models. arXiv preprint arXiv:2307.09288 (2023)
38. Vermeer, M., Kadenko, N., van Eeten, M., Gañán, C., Parkin, S.: Alert alchemy: SOC workflows and decisions in the management of NIDS rules. In: Proceedings of the ACM Conference on Computer and Communications Security (CCS) (2023)
39. Wagner, T.D., Mahbub, K., Palomar, E., Abdallah, A.E.: Cyber threat intelligence sharing: survey and research directions. Comput. Secur. **87**, 101589 (2019)
40. Zamfirescu-Pereira, J., Wong, R.Y., Hartmann, B., Yang, Q.: Why Johnny can't prompt: how non-AI experts try (and fail) to design LLM prompts. In: Proceedings of the CHI Conference on Human Factors in Computing Systems, pp. 1–21 (2023)

Pairing Security Advisories with Vulnerable Functions Using Open-Source LLMs

Trevor Dunlap[1,2(✉)], John Speed Meyers[2], Bradley Reaves[1], and William Enck[1]

[1] North Carolina State University, Raleigh, NC, USA
{bgreaves,whenck}@ncsu.edu
[2] Chainguard, Kirkland, WA, USA
{trevor.dunlap,jsmeyers}@chainguard.dev

Abstract. As the reliance on open-source software dependencies increases, managing the security vulnerabilities in these dependencies becomes complex. State-of-the-art industry tools use reachability analysis of code to alert developers when security vulnerabilities in dependencies are likely to impact their projects. These tools heavily rely on precisely identifying the location of the vulnerability within the dependency, specifically vulnerable functions. However, the process of identifying vulnerable functions is currently either manual or uses a naive automated approach that falsely assumes all changed functions in a security patch link are vulnerable. In this paper, we explore using open-source large language models (LLMs) to improve pairing security advisories with vulnerable functions. We explore various prompting strategies, learning paradigms (i.e., zero-shot vs. few-shot), and show our approach generalizes to other open-source LLMs. Compared to the naive automated approach, we show a 173% increase in precision while only having an 18% decrease in recall. The significant increase in precision to enhance vulnerable function identification lays the groundwork for downstream techniques that depend on this critical information for security analysis and threat mitigation.

Keywords: Vulnerable Function · Security Advisory · Security Database · Large Language Model

1 Introduction

In software development, leveraging open-source dependencies is fundamental, accelerating innovation and development processes. Yet, these dependencies often carry the burden of security vulnerabilities, posing significant management challenges [2]. This challenge has led to the widespread use of software composition analysis tools (e.g., OWASP Dependency-Check, GitHub Dependabot) that notify developers of potential vulnerabilities in their dependencies. One shortcoming is that these tools often overgeneralize by flagging any inclusion or use

F. Maggi et al. (Eds.): DIMVA 2024, LNCS 14828, pp. 350–369, 2024.
https://doi.org/10.1007/978-3-031-64171-8_18

of a known vulnerable dependency as potentially vulnerable, resulting in a high rate of false positives [10] and alert fatigue [1].

State-of-the-art industry tools [14,17,28] have begun to determine how those dependencies are used within projects. These tools rely on a program analysis concept known as *reachability* to determine whether functions containing vulnerable portions of code are ever called within dependencies. By assessing if vulnerable functions are called, developers can prioritize genuine threats, focusing on vulnerabilities that pose an actual risk to their projects. However, the success of reachability analysis first depends on accurately identifying vulnerable functions within dependencies. Incorrectly identifying safe functions as vulnerable causes unnecessary alerts for developers about threats that do not exist.

There are two common approaches for pairing security advisories with vulnerable functions: (1) manual curation and (2) using a naive automated approach that considers all changed functions in a security patch link as vulnerable. The Go Vulnerability Database (GoVulnDB [16]), supported by Google's Go security team, exemplifies the manual approach by releasing high-quality security reports for the Go community. In these reports, the Go security team manually identifies the root cause of vulnerabilities, accurately pinpointing the exact vulnerable functions. Despite its high accuracy and the trust it builds, this manual process is time-consuming and limited in scalability.

Alternatively, the automated approach, often seen in academic literature [4,7, 15,27,40], is intuitive but flawed. The automated approach assumes all modified functions in a security patch are vulnerable. The underlying assumption is that developers are only addressing the vulnerability and not any additional changes during the patching process. However, our preliminary analysis found that out of all the functions the automated approach labeled as vulnerable, only 22% were correct when using GoVulnDB as ground truth. This means that 78% of the functions modified in a security patch were unrelated to the actual vulnerability, underscoring the significant issue of overestimation and highlighting the need for improved automated methods.

In this paper, we hypothesize that LLMs can significantly improve the accuracy of pairing security advisories with vulnerable functions over the current naive automated approach. Our high-level approach is to take a vulnerability description from the security advisory and the associated changes from the security patch link to ask the model if the changes fix the underlying vulnerability. We explore various model sizes (i.e., CodeLlama 7 billion (B) parameters, 13B, and 34B versions [31]) using a variety of prompting strategies. We develop three prompts: a simple standard prompt, a detailed prompt, and a chain-of-thought prompt that involves the model providing an explanation [38]. We evaluate these prompts in two learning paradigms: either zero-shot or few-shot. The zero-shot approach does not have examples in the prompt, and the few-shot relies on our retrieval system to obtain similar examples to provide in the prompt to help guide the model. Additionally, we evaluate the computation times for each strategy. Finally, we explore how these techniques generalize to three other popular code-oriented LLMs (Mixtral [21], WizardCoder [26], and DeepSeek [26]).

```
func unzipFile(file *zip.File, dstDir
      string) error {
-   filePath := path.Join(dstDir, file
      .Name)
+   name := strings.TrimPrefix(
      filepath.Join(string(filepath.
      Separator), file.Name), string(
      filepath.Separator))
+   filePath := path.Join(dstDir, name
      )
```

(a) Modifications in the *unzipFile* function prevent a path traversal vulnerability by sanitizing the file name before joining the destination directory path.

```
func RemoveFile(path string) error {
-   err := os.Remove(path)
-   return err
+   return os.Remove(path)
}
```

(b) Refactoring of the *RemoveFile* non-vulnerable function to directly return the result of os.Remove(path), removing the temporary variable.

Fig. 1. Condensed updates of multiple functions in the patch for the Go utility library, *go-huge-util*, addressing CVE-2023-28105. Figure 1a are changes to the root causing vulnerable function, *unzipFile*. Figure 1b shows unrelated changes to the underlying vulnerability in function *RemoveFile*.

Our experimentation demonstrates the following key findings:

- LLMs surpass the naive automated approach that assumes all functions modified in a security patch are vulnerable. We show a 173% increase in precision (i.e., from 0.22 to 0.60) with only an 18% decrease in recall (i.e., from 0.90 to 0.74).
- Smaller models, with appropriate prompting, can match or even exceed the performance of larger models, suggesting a resource-efficient alternative.
- While chain-of-thought prompting proves the best approach in zero-shot settings, it comes at a 13x slowdown compared to more streamlined prompting with a few-shot setting that can show similar results.
- Our approach generalizes to multiple LLMs, allowing downstream users to pair security advisories with vulnerable functions.

We have released our data and implementation on GitHub.[1]

2 Background

2.1 Motivating Example

Security advisory CVE-2023-28105 highlights a path traversal vulnerability in the *go-huge-util* Go library. Within the security advisory, a reference link exists, also known as a patch link, that points to how a developer applies a fix for the vulnerability. These patch links are the basis for identifying vulnerable functions. However, the patch for CVE-2023-28105 does more than fix the vulnerable function, as described by the developer in the commit message: *"fix zip.Unzip path traversal vulnerability and add some new file utility functions."* Specifically, the

[1] https://github.com/s3c2/llm-vulnerable-functions.

```
id: GO-2023-1640
modules:
    - module: github.com/dablelv/go-huge-util
      packages:
        - package: github.com/dablelv/go-huge-util/zip
          symbols:
            - unzipFile
          derived_symbols:
            - Unzip
cves:
    - CVE-2023-28105
references:
    - fix: https://github.com/dablelv/go-huge-util/commit/0e308b0
```

Fig. 2. Condensed GO-2023-1640 report for CVE-2023-28105.

patch has modifications in four files, modifies five existing functions, adds two new functions, and totals 104 line additions and 45 line deletions. The comprehensive changes within the patch link underscore the complexity of accurately identifying vulnerable functions.

Figure 1a contains the changes to the root causing vulnerability within the patch link for CVE-2023-28105 and its fix in the *unzipFile* function. The fix was straightforward: the file path input requires sanitization to prevent the traversal vulnerability. Interestingly, the security advisory and developer mentions the function *zip.Unzip*: *"When users use 'zip.Unzip' to unzip zip files from a malicious attacker, they may be vulnerable to path traversal."* However, the *Unzip* function was unchanged during the patch link, and this is simply the public-facing function that calls the underlying vulnerable function, *unzipFile*.

In addition to fixing the underlying path traversal vulnerability, other changes occurred within the patch link. Figure 1b is an example of one of the five existing functions modified during the patch link unrelated to fixing the path traversal vulnerability. In this case, the developer refactored the function *RemoveFile* to directly return the status of removing the given file instead of saving the result in a temporary variable. Using a naive approach to label all changed functions in a patch link as vulnerable would misidentify the non-vulnerable function *RemoveFile* as vulnerable, thus creating noise for downstream data consumers.

2.2 Go Vulnerability Database (GoVulnDB)

The Go Vulnerability Database (GoVulnDB), supported by Google, is a high-quality dataset of vulnerabilities in Go [16]. The Go Security team compiles data from various datasets (e.g., the National Vulnerability Database, GitHub Advisory Database). The Go Security team then manually checks reports for precise details about the vulnerabilities, including descriptions and affected versions [18].[2]

[2] After the submission of this paper, GoVulnDB started experimenting with assigning any changed functions within the patch links as vulnerable.

Algorithm 1: Generalization of the naive automated approach used in prior work [4,7,15,27,40].

```
Input: patchLink from the security advisory
Output: functions unique set of changed functions in the patch
ChangedFuncs(patchLink):
    functions ← empty set
    for (gitHunk, fileName) ∈ PatchParser(patchLink) do
        if "test" not in ToLower(fileName) then
            for lineNumber ∈ gitHunk do
                for func ∈ Tree-Sitter(fileName) do
                    if func.startLine ≤ lineNumber ≤ func.endLine then
                        functions.add(func.name)
    return functions
```

Vulnerable Functions: One unique aspect of GoVulnDB is its curation of vulnerable functions. These vulnerable functions are in two distinct fields: *symbols* and *derived symbols*. The *symbols* field identifies the root causing vulnerable functions, such as functions or methods, that directly cause the reported vulnerability. These initial symbols are *manually* identified by the Go security team. The *derived symbols* represents a further analysis of public functions that reach the root causing vulnerable symbols. These *derived symbols* are generated automatically using portions of their static analysis tool govulncheck [17] and converted into reports using their vulnreport tool [19]. The GoVulnDB is a reliable, ground-truth foundation for understanding the nature of pairing security advisories with their associated vulnerable functions.

Figure 2 shows the GoVulnDB report, GO-2023-1640, for the prior motivating examples. The Go security team manually reviewed the report and identified the root causing vulnerable function as *unzipFile*. When automatically generating the *derived symbols*, the *Unzip* function is identified as the public-facing function initially mentioned in the security advisory description.

3 Naive Automated Approach (*ChangedFuncs*)

While manual curation of vulnerable functions delivers high-quality data (e.g., GoVulnDB), it suffers in scalability. The alternative approach, commonly seen in academic literature [4,7,15,27,40], uses an automated approach. The automated approach, while intuitive, is naive. The underlying assumption is that all modified functions within a patch link are vulnerable. However, we find the approach only correctly identifies 22% of the root causing vulnerable functions (Sect. 5.2), further emphasizing the need for an accurate automated approach.

To replicate the naive automated approach for identifying vulnerable functions, we implement Algorithm 1, referred to as *ChangedFuncs*. This process begins with a patch link from a security advisory. Such patch links lead to specific commits in a project's source code repository, highlighting the changes.

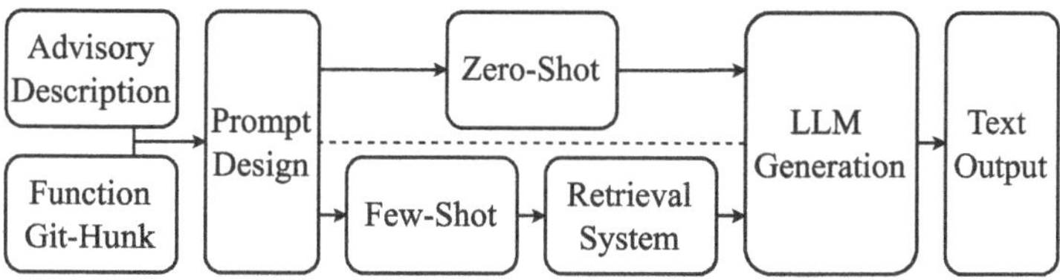

Fig. 3. Our design approach involves using either a zero-shot or few-shot technique for pairing security advisories with vulnerable functions.

These commits, also known as git-diffs, appear in *hunk* levels (git-hunks). A git-hunk represents a block of code changes within a file. Naturally, git-hunks do not specify if these changes are within functions, as modifications can occur inside and outside functions. Determining if the git-hunks are modifying code inside of a function is trivial. We first parse the patch links with PatchParser [13], a Python library that transforms commits into machine-readable data and provides the associated line numbers of git-hunk modifications. Using Tree-sitter [33], a language-agnostic parsing library, we identify the functions' start and end line numbers within the same file the git-hunk appears in. If a git-hunk's modifications are within a function's line range, the git-hunk is labeled with the function name. We also exclude files containing the word "test," ensuring only non-test functions impacted by the patch are considered. In our results, we use *Changed-Funcs* as a baseline, treating each identified function within the patch link as vulnerable.

4 LLM Approach

Figure 3 overviews our approach for using Large Language Models (LLMs) to pair security advisories with vulnerable functions. Our approach consumes a given security advisory (i.e., GoVulnDB report) and individual function code changes (i.e., git-hunks) from the associated patch link within the security advisory. We design three prompts: a standard, detailed, or chain-of-thought prompt, all aimed at determining whether the code changes address the vulnerabilities described in a security advisory. We compare these prompts in two scenarios: a "zero-shot" setting, where the LLM operates without prior examples, and a "few-shot" setting, incorporating similar examples into the prompt to guide the LLM's analysis. These similar examples rely on our retrieval system. From there, the prompt is given to the LLM to generate an output to determine if the given function changes are related to fixing the specific security advisory description.

4.1 Input Data

Our approach relies on two inputs: the description from the security advisory and associated function changes (i.e., git-hunk) from the patch link.

Security Advisory Description: We rely on the free-form text description given within the security advisory. We preprocess the description details by removing any additional spaces and new lines.

Function Git-Hunk: The input granularity within the system is at the git-hunk level based on the patch links provided within the security advisory. We chose this level as it helps separate individual changes for a commit.

4.2 System Prompt Design

LLMs rely on system prompts to instruct the model on what to do. We design three system prompts to instruct the model to determine if the git-hunk is responsible for fixing a given security advisory. 1) The standard prompt, designed to be straightforward, asks if the provided git-hunk fixes a vulnerability. 2) The detailed prompt has more specific instructions for the model, asking to consider the git-hunk's direct relevance to the vulnerability description. These prompts instruct the model to respond in a boolean fashion of either *True* or *False*. Where *True* represents the git-hunk was indeed fixing or related to the vulnerability description, and *False* would mean the fix is unrelated. 3) The third prompt uses the chain-of-thought method [38], prompting the model to explain the git-hunk's function before producing a decision. Each of our prompts is as follows:

> ***Standard Prompt:*** I want you to act as a vulnerability fix detection system. Determine if the following git-hunk fixed a vulnerability. Respond with only 'True' or 'False.'
>
> ***Detailed Prompt:*** Your task is to analyze the provided code changes (GIT-HUNK) to determine if they target a specific vulnerability (Fixed Vulnerability Description). Review each line, considering its direct relevance to the vulnerability description. Ignore any new vulnerabilities that may have been introduced. Answer 'True' if changes are directly related to fixing or even somewhat partially related to fixing the vulnerability. Only provide a 'False' conclusion if the GIT-HUNK changes are absolutely unrelated to the vulnerability. Respond with only 'True' or 'False.'
>
> ***Chain-of-Thought (CoT):*** Your task is to analyze the provided code changes (GIT-HUNK) to determine if they target a specific vulnerability (Fixed Vulnerability Description). Review each line, considering its direct relevance to the vulnerability description. Ignore any new vulnerabilities that may have been introduced. Provide a brief description of what the GIT-HUNK changes have done, then conclude by labeling 'True' if changes are directly related to fixing or even somewhat partially related to fixing the vulnerability. Only provide a 'False' conclusion if the GIT-HUNK changes are absolutely unrelated to the vulnerability. Justify your decision before ending with a clear 'True' or 'False' decision. Do not answer right away. Answer in the following format:

Explanation - {Your Explanation}
Final Decision - {True/False}.

4.3 Zero-Shot and Few-Shot Learning Paradigms

Two distinct learning paradigms exist in LLMs: zero-shot and few-shot learning. The primary difference between the two is if example data is present in the prompt. Figure 4 illustrates the high-level template for each approach. A primary benefit of these approaches is that they do not require end-users to fine-tune any models and avoid the significant effort of obtaining training data.

```
{system prompt}

# System Input #
Advisory Description:
{advisory-description}
Function GIT-HUNK:
======================
{git-hunk}
======================
```

(a) Zero-shot prompt template for pairing security advisories to vulnerable functions.

```
{system prompt}

{retrieval system examples}

# System Input #
Advisory Description:
{advisory-description}
Function GIT-HUNK:
======================
{git-hunk}
======================
```

(b) Few-shot prompt, note the addition of the retrieval-based examples.

Fig. 4. Simplified prompt templates for both the zero-shot and few-shot paradigms. The difference is the addition of the retrieval examples placed after the system prompt within the few-shot template.

Zero-Shot: The zero-shot learning paradigm prompts an LLM without examples [30]. Instruction-tuned models, such as the ones used within our experiments, have shown promising capabilities in zero-shot learning [37]. Additionally, zero-shot is useful when end-users have little to no example data.

Few-Shot: Contrary to zero-shot, few-shot learning incorporates a set of examples within the prompt [6]. These examples serve as a guide, allowing the LLM to grasp the context and specifics of the new task quickly. This approach is beneficial in limited data scenarios, but a set of known ground truth labels must exist. In the few-shot setting, these examples are added directly after the system prompts, shown in Fig. 4a. However, determining which examples to provide within the prompt is essential for few-shot learning.

4.4 Retrieval System

The retrieval system enhances few-shot learning by providing relevant examples to the git-hunk under analysis. The relevant examples are returned through

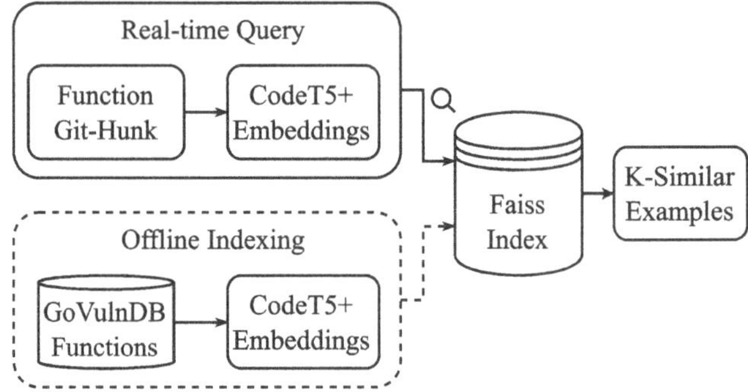

Fig. 5. The retrieval system takes the function git-hunk as input. Then, it returns k-similar examples based on the set of labeled GoVulnDB changed functions at the git-hunk level from the ground truth data.

a similarity comparison process of the known labeled examples and the git-hunk in question. We begin by building an offline datastore of known labeled examples, with the GoVulnDB serving as our primary data source (as detailed in Sect. 5.1). The core of this system lies in transforming these git-hunks into embeddings, which are compact vector representations that encapsulate their semantic meaning. These embeddings are typically generated with a specialized code embedding model (i.e., CodeT5+ [36]). The embeddings are then stored in a datastore optimized for vector retrieval (i.e., Faiss index [12]). Upon querying, the system fetches both *True* and *False* examples of git-hunks with the associated advisory description similar to the git-hunk under analysis (Fig. 5).

Populating the Retrieval System with Examples: To populate the retrieval system, we rely on a model specifically designed to create embeddings, the CodeT5+ 110M embedding model [36]. Wang et al. [36] extensively evaluated CodeT5+ and found it performed exceptionally well for retrieval tasks. The input into the embedding model is at the git-hunk level. The resulting embeddings from this process are 256-dimensional vectors that we store within a Faiss index [12].

Querying the Retrieval System: When inputs come into the retrieval system, we use the same CodeT5+ embedding model as previously described. When querying the retrieval system, the Faiss index computes a similarity metric (based on the dot product) between the input vector and all pre-populated vectors. The search returns the top-K similar git-hunks with the associated advisory description of both *True* and *False* examples in alternating order. Leading with the most similar *True* example first, then the next most similar *False* example. We show the impact of this ordering in Sect. 5.4. We also ensure that the examples come from a different advisory and git-hunk, confirming that we are not leaking data or labels at inference time. The overall retrieval system is implemented using the LangChain library [23].

4.5 LLM Text Generation

LLMs use an autoregressive generation method, where text is generated iteratively by predicting the next words, more specifically tokens, based on the context of all previously generated tokens [34]. This process involves a sequence of forward passes through the model's network, each pass calculating the probability distribution for the next possible token. The next token is selected based on this distribution and then appended to the growing text sequence. Generating longer sequences increases computational time due to the number of iterative passes required through the model. Consequently, the output format from the LLM (e.g., boolean responses vs. an explanation) significantly impacts generation time.

Standard and Detailed Prompt Responses: When we use the standard and detailed prompts, we aim for a boolean single-word response of *True* or *False* from the model. This requires just a single forward pass through the model to produce the boolean response. At the end of this step, the model produces preliminary scores, known as logits, for each potential token in its vocabulary [3]. These logits are the model's initial guesses, not yet probabilities. We then use a function called softmax to turn these logits into actual probabilities, showing how confident the model is in each outcome. By focusing on the probabilities for *True* and *False* tokens, we can identify the model's decision.

Chain-of-Thought Responses: The expected result is a longer free-form text when using the chain-of-thought (CoT) prompt. This is because the model explains what the git-hunk has changed, then concludes with a definitive *True* or *False*, categorizing the git-hunk as related or unrelated to the vulnerability fix. A consequence of the longer text will be an increase in computation time. To identify the final decision, we search the text for either *True* or *False* values.

4.6 Aggregating Results to the Function Level

Predictions from the LLM are at the git-hunk level. However, this paper aims to associate security advisories with entire functions rather than individual git-hunks. In some cases, multiple changes can occur within a single function spanning multiple git-hunks, requiring us to aggregate the results into one function. To do so, if any git-hunk of a function is predicted as *True*, indicating a relation to the vulnerability, we classify the entire function as vulnerable. This methodology assumes that the vulnerability impacts the whole function if any portion of the function is involved in the vulnerability fix.

5 Results

We evaluate the use of LLMs to pair security advisories with vulnerable functions with the goal of addressing the following questions:

Q1: How well does the naive approach work for identifying vulnerable functions?

Q2: How do various prompts impact the performance of using LLMs?
Q3: How does zero-shot and few-shot learning impact the effectiveness of LLMs?
Q4: How does the computational time compare across prompt strategies?
Q5: Can the developed prompts generalize from one LLM to another LLM?

Key Takeaways: Our analysis shows that using open-source large language models (LLMs) to link security advisories with vulnerable functions presents various insights and challenges. Current automated detection methods, like the naive *ChangedFuncs* approach, identify only 22% of actual vulnerable functions, leading to many false positives. This highlights the need for more advanced techniques. LLMs can improve precision in detecting these functions by up to 173% compared to *ChangedFuncs*; however, they have an 18% drop in recall. The study also finds that the effectiveness of LLMs varies with the quality of the input prompt, with well-designed prompts significantly boosting the performance of smaller models. Few-shot learning enhances recall but comes at the cost of reducing precision. The Chain-of-Thought prompting method excels in performance but requires significantly more computational resources, 13 times slower than other methods. Finally, our findings apply across different LLMs.

5.1 Experimental Setup

Throughout the results, we primarily focus on the CodeLlama family of models [31]. The CodeLlama family is designed explicitly for coding tasks and has instruction-following versions, each available in 7 billion (B), 13B, and 34B parameter configurations. Additionally, we also evaluate other popular coding-oriented LLMs: Mixtral 7x8B [21], DeepSeek [5], and WizardCoder [26]. We use a machine with an Intel i7-9700k CPU, 128 GB RAM, and two NVIDIA RTX 3090 Ti GPUs. During evaluation, each model is split across both GPUs with 4-bit quantization and parameters set to half-precision (float16). We use the same CodeLama parameters as Meta for text generation throughout [31]. We evaluate against the GoVulnDB data described in Sect. 5.1.

Data Collection: We rely on the GoVulnDB as ground truth data. We initially cloned GoVulnDB on October 26, 2023, and obtained 409 GoVulnDB reports. Out of these, 318 (78%) had an associated patch link. Of those 318 security advisories, 280 (88%) of the reports listed vulnerable functions. When a vulnerable function is not listed, the entire project is considered vulnerable; therefore, we do not consider those for further analysis. These 280 reports form the foundation for our subsequent analysis. Using *ChangedFuncs*, Algorithm 1, we identify 2,370 modified functions from 298 patch links across the 280 GoVulnDB reports. From GoVulnDB, 528 of those functions were labeled as vulnerable functions. We consider the remaining 1,842 functions non-vulnerable. We note that some git-hunks can change multiple functions; we split these git-hunks to ensure that only one function is processed at a time.

5.2 Comparing the *ChangedFuncs* to GoVulnDB

When comparing *ChangedFuncs* (Sect. 3) to the GoVulnDB dataset, we found the approach produces a precision of 0.22. The remaining functions (1,842 functions) changed during the patch links were not listed in the GoVulnDB reports, resulting in false positives. Furthermore, 61 reported vulnerable functions were not found within the patch link. The overall recall rate is 0.9, demonstrating the ability to capture the majority of vulnerable functions. Through a manual analysis, we found that the missing functions were either due to a missing patch link from the advisory or were mentioned in a subsequent link (e.g., a reference to the Issue). Researchers and practitioners using the naive automated approach will create significant noise for downstream consumers.

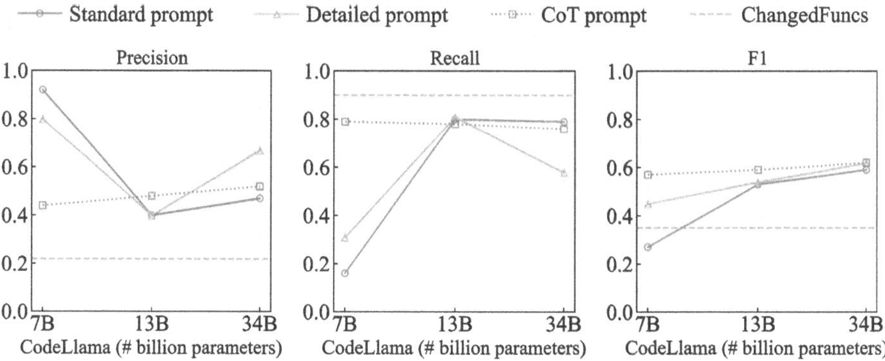

Fig. 6. Zero-shot setting between the standard, detailed, and chain-of-thought (CoT) prompt across CodeLlama 7B, 13B, and 34B. Prompt design has a greater positive impact on the smaller 7B model, allowing it to produce similar results to the larger 13B and 34B models.

5.3 LLM Prompt Impacts

Figure 6 compares the standard, detailed, and CoT prompts from Sect. 4.2 using the CodeLlama model family (7B, 13B, 34B) in a zero-shot context. The *ChangedFuncs* approach, assuming all changed functions in a patch link are vulnerable, serves as the baseline. The evaluation focuses on *True* labeled functions to ensure fairness against the *ChangedFuncs* baseline.

The prompts have the largest impact on the smaller 7B model. For instance, switching from the standard to the detailed prompt resulted in a 66.7% increase (0.27 to 0.45) in the F1 score. Then, switching from the detailed prompt to the CoT prompt resulted in another 26.7% increase (0.45 to 0.57) in the F1 score. However, these increases in F1 have tradeoffs in precision and recall. Both the detailed and CoT decreased the precision and increased the recall in the 7B model. A higher precision and lower recall indicate the model was predominately responding *False* to each input, with fewer *True* responses. However, when the recall increased, the model responded with more *True* responses.

The prompts have a less significant impact on the larger CodeLlama 13B and 34B models. On average, the 13B version has a 5.6% increase (ranging from 0.53 to 0.59) in F1 score across the prompts, and the 34B version has a 2.5% increase (ranging from 0.59 to 0.62). For the larger CodeLlama-13B model, the prompts have similar performance across precision and recall. The largest CodeLlama-34B model performed similarly using the standard and CoT prompts. However, the detailed prompt has a higher precision of 0.67 and a lower recall of 0.58. Both the detailed and CoT prompt in the CodeLlama-34B model had the highest F1 score of 0.62. Overall, the prompts have a minor impact on the larger models.

An additional benefit of the CoT prompt is the extra explanation provided. Below is an example of the response from the CodeLlama-34B zero-shot setting from Fig. 1a in the Background:

> The GIT-DIFF changes have addressed the path traversal vulnerability in the 'unzipFile' function by using the 'path.Join' and 'strings.TrimPrefix' functions to ensure that the file path is valid and does not contain any malicious traversal sequences. Final Decision: True.

Table 1. Zero-shot vs Few-Shot in terms of F1 across CodeLlama (7B, 13B, 34B) using the standard, detailed, and CoT prompt. Best metrics are bolded.

# Examples	CodeLlama-7B				CodeLlama-13B				CodeLlama-34B			
	0	3	5	10	0	3	5	10	0	3	5	10
Standard	0.27	0.38	**0.39**	0.35	0.53	**0.57**	0.54	0.53	0.59	0.53	0.54	**0.60**
Detailed	0.45	0.53	**0.58**	0.52	0.54	**0.56**	0.56	0.54	**0.62**	0.57	0.59	**0.62**
CoT	0.57	0.56	0.59	**0.60**	0.59	**0.60**	**0.60**	0.55	0.62	**0.63**	0.62	0.61

5.4 Zero-Shot vs. Few-Shot

Table 1 compares the performance of the CodeLLama 7B, 13B, and the 34B model in a few-shot learning context using the standard, detailed, and CoT prompting. The few-shot approach helps improve performance compared to the zero-shot approach, primarily in recall in the basic and detailed prompt settings.

Introducing a few-shot paradigm has the largest impact on the smaller CodeLlama 7B model. On average, each prompt in a few-shot setting increased the F1 score of the CodeLlama 7B model by 26.2%. Few-shot learning increased the CodeLlama 13B version's F1 on average by 4.1%, and the CodeLlama 34B version's by only 1.1%. The overall trend for the standard and detailed prompt after adding few-shot examples was an increase in recall and a decrease in precision. The CoT prompt had the opposite reaction, typically seeing a minor decrease in recall and an increase in precision. Interestingly, a few-shot approach in the detailed and CoT prompts allows the smaller CodeLlama 7B model to perform similarly to the 13B and 34B versions.

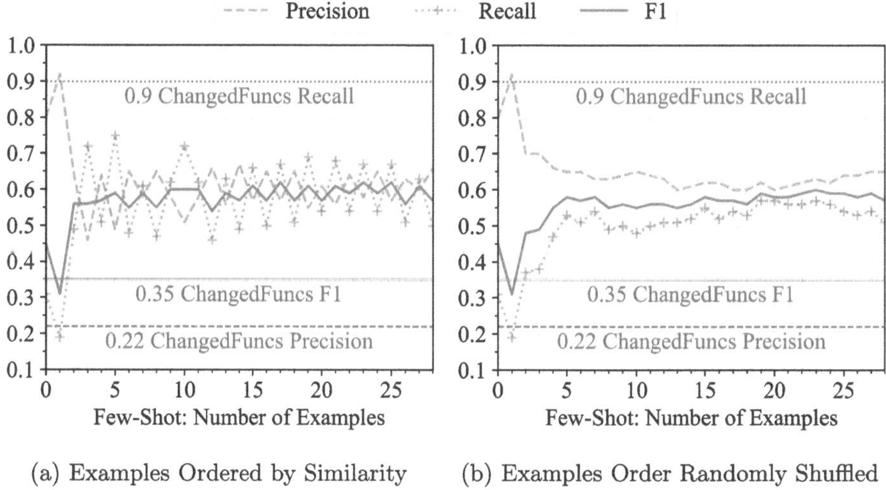

(a) Examples Ordered by Similarity (b) Examples Order Randomly Shuffled

Fig. 7. Randomly shuffling the examples in a few-shot setting of CodeLlama 7B using the detailed prompt stabilizes the overall performance of the model.

Example Order: When experimenting with the few-shot paradigm, we noticed that the order in which the examples in the prompt appear impacts the model performance, as shown in Fig. 7. For instance, our retrieval system returns both *True* and *False* examples (detailed in Sect. 4.4). Once examples are returned, we place them in an alternating order in the prompt: start with a *True* example followed by a *False* example.

Figure 7a shows the alternating impact of precision and recall depending on the number of examples in the prompt. The number of examples changes the type of example before the input. For instance, an even number of examples ends with a *False* example, and an odd number ends with a *True* example. A closing *True* example tends to influence the model towards a true prediction, enhancing recall but at the cost of precision. While a closing *False* example influences the model towards false predictions, improving precision while reducing recall. In Fig. 7b, we randomly shuffle the order of examples to counteract the observed bias. Our findings suggest that structured example sequencing creates tradeoffs between precision-recall, whereas randomization offers stable model performance.

Retrieval Impact: Additionally, we observed the similarity of returned examples impacts the model's predictive performance. Specifically, a higher mean retrieval score for a given example type (e.g., *True* examples) correlates with the model predicting that the target function addressed a vulnerability. A greater discrepancy in retrieval scores between *True* and *False* examples led to more accurate predictions by the model. However, when retrieval scores for *True* and *False* examples were similar, the model had difficulty distinguishing between the two, often mislabeling functions as *True*. This pattern was also evident in missed instances, where *False* examples had higher overall retrieval scores. The influence of retrieval scores on model accuracy was greater under an alternating example approach, though the trend continues with randomly shuffled examples.

5.5 Generalization Across Additional LLMs

Here we demonstrate the performance of prompts across the Mixtral 8x7B [21], DeepSeek 33B [5], WizardCoder 15B [26], and CodeLlama 34B [31] models. We evaluate using the 5-shot detailed prompt for each model.

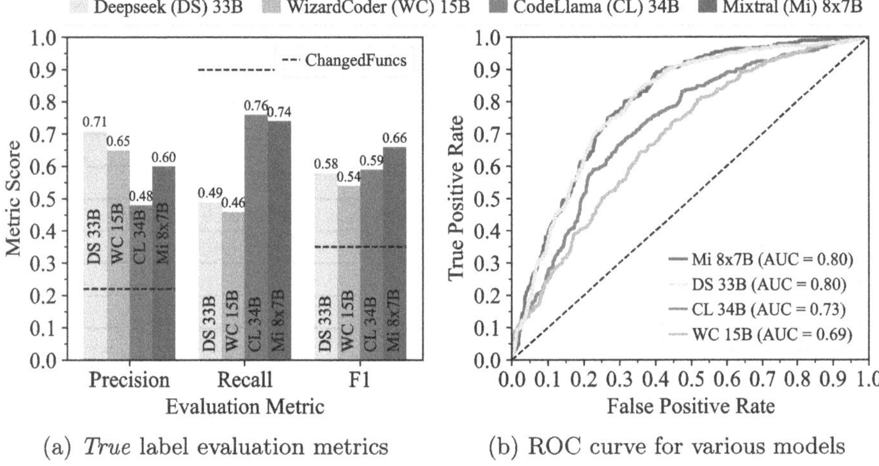

(a) *True* label evaluation metrics (b) ROC curve for various models

Fig. 8. Performance metrics for DeepSeek 33B, WizardCoder 15B, CodeLlama 34B, and Mixtral 8x7B using the detailed prompt in a 5-shot setting. The best overall performing model was Mixtral 8x7B, with a 173% increase in precision and an 18% decrease in recall compared to the naive approach.

Figure 8a specifically focuses on performance regarding *True* labels. Notably, Mixtral 8x7B has the highest overall F1 score. Specifically, its precision is a 173% improvement over the *ChangedFuncs* (0.60 vs. 0.22) and an 18% decrease in recall (0.90 vs 0.74). In contrast, CodeLlama 34B has a slightly higher recall than Mixtral 8x7B but significantly reduces precision (0.60 to 0.48). DeepSeek and WizardCoder achieved higher precision scores but at the expense of lower recall rates than CodeLlama and Mixtral. This suggests that DeepSeek and WizardCoder were more prone to producing *False* responses than the other two models. We note that the Mixtral 8x7B is the largest model (47B parameters [21]) of any code model we evaluated, which could be a factor in the higher performance.

Figure 8b focuses on the models' ability to distinguish *True* from *False* labels. The ROC curve is based on the probabilities from the *True* and *False* tokens as described in Sect. 4.5. Mixtral 8x7B and DeepSeek models achieved an AUC of 0.80, with CodeLlama 34B and WizardCoder 15B closely following. Each model demonstrates their effectiveness in differentiating between labels.

Table 2. Comparing mean generation times (seconds) at report level with detailed vs. CoT prompts across CodeLlama models (7b/13b/34b) in zero-shot and few-shot settings. The Increase Factor CoT measures the slowdown with CoT prompts.

Model	Zero-Shot Time			Five-Shot Time		
	Detailed	CoT	Increase Factor CoT	Detailed	CoT	Increase Factor CoT
CodeLlama 7B	1.30 s	38.80 s	30×	4.57 s	43.01 s	9×
CodeLlama 13B	2.87 s	46.85 s	16×	10.13 s	65.54 s	6×
CodeLlama 34B	8.08 s	88.05 s	11×	28.78 s	126.54 s	4×

5.6 Computational Time Performance

The results in Table 2 show the mean computational time the CodeLlama models took to generate a response for pairing a GoVulnDB report with a vulnerable function. Report times were calculated based on the inference times for git-hunks within each report, with an average of five git-hunks per patch link per report. For instance, CodeLlama 7B averaged 0.31 seconds (s) per git-hunk, leading to a mean report time of 1.30 s. The standard and detailed prompts have similar performance. Therefore, we only show the detailed prompt.

CoT prompts were, on average, 13 times (×) slower than the detailed prompt. Specifically, CoT prompts were 19x slower in zero-shot settings and 7× slower in few-shot settings. Overall, the few-shot approach was 2.4× slower than the zero-shot approach. Additionally, generation time doubles (2.1×) when moving from the smaller CodeLlama 7B model to the 13B model and a similar increase when moving from the 13B to the 34B model, regardless of the prompt or learning paradigm. These findings highlight the generation complexity of CoT prompts (discussed in Sect. 4.5), the inclusion of more text (i.e., few-shot), and larger models contribute to longer performance times.

6 Discussion

LLMs demonstrate a possible path forward for pairing security advisories with vulnerable functions, offering high-quality pairings with minimal manual effort, thereby enhancing industry vulnerability management efforts. Despite the potential, considerable scope exists for improvement, especially in increasing the overall recall of our analysis. Integrating program analysis with LLMs could improve recall. For example, a challenge for all models was identifying vulnerabilities addressed by introducing new functions, which are then invoked from within the existing vulnerable function. Such fixes are common in vulnerabilities requiring sanitization (e.g., CWE-20 Improper Input Validation, CWE-79 Cross-Site Scripting), where sanitization needs to be applied across multiple code locations. In such instances, additional context (e.g., surrounding code) would be required. Outside of code aspects, integrating commit messages and identifying additional links or missing patch links could enrich LLM effectiveness in the future. This could address cases where patch links are missing or vulnerable functions are located elsewhere. Another area for improvement is in the retrieval of examples. When the retrieval system finds similar examples between both *True* and *False* labels, the models have trouble discerning the correct results.

Threats to Validity: Our research is subject to both internal and external validity threats. Internally, we trust that the Google Go security team accurately identifies vulnerable functions in Go data. If these identifications are inaccurate, our results and evaluations could be inaccurate. The models we evaluated were also trained on public data, potentially including some of the Go code under investigation and GoVulnDB report data. However, our methodology, which pairs these security advisories with vulnerable functions from git-diff data, differs from the original training and testing approaches of these models, helping to mitigate the risk of data leakage. Externally, limiting our evaluation to the Go language could impact the generalizability of our findings. We expect similar performance across different languages, drawing on the generalization capabilities of LLMs [5,21,26,31]. Nevertheless, this assumption remains speculative without further empirical validation across a broader set of programming languages for pairing security advisories with vulnerable functions. This highlights a critical area for future research. However, it will take substantial effort first to accumulate high-quality data such as GoVulnDB in other languages.

7 Related Work

Vulnerability Datasets: Recent work [8,9] has found that commonly used vulnerability datasets have significant inaccuracy rates. These issues in such labels are due to automatic data collection and flawed semantic filters, underscoring the need for manual verification and comprehensive function analysis to assess vulnerability relevance accurately. For instance popular datasets such as CVEFixes [4], ReVeal [7], BigVul [15], CrossVul [27], and Devign [40] consider the changed functions within a security patch as vulnerable. Guo and Bettaieb found that vulnerability dataset quality (i.e., mislabeled data) can significantly degrade models that depend on noisy datasets [20]. Alongside our research, Wang et al. introduced ReposVul [35], a vulnerability dataset featuring CVE entries in 1,491 projects across four programming languages, detailed to the line level. ReposVul uses vulnerability untangling, using both LLMs and static analysis tools to separate vulnerability-related code changes from unrelated patch changes.

LLMs for Vulnerability Detection: The growing interest in LLMs for software vulnerability detection has significant insights, yet the field is still in its infancy. In several scenarios, Khare et al. [22] showed GPT-4's ability to outperform static analysis and other deep learning-based approaches in detecting vulnerabilities. Zhang et al. [39] assessed ChatGPT's performance on synthetic Java datasets and the CVEFixes dataset with various prompting strategies. However, Purba et al. [29] evaluated the efficacy of both open-source and proprietary LLMs in detecting SQL injection and buffer overflow vulnerabilities, noting a high incidence of false positives. Liu et al. [25] combined LLMs with binary taint analysis to uncover 37 new bugs in real-world firmware. The LLM4Vuln framework [32] focuses on identifying smart contract vulnerabilities by enhancing LLMs with additional knowledge, tool integration, and prompt engineering, uncovering nine zero-days. Other studies have explored using LLMs to generate fuzzing inputs [11] and integrating them with static analysis tools for new vulnerability discovery [24]. However, our interest lies in pairing security advisories with the corresponding vulnerable functions, differing from identifying new vulnerabilities.

8 Conclusion

In summary, this study highlights the challenge of automating the association between security advisories and vulnerable functions and demonstrates the potential of LLMs as a promising initial solution. Our findings suggest that while CoT prompting enhances accuracy, it incurs computational costs. Alternatively, strategies using few-shot learning and concise prompts achieve comparable accuracy with reduced computational overhead. These advancements suggest a path to scale the automated matching of security advisories with vulnerable functions, which is key for enhancing vulnerability management processes.

Acknowledgments. This work is supported in part by NSF grants CNS-1946273 and CNS-2207008. Any findings and opinions expressed in this material are those of the authors and do not necessarily reflect the views of the funding agencies.

References

1. Alahmadi, B.A., Axon, L., Martinovic, I.: 99% false positives: a qualitative study of SOC analysts' perspectives on security alarms. In: 31st USENIX Security Symposium (USENIX Security 2022), pp. 2783–2800. USENIX Association (2022)
2. Alomar, N., Wijesekera, P., Qiu, E., Egelman, S.: "You've got your nice list of bugs, now what?" Vulnerability discovery and management processes in the wild. In: Proceedings of the Sixteenth USENIX Conference on Usable Privacy and Security, SOUPS 2020. USENIX Association (2020)
3. Arora, A.: Understanding logits, sigmoid, softmax, and cross-entropy loss in deep learning—wandb.ai (2023). https://wandb.ai/amanarora/Written-Reports/reports/Understanding-Logits-Sigmoid-Softmax-and-Cross-Entropy-Loss-in-Deep-Learning--Vmlldzo0NDMzNTU3

4. Bhandari, G., Naseer, A., Moonen, L.: CVEfixes: automated collection of vulnerabilities and their fixes from open-source software. In: Proceedings of the 17th International Conference on Predictive Models and Data Analytics in Software Engineering, PROMISE 2021, pp. 30–39. Association for Computing Machinery, New York (2021). https://doi.org/10.1145/3475960.3475985
5. Bi, X., et al.: DeepSeek LLM: scaling open-source language models with longtermism. arXiv preprint arXiv:2401.02954 (2024)
6. Brown, T.B., et al.: Language models are few-shot learners. In: Proceedings of the 34th International Conference on Neural Information Processing Systems, NIPS 2020. Curran Associates Inc. (2020)
7. Chakraborty, S., Krishna, R., Ding, Y., Ray, B.: Deep learning based vulnerability detection: are we there yet? IEEE Trans. Softw. Eng. **48**(09), 3280–3296 (2022). https://doi.org/10.1109/TSE.2021.3087402
8. Chen, Y., Ding, Z., Alowain, L., Chen, X., Wagner, D.: DiverseVul: a new vulnerable source code dataset for deep learning based vulnerability detection. In: Proceedings of the 26th International Symposium on Research in Attacks, Intrusions and Defenses, RAID 2023, pp. 654–668. Association for Computing Machinery, New York (2023). https://doi.org/10.1145/3607199.3607242
9. Croft, R., Babar, M.A., Kholoosi, M.M.: Data quality for software vulnerability datasets. In: Proceedings of the 45th International Conference on Software Engineering, ICSE 2023, pp. 121–133 (2023). https://doi.org/10.1109/ICSE48619.2023.00022
10. Dann, A., Plate, H., Hermann, B., Ponta, S., Bodden, E.: Identifying challenges for OSS vulnerability scanners - a study & test suite. IEEE Trans. Softw. Eng. **48**(09), 3613–3625 (2022). https://doi.org/10.1109/TSE.2021.3101739
11. Deng, Y., Xia, C.S., Peng, H., Yang, C., Zhang, L.: Large language models are zero-shot fuzzers: fuzzing deep-learning libraries via large language models. In: Proceedings of the 32nd ACM SIGSOFT International Symposium on Software Testing and Analysis, ISSTA 2023, pp. 423–435. Association for Computing Machinery, New York (2023). https://doi.org/10.1145/3597926.3598067
12. Douze, M., et al.: The faiss library. arXiv preprint arXiv:2401.08281 (2024)
13. Dunlap, T.: A python package for extracting commit features - patchparser (2022). https://github.com/tdunlap607/patchparser
14. EndorLabs: Endor Labs vs. SCA (2022). https://www.endorlabs.com/endor-labs-vs-sca
15. Fan, J., Li, Y., Wang, S., Nguyen, T.N.: AC/C++ code vulnerability dataset with code changes and CVE summaries. In: Proceedings of the 17th International Conference on Mining Software Repositories, MSR 2020, pp. 508–512. Association for Computing Machinery (2020). https://doi.org/10.1145/3379597.3387501
16. Google: Go Vulnerability Database. https://go.dev/doc/security/vuln/database
17. Google: Govulncheck. https://pkg.go.dev/golang.org/x/vuln/cmd/govulncheck
18. Google: Handling Go Vulnerability Reports. https://github.com/golang/vulndb/blob/master/doc/triage.md#add-a-new-report-label-needsreport
19. Google: Vulnreport. https://pkg.go.dev/golang.org/x/vulndb/cmd/vulnreport
20. Guo, Y., Bettaieb, S.: An investigation of quality issues in vulnerability detection datasets. In: 2023 IEEE European Symposium on Security and Privacy Workshops (EuroS&PW) (2023). https://doi.org/10.1109/EuroSPW59978.2023.00008
21. Jiang, A.Q., et al.: Mixtral of experts. arXiv preprint arXiv:2401.04088 (2024)
22. Khare, A., Dutta, S., Li, Z., Solko-Breslin, A., Alur, R., Naik, M.: Understanding the effectiveness of large language models in detecting security vulnerabilities. arXiv preprint arXiv:2311.16169 (2023)

23. LangChain: Custom example selector (2023). https://python.langchain.com/docs/modules/model_io/prompts/example_selectors/custom_example_selector
24. Li, H., Hao, Y., Zhai, Y., Qian, Z.: The Hitchhiker's guide to program analysis: a journey with large language models. arXiv preprint arXiv:2308.00245 (2023)
25. Liu, P., et al.: Harnessing the power of LLM to support binary taint analysis. arXiv preprint arXiv:2310.08275 (2023)
26. Luo, Z., et al.: WizardCoder: empowering code large language models with evol-instruct. arXiv preprint arXiv:2306.08568 (2023)
27. Nikitopoulos, G., Dritsa, K., Louridas, P., Mitropoulos, D.: CrossVul: a cross-language vulnerability dataset with commit data. In: Proceedings of the 29th ACM Joint Meeting on European Software Engineering Conference and Symposium on the Foundations of Software Engineering, ESEC/FSE 2021, pp. 1565–1569. Association for Computing Machinery, New York (2021). https://doi.org/10.1145/3468264.3473122
28. Ponta, S.E., Plate, H., Sabetta, A.: Detection, assessment and mitigation of vulnerabilities in open source dependencies. Empir. Softw. Eng. **25**(5), 3175–3215 (2020). https://doi.org/10.1007/s10664-020-09830-x
29. Purba, M.D., Ghosh, A., Radford, B.J., Chu, B.: Software vulnerability detection using large language models. In: 2023 IEEE 34th International Symposium on Software Reliability Engineering Workshops (ISSREW), pp. 112–119 (2023). https://doi.org/10.1109/ISSREW60843.2023.00058
30. Radford, A., et al.: Language models are unsupervised multitask learners. OpenAI Blog **1**(8), 9 (2019)
31. Roziere, B., et al.: Code llama: open foundation models for code. arXiv preprint arXiv:2308.12950 (2023)
32. Sun, Y., et al.: LLM4Vuln: a unified evaluation framework for decoupling and enhancing LLMs' vulnerability reasoning. arXiv preprint arXiv:2401.16185 (2024)
33. TreeSitter: An incremental parsing system for programming tools - tree-sitter (2023). https://tree-sitter.github.io/tree-sitter/
34. Vaswani, A., et al.: Attention is all you need. In: Guyon, I., et al. (eds.) Advances in Neural Information Processing Systems, vol. 30. Curran Associates, Inc. (2017)
35. Wang, X., Hu, R., Gao, C., Wen, X.C., Chen, Y., Liao, Q.: ReposVul: a repository-level high-quality vulnerability dataset. arXiv preprint arXiv:2401.13169 (2024)
36. Wang, Y., Le, H., Gotmare, A.D., Bui, N.D., Li, J., Hoi, S.C.: CodeT5+: open code large language models for code understanding and generation. arXiv preprint arXiv:2305.07922 (2023)
37. Wei, J., et al.: Finetuned language models are zero-shot learners. arXiv preprint arXiv:2109.01652 (2021)
38. Wei, J., et al.: Chain-of-thought prompting elicits reasoning in large language models. In: Koyejo, S., Mohamed, S., Agarwal, A., Belgrave, D., Cho, K., Oh, A. (eds.) Advances in Neural Information Processing Systems, vol. 35, pp. 24824–24837. Curran Associates, Inc. (2022)
39. Zhang, C., Liu, H., Zeng, J., Yang, K., Li, Y., Li, H.: Prompt-enhanced software vulnerability detection using ChatGPT. arXiv preprint arXiv:2308.12697 (2023)
40. Zhou, Y., Liu, S., Siow, J., Du, X., Liu, Y.: Devign: effective vulnerability identification by learning comprehensive program semantics via graph neural networks. In: Wallach, H., Larochelle, H., Beygelzimer, A., d'Alché-Buc, F., Fox, E., Garnett, R. (eds.) Advances in Neural Information Processing Systems, vol. 32. Curran Associates, Inc. (2019)

Extended Abstract: Assessing Language Models for Semantic Textual Similarity in Cybersecurity

Arian Soltani(✉), DJeff Kanda Nkashama, Jordan Felicien Masakuna, Marc Frappier, Pierre-Martin Tardif, and Froduald Kabanza

GRIC, Université de Sherbrooke, Sherbrooke, QC J1K 2R1, Canada
{arian.soltani,nkad2101,jordan.felicien.masakuna,marc.frappier, pierre-martin.tardif,froduald.kabanza}@usherbrooke.ca

Abstract. In light of the significant strides made by large language models (LLMs) in the field of natural language processing (NLP) [5], our research seeks to evaluate and contrast their proficiency in establishing associations within the realm of cybersecurity. Our experimental framework involves juxtaposing actual connections from various cybersecurity knowledge graphs (including MITRE CAPEC, D3FEND, and CVE connections to ATT&CK) against predictions made by LLMs using semantic textual similarity (STS). These connections span a broad spectrum, encapsulating diverse abstractions of threat descriptions, attack patterns, defense strategies, and vulnerabilities. The language models chosen for this study are varied, comprising state-of-the-art models from STS leaderboards, LLMs (GPT3.5 and PaLM), and ATTACK BERT [1], a cybersecurity domain-specific language model. Our experiments provide valuable insights into the differentiation between language models and data sources, thereby facilitating the broader application of STS in cybersecurity.

Keywords: Cybersecurity · Language Models · Intrusion Detection Systems (IDS) · MITRE ATT&CK

1 Introduction

The ever-increasing reliance on computer systems and digital technologies has led to a corresponding rise in cybersecurity threats, posing a significant risk to the security and integrity of cyber assets. As organizations increasingly depend on interconnected networks and digital infrastructures, the potential vulnerabilities have expanded, necessitating robust cybersecurity measures. Due to this trend, the demand for cybersecurity specialists has surged, reflecting the critical need for expertise in safeguarding sensitive information and defending against sophisticated cyber attacks. However, this growing demand is exacerbated by a pervasive talent shortage within the field of cybersecurity [6], presenting a challenge to organizations seeking to reinforce their cyber-defenses. Recent reports indicate that cybersecurity analysts overwork, leading to potential gaps in security coverage and increased vulnerability to cyber threats [17].

F. Maggi et al. (Eds.): DIMVA 2024, LNCS 14828, pp. 370–380, 2024.
https://doi.org/10.1007/978-3-031-64171-8_19

Artificial Intelligence (AI) has emerged as a crucial pillar in advancing cybersecurity, addressing the growing complexity of cyber threats. AI-powered tools can detect attacks in real-time, automating incident response processes and streamlining threat hunting. As the amount of data generated by connected devices increases exponentially, AI becomes essential in analyzing this data to mitigate cyber threats [16].

One of the most important ways AI can help alleviate the pressure on cybersecurity analysts is making associations among concepts and entities. Consider the following examples of tasks in the workflow of a cybersecurity analyst:

- finding the underlying goal of an attack (identifying the attacker's motives and objectives)
- predicting the next possible move of an attacker (threat prediction)
- suggesting defense mechanisms for threats (threat mitigation)
- anticipating possible threats based on vulnerabilities (proactive cyber risk management)

All of these tasks include recognizing **semantic relationships**, a capability that recent language models are increasingly adept at developing [13]. This notion has been previously explored for a narrow use-case, namely connecting vulnerabilities to threats [1], yet its general usability in the broader context remains uncharted. To our knowledge, currently there is no proposed general method of determining how competent AI systems are in making associations in the context of cybersecurity.

AI holds promise for enhancing cybersecurity efficiency, yet it faces challenges such as a lack of contextual understanding, leading to misclassifications. The necessity for transparency and interpretability in critical cybersecurity operations is not met by the opaqueness of large neural networks. Additionally, LLMs are prone to "hallucination", producing plausible but incorrect information [12], which undermines their reliability in cybersecurity contexts where accuracy is paramount. However, leveraging the embeddings from language models (LMs) could offer a compromise, harnessing LM strengths while mitigating generative AI weaknesses. Embeddings represent complex data as high-dimensional vectors, facilitating various algebraic operations. In cybersecurity, embeddings have been effectively applied to tasks like linking vulnerabilities to potential exploits, providing context-sensitive insights that are crucial for threat intelligence and incident response [1,3]. Their compatibility with existing cybersecurity systems, such as intrusion detection systems (IDS) and security information and event management systems (SIEM), renders them a valuable tool for enhancing security measures.

Considering the advantages that embeddings offer for cybersecurity applications, our research focuses on evaluating the proficiency of current LMs in discerning a variety of semantic relationships pertinent to cybersecurity, including attack types, patterns, vulnerabilities, and countermeasures. We leverage relationships derived from cybersecurity knowledge bases to assess and compare the effectiveness of different LMs, thereby enabling the combination of various LMs to foster innovation in their application to cybersecurity tasks, such as ensemble

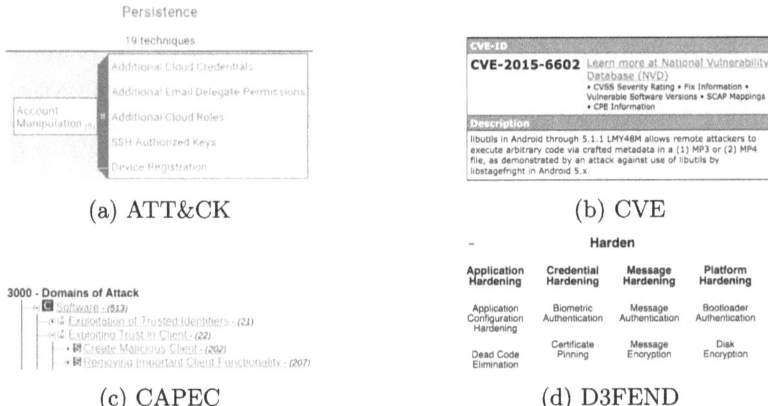

(a) ATT&CK (b) CVE (c) CAPEC (d) D3FEND

Fig. 1. Examples from threat and defence description data sources

approaches. Preliminary findings underscore the promise of LM embeddings in identifying semantic connections within the cybersecurity domain.

The subsequent sections of this paper are structured as follows: Sect. 2 provides an overview of the relevant literature and background information; Sect. 3 delineates the methodology employed in our study; Sect. 4 presents the results of our experimental evaluations; and the concluding section discusses implications for future research.

2 Background and Related Work

Background. Cybersecurity knowledge bases such as MITRE ATT&CK, CAPEC, D3FEND, and CVE play a pivotal role in the workflow of cybersecurity analysts, each covering an aspect of defense. *ATT&CK* enumerates and taxonomizes singular attack tactics, techniques, and procedures (TTPs), enabling analysts to perform cyber-threat intelligence (CTI), intrusion detection, risk assessment, and many other operations [15]. *CAPEC* catalogs known and common patterns of attack. In CAPEC, attack patterns are described and connected to related TTPs and weaknesses. *CVE* is the common vulnerability enumeration database, containing short descriptions of exposed vulnerabilities. Vulnerabilities are computer system weaknesses that enable attackers to take advantage. They are continuously discovered, exposed, exploited, and patched by developers, analysts, and attackers. *D3FEND* focuses on the opposite aspect compared to the aforementioned; it contains a top-down hierarchy of defensive strategies. Entries are linked to ATT&CK to mark the threats they can mediate. These knowledge bases contain cybersecurity domain knowledge and connections among them can be harbored as a testing ground for assessing LMs' capability in recognizing cybersecurity semantic similarities. First, previous cases of LM application in cybersecurity are presented. Then in the next sections, the generalized comparison grounds are presented contrasting LMs' general capability in this domain.

Related Work. Language models present immense opportunities for applications in cybersecurity, owing to their ability in processing natural language content. This ability can be exploited for matching threats in emails, messaging apps, social media, event logs, vulnerability descriptions, etc. with high speed. Finding matches using LMs is achieved in three main ways: using a corresponding task in MLP, general purpose language models, and specialized language models. Wåreus and Hell introduced a method that can track vulnerabilities in software versions automatically [18] by utilizing named entity recognition (NER). Kuppa et al. developed a joint word embedding space to match ATT&CK and CVE entries [11]. General purpose LMs for generating sentence embeddings such as Sentence BERT and its variations also proved useful in matching CAPEC and CVE entries [10], and expanding D3FEND-ATT&CK connections for incident response [3].

First endeavours for developing a domain-specific language models were fine-tuning BERT, a general-purpose base language model, on cybersecurity corpus using the masking technique [2,14] optimized in performing NER and sentiment analysis, and capable of generating *word* embeddings. After, Abdeen et al. introduced ATTACK BERT, a specialized cybersecurity *sentence* embedding model, and used it to match CVE and ATT&CK entries outperforming Sentence BERT. With the advent of LLMs, their generative capabilities are utilized [4,7]. However, their embedding capabilities are not yet examined.

Introducing new concepts to cybersecurity where robustness and tangibility are vital, requires investigation beyond narrow use-cases. By exploiting the information already used in operation, our approach introduces a testing ground encompassing several aspects of semantic connections in cybersecurity. This highlights the distinction of LMs for applications in cybersecurity and enables subsequent research to continuously improve, and tangibly infuse cybersecurity tools such as IDSs with the capabilities of modern LMs.

3 Methodology

3.1 Data

To assess LMs' capability in recognizing semantic connections in the context of cybersecurity, we extract descriptions and labelled connections among cybersecurity knowledge bases (ATT&CK, etc.), and then based on the description, LMs can predict relationships using STS. By comparing the ground truth with LMs' predictions, LMs' capability in recognizing these relationships can be measured.

The selected datasets are not exhaustive. There exist other semantic connections among entries, though the existing definitions and connections are rigorously reviewed by subject matter experts (SMEs) and the chosen sources are already being used by cybersecurity analysts. Therefore, we assume that the

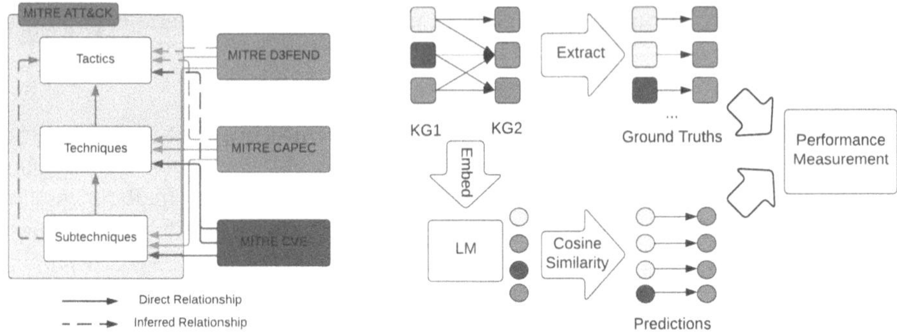

Fig. 2. Data source connections

Fig. 3. Processing pipeline

outlined relationships are evident, and they must be unmistakable for a cyber-security analyst.

After extracting, cleaning, and preprocessing definitions and relationships, definitions will be processed into embedding vectors using several language models. There are several ways to perform the preprocessing steps. For example, in [1], a subset of the descriptions matching the attack vector templates are selected. In contrast, our goal is to measure LMs ability in understanding and generalizing definitions. So instead of snippets, embeddings of the whole definition documents are processed in this work. Moreover, contrary to previous works, we expand the dataset to unambiguous adjacencies. For example, "CAPEC-25: Forced Deadlock" is mapped to subtechnique "T1499.004: Endpoint Denial of Service: Application or System Exploitation" in ATT&CK. In this instance, we expand the labels to the subtechniques parent nodes: technique "T1499: Endpoint Denial of Service", and tactic "TA0040: Impact". The validity of this expansion is supported by the hierarchical tree structure of ATT&CK. Figure 2 depicts the connections among data sources, marking the extended (indirect) connections as dashed lines.

3.2 Similarity and Performance Measures

Having processed the documents to embeddings and extracted the ground truth, LMs' predictions can be measured. To achieve this using the common STS process, cosine similarity between the embedding vectors are calculated, and the top scoring vector pairs are selected (Fig. 3). For two vectors $\vec{a}$ and $\vec{b}$:

$$\text{Cosine Similarity}(\vec{a}, \vec{b}) = \frac{\vec{a} \cdot \vec{b}}{||\vec{a}||\,||\vec{b}||}$$

So assuming the embedding function $\vec{f_e}$ calculates embeddings of a text sequence, for two text sequences a and b, the similarity score would be:

$$\text{Similarity}(a, b) = \text{Cosine Similarity}(\vec{f_e}(a), \vec{f_e}(b))$$

Table 1. Pair counts for each subtask

Group	Subtask Name	Pair Count
Inter-ATT&CK	TecTac	227
	SubTec	411
	SubTac	558
CAPEC-ATT&CK	CapTac	271
	CapTec	235
	CapSub	156
D3FEND-ATT&CK	DefTac	795
	DefTec	1271
	DefSub	2631
CVE-ATT&CK	CveTac	2484
	CveTec	2057
	CveSub	498

In this study, similar to [1], we employ a similarity score to evaluate LMs' predictive capabilities through various multilabel classification evaluation metrics: recall@k, coverage error (CE), label ranking average precision (LRAP), and label ranking loss (LRL). Each metric serves as a distinct indicator of performance from various perspectives. Coverage error quantifies the average number of top predictions needed to encompass all actual labels, with its maximum value being the total number of labels. LRAP measures the percentage of top-ranked labels that are correct, indicating the precision of the ranking process. Label ranking loss (LRL) assesses the average ratio of incorrectly to correctly ranked labels, providing insight into the model's ranking accuracy. Lastly, recall@k determines the fraction of true labels that appear among the top-k predictions, offering a measure of the model's ability to capture relevant labels.

3.3 Language Models

Selected language models for this work are from the Massive Text Embedding Benchmark (MTEB) Leaderboard [8]. This list represents state of the art sentence embedding language models in several tasks such as classification, clustering, reranking, retrieval, STS along with information about the models' size, embedding dimensions, and maximum accepted tokens. It is continuously updated, introducing more advanced models outperforming previous state-of-the-art. Our curated selection features top-tier models from the MTEB leaderboard, alongside two distinguished Large Language Models (LLMs): GPT-3.5 (the foundational model behind ChatGPT) and PaLM. Additionally, we spotlight a specialized sentence embedding model tailored for the cybersecurity domain: ATTACK BERT (AB) [1]. Beyond these individual models, we also present a series of their equally weighted ensembles, which are denoted with an "E-" prefix.

4 Experimental Evaluation

To perform the comparison, first we prepare a dataset of pairs containing relations between knowledge base entries in section. Due to the variety of the data sources, and thus their different use-cases, the data is broken to several parts to highlight their difference, and they will be investigated as different subtasks. First, each of these subtasks will be explained, then the experiment results will be presented. Finally, results will be inspected in the corresponding section.

4.1 Subtasks

For our study, we collect data pertinent to specific subtasks from their respective sources, enabling us to evaluate language models' performance on individual subtasks (as shown in Fig. 4) and their combined average (refer to Table 2). Subtasks are designated based on cross-referencing elements from various cybersecurity frameworks: Tactics (Tac), Techniques (Tec), and Subtechniques (Sub) from ATT&CK; Patterns from CAPEC (Cap); Vulnerabilities from CVE (Cve); and Defense Strategies from D3FEND (Def). Within the MITRE ATT&CK framework, we examine the structural relationships—TecTac (Techniques to Tactics), SubTec (Subtechniques to Techniques), and SubTac (Subtechniques to Tactics)—to assess LMs' proficiency in matching abstract and technical concepts. We apply a similar approach to analyze the interplay between CAPEC attack patterns (CapTac, CapTec, CapSub), CVE vulnerabilities (CveTac, CveTec, CveSub), and D3FEND defense strategies (DefTac, DefTec, DefSub). These subtasks are critical for understanding various aspects of cybersecurity. This structured approach allows us to dissect and quantify how well LMs can discern and interpret the intricate web of relationships that define the cybersecurity landscape.

The number of pairs available for each subtask is displayed in Table 1. For these subtasks, measures (LRAP, coverage error, etc.) are calculated among predicted LM connections and ground truth labels, presented in Table 2. Moreover, in Fig. 4 for each subtask, a sweep on the k parameter of recall@k has been provided, demonstrates how LMs' precision changes at according to the number of top selected suggestions.

4.2 Results

Figure 4 aims to differentiate between subtasks, whereas Table 2 sheds light on the performance from the perspective of LMs. The analysis reveals significant variability in model effectiveness across subtasks and metrics, with ensemble models surpassing both domain-specialized models such as ATTACK BERT (AB) and LLMs.

Such fluctuations emphasize the impact that differing architectures, training methodologies, and datasets have on a model's ability to comprehend and process various concepts effectively. By incorporating this diversity, ensemble models enhance performance across various tasks. The recall@k metrics further reveal

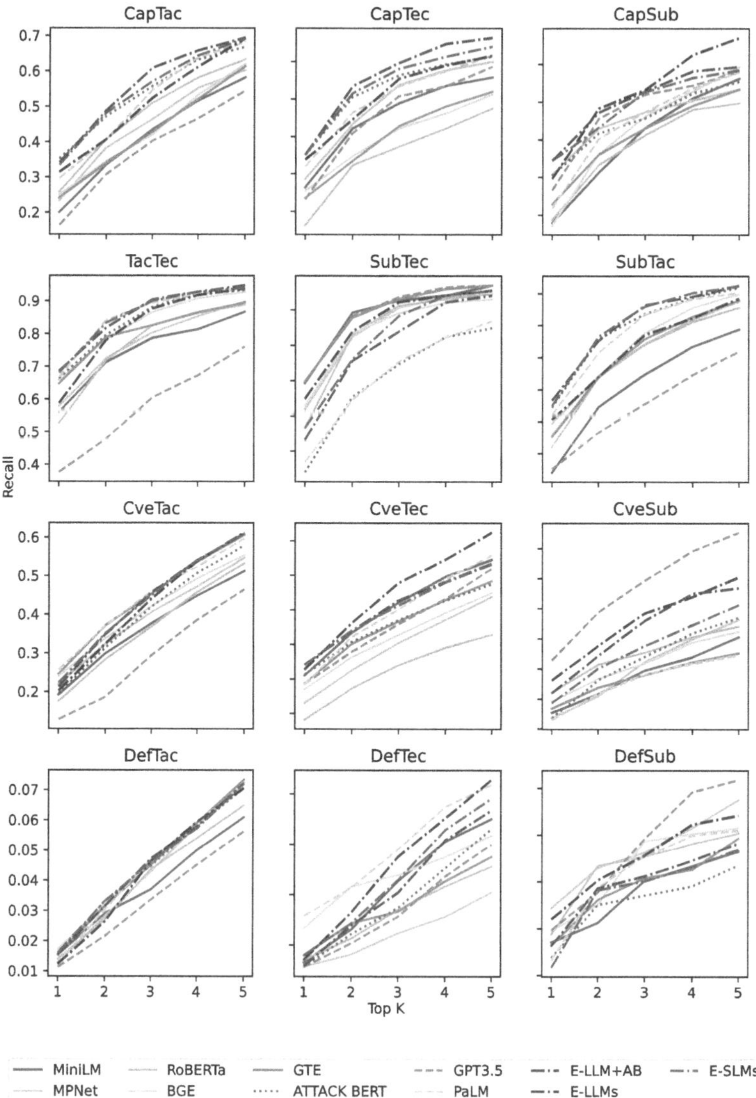

Fig. 4. Change of recall@k on top k parameter

that there isn't a single model that consistently leads in performance across all subtasks, thereby underscoring the importance of selecting models tailored to specific tasks. For instance, GPT-3.5 shows prowess in discerning connections between subtechniques but falls short in associating related tactics. The performance in defensive tasks is particularly weak, with a maximum recall of 0.08 in the top 5 selections, revealing a substantial deficiency in current models' capac-

Table 2. Language models' average results

Language Model	Variation	CE	LRAP%	LRL	Recall@5%
MiniLM	`all-MiniLM-L12-v2`	15.10	66.3%	0.123	43.9%
MPNet	`all-mpnet-base-v2`	15.18	68.0%	0.113	45.3%
RoBERTa	`all-roberta-large-v1`	15.85	65.8%	0.115	43.4%
BGE	`bge-large-en-v1.5`	15.37	68.3%	0.106	45.4%
GTE	`gte-large`	14.58	68.4%	0.103	45.2%
ATTACK BERT	`basel/ATTACK-BERT`	14.39	70.4%	0.096	46.7%
GPT3.5	`text-embedding-ada-002`	13.79	64.8%	0.132	42.5%
PaLM	`textembedding-gecko@001`	12.96	70.0%	0.090	47.3%
E-LLM+AB	GPT3.5 + PaLM + AB	12.31	**72.0%**	**0.085**	**48.7%**
E-LLMs	GPT3.5 + PaLM	**11.59**	70.7%	0.090	**48.7%**
E-SLMs+AB	BGE + GTE + AB	13.27	71.7%	0.088	48.4%

ity to automate defensive cybersecurity tasks. This aligns with prior studies that have recognized the intricate and context-sensitive nature of these tasks [3,9].

In other subtasks, the findings suggest that while the current LM embeddings might not reach complete autonomy in recognizing relationships, they may still play a crucial role in streamlining the decision-making process for analysts by helping prioritize options. Notably, smaller LMs are shown to deliver performance on par with their larger counterparts, offering a cost-effective solution for scenarios requiring the processing of large volumes of data where deploying LLMs may be financially impractical. Despite the intriguing insights these experiments offer, the occasional discrepancies in performance across different metrics signal a need for further investigation to confirm and refine these observations.

5 Conclusion and Future Work

This study evaluated LMs' effectiveness in identifying semantic relationships within the cybersecurity domain, leveraging established links between cybersecurity knowledge bases. Our findings indicate that singular and ensembled LMs possess a certain level of proficiency in executing specific tasks, potentially reducing the workload for cybersecurity analysts. However, when it comes to devising defensive strategies, further investigation is necessary. This research serves as a preliminary step, suggesting that incorporating additional data sources could enhance the robustness and reliability of the findings, thereby accelerating the integration of natural language processing (NLP) advancements into cybersecurity. Furthermore, the presented methodologies are poised for implementation in current cybersecurity frameworks to provide practical value in real-world applications.

References

1. Abdeen, B., Al-Shaer, E., Singhal, A., Khan, L., Hamlen, K.: SMET: semantic mapping of CVE to ATT&CK and its application to cybersecurity. In: Atluri, V., Ferrara, A.L. (eds.) DBSec 2023. LNCS, vol. 13942, pp. 243–260. Springer, Cham (2023). https://doi.org/10.1007/978-3-031-37586-6_15
2. Aghaei, E., Niu, X., Shadid, W., Al-Shaer, E.: SecureBERT: a domain-specific language model for cybersecurity. In: Li, F., Liang, K., Lin, Z., Katsikas, S.K. (eds.) Security and Privacy in Communication Systems. LNICST, vol. 462, pp. 39–56. Springer, Cham (2022). https://doi.org/10.1007/978-3-031-25538-0_3
3. Akbar, K.A., Halim, S.M., Hu, Y., Singhal, A., Khan, L., Thuraisingham, B.: Knowledge mining in cybersecurity: from attack to defense. In: Sural, S., Lu, H. (eds.) DBSec 2022. LNCS, vol. 13383, pp. 110–122. Springer, Cham (2022). https://doi.org/10.1007/978-3-031-10684-2_7
4. Al-Hawawreh, M., Aljuhani, A., Jararweh, Y.: ChatGPT for cybersecurity: practical applications, challenges, and future directions. Clust. Comput. **26**(6), 3421–3436 (2023)
5. Bubeck, S., et al.: Sparks of artificial general intelligence: early experiments with GPT-4. arXiv preprint arXiv:2303.12712 (2023)
6. Crumpler, W., Lewis, J.A.: The Cybersecurity Workforce Gap. JSTOR (2019)
7. Gupta, M., Akiri, C., Aryal, K., Parker, E., Praharaj, L.: From ChatGPT to ThreatGPT: impact of generative AI in cybersecurity and privacy. IEEE Access **11**, 80218–80245 (2023)
8. Huggingface: MTEB Leaderboard (2023). https://huggingface.co/spaces/mteb/leaderboard. Accessed 1 Dec 2023
9. Kaiser, F.K., Andris, L.J., Tennig, T.F., Iser, J.M., Wiens, M., Schultmann, F.: Cyber threat intelligence enabled automated attack incident response. In: 2022 3rd International Conference on Next Generation Computing Applications (NextComp), pp. 1–6. IEEE (2022)
10. Kanakogi, K., et al.: Tracing CVE vulnerability information to CAPEC attack patterns using natural language processing techniques. Information **12**(8), 298 (2021)
11. Kuppa, A., Aouad, L., Le-Khac, N.A.: Linking CVE's to MITRE ATT&CK techniques. In: Proceedings of the 16th International Conference on Availability, Reliability and Security, pp. 1–12 (2021)
12. McKenna, N., Li, T., Cheng, L., Hosseini, M.J., Johnson, M., Steedman, M.: Sources of hallucination by large language models on inference tasks. arXiv preprint arXiv:2305.14552 (2023)
13. Min, B., et al.: Recent advances in natural language processing via large pre-trained language models: a survey. ACM Comput. Surv. **56**(2), 1–40 (2023)
14. Ranade, P., Piplai, A., Joshi, A., Finin, T.: CyBERT: contextualized embeddings for the cybersecurity domain. In: 2021 IEEE International Conference on Big Data (Big Data), pp. 3334–3342. IEEE (2021)
15. Roy, S., Panaousis, E., Noakes, C., Laszka, A., Panda, S., Loukas, G.: SoK: the MITRE ATT&CK framework in research and practice. arXiv preprint arXiv:2304.07411 (2023)
16. Sarker, I.H., Furhad, M.H., Nowrozy, R.: AI-driven cybersecurity: an overview, security intelligence modeling and research directions. SN Comput. Sci. **2**, 1–18 (2021)

17. Venturebeat: Mental Health: 66% of cybersecurity analysts experienced burnout this year (2023). https://venturebeat.com/security/mental-health-cybersecurity-analysts/. Accessed 19 July 2023
18. Wåreus, E., Hell, M.: Automated CPE labeling of CVE summaries with machine learning. In: Maurice, C., Bilge, L., Stringhini, G., Neves, N. (eds.) DIMVA 2020. LNCS, vol. 12223, pp. 3–22. Springer, Cham (2020). https://doi.org/10.1007/978-3-030-52683-2_1

Extended Abstract: A Transfer Learning-Based Training Approach for DGA Classification

Arthur Drichel(✉), Benedikt von Querfurth, and Ulrike Meyer

RWTH Aachen University, Aachen, Germany
{drichel,meyer}@itsec.rwth-aachen.de,
benedikt.von.querfurth@rwth-aachen.de

Abstract. We propose a novel training method based on transfer learning (TL) and fine-tuning to improve the detection performance of domain generation algorithm (DGA) classifiers. This new training approach allows a classifier to learn nuances of domains generated by specific DGAs, which greatly improves a classifier's separation performance between benign and malicious samples. Additionally, we develop and optimize four novel models for DGA binary and multiclass classification based on more recently proposed deep learning (DL) architectures. We comparatively evaluate the resulting DGA classifiers in a unified setting for statistically significant improvements, and assess whether the classifiers generalize well between different networks and are time-robust. In a real-world setting, our best performing model improves the state of the art by 2–3.4% in true-positive rate (TPR) at the same fixed false-positive rates (FPRs).

1 Introduction

Domain generation algorithms (DGAs) are used by bots to connect to their command and control (C2) server. DGAs are pseudo-random algorithms that generate a large amount of domains which are successively queried by the bots. Most queries result in non-existent domain (NXD) responses, and only those that have been pre-registered by the botnet master are successfully resolved. This creates an asymmetric situation in which defenders must block all generated domains, while it is sufficient for an adversary if a single domain is successfully resolved. Machine learning (ML) based approaches have been proposed in the past that are able to distinguish benign from malicious domains (binary classification) and also assign malicious domains to the specific DGAs that generated them (multiclass classification).

For the binary classification task, a high true-positive rate (TPR) is essential as due to the underlying asymmetry a single successful domain resolution is sufficient for a botnet master to maintain control of a botnet. At the same time, a low false-positive rate (FPR) is indispensable in this setting due to the base-rates for benign and malicious samples: the number of benign DNS queries by far exceeds the number of queries containing DGA-generated domains. Thus, only

F. Maggi et al. (Eds.): DIMVA 2024, LNCS 14828, pp. 381–391, 2024.
https://doi.org/10.1007/978-3-031-64171-8_20

a very low FPR will lead to a feasible amount of false positives. State-of-the-art deep learning (DL) classifiers for the binary task, such as recurrent (RNN) [17], convolutional (CNN) [20], and residual (ResNet) neural networks [6] seem to promise to achieve this, providing accuracies of up to 99.92% with TPRs of 99.98% and FPRs of 0.15% [6].

However, prior work falls short when it comes to comparing different classifiers. Previous studies (e.g. [6,17,20]) do not analyze the differences in classification performance (which are often on the order of fractions of a percent) for statistical significance, making it impossible to determine the best model.

Moreover, recently, the authors of [5] analyzed the explainability of such binary DGA classifiers and uncovered several biases inherent to the classification approaches that allow attackers to evade detection with ease. The authors proposed a method to reduce the identified biases in these classifiers and showed that their biased-reduced version in a real-world evaluation setting only achieve a TPR of 89% at a fixed FPR of 0.8%. Their work thus re-opens the question on how to further improve the detection capabilities of such classifiers.

In this work, we address these problems and make the following contributions: (1) We develop and optimize four novel models for DGA binary classification and four additional variants for DGA multiclass classification based on more recently introduced DL architectures (ResNeXt [18], ConvNeXt [8], Transformer [16], and RWKV [11]). (2) We propose a new training method based on transfer learning (TL) that enables a binary classifier to learn features of domains generated by specific DGAs, which greatly improves the classifier's separation performance between benign and malicious samples. (3) We compare the novel models with the state of the art in a unified and experimental bias reduced setting, analyze the achieved improvements for statistical significance, and evaluate their generalizability and temporal robustness using real-world network data.

2 Related Work

2.1 Domain Generation Algorithm Detection

Related work on DGA detection can be divided into context-less (e.g., [6,17,20]) and context-aware (e.g., [1,3,13]) approaches. The former exclusively use information that can be extracted from a single domain for classification. The latter incorporate additional contextual information or rely on third-party data providers to further improve classification performance. Several previous studies (e.g., [6,14,17,20]) have shown that context-less approaches are less resource intensive and less privacy intrusive than context-aware approaches, while achieving similar or better classification performance. Note that these context-less approaches are not intended for single-domain based decision-making, but rather to be integrated into detection systems that allow decision-making based on multiple classifications to further reduce FPRs in practice [5].

The proposed ML approaches can additionally be divided into feature-based (e.g., [3,4,14]) and feature-less (DL) methods (e.g., [6,17,20]). Feature-based approaches require prior feature engineering to extract relevant information from an input domain that can be processed by a ML classifier. As these features are handcrafted, they may conceal important characteristics of an input domain to the classifier. This could lead to poorer classification performance compared to feature-less approaches that learn to extract relevant features on their own. In the past, several studies (e.g., [6,10,15,17]) have shown that DL-based classifiers achieve superior classification performance.

Since DL classifiers automatically learn to extract relevant features for classification and do not require integration of domain knowledge, they are often referred to as black-boxes. In the context of DGA detection, related work (e.g. [4,5]) analyzed the explainability of DGA detection classifiers. In [5], the authors critically analyzed the automatic feature extraction performed by DL classifiers, uncovered biases in state-of-the-art classifiers that allow easy detection bypass, and proposed a classification setting that reduces biases and their exploitability. In this work, we use the proposed bias-reduced classification setting to mitigate biases in novel model architectures.

Additionally, DL classifiers are known to be vulnerable to adversarial attacks, where semantic gaps in the training data can lead to blind spots in a classifier. This makes classifiers susceptible to small input perturbations that can be exploited to trick a classifier into misclassifications. Adversarial training can be used to mitigate this problem. Related work has also dealt with this issue in the context of DGA detection (e.g. [6,10,15]).

2.2 Transfer Learning

Transfer learning (TL) is a technique for training a classifier by reusing knowledge gained from training on other data to solve a different task. Thereby, TL drastically reduces training costs and is particularly useful when training data is sparse, as training based on an already trained network can significantly reduce the amount of training data required for a well-performing classifier [7]. Even initializing model weights based on a network trained on a dataset that is very distant from the target task can improve a classifier's performance compared to random initialization [19]. Recently, Iman et al. [7] extended previous studies by reviewing recent work on the application of different TL approaches for various use cases. Model-based TL approaches that reuse pre-trained network layers with different combinations of freezing and fine-tuning of model weights were identified as the most commonly used, efficient, and effective TL techniques.

In our work, we therefore focus on model-based TL approaches. However, in contrast to reducing the training costs by TL, we aim to utilize the interconnection between the DGA binary and the multiclass classification task to encourage the classifier to extract more complex features and thereby improve the classification performance. In Sect. 4, we present the details of our approach.

3 Evaluation Setup

3.1 Data

For training a supervised ML classifier for DGA detection, both benign domain names and domains labeled as generated by a DGA are required. In our work, we focus on capturing real network traffic for benign data to avoid potential biases that can arise from using artificially generated data. Related work (e.g., [1, 4,6,14]) suggests focusing on non-resolving DNS traffic (NX-traffic) for DGA detection. The reason for this is that the volume of NX-traffic is smaller and therefore easier to monitor, NXDs are less privacy-critical compared to resolving domains, and botnet activity can be detected before the next command is sent to the bots. Hence, in our work, we follow the suggestions of related work and focus on NX-traffic. Below we briefly describe our data sources:

DGArchive: To obtain DGA-labeled domains, we use the open source intelligence (OSINT) feed of DGArchive [12]. At the time of our study, DGArchive is the OSINT feed that incorporates the most DGAs and contains approximately 199 million unique samples generated by 115 different DGAs.

University Network: We obtain benign labeled data from the central DNS resolver of the RWTH Aachen University campus network which incorporates several administrative and academic networks, student residence networks, and the network of the affiliated university hospital. For our evaluation, we obtain a one-month recording of NXDs from mid-October 2017 to mid-November 2017 that contains approximately 35 million unique NXDs. Note, we deliberately selected older NX-traffic data for our evaluation. In this way, and by additionally leveraging data from a different enterprise network, we can analyze whether DGA classifiers generalize well to new environments and are consistent over time. In Sect. 6, we show that the novel classifiers perform extremely well even when there is about 17 months between the time of training and the test data, demonstrating that the limitations of using older NX-traffic data are minimal.

3.2 Models

In this work, we propose four novel DGA classification models based on the ResNeXt [18], ConvNeXt [8], Transformer [16], and RWKV [11] architectures. For each model architecture, we propose a binary classifier, which we denote with a preceding *B-*, and a multiclass classifier, which is denoted with a preceding *M-*. We use the state-of-the-art ResNet-based models [6] to compare the classification performance. For the sake of brevity, we refer the reader to the accompanied source code[1] for details of the individual models. In the following, we introduce the data preprocessing required to train and test DGA detection models.

Input Preprocessing: In order for domain names to be processed by neural networks, they must first be translated into a numerical representation. We start processing the data by filtering all benign labeled domains against DGArchive to remove potentially included malicious domains before training and testing a classifier. Then, we follow the methodology proposed in [5] for a bias-reduced classification setting to mitigate biases in novel model architectures. Here, we replace each unique character with a corresponding integer for each domain name. For multiclass classification, we pad every encoded fully qualified domain

[1] https://gitlab.com/rwth-itsec/dga-transfer-learning-based-training.

name (FQDN) to the maximum domain length of 253 characters with zeros from the left side. For binary classification, we first extract the encoded effective second-level domain (e2LD) and similarly pad it to the maximum label size of 63. In the case of a malicious domain, the e2LD corresponds to the part of the domain that must be registered by an adversary under any TLD. By focusing on the e2LD, all data that can be freely chosen by the adversary is ignored, effectively eliminating known biases [5].

3.3 Datasets

To reproduce the results of related work and compare the different models, we create two types of evaluation datasets, one for each of the classification tasks ($DS_{B\text{-}eval}$ and $DS_{M\text{-}eval}$). Both dataset types contain benign and malicious labeled samples. Focusing only on the e2LD for binary classification drastically reduces the number of available benign samples from the university network from 35 million unique FQDNs to approximately 125k e2LDs. To create a balanced dataset for $\mathrm{DS}_{\mathrm{B\text{-}eval}}$, we randomly select at most 1,450 e2LDs for each DGA in DGArchive and combine them with the benign e2LDs from the university network. We repeat this process 20 times using different seeds for the random sampling, resulting in 20 $\mathrm{DS}_{\mathrm{B\text{-}eval}}$ datasets, each containing approximately 250k e2LDs. For $\mathrm{DS}_{\mathrm{M\text{-}eval}}$, we include up to 10,000 random FQDNs for each DGA and the benign class, resulting in a dataset of 731k samples. Note, similar to related work, we use balanced datasets to reproduce their results. However, in Sect. 6, we approximate the real distribution of benign and malicious samples by classifying all benign NXDs that occurred in an enterprise network in a month, and only the malicious domains that were actually generated by DGAs in that specific time span. Thereby, the malicious samples only account for 0.31% of all classified domains, and we effectively mitigate spatial experimental bias.

3.4 Hyperparameter Optimization

Similarly to the evaluation datasets, we create two datasets to optimize the novel models. We optimize the hyperparameters of the four proposed binary and the four multiclass models using random search with $n = 60$ trials. 60 trials are often practically sufficient because the maximum of 60 random observations lies within the upper 5% of the true maximum with a probability of 95% if the near-optimal region of the hyperparameters occupies at least 5% of the total optimization surface. To significantly reduce the total optimization space to the most promising regions, we align the hyperparameter ranges of the CNN-based models with the state-of-the-art ResNet classifiers. The hyperparameters ranges for the Transformer and RWKV-based models were chosen so that they roughly correspond to the parameter complexity of their CNN counterparts.

3.5 Ethics

During our study, we ensure that we do not collect any personally identifiable information. We only monitor NX-traffic and store only the queried domains when collecting data from real-world networks. In this process, we omit all other data. As a result, we only get a list of domain names that occurred during the recording period, with no reference to users within the networks. Moreover, we focus on NX-traffic, as NXDs are less privacy sensitive compared to resolving domain names. This is because they usually do not contain user-typed domains, with the exception of typo domains.

4 Transfer Learning-Based Training for DGA Detection

One of our main contributions is the proposal of a new training scheme for DGA detection that leverages transfer learning (TL) and fine-tuning to significantly improve the classification performance of DGA binary classifiers. First, we propose to train a multiclass model conventionally on a binary balanced e2LD dataset such as $\mathrm{DS}_{\text{B-eval}}$. Here, the labels for the domains generated by the specific DGAs must be retained for training. Then, the last layer (the multiclass classification layer) must be replaced by a binary classification layer. Finally, we propose to fine-tune the model using the same dataset, but now with binary class labels and a 100 times smaller learning rate (1×10^{-5}). Note that the direct use of the initially trained multiclass model for binary classification leads to a poorer classification performance compared to a conventionally trained binary classifier. Moreover, freezing the weights of the multiclass model before fine-tuning or a learning rate that is too high also leads to poorer classification performance.

The main intuition behind this approach is to encourage the classifier to learn to extract more complex features by providing it with additional information about the origin of the domains. In the following, we show that the additional knowledge learned through this approach can be transferred to binary models to improve classification performance in a statistically significant way.

5 Evaluation

In this section, we use the datasets described in Sect. 3.3 and follow the methodology proposed in [5] to comparatively evaluate the classifiers' performances in a bias-reduced classification setting.

In Table 1, we report the averaged results of 20 repetitions of a five-fold cross-validation. In addition, we report the binary classification results for the RWKV model (which performs best on average) and the Transformer model (which performs best at lower FPR ranges) trained with our newly proposed training scheme, and label these classifiers with a trailing. *TL*. In Table 2, we present the averaged multiclass classification results of the five-fold cross-validation.

B-RWKV achieves the best results among the conventionally trained binary classifiers, while M-ResNeXt achieves the best results for multiclass classification. TL-based training has only a minor influence on the average classification performance of the Transformer model. On the other hand, the RWKV model benefits greatly from TL-based training, which improves the TPR by 0.55% at

Table 1. Binary Results

Model	TL	ACC	TPR	FPR
B-ResNet [5,6]	✗	0.8899	0.8951	0.1152
B-ConvNeXt	✗	0.8960	0.9011	0.1091
B-ResNeXt	✗	0.8954	0.9039	0.1131
B-RWKV	✗	0.8978	0.9070	0.1114
B-RWKV.TL	✓	**0.9088**	**0.9125**	**0.0949**
B-Transformer	✗	0.8940	0.8939	0.1059
B-Transformer.TL	✓	0.8948	0.8971	0.1076

Table 2. Multiclass Results

Model	F1-Score	Precision	Recall
M-ResNet [5,6]	0.7952	0.8123	0.8054
M-ConvNeXt	0.7935	0.8128	0.8029
M-ResNeXt	**0.8022**	**0.8218**	**0.8070**
M-RWKV	0.7965	0.8140	0.7963
M-Transformer	0.7904	0.8108	0.7984

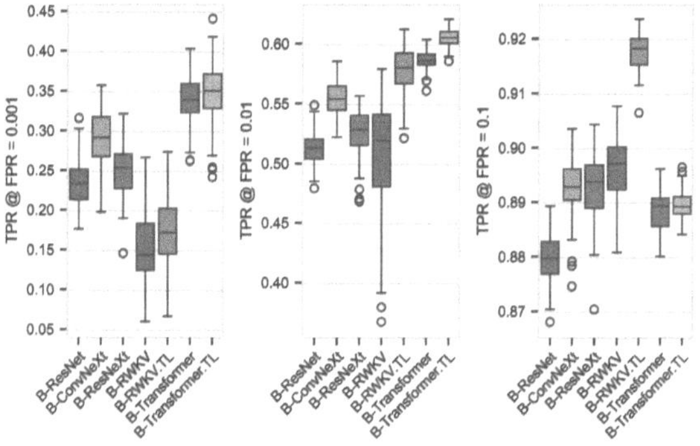

Fig. 1. Box Plots of Statistical Analysis

Table 3. Statistical Analysis of TPRs at three different fixed FPRs

Model	TL	FPR = 0.001			FPR = 0.01			FPR = 0.1		
		TPR	*p*	*P*	TPR	*p*	*P*	TPR	*p*	*P*
B-ResNet [5,6]	✗	0.2334	-	-	0.5122	-	-	0.8797	-	-
B-ConvNeXt	✗	0.2912	**0.0008**	**0.9957**	0.5539	**0.0000**	**1.0000**	0.8926	**0.0000**	0.8201
B-ResNeXt	✗	0.2503	0.1571	0.6596	0.5255	0.1192	0.6147	0.8929	**0.0002**	0.8145
B-RWKV	✗	0.1524	0.9995	0.0001	0.5075	0.5771	0.2716	0.8959	**0.0000**	**0.9638**
B-RWKV.TL	✓	0.1727	0.9941	0.0018	0.5783	**0.0000**	**1.0000**	**0.9176**	**0.0000**	**1.0000**
B-Transformer	✗	0.3385	**0.0000**	**1.0000**	0.5863	**0.0000**	**1.0000**	0.8883	**0.0003**	0.2868
B-Transformer.TL	✓	**0.3467**	**0.0000**	**1.0000**	**0.6050**	**0.0000**	**1.0000**	0.8893	**0.0000**	0.4400

a 1.65% lower FPR. Compared to the state of the art, B-RWKV.TL improves the TPR by over 1.7% and the FPR by over 2%.

In the following, we analyze the achieved improvements in classification performance for statistical significance. Here, we focus on binary classification, as this is where the improvement in detection performance has the greatest practical benefit. To this end, we first compute the averaged TPRs for the three fixed FPRs of [0.001, 0.01, 0.1]. Note that the highest FPR value of 0.1 is only used for reasons of comparability with the state of the art. Then, we perform a frequentist correlated t-test [9], which takes into account the correlation due to the overlapping training sets within the cross-validation. In Table 3, we report the averaged TPRs together with the *p*-values for each model which reflect the degree of data compatibility with the null hypothesis. We mark those entries in bold that are below the strict significance threshold $p \leq 0.005$, which indicates a significantly better model compared to the state-of-the-art model B-ResNet. In addition, we perform a Bayesian correlated t-test [2] to mitigate the known problems of the null hypothesis significance test for the comparison of classifiers. Here, we use a rope of 0.01 to specify the region of practical equivalence (i.e., we consider two classifiers equivalent if the difference in their performance is smaller than 1%). We present the probability P that a model performs significantly better than the state of the art alongside the *p*-values in Table 3 and highlight all probabilities above 90%.

In Fig. 1, we present the box plots for all three fixed FPRs and all models. Considering the medians, all novel models have a greater median than the state of the art, with the exception of the RWKV models at a FPR of 0.001. However, the RWKV models improve with increasing FPR, and at FPR = 0.1, B-RWKV.TL is the best, outperforming the state of the art by 3.8%. Statistically, the results of both t-tests in Table 3 show that the ConvNeXt and Transformer models are significantly better than the state of the art at the FPR values of 0.001 and 0.01. Both RWKV models are statistically better at the highest FPR of 0.1. Moreover, our proposed TL-based training scheme significantly improves the RWKV model at the FPR of 0.01, making it statistically better than B-ResNet. On average, B-RWKV.TL performs best at a FPR of 0.1, while B-Transformer.TL performs best at the two smaller FPRs, outperforming the state of the art by 11.3% at the FPR of 0.001. Across all FPRs, the RWKV model benefits more from TL-based training than the Transformer model.

6 Real-World Study

The classification results obtained by the classifiers and reported in the previous section come from an artificial classification setting, which is mandatory to obtain bias-reduced classifiers. However, these results cannot be transferred to a real-world setting. Moreover, the evaluation performed to reproduce the results of related work is affected by spatial and temporal experimental biases. In the following, we assess the true classification performance of the new models in an experimental bias reduced evaluation and examine whether the classifiers generalize well between different networks and are time-robust. To this end, we make use of a second source of benign data:

Enterprise Network: Several central DNS resolvers of Siemens AG, covering the regions Asia, Europe, and USA, are used as a second source for benign labeled data. As a result, the data obtained from this source is particularly diverse. For our evaluation, we obtain a one-month record of benign NXDs from April 2019, which contains around 373 million NXDs.

Datasets: We create two additional datasets to measure the real-world performance of the novel models. First, we create $DS_{RW\text{-}train}$, which we use to train the various models for our real-world study. Here, we select the 125k benign e2LDs from the university network and the same number of malicious samples available in DGArchive at the end of the benign data recording period (November 2017), stratified across the included DGAs. Up to this point in time, DGArchive contains approximately 53 million unique domains generated by 85 different DGAs. To evaluate the trained models, we create $DS_{RW\text{-}test}$, which contains all 373 million benign NXDs from the enterprise network of the recording period April 2019. Additionally, we include 1.2 million unique malicious domains that were generated by 46 DGAs in April 2019 (four of which were unknown at the time of training). This results in a time difference of around 17 months between the training and test samples.

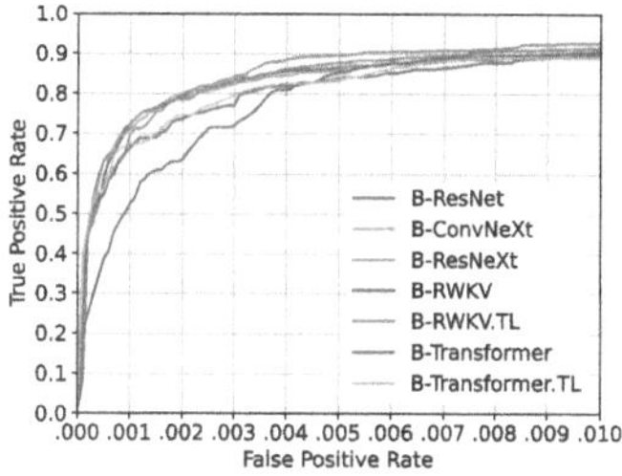

Fig. 2. Real-World Study

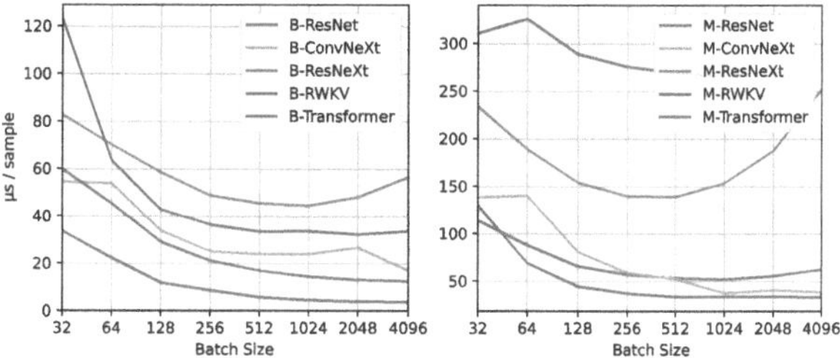

Fig. 3. Inference Times as a Function of Batch Size

Results: We present the receiver operating characteristic (ROC) curves for the FPR range of [0, 0.01] in Fig. 2. In general, the huge improvements in classification performance over the state of the art achieved by the novel models in the artificial evaluation setting do not transfer to the real-world scenario. The CNN-based models (B-ConvNeXt and B-ResNeXt) achieve similar results to B-ResNet, while the ROC curves of the Transformer models are below the curve of the state of the art. This could indicate that the Transformer models are possibly overparameterized and adapt too strongly to the training data distribution of the university network. The B-RWKV model performs the worst. However, when B-RWKV is trained using our proposed TL-based approach, the model behaves similarly to the state of the art up to a FPR of 0.0025. From then on, B-RWKV.TL is the best model overall and even outperforms B-ResNet from a FPR of 0.0033 by 2–3.4%. Although the improvements in classification performance from the artificial scenario do not translate to the real world, the CNN-based models and B-RWKV.TL in particular prove to be robust over time and generalize well between networks.

Training and Inference Time: In Fig. 3, we show the inference times for the different models as a function of batch size using an NVIDIA Tesla V100 GPU.

All binary models are slower compared to the state-of-the-art B-ResNet model. For certain models, an increase in inference time can be measured with increasing batch size, which is due to quantization effects. We account all models as real-time capable since within the enterprise network there are on average 148 NXDs and a maximum of 2471 NXDs queried per second. Thus, even at the worst inference time of the binary classifiers of 123 µs/domain (8130 domains/s), the models would be able to keep up with domain query peaks.

The training times of the detection models are negligible for practical use and are between 70 s and 15 min, depending on the model.

7 Conclusion

In this work, we proposed a novel training method based on TL and fine-tuning, which improves detection performance of DGA binary classifiers. Moreover, we developed and optimized four novel model architectures for DGA binary classification and four additional variants for DGA multiclass classification. We statistically compared the newly developed models in a unified framework with the state of the art and showed that caution is needed when comparing classifiers in an artificial environment, which is necessary for training bias-reduced classifiers, as the classification results obtained may not be transferable to the real world. In our real-world study, we avoided experimental bias and assessed the generalizability and temporal robustness of the novel models. The best model B-RWKV.TL is able to improve the state of the art from a FPR of 0.0033 by 2–3.4%.

Acknowledgments. We thank the Cyber Analysis & Defense department of Fraunhofer FKIE, Siemens AG, and the IT Center of RWTH Aachen University for providing data for our research.

Data Availibility Statement. We make the source code of the ML models and the TL-based training procedure publicly available (See footnote 1) to facilitate future work.

References

1. Antonakakis, M., et al.: From throw-away traffic to bots: detecting the rise of DGA-based malware. In: USENIX (2012)
2. Benavoli, A., et al.: Time for a change: a tutorial for comparing multiple classifiers through Bayesian analysis. JMLR **18**(77), 1–36 (2017)
3. Bilge, L., et al.: Exposure: a passive DNS analysis service to detect and report malicious domains. Trans. Inf. Syst. Secur. **16**(4), 1–28 (2014)
4. Drichel, A., Faerber, N., Meyer, U.: First step towards EXPLAINable DGA multiclass classification. In: ARES. ACM (2021)
5. Drichel, A., Meyer, U.: False sense of security: leveraging XAI to analyze the reasoning and true performance of context-less DGA classifiers. In: RAID (2023)
6. Drichel, A., et al.: Analyzing the real-world applicability of DGA classifiers. In: ARES. ACM (2020)
7. Iman, M., Arabnia, H.R., Rasheed, K.: A review of deep transfer learning and recent advancements. Technologies **11**(2), 40 (2023)
8. Liu, Z., et al.: A ConvNet for the 2020s. In: CVPR (2022)
9. Nadeau, C., Bengio, Y.: Inference for the generalization error. In: Advances in Neural Information Processing Systems, vol. 12. MIT Press (1999)
10. Peck, J., et al.: CharBot: a simple and effective method for evading DGA classifiers. IEEE Access **7**, 91759–91771 (2019)
11. Peng, B., et al.: RWKV: reinventing RNNs for the transformer era. arXiv:2305.13048 (2023)
12. Plohmann, D., et al.: A comprehensive measurement study of domain generating malware. In: USENIX (2016)
13. Schiavoni, S., Maggi, F., Cavallaro, L., Zanero, S.: Phoenix: DGA-based botnet tracking and intelligence. In: Dietrich, S. (ed.) DIMVA 2014. LNCS, vol. 8550, pp. 192–211. Springer, Cham (2014). https://doi.org/10.1007/978-3-319-08509-8_11

14. Schüppen, S., et al.: FANCI: feature-based automated NXDomain classification and intelligence. In: USENIX (2018)
15. Spooren, J., et al.: Detection of algorithmically generated domain names used by botnets: a dual arms race. In: SAC. ACM (2019)
16. Vaswani, A., et al.: Attention is all you need. In: NeurIPS, vol. 30 (2017)
17. Woodbridge, J., et al.: Predicting domain generation algorithms with long short-term memory networks. arXiv:1611.00791 (2016)
18. Xie, S., et al.: Aggregated residual transformations for deep neural networks. In: CVPR (2017)
19. Yosinski, J., et al.: How transferable are features in deep neural networks? In: NIPS. MIT Press (2014)
20. Yu, B., et al.: Character level based detection of DGA domain names. In: IJCNN. IEEE (2018)

Hardware and Firmware Security

Seum Spread: Discerning Security Flaws in IoT Firmware via Call Sequence Semantics

Anis Lounis(✉), Anthony Andreoli, Mourad Debbabi, and Aiman Hanna

Security Research Centre, Concordia University, Montreal, QC H3G 1M8, Canada
{anis.lounis,anthony.andreoli,mourad.debbabi,aiman.hanna}@concordia.ca

Abstract. Internet-of-Things (IoT) devices play a central role in the modern digital landscape by providing uninterrupted convenience, but their history has been marked by the opportunities they provide bad actors desiring to exploit them. Vulnerable firmware remains a major factor in exploiting IoT devices. Persistent storage interfaces, which allow data retention even after power loss, can exacerbate this problem. This work examines and informs on the security impact of persistent storage through the lens of security-critical sanitization bypass flaws, computed attack surface spread, and the presence of command injection vulnerabilities; three facets of firmware code that can influence its susceptibility to exploitation. The assessment is established upon manually annotated call sequences that allow reasoning about path sanitization in order to formulate well-founded observations. We assemble a dataset of 100 IoT firmware images from four well-known vendors and initially find 68 devices whose persistent storage receives attacker-controllable input in over 4800 unique unsanitized paths. Furthermore, we discovered 77 instances of sanitization bypass flaws in 32 devices. In addition, we create the *taint spread* metric to assess the impact of tainted persistent storage on a firmware's attack surface; in one alarming instance we find tainted data to affect over 1500 firmware code locations. Finally, we leverage the modeled call sequences to detect and exploit seven 0-day command injection vulnerabilities in five acquired devices and five assigned PSVs.

Keywords: Internet of Things · Vulnerability Detection · Command Injection · Sanitization Bypass · Attack Surface

1 Introduction

Internet-of-Things (IoT) devices are seamlessly integrated into the modern digital landscape, fostering an era of convenience and interconnectivity in wearables,

A. Lounis and A. Andreoli—The first two authors contributed equally to this work.
The authors gratefully acknowledge the support and collaboration of the Communication Security Establishment (CSE).

F. Maggi et al. (Eds.): DIMVA 2024, LNCS 14828, pp. 395–415, 2024.
https://doi.org/10.1007/978-3-031-64171-8_21

smart homes, public infrastructure, enterprise environments, and industrial control systems. Despite these apparent gains, IoT devices are often moored to security-critical vulnerabilities exploited by global threat actors. Compounding the issue are the projected 29.42 billion devices deployed by 2030 [27], surfacing an unprecedented global vulnerability footprint ripe for exploitation. In 2016, the notorious *Mirai* botnet spread by infecting over 200,000 insecure devices in more than 10 countries, orchestrating mass-scale Dynamic Denial-of-Service attacks [3]. Its most recent V3G4 variant was detected in 2022, weaponizing 13 Remote Code Execution (RCE) vulnerabilities in IoT devices [4]. In October of 2023, an exploit chain of two 0-day vulnerabilities was leveraged against devices running Cisco's IOS XE operating system by compromising a WAN exposed web interface and subsequent privilege escalation ultimately leading to a non-persistent implant affecting over 60,000 devices [1]. Attacks of such magnitude serve as foreboding reminders of the necessity to understand and neutralize threats by recognizing security-susceptible facets of IoT devices.

The primary objective of this work is to investigate and inform on the security of embedded systems by studying the utilization of persistent storage native to these platforms. This is accomplished through the lens of critical sanitization security bypass flaws, the propagation of unsanitized input reflecting the spread of attack surface, and the presence of exploitable command injection vulnerabilities. To address these concerns, we propose an approach for modeling security-relevant call sequences captured through static taint analysis and annotated with manually verified call semantics. These facilitate reasoning about the sanitized nature of tainted paths and lend credibility to the presented findings. Emphasis is placed on command injection vulnerabilities due to their high severity, and potential to serve as a precondition for remote code execution and device takeover [19,34]. We further provide a procedure for flagging security-critical sanitization bypass flaws, and a metric quantifying the effective attack surface marked by unsanitized input spreading across device firmware via persistent storage.

We conduct our study on a dataset of 100 IoT firmware images from four vendors, and discover command injection vulnerabilities reported by our semantic model. We obtain tainted call sequences (paths) and initially observe 4800 unique occurrences of unsanitized input tainting persistent storage over 68 firmware devices. We then find 77 instances of sanitization bypass flaws characteristic of three of our exploited vulnerabilities. We further exploit seven 0-day vulnerabilities for full device takeover and responsibly disclose them to their appropriate vendor; five PSVs were assigned to these paths. Finally, we implement a novel attack surface spread metric, and in one device found it to reveal an increase in attack surface by spreading tainted data to 1750 code locations. Our contributions are:

1. A novel, architecture-agnostic approach for vulnerability identification that leverages annotated security-relevant calls and considers tainted data across persistent storage
2. Seven unique 0-day vulnerabilities evidenced by internally developed Proof-of-Concept (PoC) exploits subsequently disclosed to respective vendors.

3. A procedure that identifies critical instances of sanitization bypass flaws based on sanitization discrepancy analysis.
4. Attack surface spread (also referred to as taint spread) – a novel metric tailored to IoT systems that quantifies device attack surface in terms of the propagation of unsanitized input spread through persistent storage to firmware code locations.
5. A annotated dataset of security-relevant calls with precise code locations and semantics serving as groundtruth for further research endeavors[1].

The remainder of the paper is organized as follows. Section 2 covers the IoT firmware environment and security-relevant calls in relation to taint-based vulnerabilities. Section 3 depicts our approach for modeling mishandled input leading to command injection vulnerabilities, Sect. 4 defines sanitization discrepancy analysis and the taint spread metric, Sect. 5 describes the discovery and exploitation of 0-day vulnerabilities, remarks on sanitization bypass, and elaborates on the applicability of spread. Section 6 describes the current related work in IoT vulnerability discovery, and finally Sect. 7 summarizes our findings and expands on future avenues of interest.

2 Background

This section explains the IoT firmware environment targeted in this work, defines taint-based vulnerabilities, elaborates on the semantics categories of security-relevant calls, and details the toolset leveraged for capturing dataflow paths.

2.1 Expected IoT Firmware Environment

Firmware refers to the software running on embedded systems. It includes both the operating system and the suite of services controlling device functionality, typically delivered with binary executables and pre-compiled libraries for interacting with hardware modules. This work assumes two environmental preconditions; an embedded web application as the chosen attack vector, as well as reliance on persistent storage.

Embedded Web Application. IoT firmware is often released with embedded web application(s), and one such example is the administrative interface provided for device configuration [11] running atop highly privileged server processes [20]. Given such properties, applications constitute an ideal attack vector addressed by prior work on vulnerability detection, either through static analysis [7–9,22], or fuzzing emulated firmware [5,10,13,15,16,21,33].

Persistent Storage. Non-Volatile Random Access Memory (NVRAM) is native to IoT devices and used for storing device configurations [32]. `Libnvram` is the de-facto library for accessing NVRAM in embedded systems through key-value storage getters and setters. For example, `nvram_set("key", "value")`

[1] Dataset available through email correspondence with first two listed authors.

writes "value" to a location indexed by "key", whereas nvram_get("key") returns this value. Dependency on Libnvram is evidenced via the native support by modern firmware emulators providing a mocked version to improve emulation success [5,13,15,16,32]. Prior work has leveraged Libnvram for tainting data across binary boundaries within firmware [7]. Our work includes Libconfig within the persistent storage model as it provides an identical programming interface to Libnvram through config_{get,set} functions.

2.2 Taint-Based Vulnerabilities

The concept of *taint* refers to tracking the propagation of unsafe data across the system under inspection and taint-based vulnerabilities arise when such data reaches security-critical operations [8,9,31].

Command Injections. Such vulnerabilities occur when unsanitized data reaches the command line interface allowing execution of arbitrary system commands [25]. Focus on command injections in embedded systems is motivated by their high severity and low attack complexity since their exploitation often meets no further inherent protection, in contrast to memory corruption RCEs susceptible to neutralization via Data Execution Prevention (DEP) or Address Space Layout Randomization (ASLR).

Proper Input Handling. Industry standards emphasize input sanitization to ensure only safe values are processed to avoid bugs and security flaws. OWASP provides comprehensive secure development guidelines, and prescribes two forms of input handling: *input validation* and *data escaping* [14]. The former enforces syntactic and semantic correctness of data, while the latter is a data representation requirement that prevents misinterpretation. Proper input handling thus ensures user input is treated strictly as data and avoids unintended interpretations such as executable code.

2.3 Binary Query System

Our work relies on the binary code query system proposed by the authors of [2] for specifying and capturing dataflow paths of interest. It is used throughout this work to express needed primitives such as `reach(s, f)` that indicates whether string s connects to any parameter of function f through data-flow. The system was chosen as it contains features necessary to this work, including: (a) architecture agnostic data-flow analysis, (b) data-flow pattern matching capabilities, (c) a rich syntax for defining patterns of interest, and (d) a set of built-in memory transfer summaries (e.g., `sprintf`). For these reasons, the paths specified in this work are captured through queries.

3 Vulnerability Modeling

In this section, we define the security-relevant semantic categories that constitute the building blocks for our modelled command injection vulnerabilities.

3.1 Security-Relevant Call Semantics

We abstract operations encountered by tainted data at the call level by annotating them with semantic categories. This technique allows for a two-fold advantage: (a) it simplifies path constraint concerns, often requiring costly symbolic execution [23,24,26,28], and (b) it enables a high-level interpretation of call sequences facilitating a multitude of security analyses outcomes, such as detecting vulnerabilities, recognizing flaws, and gauging attack surface; the threefold purpose of our work. Below, we describe the four semantic categories:

1. **Storage Accessors** correspond to *getters* or *setters* of persistent storage that read/write storage locations referred to by key k (e.g., `nvram_{get,set}`).
2. **Taint Sources** fetch user-controllable data within web server requests (e.g., `get_cgi("user_ssid")`).
3. **Command Executors** provide the ability to execute system commands through string parameters. (e.g., `system`, `popen`).
4. **Sanitizers** include *input validators* and *data escapers* (see Sect. 2.2).

In what follows, we elaborate on the usage of the above semantic categories to construct causal components which are necessary subpaths to the assembly of transitive vulnerabilities.

3.2 Transitive Vulnerabilities

We model transitive vulnerability paths as depicted in Fig. 1. Each box represents a call's semantic category, and blue arrows indicate the direction of tainted data between them. Paths are thus composed of three connected *causal components* defined in terms of our semantic categories:

1. **Source Path**: Bookended by a taint source and storage setter.
2. **Storage Transition**: Calls that transfer data between storage locations.
3. **Sink Path**: Bookended by a storage getter and command executor.

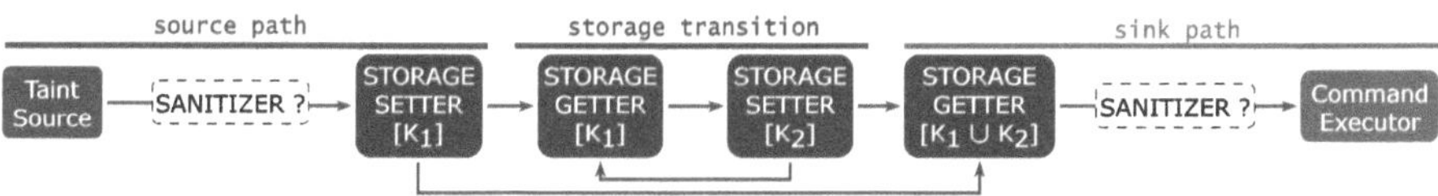

Fig. 1. Transitive Vulnerability Paths

This work holds a conservative definition of taint propagation to quantify an upper-bound of its spread across binary boundaries over persistent storage t (e.g., `Libconfig`) by assuming storage key k will load value v if v was written to k at any execution point; $\texttt{storage\_setter}_t\texttt{(k, v)} \rightarrow \texttt{(storage\_getter}_t\texttt{(k)=v)}$.

3.3 Path Definition

Path Assumptions This model makes two fundamental assumptions: (1) the presence of an annotated sanitizer is sufficient to mark a call sequence as sanitized; we do not distinguish incorrectly implemented sanitizers (false negative paths may thus result, but sanitizer correctness is out of scope), and (2) sanitizers are likely only imposed at source and sink paths as widely accepted in practice [14,18], while storage transitions are presupposed unsanitized.

The BNF grammar below defines the possible sequence of semantic categories that produce causal components, where *sanitized* specifies the grammar for a sanitized causal component.

$$\begin{aligned} source\_path &::= taint\_source \;\; sanitizer^{*} \;\; storage\_setter \\ storage\_transition &::= storage\_getter \;\; storage\_setter \\ sink\_path &::= storage\_getter \;\; sanitizer^{*} \;\; command\_executor \\ sanitized &::= taint\_source \;\; sanitizer^{+} \;\; storage\_setter \mid \\ &\qquad storage\_getter \;\; sanitizer^{+} \;\; command\_executor \end{aligned}$$

We define the predicate `isSani(cc)` to evaluate whether the annotated call sequence of causal component `cc` adheres to the *sanitized* grammar. As an example, it evaluates to true if `cc` consists of call sequence ⟨`getUserInput`, `isValidIp`, `nvram_set`⟩ whose annotated counterpart is ⟨`taint_source`, `sanitizer`, `storage_setter`⟩. A transitive vulnerability candidate is retained if and only if both its source and sink causal components are unsanitized.

4 Approach

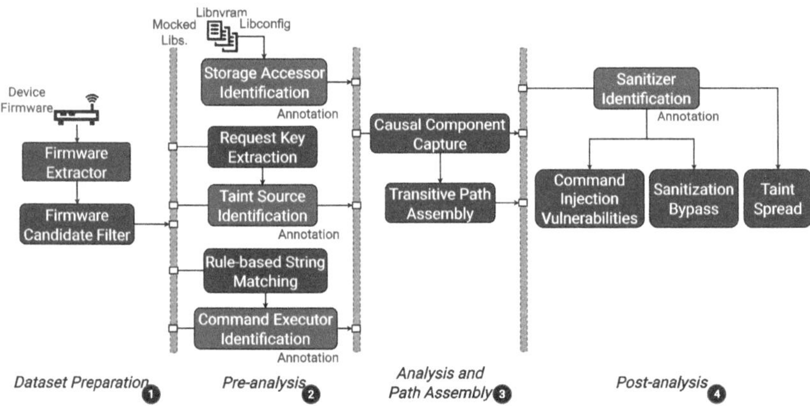

Fig. 2. Diagram of components composing our approach

Our approach is divided into four distinct phases depicted in Fig. 2. In dataset preparation (1), firmware are collected and unpacked using the FACT extraction tool[2]. The pre-analysis phase (2) produces annotated calls in three of the

[2] https://github.com/fkie-cad/fact_extractor.

four semantic categories, namely storage accessors, taint sources, and command executors. The analysis and path assembly phase (3) identifies causal components and fully assembles transitive vulnerability paths. In post-analysis (4), unlabelled calls are manually annotated with the fourth semantic category (i.e., sanitizer), completing the process of defining the semantics of each call captured within causal components. Fully annotated components are now easily interpreted through the `isSani(...)` predicate backing security-relevant findings such as exploitable vulnerabilities, critical sanitization bypass flaws and the taint spread metric.

4.1 Dataset Preparation

The dataset is prepared by collecting 158 firmware from four vendors, namely TP-Link, Netgear, ASUS, and Linksys, selected by their prevalence in contemporary IoT vulnerability works [7,22,30]. Products include WiFi range extenders, powerline adapters, mesh systems, access points, and conventional routers. Only 100 of these firmware depend on persistent storage and compose the final dataset.

4.2 Pre-analysis

The pre-analysis phase is responsible for annotating three of the four semantic categories of security-relevant calls required for analysis, consisting of: storage accessors, taint sources, and command executors. The benefit of manual annotation is its certainty; for example, annotated taint sources assure data is in-fact user-controllable, reducing false-positive results.

Annotating Storage Accessors. In addition to native library exports from `Libnvram` and `Libconfig`, embedded firmware employs other functions (e.g., `acosNvramConfig_{get,set}`). To cover alternative storage accessors, we consult mocked libraries from firmware emulators [5,16,33]; the final set includes 87 annotated storage accessors.

Annotating Taint Sources. We are inspired by the approach taken by SaTC for identifying taint sources through shared request keys between backend binaries and frontend files [7]. Our approach retains a reduced set of keys, prevents false-positive taint sources through the removal of ambiguous keys, and confirms the selection of taint sources through manual inspection.

We parse HTML `input` elements and URL query parameters within firmware web files for request keys forming the set K. A function f is deemed a candidate taint source in a backend binary if a key $k \in K$ reaches an argument of f (i.e., `reaches(k, f)`), suggesting that the function may extract a parameter value from a request object. We observe keys in K also reach persistent storage accessors denoted A. To avoid ambiguous keys, we retain functions without such overlap. The set of taint sources T for an inspected binary containing functions F is denoted by Eq. 1:

$$T = \{f \in F : \exists k \in K\,(reaches(k,f) \wedge \forall a \in A\ \neg reaches(k,a))\} \quad (1)$$

For all firmware inspected through this approach, we produce a final set of 116 annotated taint sources.

Annotating Command Executors. We employ dataflow and rule-based string matching to find custom command executors to ensure thorough path coverage. The string predicate `isLikelyCmd(s)` evaluates to true if s is a command-line string according to string matching on Linux commands (e.g., `"ifconfig"`), binary directories (e.g., `"/usr/sbin/"`), and recognizable shell sequences (e.g., common redirections like `"2>/dev/null"`). A function f is a candidate command executor if reached by such a string. Each binary is analyzed and the set of retained candidates E is computed using Eq. 2.

$$E = \{f \in F : \exists s \in S\ isLikelyCmd(s) \wedge reaches(s,f)\} \quad (2)$$

We retain 194 annotated command executor functions after manual inspection, including `_exec_cmd`, `xsystem`, and `do_command` to name a few.

4.3 Analysis and Path Assembly

Three separate queries are written to capture 1) source paths, 2) transitive paths, and 3) sink paths. Paths captured by the query system involve *may* semantics, in that the sequence of calls tainted from source to sink may be called based on control flow branching. Our work is concerned with the set of functions that *must* be called on such a path as it provides stronger guarantees for whether a path must be sanitized prior to reaching a sink. The piecewise function in Eq. 3 demonstrates the worklist algorithm over the set of basic blocks containing tainted call locations to construct *must* semantic paths. At each call to f, basic block b is instantiated to the top-of-queue element $dequeue(W)$, where W is a queue of basic blocks initialized to the reverse postorder traversal of the control-flow graph.

$$f(W) = \begin{cases} out_b = \left(\bigcap_{p \in pred_b} out_p\right) \cup calls_t(b) & |W| > 0 \\ f(enqueue_b(W, succ_b)) & \end{cases} \quad (3)$$

Operation $calls_t(b)$ returns the set of calls made in b that receive tainted input, and out_b is the set of calls propagated forward to successor block $succ_b$ computed as the meet "$\bigcap$" of the calls propagated by predecessor blocks $pred_b$. The special operator $enqueue_b(W, succ_b)$ adds $succ_b$ to W if and only if out_b differs from its prior computed set $out_{b'}$. Following the computation of $f(W)$ for a particular path under analysis, each basic block is associated with a set of calls that *must* be invoked by its predecessors and such are the calls that compose the causal components inspected in our work.

4.4 Post-analysis

This phase involves annotating sanitizers from the set of unlabelled functions found in causal components assembled in the prior phase. Transitive vulnerability paths can then be credibly filtered according to their fully annotated call sequences via the `isSani` predicate. In addition, we discover sanitization bypass flaws and compute the taint spread metric to assess the growth of device attack surface owing to tainted persistent storage.

Annotating Sanitizers. Sanitization is the final semantic category that must be annotated to: (a) validate vulnerability candidates, (b) capture instances of critical security flaws, and (c) measure taint spread. Annotation is conducted by manually inspecting calls within the collected causal components that have not been annotated as either taint sources, storage accessors, or command executors. We find 203 unlabelled calls and of these, 113 are annotated as sanitizers. Manual annotation in this study took a total of five work-hours to complete over the *entire dataset*. The time taken is short relative to state-of-the-art works in automatic static detection of IoT vulnerabilities that take an average of 1 h 7 min–8 h 25 min *per firmware sample* [12].

As an aside, if the sole interest is to validate vulnerability candidates as opposed to analyzing sanitization discrepancies and computing taint spread, only the causal components that contribute to assembling a full transitive vulnerability are considered; thus the number of components that might contain sanitizers are reduced. This yields a 10-fold reduction in vetting efforts as 26 calls would be retained (vs. 203), of which 17 would be annotated as sanitizers (vs. 113).

Validating Command Injection Vulnerabilities. With fully annotated call sequences, the constructed transitive paths $TransPaths$ are retained in R if neither their source ($source\_cc$) nor sink ($sink\_cc$) causal components are sanitized:

$$R = \{\, p \in TransPaths \,:\, \neg\,(isSani(source\_cc(p)) \vee isSani(sink\_cc(p)))\,\} \quad (4)$$

Vulnerability results and underlying discussion are detailed in Sect. 5.

Identifying Sanitization Bypass Flaws. The mere presence of unsanitized input reaching a storage location k in a source path does not inherently pose a security threat, as its usage may not involve operations with security implications. However, an intriguing scenario arises when sanitized input from path p_1 is directed to k that is elsewhere written by an unsanitized path p_2 and signifies a potential sanitization bypass. This flaw reveals the presence of sanitization discrepancy about key k. Such instances effectively compromise the guarantees that would have otherwise been provided by the sanitized path(s), and a real-world example from our dataset is expounded in Sect. 5.3. Our proposed technique for identifying sanitization bypass flaws through *sanitization discrepancy analysis*, where the key k is the $setter\_key$ of some source path p from the set of $SourcePaths$, is described by:

$$\begin{aligned} D = \{\, k = setter\_key(p_1) \,:\, p_1, p_2 \in SourcePaths \wedge \\ k = setter\_key(p_2) \,\wedge\, isSani(p_1) \wedge \neg isSani(p_2)\,\} \end{aligned} \quad (5)$$

Attack Surface Spread Metric. Although the absence of sanitization does not necessarily pose a threat, measuring its system-wide propagation to code locations dependent on persistent storage is one proactive approach to quantifying the attack surface. This metric warrants monitoring to ensure the system is inherently resilient against potential vulnerabilities, and can help prioritize patching to reduce the attack surface spread as elaborated in Sect. 5.4.

The instantiation of spread in this work is tied to persistent storage native to IoT firmware, and is designed upon four concepts:

- **Storage Key**: A key directly or transitively affected by unsanitized input.
- **First-contact (fc) Key**: The storage key directly receiving unsanitized input in a source path.
- **Code Location**: The instruction address where tainted storage is loaded.
- **Storage Transition**: Taint transfer between two tainted storage keys.

The metric measures the number of code locations loading from tainted storage keys emanating from a first-contact key. Its value reports at two levels of resolution: (1) at the key level as the number of impacted code locations reached by an fc key, and (2) at the firmware level as the number of impacted code locations over the entire system.

Spread is computed through a reachability graph G where nodes represent storage keys, and edges connect storage keys (k_1, k_2) if a storage transition exists where tainted data is transferred from k_1 to k_2. A sample spread graph for an anonymized device is displayed in Fig. 3.

$$\begin{aligned} G = \{(fc) : p \in SourcePaths \wedge \neg\, isSani(p) \wedge fc = setter\_key(p)\} \cup \\ \{(k_1, k_2) : p \in StorageTransitions \wedge \neg\, isSani(p) \wedge \\ k_1 = getter\_key(p) \wedge k_2 = setter\_key(p)\} \end{aligned} \tag{6}$$

Each node maintains a set of code locations where its representative key is loaded, provided by $locs_k$. The spread induced by a first-contact key k is computed by capturing the unique code locations $rlocs_k$ where its tainted data may propagate in terms of recursively reachable descendant keys $rdesc_k$ in G. The spread *value* of k is $spread_k$. The total spread of tainted data propagated from all first-contact keys FC in a firmware image is $rlocs$. The total spread *value* is $spread$.

$$\begin{aligned} rlocs_k &= \bigcup_{v \in rdesc_k} locs_v;\; spread_k = |rlocs_k| \\ rlocs &= \bigcup_{k \in FC} rlocs_k;\; spread = |rlocs| \end{aligned} \tag{7}$$

It is worth noting that merely sanitizing a first-contact key k_1 might not suffice to reduce the spread if $k_2 \in FC$ spreads to overlapping code locations. It is thus useful to compute the spread dependencies $sdeps_{k_1}$ as the set of first-contact keys that must also be sanitized so that the spread of k_1 may be neutralized

$$sdeps_{k_1} = \{\, k_2 \in FC : (k_1 \neq k_2) \wedge rlocs_{k_1} \cap rlocs_{k_2} \neq \varnothing \,\}. \tag{8}$$

In Fig. 3, the spread dependencies of `wla_ssid` are the following: {`wla_ssid_2`, `wla_temp_ssid_2`, `wla_temp_ssid`}.

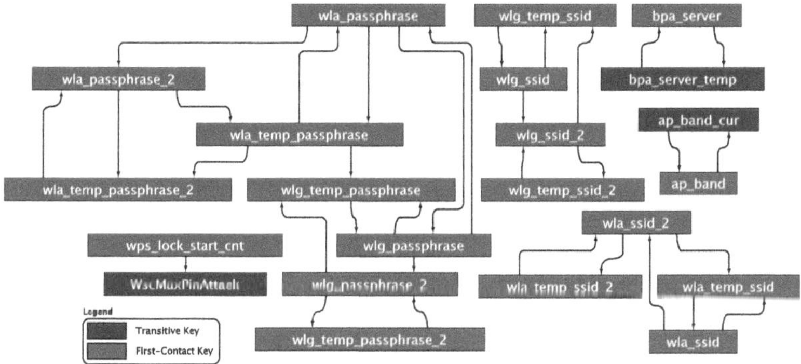

Fig. 3. Spread Subgraph from Sample Firmware

5 Results

In what follows we showcase how annotating sanitizer semantics shortlists candidate vulnerabilities, a subset of which are selected for exploitation. We then report sanitization bypass flaws discovered through sanitization discrepancies, and conclude by providing sample measurements of the proposed spread metric along with a pertinent discussion of its use-cases.

5.1 Assembled Vulnerability Paths

```
// Usage of strcmp in Validation
char *user_input = get_cgi("bandwidth_radio_choice");
// Compare for radio button value sent from the front-end
if (strcmp("2.4Ghz", user_input) == 0 || strcmp("5Ghz", user_input) == 0)
    nvram_set("bandwidth", user_input); // <—Safe

// Usage of strcmp in Non-Validation
char *user_input = get_cgi("user_ssid");
char *prior_ssid = nvram_get("ssid");
// Verify new SSID
if (strcmp(user_input, prior_ssid) != 0) {
    nvram_set("ssid", user_input); // <—Unsafe
    hdl_ssid_change();
}
```

Listing 1.1. Various usages of `strcmp`

The filtering of assembled vulnerability paths to retain candidates for exploitation is conducted as described in Sect. 4.4 with a noteworthy observation;

while most sanitizers exhibit consistent usage, some ambiguously serve both sanitization and non-sanitization purposes. For instance, `strcmp` may be purposed for input validation *or* non-security logic.

In the code of Listing 1.1, we observe `strcmp` in validation, as its usage verifies user input *must* equal a white-listed value, sanitizing it prior to storage in NVRAM. In the follow-up example, `strcmp` is used to verify user input *must not* equal to some value, allowing unsanitized values to reach NVRAM. These two scenarios produce the noted ambiguity. We thus delineate the two by categorizing *certain* sanitizers as those purely involved in sanitization, while the remaining (e.g., `strcmp`) are marked *ambiguous*. The effect of filtering causal components according to both categories is depicted in Fig. 4.

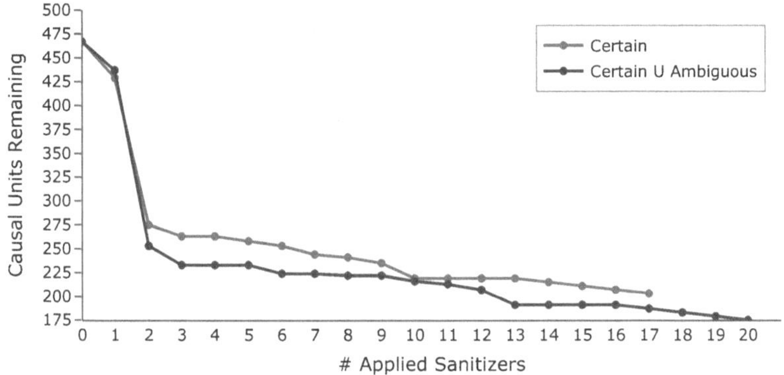

Fig. 4. Comparative view of causal components filtered by varying the inclusion of ambiguous calls as sanitizers over all firmware images

By including ambiguous calls as part of the sanitization set, 28 additional causal components are removed in comparison to the *certain* sanitizer category. This is a trade-off decision in filtering sanitized paths, and is a function of the type of certainty aimed for; ambiguous calls considered sanitizers yields sound-leaning unsanitized paths, while assuming their presence has no sanitization effect yields complete-leaning unsanitized paths. We also observe the sanitizer `inet_addr`[3] to filter 35.8% of all causal components involved in vulnerability candidates from 429 to 275. From this we draw two important takeaways: (1) knowledge of a single manually vetted sanitizer can significantly reduce the number of paths to inspect, and (2) sanitizers might have varied influence on the filtration process based on their relevance to the system context (e.g., verifying IP addresses in IoT devices and would filter more paths than email validation).

This work takes the more *complete* leaning approach and considers ambiguous functions as non-validators since an additional 28 causal components has relatively insignificant implications on validation efforts. Thus the comparative view

[3] Converts human-readable IPv4 string into an address structure.

Table 1. Comparative view of vuln. candidates (trans. paths) and the unique causal units to be validated before and after the sanitizer-based filtration.

		Before Filtering				After Filtering			Units Avoided	
Vendor	Model	Binaries	Causal Units	Trans. paths	Sanit.	Binaries	Causal Units	Trans. paths	#	%
linksys	e1200	2	2	1	1	0	0	0	2	100%
	e900	2	2	1	1	0	0	0	2	100%
netgear	eax80	5	11	14	1	0	0	0	11	100%
	ex3700	2	31	36	1	2	22	23	9	29%
	ex6100	2	23	29	1	2	15	17	8	35%
	ex6120	2	31	36	1	2	22	23	9	29%
	ex6150	2	23	40	1	2	15	28	8	35%
	ex7000	5	22	37	2	0	0	0	22	100%
	ex7500	4	9	10	1	0	0	0	9	100%
	lax20	4	4	4	3	0	0	0	4	100%
	ms60	2	4	2	2	0	0	0	4	100%
	nbk752	9	44	66	2	2	19	17	25	57%
	r6120	1	4	3	0	1	4	3	0	0%
	r6220	2	3	2	0	2	3	2	0	0%
	r6350	1	4	3	0	1	4	3	0	0%
	r6400	5	38	38	4	4	17	22	21	55%
	r6850	1	4	3	0	1	4	3	0	0%
	r7000	7	53	103	3	3	16	21	37	70%
	r7000p	7	74	131	2	4	40	43	34	46%
	r8000	3	21	18	3	2	6	9	15	71%
	r8000p	2	4	2	1	0	0	0	4	100%
	rax20	5	7	7	3	2	2	1	5	71%
	rax200	2	4	2	1	0	0	0	4	100%
	rax43	5	7	7	3	2	2	1	5	71%
	rax50	5	7	7	3	2	2	1	5	71%
	rax80	2	4	2	1	0	0	0	4	100%
	raxe500	4	10	7	1	2	6	3	4	40%
	xr1000	6	17	13	5	2	4	2	13	76%
asus	*	-	-	-	-	-	-	-	-	-
tp-link	*	-	-	-	-	-	-	-	-	-

of Table 1 reports the effect of filtering causal components solely through *certain* sanitizers. Sanitization filtered 100% of transitive vulnerability candidates in 10 of the 28 devices, leaving 18 for selection, validation, and exploitation. The flexibility provided by annotated semantic calls is visible here, whereby alternating between desired soundness does not require a complete re-run of our taint analysis, but merely requires re-applying the predicate `isSani(...)` on already-captured paths after making appropriate modifications to call annotations.

5.2 Validated Command Injections

To properly confirm the applicability of our semantic call sequence-based model for vulnerability detection, we validate vulnerability paths beyond the static perspective by acquiring and exploiting a sampling of devices. These are chosen based on three successive criteria. First, vulnerability candidates must remain after sanitization filtration. Second, evidence of sanitization discrepancy indicates a protection imposed to secure some downstream sink path that may be bypassed. Lastly, higher values of taint spread (above 80th percentile, see

Sect. 5.4) indicates opportunity for tainted data to incite confusion and incorrect assumptions on its provenance and sanity in downstream sink paths.

Table 2. Device selection and exploitation according to remaining vulnerability paths after sanitization filtration, discrepancy analysis, high spread yield, and static path validation.

Device	Vuln. Paths	Discrep.	Spread$^{80^{th}}$	Statically Validated	Acquired	Exploited	Patched?
raxe500	✓	✓	✓	**1**	✓	**1**	✓
ex3700	✓	✓	✓	**2**	✓	**2**	
ex6120	✓	✓	✓	**2**	✓	**2**	
xr1000	✓	✓	✓	**1**	✓	**1**	
r7000	✓	✓	✓	**4**	✓	**1**	
ex6100	✓	✓	✓	1			
r6400	✓	✓	✓	2			
r7000p	✓	✓	✓	2			
ex6150	✓	✓	✓	-			
r8000	✓	✓					
rax20	✓	✓					
rax43	✓	✓					
nbk752	✓						
r6120	✓						
r6220	✓						
r6350	✓						
r6850	✓						

The selection process begins with the 18 remaining devices after filtration, listed under the first column in Table 2. Five additional devices are removed that do not contain sanitization discrepancies. An additional three devices are removed due to relatively low spread. A random sample of paths was selected from each of the eight remaining devices and statically validated by opening up affected binaries and inspecting paths. Five devices with the most promising validated paths were acquired and PoC exploits were crafted to demonstrate how to acquire a remote root shell on each device. In particular, the exploited vulnerability on `raxe500` was patched in a subsequent release. Exploited instances were responsibly disclosed to the vendor, and five PSVs were assigned[4].

The redacted code in Listing 1.2 illustrates one of the exploited transitive vulnerabilities. The path is triggered across three separate binaries and transfers taint across two storage transition paths before reaching a `system` call. It is guaranteed unsanitized by our semantic call sequence annotations which, for this specific path, evaluates to ⟨`taint_source, storage_setter, storage_getter, storage_setter, storage_getter, storage_setter, storage_getter, command_executor`⟩.

[4] PSV-2023-0150, PSV-2023-0151, PSV-2023-0152, PSV-2023-0153, PSV-2023-0154.

```
// [Binary 1] Source Path
websGetVar(request_obj, "redacted", user_input);
snprintf(dest, 400, "%s_%s", user_input, other_buff);
acosNvramConfig_set("storage_key1", dest);
// [Binary 1] Storage Transition A
buff = acosNvramConfig_get("storage_key1");
strlcpy(dest, buff, 400);
first_tok = strtok_r(dest, "_", &tok_ptr);
acosNvramConfig_set("storage_key2", first_tok);
// [Binary 2] Storage Transition B
buff = acosNvramConfig_get("storage_key2");
acosNvramConfig_set("storage_key3", buff);
// [Binary 3] Sink Path
buff = acosNvramConfig_get("storage_key3");
sprintf(dest, "<redacted>␣%s␣<redacted>", buff);
system(dest);
```

Listing 1.2. Redacted Transitive Command Injection Vulnerability

5.3 Critical Sanitization Bypass Flaws

Discrepancy in the expected sanitized treatment of a storage key indicates a sanitization bypass flaw. Such cases should not occur because the presence of sanitized input reaching a storage key indicates a deliberate effort to protect storage from arbitrary values, and thus all instances writing to this key should be similarly protected.

The discrepancies detected are between paths that are sanitized from the *certain* set of sanitizers, while unsanitized paths containing ambiguous sanitizers are omitted. These considerations assure that reported cases of discrepancies are guaranteed to contain only true-positive sanitized and unsanitized path pairs. We discover 77 unique instances of this security flaw affecting 32 different devices; 22 from Netgear, and 10 from ASUS as illustrated in Table 3.

Table 3. The number of sanitization bypass instances detected through discrepancy analysis over our firmware dataset

ex3700	1	ex7000	3	r7000	13	r8000p	2
ex6100	1	gs-ax3000	1	r7000p	2	rax120	3
ex6120	1	lax20	1	r7800	1	rax20	4
ex6150	1	r6400	4	r8000	5	rax200	1
rax43	4	rbs50y	2	rt-ax55	1	rt-ax82u1	1
ax50	4	rt-ac3100	1	rt-ax56u	1	rt-ax89x	1
rax80	1	rt-ax1800s	1	rt-ax58u	1	xr1000	4
raxe500	7	rt-ax3000	1	rt-ax82u	1	xr500	2

The redacted code in Listing 1.3 illustrates a real-world example of this security flaw we identified through sanitization discrepancy. Interestingly, this instance is found in the causal component of an exploited vulnerability path. The storage location `"discrep_key"` is filled with raw user input in function `process_request_A`, while in `process_request_B` the input is validated using `contains_shell_characters` before storing it to the same key. This indicates a possibility to circumvent the sanitization constraint that `process_request_B` tries to enforce on `"discrep_key"` by exploiting the unsanitized logic in `process_request_A`.

```
void process_request_A(...) {
    int size = 0x800;
    char *user_input[size];
    get_user_input(request, "redactedA", user_input, size);
    acosNvramConfig_set("discrep_key", user_input);
}
void process_request_B(...) {
    int size = 0x800;
    char *user_input[size];
    get_user_input(request, "redactedB", user_input, size);
    int flag = contains_shell_characters(user_input);
    if (flag == 1)
        return;
    acosNvramConfig_set("discrep_key", user_input);
}
```

Listing 1.3. Redacted Sanitization Bypass for Exploited Vulnerability

5.4 Attack Surface Spread

We design taint spread as a measurement of attack surface induced by the propagation of unsanitized input to code locations reached through persistent storage which reflects its system-wide potential impact. The metric was conceived after we observed 4832 unique unsanitized source paths within our firmware dataset, as it quantifies an obvious need to track the influence of tainted data.

The attack surface spread impact is computed for each firmware and binned by intervals of 250 code locations as depicted in Fig. 5a. The majority of firmware (80%) fall into the lowest bin, thus devices with higher spread are considered to fall above the 80th percentile and are prioritized for device exploitation as explained in Sect. 5.2.

The spread of each first-contact key involved in device exploitation is computed and displayed in Fig. 5b. An intriguing observation is the tendency of the first-contact keys involved in exploitation to be ranked in the top 5–10% of all keys within their respective firmware when rank-ordered by largest spread; if sanitization would have been applied proactively to these top 10% keys (i.e., even before vulnerability detection), each instance of exploitation would be thwarted.

A transitive vulnerability path patched at its source causal component does not prevent other transitive paths from reaching the same system call for exploitation. The reason being that though these two paths may not write to

(a) Attack Surface Impact by Spread

Spread Range	No. FW	Sel. Devices
[1, 250[	55	
[250, 500[	3	✓
[500, 750[	4	✓
[750, 1000[	1	
[1000, 1250[	0	
[1250, 1500[	3	✓
[1500, 1750[	2	✓✓
[1750, 2000[	1	

(b) Exploited fc Key Spread Rank

Expl. Device	fc Spread	Spread Rank	Total Keys
XR1000	11	10	97
R7000	14	12	246
RAXE500	9	13	138
EX6120	34	8	182
	23	10	182
EX3700	23	10	182

Fig. 5. (a) Binned spread for device exploitation selection, (b) The spread of each first-contact key involved in device exploitation

the same first-contact key, they may transitively reach the same storage location loaded in a common sink path. This is caused by having their spread overlap in some code location(s). Owing to the difficulty in assessing the success of a patch because of these obscure downstream effects, we construct a **spread table** to visualize and prioritize vulnerability patching centered around first-contact keys and their spread impact as seen in Table 4. The table provides two important perspectives: (a) for each first-contact key k, the set of dependencies which must *also* be sanitized to neutralize the spread of k, and (b) a rank-ordering of total dependency spread for prioritizing highly entangled keys.

We provide an example for interpreting the spread table using Netgear's R7000P. If a patch is applied to the unsanitized source paths writing to fc-key k_1, its impacted system-wide spread of 107 code locations would be neutralized to zero if and only if every dependency with overlapping code locations $\{k_2, ..., k_{12}\}$ is also sanitized. Such an endeavour would result in a total effective protection of 466 code locations over 5 binaries and 79 persistent storage keys from being impacted by tainted data. Applying patches to a portion of first-contact keys starting from the top of the table will yield the most significant reduction in spread, and will remove the impact from keys with heavily entangled dependencies.

As a final note, though this work's application of spread is modeled around native IoT persistent storage, it allows for extension to other communication paradigms that transfer data across function boundaries such as those leveraged in Karonte [22] (i.e., sockets, files, environment variables, and shared memory). It would only require modifying the node semantics (e.g., from persistent storage keys to socket ports), edge semantics (e.g., to socket communications between ports), and code location semantics (e.g., basic block addresses).

Table 4. Truncated spread table for R7000P, depicting the first-contact key and its code location spread, followed by the spread dependencies, the number of binaries receiving their spread, and the total dependency spread.

fc-key	Spread	Tainted Storage Keys	Spread Dependencies	Binaries Affected	Deps. Spread	Deps. Tainted Storage Keys
k_1	107	8	$\{k_2, k_3, ..., k_{10}, k_{11}, k_{12}\}$	5	466	79
k_{13}	105	5	$\{k_{14}, k_{15}, ..., k_{18}, k_{19}, k_{20}\}$	5	208	13
k_{21}	13	1	$\{k_{22}, k_{23}, k_{24}, k_{25}, k_{26}\}$	5	91	6
k_{27}	23	1	$\{k_{28}, k_{29}\}$	3	40	3
k_{30}	5	1	$\{k_{31}, k_{32}, ..., k_{34}, k_{35}, k_{36}\}$	4	34	7
k_{37}	34	2	$\emptyset$	1	-	-
k_{38}	31	3	$\emptyset$	3	-	-
k_{39}	2	1	$\{k_{40}, k_{41}\}$	6	26	3
k_{42}	21	1	$\emptyset$	3	-	-
k_{43}	4	1	$\{k_{44}, k_{45}, ..., k_{62}, k_{63}, k_{64}\}$	3	17	22
k_{65}	16	2	$\emptyset$	1	-	-
k_{66}	16	1	$\emptyset$	6	-	-
k_{67}	15	2	$\emptyset$	1	-	-
k_{68}	11	2	$\{k_{69}\}$	2	13	3
k_{70}	7	1	$\{k_{71}\}$	2	13	2

6 Related Work

Vulnerability detection in firmware is approached with both static and dynamic analysis, thus we present related work considering these two aspects.

In general static analysis is favored when coverage and scalability are pursued [7–9,17,22,23]. DTaint [8] uses taint analysis with pointer-aliasing and data structure recovery, but is limited to a single binary. Karonte [22] handles cross-binary tainting via a dependency graph modeling inter-process communication, however it does not consider `Libnvram` and consequently misses transitive vulnerabilities. SaTC [7] handles `Libnvram` and incorporates taint sources inferred via shared keyword between frontend and backend code, yet makes tentative assumptions for flagging sanitizers by tracking its the longest control-flow path to gather input constraints. EmTaint [9] resolves indirect calls for better coverage with the drawback of reporting infeasible paths. FLuaScan [17] addresses embedded Lua web servers but is tailored to a specific library and would need adjustments to support others. Firmalice [23] identifies authentication bypass flaws, yet requires human-generated security policies to operate.

Dynamic analysis, notably through fuzzing, is extensively explored in IoT firmware and their embedded web interfaces [5,6,10,11,13,15,16,21,23,26,29,33]. State-of-the-art emulators include Firmadyne [5] and it's successor FirmAE [16], both highly leveraged for fuzzing purposes. KVFL [29] proposes to embrace the key-value nature of requests processed by web servers to help prioritize mutations for better fuzzing coverage which shares the insights with our work. Although web interfaces have been targeted by network fuzzers [10,13,21,33], few of them were adapted to trigger and monitor command injection vulnerabilities.

7 Conclusion

Advancing the tradecraft of IoT security to outpace threat actors demands innovative, pragmatic, and actionable solutions. For these purposes, we provide an approach that models call sequences and persistent storage to facilitate: 1) detection of critical security vulnerabilities, 2) investigation of security-related information-flows like sanitization discrepancies, and 3) assessment of attack surface through firmware-wide taint spread. Our experimentation shows that the process of manual annotation does not have significant time overhead relative to state-of-the-art IoT vulnerability detection systems. By manually annotating sanitizers, we facilitate adjustments to the soundness of our analysis.

These three objectives were reached by first identifying and exploiting seven 0-day vulnerabilities and promptly disclosed them to their respective vendors. Moreover, through sanitization discrepancy analysis, we identified 77 instances of critical sanitization bypass flaws affecting 32 unique devices. We also demonstrated how to assess the attack surface reflected by IoT devices through the lens of the taint spread metric, as well for shrinking it using the spread table. Future work could hybridize the annotation phase, however the type of analysis applied would require a certain level of soundness to ween off the certainties provided by manual validations. Finally, the attack surface spread metric can be leveraged in various fields of embedded firmware security including vulnerability triaging, patch prioritization, and device security posture auditing.

References

1. Hackers update Cisco IOS XE backdoor to hide infected devices—bleepingcomputer.com. https://www.bleepingcomputer.com/news/security/hackers-update-cisco-ios-xe-backdoor-to-hide-infected-devices/. Accessed 06 Dec 2023
2. Andreoli, A., Lounis, A., Debbabi, M., Hanna, A.: On the prevalence of software supply chain attacks: empirical study and investigative framework. Forensic Sci. Int. Digit. Invest. **44**, 301508 (2023)
3. Antonakakis, M., et al.: Understanding the mirai botnet. In: 26th USENIX Security Symposium (USENIX Security 2017), pp. 1093–1110 (2017)
4. Lei, C., Zhang, Z., Hu, C., Das, A.: Mirai variant V3G4 targets IoT devices—unit42.paloaltonetworks.com. https://unit42.paloaltonetworks.com/mirai-variant-v3g4/. Accessed 06 Dec 2023
5. Chen, D.D., Woo, M., Brumley, D., Egele, M.: Towards automated dynamic analysis for Linux-based embedded firmware. In: NDSS, vol. 1, p. 1 (2016)
6. Chen, J., et al.: IoTFuzzer: discovering memory corruptions in IoT through app-based fuzzing. In: NDSS (2018)

7. Chen, L., et al.: Sharing more and checking less: leveraging common input keywords to detect bugs in embedded systems. In: 30th USENIX Security Symposium (USENIX Security 2021), pp. 303–319 (2021)
8. Cheng, K., et al.: DTaint: detecting the taint-style vulnerability in embedded device firmware. In: 2018 48th Annual IEEE/IFIP International Conference on Dependable Systems and Networks (DSN), pp. 430–441. IEEE (2018)
9. Cheng, K., et al.: Detecting vulnerabilities in Linux-based embedded firmware with SSE-based on-demand alias analysis. In: Proceedings of the 32nd ACM SIGSOFT International Symposium on Software Testing and Analysis, pp. 360–372 (2023)
10. Clements, A.A., et al.: HALucinator: firmware re-hosting through abstraction layer emulation. In: 29th USENIX Security Symposium (USENIX Security 2020), pp. 1201–1218 (2020)
11. Costin, A., Zarras, A., Francillon, A.: Automated dynamic firmware analysis at scale: a case study on embedded web interfaces. In: Proceedings of the 11th ACM on Asia Conference on Computer and Communications Security, pp. 437–448 (2016)
12. Gao, Z., et al.: Faster and better: detecting vulnerabilities in Linux-based IoT firmware with optimized reaching definition analysis (2024). https://doi.org/10.14722/ndss.2024.24346
13. Gui, Z., Shu, H., Kang, F., Xiong, X.: FIRMCORN: vulnerability-oriented fuzzing of IoT firmware via optimized virtual execution. IEEE Access **8**, 29826–29841 (2020)
14. Anton, K., Bird, J., Manico, J.: OWASP top ten proactive controls 2018. About this Project. OWASP foundation—owasp.org (2018). https://owasp.org/www-project-proactive-controls/v3/en/0x02-about-project.html. Accessed 06 Dec 2023
15. Kim, J., Yu, J., Kim, H., Rustamov, F., Yun, J.: FIRM-COV: high-coverage greybox fuzzing for IoT firmware via optimized process emulation. IEEE Access **9**, 101627–101642 (2021). https://doi.org/10.1109/ACCESS.2021.3097807
16. Kim, M., Kim, D., Kim, E., Kim, S., Jang, Y., Kim, Y.: FirmAE: towards large-scale emulation of IoT firmware for dynamic analysis. In: Annual Computer Security Applications Conference, pp. 733–745 (2020)
17. Li, X., Wei, Q., Wu, Z., Guo, W.: Finding taint-style vulnerabilities in Lua application of IoT firmware with progressive static analysis. Appl. Sci. **13**(17), 9710 (2023)
18. Livshits, B., Chong, S.: Towards fully automatic placement of security sanitizers and declassifiers. Acm Sigplan Not. **48**(1), 385–398 (2013)
19. MITRE: CWE - CWE-78: Improper Neutralization of Special Elements used in an OS Command ('OS Command Injection') (4.13)—cwe.mitre.org. https://cwe.mitre.org/data/definitions/78.html. Accessed 06 Dec 2023
20. Neshenko, N., Bou-Harb, E., Crichigno, J., Kaddoum, G., Ghani, N.: Demystifying IoT security: an exhaustive survey on IoT vulnerabilities and a first empirical look on internet-scale IoT exploitations. IEEE Commun. Surv. Tutor. **21**(3), 2702–2733 (2019)
21. Qin, C., et al.: UCRF: static analyzing firmware to generate under-constrained seed for fuzzing SOHO router. Comput. Secur. **128**, 103157 (2023)
22. Redini, N., et al.: Karonte: detecting insecure multi-binary interactions in embedded firmware. In: 2020 IEEE Symposium on Security and Privacy (SP), pp. 1544–1561. IEEE (2020)
23. Shoshitaishvili, Y., Wang, R., Hauser, C., Kruegel, C., Vigna, G.: Firmalice-automatic detection of authentication bypass vulnerabilities in binary firmware. In: NDSS, vol. 1, p. 1 (2015)

24. Shoshitaishvili, Y., et al.: SoK: (state of) the art of war: offensive techniques in binary analysis. In: 2016 IEEE Symposium on Security and Privacy (SP), pp. 138–157. IEEE (2016)
25. Stasinopoulos, A., Ntantogian, C., Xenakis, C.: Commix: automating evaluation and exploitation of command injection vulnerabilities in web applications. Int. J. Inf. Secur. **18**, 49–72 (2019)
26. Stephens, N., et al.: Driller: augmenting fuzzing through selective symbolic execution. In: NDSS, vol. 16, pp. 1–16 (2016)
27. Topics, T.I.E.: IoT connected devices worldwide 2019–2030. Statista—statista.com (2023). https://www.statista.com/statistics/1183457/iot-connected-devices-worldwide. Accessed 06 Dec 2023
28. Vadayath, J., et al.: Arbiter: bridging the static and dynamic divide in vulnerability discovery on binary programs. In: 31st USENIX Security Symposium (USENIX Security 2022), pp. 413–430 (2022)
29. Wang, C., Zhao, S., Peng, J., Zhu, J.: KVFL: key-value-based persistent fuzzing for IoT web servers. Comput. J. bxad110 (2023)
30. Wu, Y., et al.: Your firmware has arrived: a study of firmware update vulnerabilities (2023)
31. Yamaguchi, F., Maier, A., Gascon, H., Rieck, K.: Automatic inference of search patterns for taint-style vulnerabilities. In: 2015 IEEE Symposium on Security and Privacy, pp. 797–812. IEEE (2015)
32. Yun, J., Rustamov, F., Kim, J., Shin, Y.: Fuzzing of embedded systems: a survey. ACM Comput. Surv. **55**(7), 1–33 (2022)
33. Zheng, Y., Davanian, A., Yin, H., Song, C., Zhu, H., Sun, L.: FIRM-AFL: high-throughput greybox fuzzing of iot firmware via augmented process emulation. In: 28th USENIX Security Symposium (USENIX Security 2019), pp. 1099–1114 (2019)
34. Zhong, W.: Command Injection. OWASP Foundation—owasp.org. https://owasp.org/www-community/attacks/Command_Injection. Accessed 06 Dec 2023

GLUEZILLA: Efficient and Scalable Software to Hardware Binding using Rowhammer

Ruben Mechelinck[1(✉)], Daniel Dorfmeister[2], Bernhard Fischer[2], Stijn Volckaert[1], and Stefan Brunthaler[3]

[1] DistriNet, KU Leuven, Leuven, Belgium
{ruben.mechelinck,stijn.volckaert}@kuleuven.be
[2] Software Competence Center Hagenberg, Hagenberg, Austria
{daniel.dorfmeister,bernhard.fischer}@scch.at
[3] μCSRL, CODE Research Institute, University of the Bundeswehr Munich, Munich, Germany
brunthaler@unibw.de

Abstract. Industrial-scale reverse engineering affects the majority of companies in the mechanical and plant engineering sector and imposes significant economic damages. Although reverse engineering mitigations exist, economic damage has not been impacted, indicating that they have failed to address the problem. A closer investigation shows that industrial-scale reverse engineering typically only expends efforts on replicating hardware, since software can often be copied verbatim—no reverse engineering effort required.

We present GLUEZILLA, a system that binds software to hardware through user-space rowhammer PUFs. GLUEZILLA transforms programs such that they only exhibit their intended behavior on the single machine they are bound to at compile time. When run on any other machine, the programs will exhibit a different functionality. GLUEZILLA relies on unclonable machine features and thereby forces counterfeiters to not clone just the hardware but also the software. Cloning both hard- and software drives up reverse engineering costs, thereby also decreasing the economic viability of industrial-scale reverse engineering.

GLUEZILLA works on commodity hardware and does not rely on expensive hardware components. Our evaluation shows that GLUEZILLA is effective and incurs 16% run-time performance overhead in a practical case.

1 Introduction

Estimated at 6.4 billion euros in Germany alone, industrial-scale reverse engineering poses a significant risk that endangers economic prosperity [34]. Mechanical engineering companies traditionally invest substantial amounts into research and development, resulting in high-tech products with correspondingly high premiums. To offset these premiums and leapfrog lacking prior research and development costs, industrial-scale reverse engineering has become a billion-dollar business. Based on industry findings, industrial-scale reverse engineering is financially

F. Maggi et al. (Eds.): DIMVA 2024, LNCS 14828, pp. 416–438, 2024.
https://doi.org/10.1007/978-3-031-64171-8_22

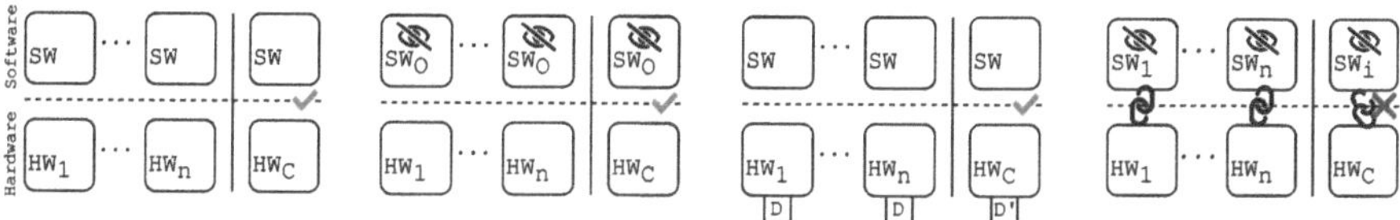

(a) No protection: An attacker can simply copy the software to a cloned device (HW_C). (b) Obfuscation: An attacker cannot easily reverse-engineer the software but can copy it to a cloned device. (c) Dongle: An attacker can clone or steal a dongle to run the software on a cloned device. (d) GLUEZILLA: An attacker can neither run the software on a cloned device nor easily reverse-engineer it.

Fig. 1. Overview of different protection modes of intellectual property.

sound and viable. This financial viability, however, rests crucially on the costs involved in reverse engineering versus actual development costs. If stealing intellectual property through reverse engineering costs less than actual development, then reverse engineering pays off; otherwise actual development is preferable.

The costs of reverse engineering can roughly be divided into two parts: (i) costs to reverse-engineer hardware machinery, and (ii) costs to reverse-engineer software components. At present, most reverse engineering efforts are spent on hardware components. A fundamental problem favoring industrial reverse engineering practices is that a machine's operational software runs on both the original and the reverse-engineered machine.

Thus, we propose an approach, called GLUEZILLA, which protects against reverse engineering of software. In contrast to other protection techniques—such as dongles or obfuscation [23], whose protective power remains low [3,25]—, GLUEZILLA also protects against unauthorized copying of software, as shown in Fig. 1. Software obfuscation thwarts the reverse engineering process but poses no obstacle to software cloning. Dongles, on the other hand, hinder software running across machines, but an attacker can clone them or remove the checking mechanism, which enables copies of the associated software instance to run on any machine.

Forcing industrial-scale reverse engineering companies to reverse-engineer both hard- *and* software components increases overall costs, which is the key goal of GLUEZILLA. To this end, GLUEZILLA glues hardware and software together in a novel, cost-efficient, and effective way. The resulting programs behave differently on different machines, having an intentional and an unintentional execution mode. Both modes are fully functional but perform different actions. For each software instance, there is only one associated machine instance which lets the program run its intentional execution mode. If the software runs on any other machine, including exact clones of the associated machine, the program remains in unintentional execution mode. GLUEZILLA achieves this goal by way of combining user-space-only, rowhammer-based [17] physically unclonable functions (PUFs) [13] with techniques borrowed from software diversity [21]. To the best of our knowledge this combination has not yet been proposed. Through its unique design, binaries compiled with GLUEZILLA need not carry any anti-theft protection techniques, which more often than not help adversaries during reverse engineering. GLUEZILLA works on commodity hardware without the need for expensive hardware components.

Summing up, this paper makes the following contributions:

- We present GLUEZILLA, a system that glues software to hardware through user-space rowhammer PUFs (see Sect. 6). The resulting glued software poses substantial obstacles to industrial-scale reverse engineering, thus driving up costs.
- We break with the prevalent hardware-oblivious paradigm of prior software defenses by introducing *hardware-dependent* software diversification (see Sect. 7).
- We discuss how our approach secures programs against reverse engineering and report findings of our experimental evaluation of GLUEZILLA (see Sects. 8 and 9).

2 Background

In this section, we describe the fundamentals of our approach, i.e., rowhammer. Rowhammer is a disturbance error in DRAM modules that allows unprivileged malicious actors to flip bits in physical memory, including in page frames allocated to privileged processes [17]. Each DRAM cell stores a single bit of data, whose value is determined by the charge of the capacitor. Cells naturally lose their charge over time, so the memory controller periodically recharges all rows to preserve their values. Because of implicit electromagnetic couplings in the DRAM chip, one can increase the charge leakage of some memory cells in the chip, i.e., the victim cells, by accessing a certain pattern of neighboring DRAM rows, i.e., the aggressor rows, at a very high frequency, called hammering. In the extreme case, the leftover charge in the victim cell falls below a threshold value and results in a flip of the logical value originally stored in the victim cell, i.e., a bit flip.

Rowhammer affects most commodity DDR3 and DDR4 DRAM chips [14,16, 24], including high-end ECC RAM [7], and can be reliably exploited. According to Kim *et al.*, more than 70% of rowhammer-susceptible cells flipped in ten out of ten iterations on DDR3 [17]. Schaller *et al.* show that the number of bit flips increases with temperature, however, the noise level stays constant [29]. Furthermore, the set of flippable bits depends on variances in the manufacturing process of the DRAM chip. This makes the set of flippable bits and the rows that contain these bits a unique and unclonable identifier for the DRAM chip. For this reason, Schaller *et al.* proposed to use rowhammer as an intrinsic PUF [29]. Recently, Jattke *et al.* showed rowhammer-induced bit flips on one out of ten tested DDR5 modules [15]. They conclude, however, that more work is required to reliably circumvent modern rowhammer mitigations, on-die ECC, and higher refresh rates available in DDR5 modules.

After its initial discovery in 2014, several security researchers demonstrated rowhammer-based exploits, for example to gain kernel privileges [30], extract sensitive information from a victim VM using a malicious co-hosted VM [5], gain unauthorized access to co-hosted VMs [27], or to escape from browser sandboxes [5,8,12]. Kwong *et al.* even showed that rowhammer can be used as a read primitive to leak sensitive information of other processes [20].

Most successful rowhammer exploits take three steps. During *memory templating*, the attacker constructs a list of vulnerable cells by probing the DRAM [35]. Next, the attacker loads a virtual memory page of the victim process into a physical page frame with a vulnerable cell at the desired offset. Attackers commonly resort to *memory massaging* techniques to bring the operating system's page frame allocator into a predictable state, and subsequently trigger a page allocation in the victim process. Finally, the attacker starts *hammering* the correct aggressor rows and causes the expected bits to flip in the page of the victim process's page.

3 Rowhammer Properties

In this section, we lay out a number of properties of the rowhammer effect, which we will later use as building blocks for stealthy, machine-aware computations. In the past, rowhammer was primarily used for targeted exploitation, because it allows stealthy cross-process memory modifications. Apart from this property, the rowhammer effect has many other interesting properties, including:

1. **Unclonable and unique bit flip pattern.** Whether a DRAM cell is susceptible to rowhammer or not depends on many uncontrollable variations during the manufacturing process of the DRAM chip. The distribution and sensitivity of rowhammer-susceptible memory cells is, therefore, unclonable and practically unique for large memory regions.
2. **Bit flips in the full memory range.** Rowhammer-susceptible cells and their corresponding aggressor rows are distributed throughout the whole physical memory. This allows programmers to use rowhammer to perform intra- and inter-process alterations of code and data. Inter-process bit flips, in particular, are well documented in the literature on rowhammer attacks.
3. **No static knowledge of victim cell locations.** For many rowhammer patterns, one can deduce the victim row based on knowledge of the location of the aggressor rows. For example, in the single-sided, double-sided, one-location, many-sided, and half-double hammer patterns, the victim rows are located next to the aggressor row(s) [9,11,17,18,28,30]. However, the precise location of rowhammer-susceptible cells is still unknown because it can be any variation of the 65,536 bits in the row (assuming a DRAM row is 8 KiB long). In more complex hammer patterns, e.g., non-uniform hammering [14], one cannot even deduce the precise location of the victim row(s).
4. **Bit flip success depends on execution speed.** If the aggressor accesses happen too slowly, e.g., because the CPU executes different instructions (of any process) in between, the victim cells will not lose enough charge in between row refreshes, resulting in no bit flips.
5. **Bit flip information isolated from OS and CPU.** Bit flips in memory happen directly in DRAM without memory write instructions. They are the result of successive memory reads to different locations in physical memory.

The OS and CPU are, therefore, unaware of any modifications in memory due to bit flips. One important consequence is that the OS will not trigger any faults, for example, when flipping bits in pages mapped as read-only or pages that are mapped in another process's address space.

We introduce the idea of using these rowhammer properties for purposes other than exploitation. Schaller *et al.* already explored the application of property 1 to build a rowhammer PUF [29]. Properties 2 and 5 allow programs to make modifications to any process, including their own, without the OS or the CPU noticing. These properties, furthermore, enable a new form of self-modifying code, even in non-writable code pages. Due to property 3, the static image of the binary does not contain any direct information on which modifications are performed at run time. These properties limit the amount of information one can extract from the static image about the run-time behavior. Properties 4 and 5 frustrate several dynamic analysis techniques.

4 Case Study: GlueZilla

We now present GlueZilla[1], a new approach to prevent software theft by binding a software instance to a single associated machine instance. Hackers who engage in industrial-scale reverse engineering specialize in creating replicas of expensive pieces of machinery, including their software. Currently, these hackers spend most of their efforts on reverse engineering and copying the machine's hardware components. Once they have copied these components, they can run the original machine's software on the replica machine as-is. GlueZilla can make reverse engineering much more difficult and costly by forcing hackers to reverse-engineer the hardware and subsequently modify the original software before it can run on a copied machine. One of the unique properties that sets GlueZilla apart from most traditional hardware-software binding mechanisms in use today (e.g., those that use dongles or trusted platform modules (TPMs)) is the stealthiness of the interaction between the GlueZilla-protected software and the hardware it is bound to. Whereas protected user-space software interacts with dongles and TPMs over well-defined and easily identifiable interfaces, GlueZilla-protected software uses standard memory access instructions that can hide in plain sight.

The protection system we present in this section is just one implementation of our hardware-software binding approach. This implementation does not offer bulletproof protection, as we will show in Sects. 8 and 10. GlueZilla is vulnerable to attackers who can create full-memory snapshots directly on a machine that runs the associated software instance.

5 Threat Model

For this case study, we make the following assumptions about the host system, the attacker, and the protected applications.

[1] https://github.com/COMET-DEPS.

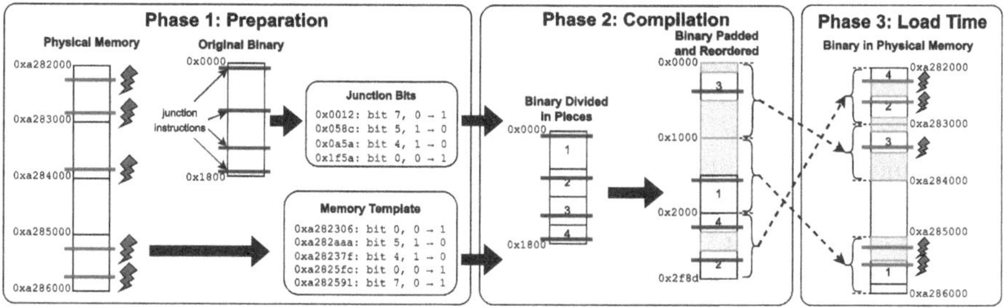

Fig. 2. GLUEZILLA generates a memory template on the target device and compiles the protected program such that all developer-specified junction instructions can be loaded into rowhammer-susceptible DRAM cells at run time.

We assume the host system uses commodity rowhammer-susceptible DRAM modules. Related work shows that this includes most of the DDR3 and DDR4 DRAM modules in use today [14,16,24], including those that use ECC [7]. Given the fact that no general rowhammer mitigation exists as of yet [7,9], we consider this a realistic assumption.

We assume the protected application is a user-space program, deployed on a machine that is part of industrial machinery, such as an assembly line. We consider an attacker performing industrial espionage, i.e., aiming to obtain a copy of the binary image of the protected software, for example, by stealing it from the storage of a genuine machine or by intercepting software updates. The attacker is, furthermore, able to acquire or create machine clones that meet the same specifications as the original machine. They will, however, show different results compared to the original machine when querying physically unclonable components, such as the DRAM modules. The attacker's goal is to deploy fully functional copies of the protected software on clones of the original machine. The attacker does not have access to the original source code.

6 GLUEZILLA: Overview and Design

Overview. GLUEZILLA programs contain two functional operation modes: an *intentional execution mode* and an *unintentional execution mode*. The intentional execution performs the original, intended purpose of the program. The unintentional execution performs actions differing from the intended purpose without obvious signs that something is wrong with the program, e.g., program crashes. These two operation modes only differ at unsuspicious-looking program points. For example, a branch condition in the intentional execution could be inverted in the unintentional execution, or a call in the intentional execution to one function could be changed to a call to another function in the unintentional execution. If chosen specifically so as to not arouse suspicion, it becomes difficult for an attacker to distinguish unintentional from intentional behavior. Reliable distinction of both behaviors requires expert understanding of the software's intentional behavior. Adversaries typically lack this knowledge since gaining this knowledge is the primary reason for reverse engineering the program in the first place.

Once our GLUEZILLA compiler has built the protected GLUEZILLA program, it executes the unintentional behavior by default. The program uses rowhammer-induced bit flips to switch to its intentional execution mode at run time. GLUEZILLA leverages several of the rowhammer properties we laid out in Sect. 3 to implement the run-time switch from the unintentional to intentional execution mode. Property 2 allows GLUEZILLA to use rowhammer-induced bit flips to morph the unintentional code into the intentional code. For the transition to intentional code to be successful, the protected program requires specific bits to flip. Due to property 1, the desired bit flip pattern is unclonable and only present at the associated machine for which we compiled the software instance. This ensures the intentional execution mode is only reconstructed on this physical machine. The static binary image, furthermore, does not contain any information about the intentional code, as explained by property 3. Finally, property 5 states that the CPU does not explicitly perform the required memory modifications. There is, thus, no need for any unstealthy branching instruction inside the protected program itself that conditionally reconstructs the intentional code. As explained in property 5, these memory modifications also work in non-writable code pages.

Design. Conceptually, GLUEZILLA operates in several phases shown in Fig. 2 that are required to create binaries glued to a specific hardware instance, as described above:

1. *Pre-Compilation Phase*:
   (a) Memory Templating Stage: In this stage, we gather information about the specific location of bit flips in the physical memory of the target device.
   (b) Junction Instruction Selection Stage: A junction instruction is a pair that consists of (i) the original, intentional instruction, and (ii) its unintentional instruction counterpart. During this stage, the compiler selects suitable pairs, either automatically through random selection or manually.
2. *Compilation Phase*: In the compilation phase, the compiler fuses the information of the memory templating with the selected junction instructions. Since both instructions and possibly even functions need to be put into specific locations, the compiler needs to perform advanced code scheduling techniques.
3. *Load Time Phase*: Before we can execute a program, we need to perform the required memory massaging at load time. This phase ensures that all required physical page frames from user space are allocated. Then, we load all of the code pages into the expected page frames.
4. *Run-Time Phase*: Rowhammering is used on junction instructions, such that the program's control flow is diverted from unintentional to intentional execution mode on the target device.

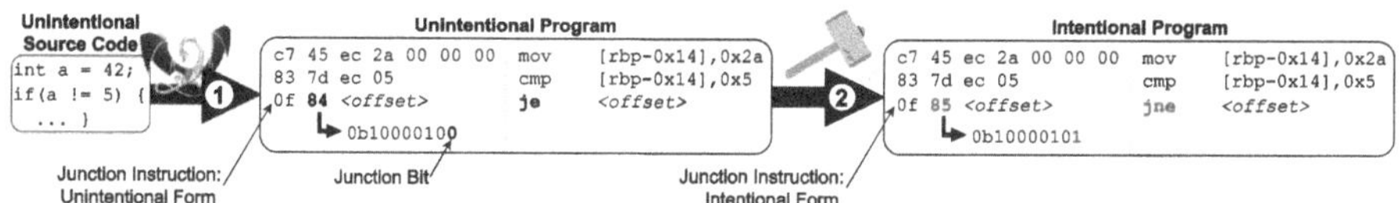

Fig. 3. At run time, GLUEZILLA uses rowhammer to transform junction instructions from their unintentional into their intentional form.

7 The Implementation of GLUEZILLA

In this section, we describe the previously mentioned phases in greater detail.

7.1 GLUEZILLA Memory Templating

Our approach requires a priori information about the behavior of rowhammer on the target device. We gather this information by measuring the rowhammer-susceptibility of the device's physical memory using a number of hammer patterns that are known to induce bit flips [9,11,14,17,18,28,30]. For all vulnerable cells we discover, we collect the parameters required to reproduce the bit flip later on, including (i) the physical address and (ii) the bit index of the vulnerable byte, (iii) the corresponding aggressor row addresses, (iv) the hammer pattern, and (v) the flip direction. We repeat this scan several times to assess the repeatability of each bit flip. Our memory templater and additional analysis tools consist of approximately 3,500 lines of C++ code.

7.2 GLUEZILLA Junction Instruction Selection

Junction instructions define the points where the intentional and unintentional execution of the program differ. The effort to reverse-engineer the protected program increases with greater difference between both execution forms and, therefore, the number of junction instructions. Selecting these instructions is hard to automate because we want to ensure that these differences are not obvious to reverse engineers. We want to guarantee that the program is not actively harmful, even when it executes its unintentional code.

An additional challenge lies in the sparse distribution of rowhammer-susceptible cells across the DRAM. In practice, this distribution limits us to a single bit flip per junction instruction. We refer to this bit as the junction bit. Fortunately, if chosen well, flipping one bit in a few junction instructions often suffices to make the intentional and unintentional executions of a program differ substantially. This is especially true on architectures with dense instruction sets (e.g., x86). On such architectures, there are many valid machine instruction sequences that differ in only one bit from several distinct sequences with different semantics. The individual instructions in the new interpretation of the sequence can have a different encoding length as long as the total sequence is still valid, i.e., the bit flip did not introduce invalid instructions in the code sequence.

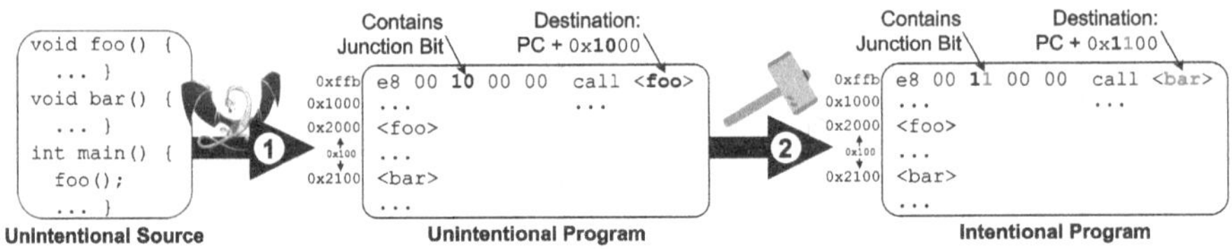

Fig. 4. Junction bits inside control transfer destinations add constraints to the relative position of the destinations.

This allows us to create a different control flow (e.g., by changing loop predicates, jump or call targets, etc.) and data flow (e.g., by changing arguments, constants, etc.) between the intentional and unintentional executions.

We classify junction instructions w.r.t. their position-related constraints into the following two classes:

1. Unmovable junction instructions (U-JI) only require that their junction bit be aligned with a suitable rowhammer-susceptible DRAM cell. Figure 3 illustrates an unintentional `je` instruction, which becomes an intentional `jne` instruction after the bit flip ②. This instruction should be placed into a DRAM cell in which we can flip bit 0 of the instruction.
2. Movable junction instructions (M-JI) create inter-instruction placement constraints. Figure 4 illustrates an unintentional call to `foo`, which becomes an intentional call to `bar` after the bit flip ②. If we choose bit index 8 of the second byte of the branch target operand as our junction bit, the target changes from 0x1**0**00 to 0x1**1**00 during the run-time transition from the unintentional to the intentional execution mode. This puts a relative placement constraint on the two call targets (`foo` and `bar`) as they must be exactly 0x100 bytes apart.

To select junction instructions, we explored an automatic and a manual operation mode. The automatic operation mode selects a provided number of unmovable and movable junction instructions randomly and uniformly across the binary. This mode, however, provides no guarantee about the behavior of the software on cloned machines. It may, in the worst case, cause harm to the machine, or its environment. Instead of ignoring such a security-critical scenario, we also support a manual operation mode.

In manual operation mode, we rely on a developer to modify their source code to unintentional form and keep track of the changed instructions, i.e., the junction instructions. To facilitate this task, the developer can construct the unintentional behavior by adding subtle alterations to the existing control and data flow graph of the intentional behavior. To increase the number of available alterations, a developer can perform semantic-preserving transformations, such as data structure or function parameter permutation.

7.3 Compiling GLUEZILLA Binaries

The transition from intentional to unintentional execution mode only works if we can position the program's code in physical memory such that every junction bit is stored in a rowhammer-susceptible DRAM cell. This poses several challenges:

Challenge 1: Bit Alignment. Machine instructions are aligned to byte boundaries in physical memory. Thus, if the X-th bit within a junction instruction is a junction bit, then it is only possible to store this bit in the Y-th cell of a DRAM row if X and Y are congruent modulo 8 (i.e., $X \equiv Y \pmod 8$).

Challenge 2: Page Alignment. For practical reasons, and although not strictly required, code pages within a program binary should be aligned to page boundaries when loaded into virtual and physical memory. This means that if a junction instruction is at page offset X within a code page of a program binary, said instruction can only be stored at byte offset Y in a page frame if X and Y are congruent modulo the page size.

GLUEZILLA: Compiler + Loader. We developed a compiler toolchain and a binary loader to tackle these challenges. To overcome the first challenge, our compiler consults the memory template of the target device to find candidate cells for each junction bit that complies with the required bit alignment. We use this list of candidates to constrain our layouting passes. These layouting passes assign each code page to a physical page frame with a suitable candidate DRAM cell, where that page's contents must be loaded. We emit information about these assignments alongside the program binary in the form of a *loader map*. Afterwards, they apply dead `nop` insertion and instruction/basic block reordering [21] to move each junction instruction to the page offset of the candidate cell, which solves the page alignment problem. We implemented all logic related to the custom binary layout in the Machine Code (MC) sub-project of LLVM 11 in almost 1,200 lines of C++ code.

7.4 Loading GLUEZILLA Binaries into Memory

GLUEZILLA only works if it can guarantee that every junction instruction is stored in its intended physical memory cell. However, user-space programs have no direct control over where their code pages get loaded into physical memory by the operating system. We tackle this challenge using *memory massaging* from user space, similar to prior rowhammer attacks [11,20,27,28,30,35]. In detail, we massage the operating system's page frame allocator so we can load every code page that appears in our loader map into its intended physical frame.

The massaging techniques presented in related work are built for attack scenarios where the attacker controls the massaging process, and tries to load pages of the victim process in a specific physical page frame [11,20,27,28,30,35]. These techniques typically leave a time window between the freeing of rowhammer-susceptible page frames by the massaging process and the reallocation of said frames by the victim program. In this time window, the operating system could allocate the desired frame to other programs running in the background. Most attacks tolerate this probabilistic behavior, as they typically require only very few rowhammer-susceptible page frames and one can simply relaunch the attack without anyone noticing.

In contrast, programs protected by GLUEZILLA might require rowhammer-susceptible cells in many different page frames and, therefore, probabilistic massaging is inadequate. To address this challenge, we design a *deterministic* massaging technique by performing the massaging step from within the same process whose page frames we want to hammer. This decision closes the time window, because we do not have to deallocate frames only to trigger their reallocation in a different process. Our technique allocates pages until we find all physical page frames in the loader map and their corresponding aggressors.

We implemented our binary loader and supporting tools in approximately 1,200 lines of Rust code.

7.5 Rowhammering GLUEZILLA Binaries at Run-Time

After loading all code pages into their associated page frames, we can execute the program and use rowhammer to transition between execution modes. We flip the junction bit by rapidly hammering the junction instruction's corresponding aggressor rows with the chosen hammer pattern.

In most cases, this operation works only in one direction: it can transform a junction instruction from its original unintentional form into its intentional form but not vice versa. Thus, after executing a junction instruction once, it would remain in its intentional form until the program stops. However, it is also possible to reverse the transformation after executing the junction instruction. For example, we could reset the contents of page frames containing junction instructions by either explicitly reloading the code from disk or by forcing the operating system's pager to evict the frame and, subsequently, page it back in. In this case, we could restore the junction instructions' unintentional forms, either periodically or immediately after executing them.

Other degrees of freedom lie in the timing of the transition and the placement of the hammering instructions. We could invoke the transformation at program startup, just before executing the junction instruction, or anywhere in between. Similarly, we can invoke the transformation from within the GLUEZILLA-protected process or from an external process. By separating the transformation code in time and space from the junction instruction it transforms, we can increase stealthiness and resilience against reverse engineering. However, too much separation, especially in time, can facilitate memory snapshotting attacks, as we explain in Sect. 10.

While rowhammering, the protected program uses approximately all available memory bandwidth. To enable a successful transition, no other program running on the system should consume a significant portion of the memory bandwidth while the protected program is hammering. Multiple GLUEZILLA protected programs can run on the same system simultaneously as long as they adhere to this constraint, i.e., they should synchronize their rowhammer-based transitions.

8 Security Evaluation

An attacker can try to reverse-engineer the protected binary to extract secrets or to remove the machine dependence. We describe two attackers that have access to increasingly powerful analysis techniques: (i) static analysis and (ii) dynamic analysis on a cloned machine.

Static Analysis. Due to rowhammer properties 2 and 3, the binary is only an image of the unintentional program and does not contain any direct information about the transformations performed by GLUEZILLA to transition to the intentional execution mode.

The attacker can perform a differential attack on multiple binary instances that are compiled for different machines. However, the unintentional behavior and the set of junction bits are exactly the same for all instances. Only the order of the code fragments and the amount of padding between them differs. These reorderings do not provide any information disclosing junction instructions.

By inspecting the loader map, the attacker can learn which page frame is assigned to each code page. However, due to property 1, the attacker knows neither the location nor the number of rowhammer-susceptible cells these page frames contain. Additionally, due to rowhammer property 3, even if the attacker learns the location of the victim row based on the aggressor rows, they cannot reliably infer which bit flips are induced by them. Assuming a row is 8 KiB long (65,536 bits), the total number of possible variations is given by: $\sum_{p=1}^{M} \binom{N}{p}$ with M being the attacker-assumed maximum number of bit flips in a row and N being the number of bits giving valid instructions when flipped. Let us assume the compiler emits code fragments with 35% code and 65% `nop` instructions[2] and that 50% of the code bits result in valid instructions when flipped. On a single channel system, this gives 11,469, 131,537,961, or 1,508,345,821,725 possible bit flip variations if the attacker assumes a maximum of one, two or three flips, respectively. For every considered bit flip, the attacker has to reason about its effect on the program semantics, which is difficult to automate. This difficulty renders a brute-force attack impractical due to prohibitive time requirements. When using complex hammer patterns, the victim row is not adjacent to the aggressor row and bit flips can happen anywhere in the program, providing additional protection.

We conclude that a static adversary cannot obtain sufficient information regarding the intentional behavior using static analysis alone.

Dynamic Analysis on Cloned Hardware. In this scenario, the attacker has a clone of the software instance and a clone of the machine associated with the software instance. The protected program is always loaded at the same physical addresses, dictated by the loader map. Property 1 states that the number and location of rowhammer-susceptible DRAM cells are different between any two

[2] Average size overhead for 100 junction instructions in SPEC CPU 2017 (see Sect. 9).

systems. Therefore, on a cloned machine, the junction bits will not reside in the required rowhammer-susceptible cells, leading to either no bit flips or unexpected bit flips. The attacker, therefore, has no incentive to perform any type of dynamic analysis on this program execution since they cannot learn anything about the intentional behavior.

9 Empirical Evaluation

We performed our experiments on both DDR3 and DDR4 systems. For DDR3 memory, we used the double-sided hammer pattern to induce bit flips [30]. Due to the presence of rowhammer mitigations on DDR4, e.g., Target Row Refresh (TRR), we required more advanced hammer patterns, such as half-double sided hammering [18], many-sided hammering [9,28], or non-uniform hammer patterns [14]. First, we evaluated the repeatability of the rowhammer-susceptible DRAM cells and showed that these can be used as a foundation for GlueZilla. Second, we evaluated the functionality of GlueZilla by protecting `coreutils` programs and performance on the SPEC CPU 2017 benchmark suite.

Bit Flip Repeatability. As we require individual bits to flip reliably, we must ensure a high repeatability. In several experiments conducted on multiple DRAM/machine combinations, a share of rowhammer bit flips is highly reliable, i.e., flips in all tests, under specific operating conditions.

For example, Fig. 5a shows the results of two rowhammer experiments, executed using the same 54 MiB of consecutive DDR3 memory of vendor B. Per experiment, we executed 1,000 iterations of a double-sided rowhammer test, with the only difference being the initialization of the victim and aggressor rows so that we can find bit flips in both directions ($0 \rightarrow 1$ and $1 \rightarrow 0$). While some bit flips occur in a very small number of tests, which can be considered as noise, several bits flipped in all iterations and thus are suitable for use in GlueZilla. Figure 5a shows that there are several reliable bit flips for each bit index in both flip directions.

We executed the same experiment on DDR4 memory of vendor A, although for just 800 KiB of consecutive memory and using a Blacksmith-generated non-uniform hammer pattern. Since DDR4 is more susceptible to bit flips than DDR3, the smaller memory area was sufficient to induce a comparable number of bit flips. As evidenced by the results shown in Fig. 5b, the results align with the ones obtained for DDR3. In consequence, we find that both DRAM types offer sufficiently reliable bit flips.

Among other factors, temperature and aging can influence the repeatability of rowhammer bit flips [4,17,24,29,33]. We do not consider these factors in this paper and want to refer to future work for possible error correction mechanisms.

Functionality Evaluation. To evaluate the practicality of GlueZilla, we applied GlueZilla to the `coreutils` programs `ls` and `cp`. We constructed an

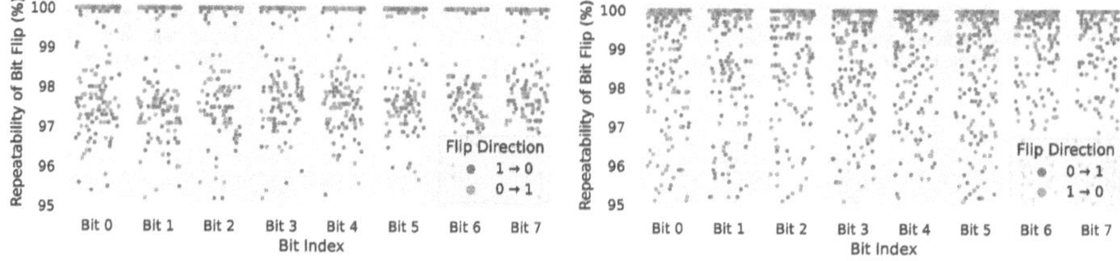

(a) Double-sided pattern on DDR3 DRAM (b) Non-uniform pattern on DDR4 DRAM

Fig. 5. Bit flip repeatability with 1,000 iterations per flip direction. Each dot represents a bit flip at a specific location in DRAM. They are grouped by their bit index within the respective directly addressable byte.

unintentional behavior for both programs and used the manual operation mode described in Sect. 7.2 to select the junction instructions.

For cp, we made the unintentional variant of the program delete rather than copy the source file. We implemented the unintentional behavior by modifying:

- Two function signatures to add an extra pointer indirection for some arguments.
- The struct layout of two structs to put common fields at the same offset.
- A function call that serves as movable junction instruction with the remove function as unintentional destination and the copy function as intentional destination.

For ls, we built the unintentional behavior by adding disruptive operations to the intentional behavior so it does not print the current directory correctly. We implemented this behavior by adding:

- A conditional function call that increments the pointer to the element to print. The condition always evaluates to true, so the unintentional form effectively skips an entry in each iteration. This condition functions as an unmovable junction instruction in which we flip an operand, so it evaluates to false in the intentional execution.
- A time dependence that jumps over the print code between 3 p.m. and 6 p.m. This jump is a movable junction instruction with the print code as intentional destination and the code thereafter as unintentional destination.
- An increment of two bytes to the char pointer pointing to the current entry's name. Each entry will thus miss the two starting letters in its name. The increment instruction is an unmovable junction instruction in which we flip the constant from two to zero.

We performed this experiment on both a DDR3 and a DDR4 system. For each system, we took two different DRAM modules that contain rowhammer-susceptible cells of which one is associated with the program by the GLUEZILLA compiler. At run time, we verified that the intentional execution is only unlocked on the associated module while on the other module the program performs its unintentional execution.

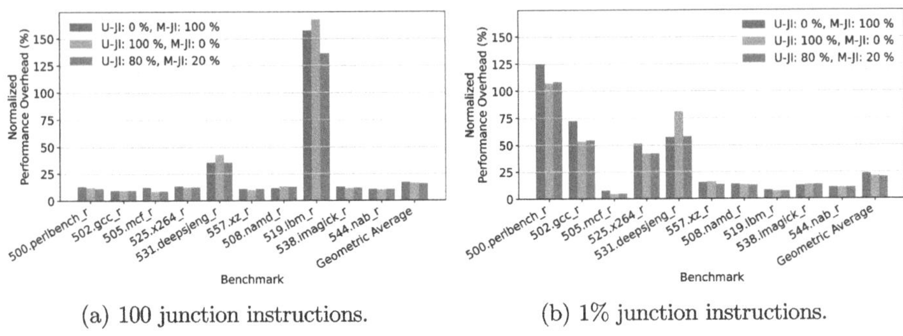

(a) 100 junction instructions.

(b) 1% junction instructions.

Fig. 6. Performance impact of GlueZilla on SPEC CPU 2017.

Performance Evaluation. Our current prototype performs all massaging and hammering at program launch, so these steps only add overhead when the protected program starts. The start-up overhead heavily depends on the current state of the memory allocator. For the average case, when most of the required page frames are free, this takes about 2.7 s for a program requiring about 1,000 page frames.

At run time, the custom binary layout adds performance overhead because of reduced cache utility due to reduced code locality. We evaluated this overhead on all C benchmarks of the SPEC CPU 2017 benchmark suite (we omit C++ benchmarks as our prototype does not currently support C++ exceptions). Our machine has an Intel Core i7-2600 CPU and runs Fedora 39 with Linux 6.6. We used four DRAM modules with 12 GiB of DDR3 memory in total, operating at 1333 MT/s. To reduce interference with other processes, we disabled SMT, turbo boost and isolated one core to which we pinned the protected program and set the scaling governor to performance.

For each benchmark, we used the automatic operation mode, which selects a provided number of unmovable and movable junction instructions randomly and uniformly across the binary. This mode allows us to measure the effect of the custom layout on large quantities of junction instructions. Furthermore, we defined the unintentional form equal to the intentional form so we do not require the rowhammer step in our layout-related overhead measurements.

Figure 6a shows the run-time performance overhead of binaries containing 100 uniformly distributed junction instructions, each requiring one junction bit. We divide these in unmovable and movable junction instructions in three configurations, represented by the three bars per benchmark. For each configuration, we ran three experiments per benchmark program, each with a different set of junction bits and report the geometric average. Figure 6b shows the overhead when 1% of all machine instructions in the binary are junction instructions. This setting is the maximum amount of junction instructions we could select with a memory template containing about 200,000 rowhammer-susceptible DRAM cells.

Our current compiler prototype does not yet take into account any optimization regarding code placement and alignment, cache prefetching and utilization, branch probabilities, etc. It may even deteriorate the optimizations performed by earlier compiler passes, resulting in an almost worst-case overhead. This explains why both figures show a relatively high impact for the modified binary layout. However, in the realistic scenario of 100 junction instructions, the run-time performance overhead is limited to 16%. The geometric average of the size overhead of all benchmarks over the three configurations is 185% for 100 junction instructions and 843% when 1% of all instructions are junction instructions.

We investigate these results using the hardware performance counters available in our benchmark machine. In our experiments using 100 junction instructions, all measured performance loss is caused by a variety of effects related to suboptimal code layout, including instruction cache misses, iTLB misses, and CPU frontend stalls. For 519.lbm_r, however, we found that almost all overhead is caused by branch resteers in the CPU frontend. 519.lbm_r is the smallest benchmark program and only includes around 1,800 assembly instructions that fit on three 4 KiB pages in the baseline binary. Adding 100 junction instructions to these small binaries results in many small code fragments, often containing only one or two assembly instructions separated by large amounts of padding. To fit the spare distribution of rowhammer-susceptible cells in physical memory, the padding can be multiple KiB long. As suggested by Godbolt, this greatly underuses the available entries in the branch target buffer (BTB) leading to many branch mispredictions [10]. We found that the protected program had approximately 12,000 times more branch resteers compared to the baseline version. During a branch resteer, the branch address calculator updates an entry in the BTB that caused a branch misprediction or misidentification and restarts instruction fetching. These additional resteers account for 10% of all reference clock cycles.

In our experiments where 1% of all instructions are junction instructions, we found that more benchmarks are affected, albeit not as much as 519.lbm_r with 100 junction instructions, by branch resteers, most notably 500.perlbench_r, 502.gcc_r, 525.x264_r, and 531.deepsjeng_r. 519.lbm_r in these experiments only has approximately 18 junction instructions (1% of its 1,800 assembly instructions) and, therefore, has less run-time overhead compared to our evaluation using 100 junction instructions.

10 Discussion

Implementation Limitations. Our prototype presented in Sect. 7 only focuses on the functional requirements and, therefore, has some limitations related to security and performance.

We choose to construct the required binary layout during the compilation phase, because all necessary information is available at this stage. This approach, however, results in large binaries of which a large portion is `nop` padding. Alternatively, we could put each page in a separate ELF segment and directly load it in the assigned page frame. This approach only has intra-page padding but requires more information to transfer between the compiler and the linker.

The padding also identifies the edges of each code fragment which either contains a single junction instruction or is the destination of a movable junction instruction. This information slightly narrows down the adversary's search space for junction instructions, which makes it easier to reconstruct the intentional program. For example, the attacker can identify a potential intentional destination when it is preceded by a large amount of padding in the unintentional binary. Subsequently, they scan the unintentional binary and match all target addresses of direct jump instructions with the address of the potential intentional destination. If both addresses differ only in a single bit, the adversary can with great certainty assume the jump is a movable junction instruction with the identified code fragment as its intentional target.

We can eliminate this attack by limiting the attacker's capability to identify potential intentional targets based on the left-over intra-page padding. One approach is to substitute the padding with dead code or with used code that does not have any positional constraints. In both cases, it should not be obvious where the substituted padding ends and the intentional target starts by making the substituted padding as benign looking as possible, for example:

1. Whenever the padding ends with a return instruction, there should be a function symbol associated with the succeeding code, i.e., the intentional target.
2. If the padding ends with an unconditional jump, the padding should also include (dead) jumps to the succeeding code.
3. When the padding consists of dead code, we can include apparent data dependencies between the dead padding code and the intentional target's code.
4. If the padding consists of used code, we can include data dependencies delimited by an opaque predicate between the padding and the intentional target.

Our prototype requires access to the Linux pagemap, for which we launch the program with `sudo` and later switch to an unprivileged user. Because this is bad security practice, we could leverage a kernel module that exposes a memory massaging interface limited to the memory of the current process. Such an interface would not give new capabilities to the attacker because it provides the same functionality as massaging from user space with the information in the loader map.

To reduce run-time overhead, we could define the binary layout problem as an optimization problem to increase code locality. The optimization variables include: the size and order of the code fragments, the location of the junction instruction within the fragment, and the selection of the rowhammer-susceptible cell w.r.t. its page frame offset.

Because we merely want to demonstrate the general idea of using rowhammer for stealthy computations, we rely on the developer to manually analyze the code base and select enough junction instructions (see Sect. 7.2). This could be an enormous effort depending on the desired level of protection, i.e., the distance between the intentional and unintentional form. Automating this process requires multiple analyses, including a syntactical analysis of the source code to explore neighboring valid candidate programs, and a semantical analysis to evaluate the conspicuousness of each candidate and guarantee the safety for machine and environment of the unintentional form. Automation gets easier if we relax the conspicuousness requirement and allow, for example, program crashes or infinite loops without dangerous side effects. In this case, we could select condition statements as junction instructions.

Reliability. In Sect. 9, we showed the existence of rowhammer-susceptible cells that flip reliably in all our tests. However, we cannot guarantee 100% reliability because the rowhammer effect is still a hardware disturbance error and numerous factors might undermine the reliability, e.g., hardware aging and interference with other running software. The intentional execution is not fully reconstructed in case some junction bit does not flip as expected.

To maximize reliability, we assume the software and hardware vendor do not include any constructs that actively counteract our rowhammer-based program transitions. These include advanced error correction codes for memory transactions, which could correct the bit flips in junction instructions, or anti-tamper protections, which could detect changes in the executing program code.

As many industrial systems are considered critical, our approach might need additional measures to increase reliability. For example, we could leverage multiple distinct junction instructions to enable each functionality change, rather than just one. Whether the changed functionality is enabled at run time then depends on whether the majority of its junction instructions flipped.

In the absence of suitable error correction mechanisms, our method is still applicable to non-critical applications. An alternative approach is to make the intentional and unintentional execution functionally equivalent but make the unintentional form run less optimally. We could achieve this, e.g., by using less optimal algorithms, using more system resources, and requiring more user input. This way, the program will always execute its critical behavior, but it will only execute efficiently on the associated machine.

For systems demanding absolute reliability and with no room for occasional suboptimal code execution, the software vendor may find a different, potentially weaker, protection approach more suitable.

Potential Attacks and Further Mitigations. So far, we have assumed that the reverse engineer cannot run analyses directly on a coupled software and hardware pair. This assumption is in line with our threat model and is not unrealistic given that industrial machinery is often locked down using software-based access control mechanisms and physical sealing. However, given that no protection mechanism is perfect, we now discuss which types of attacks the adversary could attempt given full and unrestricted access to an associated device.

One threat that immediately comes to mind is Dynamic Binary Translation (DBT). An attacker could learn about the protected program's behavior by running it in a tool such as Valgrind to dynamically add instrumentation code or to monitor interesting features such as the execution trace, data dependencies, etc.

DBT tools preserve neither the input program's binary layout nor the intended mapping between virtual pages and physical memory frames. Therefore, DBT tools would be unable to observe the intended bit flips and thus give the attacker very little useful information about the intentional behavior.

Layout-preserving Dynamic Binary Analysis (DBA) tools or debuggers could resort to hardware features such as performance counters, processor tracing, or hardware breakpoints to analyze an instance of the protected program. These tools can, theoretically, observe transitions between the unintentional and intentional execution modes. However, due to property 5 of rowhammer, detecting these transitions is difficult because there are no obvious, dedicated instructions that explicitly modify the code. At best, the adversary could detect the rowhammering code based on its signature, i.e., a sequence of memory read instructions and a cache line flush of the same address and observe the program's memory afterwards. Prior work, however, already explored numerous other stealthy rowhammer primitives, for example, using non-temporal and implicit loads, uncached memory regions, and cache eviction sets [1,2,12,26,35,37,38]. These techniques remove the visual clues that prompt the adversary to inspect the code pages for bit flips. The attacker can, at best, check for bit flips at regular time intervals. If this interval is big, the attacker risks missing many bit flips, while a small interval is very time-consuming and will frustrate the attacker.

Another problem with DBA tools and debuggers is their performance impact. Due to property 4 of rowhammer, analysis tools cannot meaningfully interfere with the memory controller while the protected program executes its transition code. Even minimal interference could slow down the protected program to a point where the transition fails.

One particularly dangerous threat that *would* be successful against our current prototype is memory snapshotting. Given that our implementation only supports one-time transitioning from the unintentional to intentional execution mode immediately after the program starts, the attacker could simply create a full-memory snapshot when the program reaches its original entry point. This snapshot would show the junction instructions in their intentional form. We could defend against this attack by making the program transition back and forth between its unintentional and intentional execution mode (see Sect. 7.5).

Attackers could perform their own templating step using the full set of rowhammer parameters (e.g., the aggressors, the hammer pattern, and access counts and speeds) they extracted from the binary. Using the loader map, they could identify the junction instruction that gets loaded into a rowhammer-susceptible cell. We can mitigate this attack by inserting many bogus hammer sequences that either flip unimportant bits or no bits at all. With a sufficiently high number of genuine and bogus hammer sequences, this attack becomes a time-consuming and costly endeavor.

11 Related Work

To the best of our knowledge, there is no directly related work. Thus, we explore the two closest concepts, physically unclonable functions and software diversity.

Physically Unclonable Functions. Physically Unclonable Functions (PUFs) are hardware primitives that provide component-specific responses to user-provided challenges [13,32]. PUFs derive their responses from physical hardware characteristics that are extremely difficult to clone and that are expected to be unique because they stem from the many random variations in the hardware's production process. Over the past two decades, researchers have proposed several uses for PUFs, including identification and authentication [32], cryptographic key storage [22,31], and hardware-software binding [19,36]. Using PUF-based encryption of data or code for obfuscation purposes, however, is less stealthy than GLUEZILLA because all queries to the PUF and the encryption operations are visible and prone to dynamic analysis.

Schaller *et al.* first proposed a rowhammer PUF in 2017 [29]. They validated the PUF's uniqueness and robustness and showed that its entropy is high enough to derive cryptographic keys from. Their design uses a fixed memory area and requires a kernel module for PUF interaction. GLUEZILLA's design, on the other hand, can operate at any location in physical memory, has no fixed memory size requirement, and does not need a kernel model to safeguard the reserved memory area.

Software Diversity. A side effect of our approach is that it results in a form of software diversity [6,21]. Modern software diversity addresses the problem of the software monoculture by generating a unique version of a program for each user. GLUEZILLA, however, requires specific locations for the junction instruction in memory and, therefore, generates a unique binary layout for each associated machine. Furthermore, the prevalent paradigm in software diversity is to be *hardware agnostic* [21]. GLUEZILLA is, to the best of our knowledge, the first system breaking with this paradigm.

12 Conclusions

We present GLUEZILLA, a system that binds software to hardware to prevent industrial-scale reverse engineering. To this end, we combine new compile-time transformations on the software side with DRAM PUFs on the hardware side.

Through GLUEZILLA, a program now has two different modes of operation: an intentional and an unintentional mode. The intentional mode requires the correct bit flips at the correct locations. The unintentional mode executes the program in its statically encoded representation.

We present the relevant design decisions and implementation details for a fully-fledged GLUEZILLA prototype. Our evaluation shows that the performance impact of GLUEZILLA depends primarily on the required number of instructions to modify by bit flips. Although prohibitive performance impact may arise in some configurations, we find that these occur only in compute-intensive programs of the chosen benchmark suite. Software in our target area of industrial computing systems, however, is typically not compute-intensive. As a result, we expect negligible to moderate performance impacts in the target environment.

Our security evaluation shows that the software bound to hardware by GlueZilla poses significant obstacles to reverse engineering. Reverse engineering based on static analysis is frustrated by software diversity, amplified by not knowing which candidate instructions will be modified by rowhammering. Reverse engineering based on dynamic analysis is frustrated by both not having a hardware environment with identical bit flips and not knowing *when* all of the program's candidate instructions have been modified.

Acknowledgements. The research was supported by the Austrian ministries BMK and BMAW and the State of Upper Austria in the frame of the COMET Module DEPS (FFG 888338) and the SCCH COMET competence center INTEGRATE (FFG 892418).

References

1. Aga, M.T., Aweke, Z.B., Austin, T.: When good protections go bad: exploiting anti-DoS measures to accelerate Rowhammer attacks. In: HOST (2017)
2. Aweke, Z.B., Yitbarek, S.F., Qiao, R., Das, R., Hicks, M., et al.: ANVIL: software-based protection against next-generation Rowhammer attacks. In: ASPLOS (2016)
3. Barak, B., et al.: On the (im)possibility of obfuscating programs. In: Kilian, J. (ed.) CRYPTO 2001. LNCS, vol. 2139, pp. 1–18. Springer, Heidelberg (2001). https://doi.org/10.1007/3-540-44647-8_1
4. Bepary, M.K., Talukder, B.M.S.B., Rahman, M.T.: DRAM retention behavior with accelerated aging in commercial chips. Appl. Sci. **12**(9), 4332 (2022)
5. Bosman, E., Razavi, K., Bos, H., Giuffrida, C.: Dedup est machina: memory deduplication as an advanced exploitation vector. In: S&P 2016 (2016)
6. Cohen, F.B.: Operating system protection through program evolution. Comput. Secur. **12**(6), 565–584 (1993)
7. Cojocar, L., Razavi, K., Giuffrida, C., Bos, H.: Exploiting correcting codes: on the effectiveness of ECC memory against Rowhammer attacks. In: S&P (2019)
8. Frigo, P., Giuffrida, C., Bos, H., Razavi, K.: Grand pwning unit: accelerating microarchitectural attacks with the GPU. In: S&P (2018)
9. Frigo, P., et al.: TRRespass: exploiting the many sides of target row refresh. In: S&P (2020)
10. Godbolt, M.: The BTB in contemporary Intel chips—Matt Godbolt's blog (2016). https://xania.org/201602/bpu-part-three. Accessed 18 Apr 2024
11. Gruss, D., et al.: Another flip in the wall of Rowhammer defenses. In: S&P (2018)
12. Gruss, D., Maurice, C., Mangard, S.: Rowhammer.js: a remote software-induced fault attack in JavaScript. In: Caballero, J., Zurutuza, U., Rodríguez, R.J. (eds.) DIMVA 2016. LNCS, vol. 9721, pp. 300–321. Springer, Cham (2016). https://doi.org/10.1007/978-3-319-40667-1_15
13. Herder, C., Yu, M.D., Koushanfar, F., Devadas, S.: Physical unclonable functions and applications: a tutorial. Proc. IEEE **102**(8), 1126–1141 (2014)
14. Jattke, P., van der Veen, V., Frigo, P., Gunter, S., Razavi, K.: Blacksmith: scalable rowhammering in the frequency domain. In: S&P (2022)
15. Jattke, P., Wipfli, M., Solt, F., Marazzi, M., Bölcskei, M., Razavi, K.: ZenHammer: Rowhammer attacks on AMD Zen-based platforms. In: USENIX Security (2024)

16. Kim, J.S., Patel, M., Yağlıkçı, A.G., Hassan, H., et al.: Revisiting RowHammer: an experimental analysis of modern DRAM devices and mitigation techniques. In: ISCA (2020)
17. Kim, Y., et al.: Flipping bits in memory without accessing them: an experimental study of DRAM disturbance errors. ACM SIGARCH Comput. Archit. News **42**(3), 361–372 (2014)
18. Kogler, A., et al.: Half-double: hammering from the next row over. In: USENIX Security (2022)
19. Kohnhäuser, F., Schaller, A., Katzenbeisser, S.: PUF-based software protection for low-end embedded devices. In: Conti, M., Schunter, M., Askoxylakis, I. (eds.) Trust 2015. LNCS, vol. 9229, pp. 3–21. Springer, Cham (2015). https://doi.org/10.1007/978-3-319-22846-4_1
20. Kwong, A., Genkin, D., Gruss, D., Yarom, Y.: RAMBleed: reading bits in memory without accessing them. In: S&P (2020)
21. Larsen, P., Homescu, A., Brunthaler, S., Franz, M.: SoK: automated software diversity. In: S&P (2014)
22. Lim, D., Lee, J.W., Gassend, B., Suh, G.E., Van Dijk, M., Devadas, S.: Extracting secret keys from integrated circuits. VLSI **13**(10), 1200–1205 (2005)
23. Nagra, J., Collberg, C.: Surreptitious Software: Obfuscation, Watermarking, and Tamperproofing for Software Protection. Pearson Education (2009)
24. Orosa, L., et al.: A deeper look into RowHammer's sensitivities: experimental analysis of real DRAM chips and implications on future attacks and defenses. In: MICRO (2021)
25. Piazzalunga, U., Salvaneschi, P., Balducci, F., Jacomuzzi, P., Moroncelli, C.: Security strength measurement for dongle-protected software. Secur. Priv. **5**(6), 32–40 (2007)
26. Qiao, R., Seaborn, M.: A new approach for rowhammer attacks. In: HOST (2016)
27. Razavi, K., Gras, B., Bosman, E., Preneel, B., Giuffrida, C., Bos, H.: Flip Feng Shui: hammering a needle in the software stack. In: USENIX Security (2016)
28. de Ridder, F., Frigo, P., Vannacci, E., Bos, H., Giuffrida, C., Razavi, K.: SMASH: synchronized many-sided Rowhammer attacks from JavaScript. In: USENIX Security (2021)
29. Schaller, A., Xiong, W., Anagnostopoulos, N.A., et al.: Intrinsic Rowhammer PUFs: leveraging the Rowhammer effect for improved security. In: HOST (2017)
30. Seaborn, M., Dullien, T.: Exploiting the DRAM rowhammer bug to gain kernel privileges (2015). https://googleprojectzero.blogspot.com/2015/03/exploiting-dram-rowhammer-bug-to-gain.html. Accessed 30 Apr 2024
31. Škorić, B., Tuyls, P., Ophey, W.: Robust key extraction from physical uncloneable functions. In: Ioannidis, J., Keromytis, A., Yung, M. (eds.) ACNS 2005. LNCS, vol. 3531, pp. 407–422. Springer, Heidelberg (2005). https://doi.org/10.1007/11496137_28
32. Suh, G.E., Devadas, S.: Physical unclonable functions for device authentication and secret key generation. In: DAC (2007)
33. Tehranipoor, F., Karimian, N., Yan, W., Chandy, J.A.: DRAM-based intrinsic physically unclonable functions for system-level security and authentication. VLSI **25**(3), 1085–1097 (2017)
34. VDMA: VDMA study product piracy 2022 (2022). https://www.vdma.org/documents/34570/51629660/VDMA+Study+Product+Piracy+2022_final.pdf
35. van der Veen, V., Fratantonio, Y., Lindorfer, M., Gruss, D., Maurice, C., et al.: Drammer: deterministic Rowhammer attacks on mobile platforms. In: CCS (2016)

36. Xiong, W., Schaller, A., Katzenbeisser, S., Szefer, J.: Software protection using dynamic PUFs. IEEE Trans. Inf. Forensics Secur. **15**, 2053–2068 (2019)
37. Zhang, Z., Cheng, Y., Liu, D., Nepal, S., Wang, Z., Yarom, Y.: PThammer: cross-user-kernel-boundary rowhammer through implicit accesses. In: MICRO (2020)
38. Zhang, Z., et al.: Implicit hammer: cross-privilege-boundary rowhammer through implicit accesses. IEEE Trans. Dependable Secure Comput. **20**(5), 3716–3733 (2023)

SmmPack: Obfuscation for SMM Modules with TPM Sealed Key

Kazuki Matsuo[1], Satoshi Tanda[4], Kuniyasu Suzaki[2], Yuhei Kawakoya[3], and Tatsuya Mori[1,5,6](✉)

[1] Waseda University, 3-4-1 Okubo, Shinjuku, Tokyo 169-8555, Japan
k4zuk1@nsl.cs.waseda.ac.jp
[2] Institute of Information Security, 2-14-1 Tsuruyacho, Kanagawa-ku, Yokohama-shi, Kanagawa 221-0835, Japan
[3] NTT Security (Japan) KK, 4-14-1 Sotokanda, Chiyoda-ku, Tokyo 101-0021, Japan
[4] Satoshi's System Programming Lab, Vancouver, Canada
[5] NICT, 4-2-1 Nuki-kitamachi, Kokubunji-shi, Tokyo 184-0015, Japan
[6] RIKEN AIP, 15th Floor, Nihonbashi 1-4-1, Nihonbashi, Chuo-ku, Tokyo 103-0027, Japan

Abstract. System Management Mode (SMM) is the highest-privileged operating mode of x86 and x86-64 processors. Through SMM exploitation, attackers can tamper with the Unified Extensible Firmware Interface (UEFI) firmware, disabling the security mechanisms implemented by the operating system and hypervisor. Vulnerabilities enabling SMM code execution are often reported as Common Vulnerabilities and Exposures (CVEs); however, no security mechanisms currently exist to prevent attackers from analyzing those vulnerabilities. To increase the cost of vulnerability analysis of SMM modules, we introduced SmmPack. The core concept of SmmPack involves encrypting an SMM module with the key securely stored in a Trusted Platform Module (TPM). We assessed the effectiveness of SmmPack in preventing attackers from obtaining and analyzing SMM modules using various acquisition methods. Our results show that SmmPack significantly increases the cost by narrowing down the means of module acquisition. Furthermore, we demonstrated that SmmPack operates without compromising the performance of the original SMM modules. We also clarified the management and adoption methods of SmmPack, as well as the procedure for applying BIOS updates, and demonstrated that the implementation of SmmPack is realistic.

Keywords: UEFI · SMM · TPM2.0 · Packing

F. Maggi et al. (Eds.): DIMVA 2024, LNCS 14828, pp. 439–459, 2024.
https://doi.org/10.1007/978-3-031-64171-8_23

1 Introduction

System Management Mode (SMM) is the highest-privileged operating mode in x86 and x86-64 processors. The SMM module, a type of Unified Extensible Firmware Interface (UEFI) module, operates in a memory area called the System Management RAM (SMRAM), which is accessible only during SMM. Attackers can arbitrarily modify the BIOS image and insert malware or bypass security mechanisms, such as Virtualization Based Security (VBS) [11,32], by escalating privileges to SMM. Consequently, vulnerability researchers are actively exploring SMM modules for exploits, as shown by rising CVE reports [25].

Many vulnerabilities in CVEs are caused by implementation errors, such as memory access outside the SMRAM [6,8,26]. To exploit these vulnerabilities, attackers first need to analyze the SMM module. Unfortunately, there are currently no security mechanisms that hinder attacker's vulnerability analysis. Consequently, developers of SMM modules are forced to implement these modules without vulnerabilities to defend against the exploits. Existing security mechanisms, such as secure boot, cannot prevent such attacks because attacks utilizing these vulnerabilities do not require modification to the UEFI firmware. Therefore, security mechanisms that increase the cost for attackers are necessary.

We propose SmmPack as an obfuscation system to increase the cost of vulnerability analysis of SMM modules to attackers. To the best of our knowledge, this is the first study to cautiously explore the application of obfuscation techniques to UEFI firmware while comprehensively elucidating the required technologies, specific implementation procedures, feasibility, and constraints, thus pioneering a new research domain in this area. Note that, in this paper, the act of increasing the cost of vulnerability analysis is defined as "obfuscation."

The key technical concept of SmmPack involves sealing the key used for encrypting the SMM modules in the TPM, preventing attackers from extracting the key even on their own terminals. Key retrieval is restricted by the Platform Configuration Register (PCR), making it impossible to obtain the key after the OS has booted. Additionally, during the boot phase, if such SMM module performing SMRAM dump is added, the PCR value changes, preventing key acquisition. As a result, attackers are limited to methods such as cold boot attacks involving memory transplantation and DMA within an extremely small time frame, forcing them to resort to costly methods[1]. Furthermore, these methods may not be feasible depending on the presence of other security mechanisms.

In this study, we first defined the threat model for SmmPack. Then, we implemented a prototype and evaluated its effectiveness[2]. We conducted an assessment to determine whether SmmPack can effectively prevent various methods of acquiring SMM modules that attackers might attempt. Additionally, we demonstrate that the boot time and BIOS size overheads are both within practical limits. Finally, we explain the procedures for managing and adopting SmmPack

[1] The magnitude of these costs is discussed in Sect. 5.4.

[2] As an artifact of this paper, the implementation code of SmmPack is shown at [21].

along with the process entailed in implementing BIOS updates and substantiate the feasibility of incorporating SmmPack.

The contributions of this study have been summarized as follows:

- We presented SmmPack, the first obfuscation framework developed for platform firmware, which is uniquely positioned as a pioneering solution that protects SMM modules from attacker vulnerability analysis.
- We demonstrated the effectiveness of SmmPack as a defense against various means of SMM module acquisition.
- We showed that the impact of SmmPack on the system's performance is minimal, indicating the practicality of its implementation.
- We clarified the management and deployment methods of SmmPack, as well as the procedure for applying BIOS updates, and demonstrated that the implementation of SmmPack is realistic.

2 Background

In this section, we first provide an overview of UEFI, SMM, and TPM. Next, we explain what attackers can achieve by attacking SMM and discuss the types of vulnerabilities in SMM. Finally, we explain the security mechanisms that currently exist in UEFI.

2.1 UEFI

The Unified Extensible Firmware Interface (UEFI) is a set of specifications maintained by the Unified EFI Forum [2] that defines the interface between the platform firmware and the OS. A UEFI-compliant BIOS comprises numerous UEFI modules, most of which reside within a Serial Peripheral Interface (SPI) flash memory chip. The UEFI specifications partition the boot process into seven distinct phases, with the following five phases representing the most important stages of the procedure:

- SEC (Security): The initial code running and is the root of trust of the system.
- PEI (Pre-EFI Initialization): Initializes permanent memory and handles the different states of the system.
- DXE (Driver Execution Environment): Execute drivers that initialize platform components.
- BDS (Boot Device Selection): Select the boot device and run the boot loader.
- RT (Runtime): The phase when OS executes. UEFI environment except runtime services are discarded.

Firmware that runs earlier in the boot phase has a more platform-dependent implementation. The phase until BDS is called the boot phase, and the phase after the OS starts is called the runtime phase. In this study, we specifically focused on the DXE phase, as this is where the SMM environment is also set up. Most UEFI modules running in the DXE phase are known as DXE modules. UEFI modules running after the PEI phase are usually in the PE format.

UEFI defines a standardized format for EFI firmware storage devices abstracted into Firmware Volumes (FVs). Typically, a BIOS contains multiple FVs, with one FV containing DXE modules, including SMM modules, and another FV containing PEI modules. UEFI modules in the DXE phase are individually dispatched by the dispatcher, which implies that no modules run in parallel. The dispatcher enumerates through the FVs to locate UEFI modules and execute them. FVs can include apriori files listing module GUIDs. Modules in this list run first, following the order of the file.

EDK2 is a full implementation of the UEFI specification developed by the open-source Tianocore project [29]. It is also a UEFI module development environment, which we used to build SmmPack.

2.2 SMM

The System Management Mode (SMM) is the highest-privileged operating mode of x86 and x86-64 processors, intended for use by BIOS firmware to handle low-level system management operations such as power management, system hardware control, or proprietary OEM-designed code. SMM can only be entered through a System Management Interrupt (SMI) and can only be exited with an RSM instruction or reboot. UEFI modules that run in SMM are called SMM modules, and these modules prepare SMI handlers that perform low-level system management operations.

SMRAM. SMM modules are loaded into a separate memory region called SMRAM, which can only be accessed during SMM and cannot be read via Direct Memory Access (DMA). SMM is typically initialized during the DXE phase. In this phase, SMM modules are also loaded and executed from their entry points. At the end of the DXE phase, the SMRAM is locked by setting Model Specific Registers (MSRs), which control the SMRAM. This locking makes these MSRs read-only and restricts the visibility of SMRAM solely to SMM.

SMM System Table and Protocols. SMM has a data structure called the SMM system table that allows access to the various functions used in most SMM modules. One of the functionalities provided by the SMM system table is the protocol. When a certain SMM module registers its provided function as a protocol in the system, other SMM modules can utilize that function. Protocols primarily consist of a GUID and a protocol interface structure. When the protocol is used, functions such as SmmLocateProtocol of the SMM system table first find the protocol interface structure based on the GUID. The specified function is then called, referencing the function pointers in the protocol interface structure. The protocol interface structure is usually defined in the data section of the SMM module that installed that protocol. The actual functions pointed to the protocol interface structure are also defined inside the SMM module.

CommBuffer. There are several ways to communicate data between non-SMM and SMM, the most common of which is to use the SMM communication buffer

(CommBuffer) via the SMM communication protocol or the Advanced Configuration and Power Interface (ACPI) table. When using CommBuffer, SMI handlers should call the SmmIsBufferOutsideSmmValid function to validate the data. This is because if the data passed from the CommBuffer point to the data inside SMRAM, the non-SMM code can manipulate the data inside SMRAM.

2.3 TPM

Trusted Platform Module (TPM) is a secure cryptoprocessor designed to safeguard the artifacts used for platform authentication. TPM specification has been developed by the Trusted Computing Group (TCG) [31]. Currently, TPM 2.0 is the latest version of the specification. In this study, we focused primarily on TPM 2.0, with no discussion of TPM 1.2, its predecessor. TPM is equipped with numerous anti-tampering mechanisms that enable it to effectively resist a wide array of hardware and side-channel attacks.

PCR. One of the essential features of TPM is the Platform Configuration Register (PCR), which provides a method for measuring the state of software. The value stored in the PCR is a hash of the software code or configuration data and can only be updated through an extend (or reset) operation. By referencing the PCR value, the platform firmware ensures that no code executed during the boot phase is altered. If an attacker attempts to run a tampered UEFI module during the boot chain, the change in PCR values can detect tampering. The extend operation hashes the concatenated value of the new measurement and the current PCR value. A standard TPM comprises 24 PCRs, each designated to store distinct types of measurements. In this study, we focused solely on PCR0, where the system firmware measurements are stored. The DXE phase measurement in the EDK2 framework was conducted at Firmware Volume (FV) granularity using the Tcg2Pei module during the Pre-EFI Initialization (PEI) phase [28].

Session. Many TPM commands require the use of sessions that can be divided into the following three types: (1) Password, (2) HMAC, (3) Policy. Among these three, policy session is the most powerful, allowing authentication based not only on passwords but also taking into account PCR values and other external information. In this study, we adopt a policy session.

Sealing. Storing secrets in TPM with the policy session using PCR is called sealing. Sealing can confine the accessibility of stored data to specific instances during the boot phase when platform integrity is assured.

NV Space. A certain amount of nonvolatile (NV) space is available in the TPM for user-configured storage. Authorization can be applied to the data stored in NVRAM, allowing the data to be sealed inside. The data placed in NVRAM can be accessed through a handle called the NV index.

2.4 Vulnerabilities and Attacks in SMM

By escalating privileges to SMM, attackers can arbitrary write to SPI flash, bypass hypervisor-based security mechanisms [11,32], and provide stealthy functions at OS runtime [4]. Moreover, all attacks that can be done by infecting other UEFI modules are also possible from SMM modules.

Various types of vulnerabilities that can trigger the above attacks have been identified in SMM modules, including SMM callout [6,8], confused deputy [27], and buffer overflow [7,10]. Most CVEs are caused by implementation errors in SMM modules. However, attacks against SMM are not widely known and obfuscation to hide these vulnerabilities is not currently performed.

2.5 Existing Security Mechanisms

Existing security mechanisms for preventing attacks on UEFI include secure boots and an Intel Boot Guard.

Secure Boot or Verified Boot. These security mechanisms are designed to ensure the system integrity. The central idea is to ensure that the system firmware has not been tampered with by verifying the hash of the firmware. Therefore, it is impossible to prevent attacks that do not tamper with the code. Secure boot mainly checks the integrity of the firmware after PEI and operates with trust in the OEM platform firmware before DXE.

Intel Boot Guard. Intel Boot Guard [15] verifies the integrity of both the SEC and PEI phases. The Intel Boot Guard is a hardware-based technology that cannot be disabled by users. In AMD, similar functionality is implemented as a Platform Secure Boot (PSB) [22], which also cannot be disabled by the user.

3 Threat Model

The purpose of SmmPack is to increase the cost of vulnerability analysis of SMM modules by attackers. Therefore, the threat model of SmmPack is defined as "attackers aiming to discover vulnerabilities in SMM modules."

SmmPack is not intended to prevent attacks towards SMM. Its purpose is to increase the cost of vulnerability analysis conducted prior to those attacks. To the best of our knowledge, no research has been conducted on the encryption or obfuscation of system firmware, including SMM modules. Consequently, there are no existing threat models focused on the vulnerability analysis of SMM modules. Therefore, there is a need to define a threat model targeting attackers aiming to conduct vulnerability analysis on SMM modules.

Attackers who intend to analyze SMM modules to identify vulnerabilities can obtain the modules from their own PC. Therefore, the threat model must consider the Man-at-the-End (MATE) model [13]. In general, research considering this model finds it challenging to completely prevent attackers from conducting vulnerability analysis. Instead, the focus is on increasing the costs for attackers.

It is necessary not only to address attacks on SMM but also to take measures for the vulnerability analysis of SMM modules. This is because many attacks that exploit vulnerabilities in the SMM modules cannot be prevented using existing security mechanisms. Attacks such as buffer overflow through CommBuffer [7,10] and SMM callout attacks [6,8] do not require tampering with the SMM module. Thus, such security mechanisms like secure boot is ineffective. It is highly challenging for developers to comprehend all SMM-related attacks and implement all modules without any vulnerabilities. Furthermore, even if vulnerabilities are identified, BIOS has the characteristic of having a significantly longer period until a patched version is released compared to regular software. Therefore, increasing the cost of vulnerability analysis is an important countermeasure.

In this study, the threat model was classified into the following three classes based on the attacker's accessible scope. In this threat model, attackers aim to obtain decrypted SMM modules and the key. Vulnerability analysis is also prevented by preventing the acquisition methods. Class 3 attackers have greater access than class 1 attackers; however, they must employ relatively expensive methods to obtain SMM modules.

Class 1 attacker: attacker with access to only BIOS update file.
Class 2 attacker: attacker with arbitrary software code execution above OS.
Class 3 attacker: attacker with hardware access and equipment.

Class 1 attacker. The lowest cost way to obtain SMM modules is to acquire them from BIOS update files. These BIOS update files can be freely downloaded from various manufacturer websites, making it cheaper than purchasing a device.
Class 2 attacker. A Class 2 attacker aims to obtain SMM modules by purchasing a PC containing the BIOS image they want to analyze and obtaining SMM modules from the software running on the OS. This attacker can attempt to dump SPI flash from the software, attempting to access keys through software-based access to the TPM using tools such as tpm2-tools [24], and trying to obtain decrypted SMM modules loaded in memory through memory dumps.
Class 3 attacker. A Class 3 attacker can attempt to acquire SMM modules using various methods, ranging from low-cost approaches, such as dumping the BIOS image from the SPI flash chip, to more costly methods, such as acquiring the decrypted SMM modules from memory via cold boot attacks, which require considering various hardware-specific security mechanisms. Attackers can also execute arbitrary SMM modules by flashing the SPI flash.

When evaluating using this threat model, the following assumptions are made: (1) the TPM module is tamper-resistant, (2) attacks on cryptographic algorithms themselves are considered out-of-scope, and (3) the integrity of the SEC and PEI phases is hardware-protected by the Intel Boot Guard or AMD PSB. Therefore, in this paper, we consider vulnerabilities in TPM itself, weaknesses in cryptographic algorithms, and tampering of SEC and PEI phases to be out of scope.

4 Implementation

In this section, we explain the implementation of SmmPack. First, we provide an overview of SmmPack, followed by a description of the preparation of the protocol used for the decryption, sealing, and unsealing of keys, and details of the packer. In Sect. 7, we discuss the validity of the design.

4.1 SmmPack Overview

Fig. 1 illustrates an overview of SmmPack. SmmPack packs the SMM module by encrypting its code section and adding the decrypt stub at its end. When the packed SMM module is executed, the execution starts from the decrypt stub, which calls the Unpack function of SmmPackProtocol that the SmmPackSmm module installs. Subsequently, SmmPackSmm will decrypt the code section of the packed SMM module using a key stored inside TPM.

Figure 2 shows the key operations related to SmmPack at a high level. SmmPackSmm loads the decryption key from the TPM in the early DXE phase. Because the key is sealed, the PCR value must be a specific value. The PCR value was calculated in FV units in the PEI phase. After SmmPackSmm installs the SmmPackProtocol, the packed SMM modules are executed, and once executed, they remain decrypted in SMRAM throughout the runtime.

Table 1 lists the components used in SmmPack, and Table 2 presents the environmental information. It is assumed that the packing of SMM modules will be performed by the OEM before shipment by executing the packer program only once. The keys are unique values for each BIOS implementation and are sealed before shipment to the non-volatile area of the TPM.

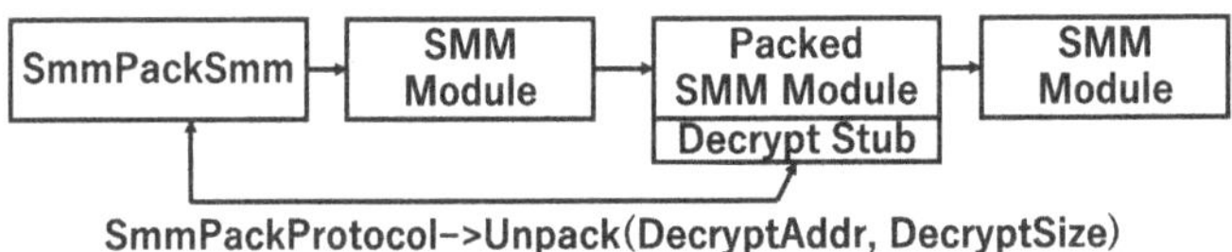

Fig. 1. SmmPack overview.

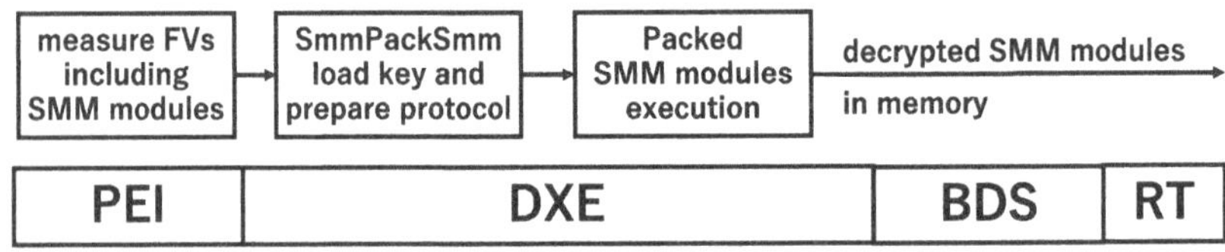

Fig. 2. High level core operations in SmmPack.

Table 1. SmmPack components.

name	description
SealKeyDxe	Seal key into the TPM
SmmPackSmm	Provides protocols for unpacking
smm-packer	Packer program
tiny-AES-c [19]	Library for AES

Table 2. Experiment setup.

name	description
Visual Studio 2019	for building packer program/C++
EDK2	for building UEFI modules
UP Squared Pro Atom 04/64 [23]	for executing UEFI modules
CPU	Intel Atom® x7-E3950
BIOS	UNAPAM22
TPM	Infineon SLB9665

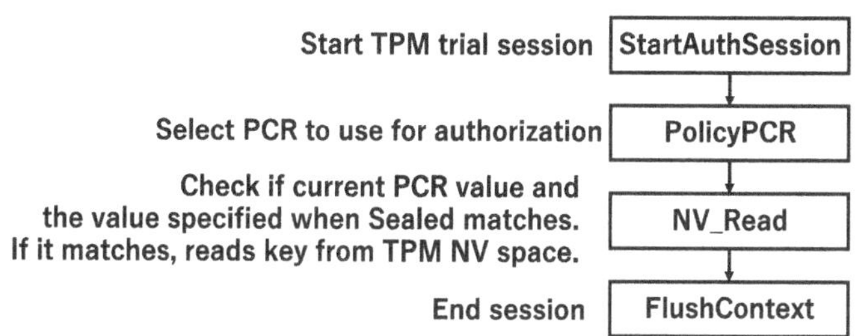

Fig. 3. Key unsealing commands.

4.2 SmmPackSmm

SmmPackSmm mainly performs the following two processes: (1) Unsealing the key from TPM and (2) Registering a protocol for decryption. SmmPackSmm is an SMM module that installs an SMM-specific protocol called SmmPackProtocol for unpacking packed SMM modules. When the packed SMM module is dispatched, it starts execution from the decrypt stub added by our packer, and the decrypt stub calls the Unpack function of SmmPackProtocol to decrypt itself. To pack more SMM modules, the protocol must be installed before any packed SMM module is executed. For this purpose, the GUID of SmmPackSmm should be written to the apriori file of its FV. Despite multiple FVs, the protocol can be used across FVs. Therefore, if SmmPackSmm is registered in the apriori file of the initially executed FV, the modules in other FVs can also be packed.

Unsealing Key from TPM. To unseal the decryption key, SmmPackSmm sends commands to the TPM as shown in Fig. 3, using the Tpm2DeviceLib library in EDK2. Tcg2Protocol and similar protocols cannot be used with SmmPack because they are not available until the DXE module that installs them is executed. First, TPM2_StartAuthSession is used to start a policy session, followed by TPM2_PolicyPCR to select the PCR to use for authorization for this session. Finally, TPM2_NV_Read is used to read the key from the nonvolatile storage of the TPM, with verification of the current PCR value to ensure that it matches the enrolled PCR value specified when the key was sealed. The key is then saved as a global variable in SmmPackSmm. It is also possible to perform decryption inside the TPM without loading the key into memory. However, this method requires SmmPackSmm to send the encrypted data (the entire text section of the packed SMM modules) to the TPM, which significantly increases the execution time. Loading the key into memory does not increase the attack vectors in the context of this study, if the key could be read from memory, so too can the unpacked SMM modules.

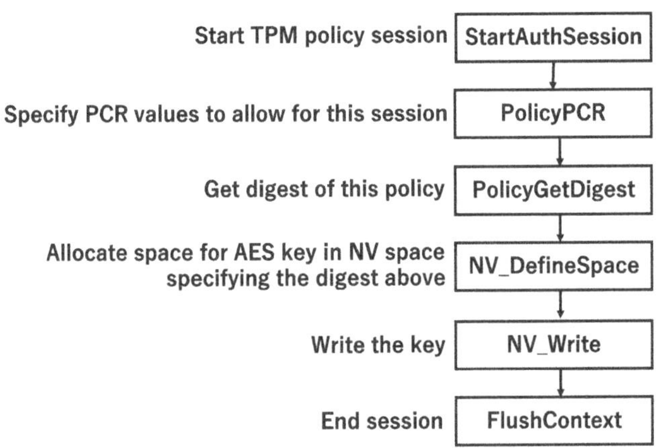

Fig. 4. Key sealing commands.

Registering a Protocol for Decryption. After the key is unsealed, SmmPackSmm installs a protocol that decrypts the packed SMM modules. The protocol is a table of function pointers, and in SmmPack, only one function, Unpack, is required. The Unpack function takes the base address and the size of the encrypted text section of SMM modules and decrypts it using the key stored in the global variable. The encryption algorithm used in SmmPack is AES-128 in CBC mode, which is implemented by using the open-source library tiny-AES-c. However, this can be any similar-strength symmetric algorithm or implemented using any other library. After the protocol is installed, the subsequent SMM modules can locate it using its GUID and call the Unpack function.

4.3 Key Sealing

The key used by SmmPack must have a unique value for each BIOS implementation. SmmPack aims to prevent the analysis of SMM modules; therefore, if the BIOS code is the same, encryption using a different key is not required. Figure 4 presents an overview of sealing a key to a TPM. TPM2_NV_DefineSpace allocates space to place a key inside the nonvolatile storage of the TPM. This command is sent with a digest of the policy obtained from TPM2_PolicyGetDigest, which contains the information on the PCR value that can access this space. This information was specified by TPM2_PolicyPCR on a trial session initiated by TPM2_StartAuthSession. SealKeyDxe, which performs this operation, should be executed before packing any SMM module with SmmPack. Alternatively, this process can be performed using other tools, such as tpm2-tools [24].

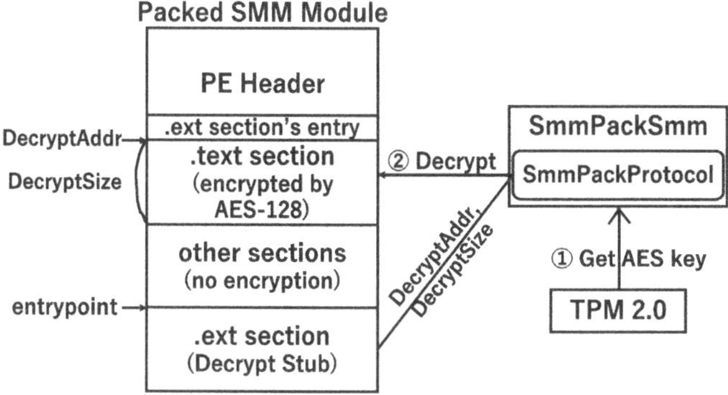

Fig. 5. Structure of packed SMM module.

4.4 Packer Design

The packer program was implemented separately from the other components as an application running on top of the OS. This is also implemented using tiny-AES-c because the algorithms used for encryption and decryption must be the same. Figure 5 shows the structure of the packed SMM module. The packer first encrypts the text section of the SMM module with AES. Then, the packer adds the ext section (decrypt stub) containing the decryption code at the end of the SMM module. The entry point of the module should also be moved to the decrypt stub. SmmPack encrypts only the text section because the purpose of SmmPack is to prevent vulnerabilities from being discovered by analyzing the codes of the SMM module.

5 Security Analysis

In this section, we assess the extent to which SmmPack increases the cost of vulnerability analysis, utilizing the threat model defined in Sect. 3. We presented the overall results in Table 3.

In summary, SmmPack can completely prevent all acquisition methods for Class 1 and Class 2 attackers and can restrict Class 3 attackers' acquisition methods to only cold boot attacks and DMA. However, neither of these acquisition methods may be applicable because of the existence of specific security mechanisms. Even if these security mechanisms were not present, performing these acquisition methods against SmmPack would still be a highly cost-intensive task. Detailed explanations are provided in Subsect. 5.4. Hence, using SmmPack, the avenues for attackers to acquire SMM modules are narrowed down to methods with a significantly higher cost.

Table 3. Threat models and attacks SmmPack can address (O represents acquisition methods that can be prevented by SmmPack, while X represents methods that cannot be prevented by SmmPack.).

class 1	BIOS update file	O
class 2	SPI flash dump (software)	O
	key read from TPM (software)	O
	memory dump from OS	O
class 3	SPI flash dump (hardware)	O
	key read from TPM (hardware)	O
	SMRAM read from SMM module	O
	DMA	X
	coldboot attack (same device)	O
	coldboot attack (memory transplant)	X

5.1 Class 1 Attacker

BIOS Update File. Although changes to the BIOS update method are necessary, distributing a BIOS image file containing SMM modules in an encrypted state would prevent attackers from analyzing them because they would not possess the decryption key. Further details regarding the BIOS update method upon adoption of SmmPack are discussed in Sect. 7.

5.2 Class 2 Attacker

SPI Flash Dump (Software). In this case, the most reliable method for obtaining SMM modules is to dump the BIOS from the SPI flash, which can be performed using tools such as CHIPSEC [12] or RWEverything [1]. With SmmPack, the obtained modules are encrypted; thus, attackers cannot analyze them.

Key Read from TPM (Software). Even if the modules are encrypted, an attacker capable of retrieving the key can still analyze their content. It is possible to access the TPM from the software running on top of the OS using tools such as tpm2-tools [24]. However, it is not possible to unseal the key from the TPM on top of the OS because the PCR value has changed by then[3]. Therefore, even if a packed SMM module is obtained on top of the OS, it cannot be analyzed.

Memory Dump from OS. Because the packed SMM modules are placed in memory in a decrypted state after execution, attackers can obtain the decrypted modules through a memory dump without the need to acquire both the encrypted SMM module and the key. Although SMM modules exist in memory during runtime, they are located in SMRAM, making it impossible for the software on the OS to read. SMRAM is accessible only by SMM and SMM is entered only by SMI. Consequently, SMRAM can only be dumped from SMM modules. Hence, Class 2 attackers cannot obtain SMM modules using memory dumps.

5.3 Class 3 Attacker

SPI Flash Dump (Hardware). Using SmmPack, SMM modules within the SPI flash are always stored in an encrypted state at any point in time. Therefore, this method by itself does not allow an attacker to analyze the SMM module.

Key Read from TPM (Hardware). To decrypt the encrypted module, the attacker must obtain the key from the TPM. Note that the vulnerability of TPM's tamper-resistance feature is considered beyond the scope of this paper. Therefore, we assume that the physical retrieval of the key is prevented.

SMRAM Read from SMM Module. While attackers can disable secure boot on their PC and write an SMM module that performs an SMRAM dump, introducing any SMM modules will alter the PCR value. As a result, neither the key nor the decrypted SMM modules are loaded in the SMRAM by SmmPackSmm. PCR values are usually measured in the PEI phase on a per-FV basis before transitioning to the DXE phase [28]. Therefore, even when running SMM modules in the early DXE phase, the PCR value changes.

DMA. An attacker can also read data from the memory via Direct Memory Access (DMA) without requiring any additional modules to write. SMRAM is an area that cannot be accessed via DMA, so generally, this is not feasible. However, the exact point at which SMRAM becomes inaccessible is after the SMRAM is locked. Consequently, performing DMA during the DXE phase after

[3] To ensure certainty, it is also possible to extend PCR0 in SmmPackSmm after unsealing the key.

the execution of SmmPackSmm is possible until SMRAM is locked. This allows for the retrieval of decrypted SMM modules or keys.

Another possible method is a cold boot attack [3], which takes advantage of the delay until the DRAM content is completely erased and the residual data are read out. Cold boot attacks can be further categorized into two methods based on their approach. The first method involves performing a cold reset within the same device and then reading the residual data. The second method involves freezing memory modules, transplanting them to another device, and then reading the residual data on that device.

Cold Boot Attack (Same Device). In this case, it is impossible to obtain the decrypted SMM module for two reasons. To carry out this attack, an attacker must follow steps similar to the following: (1) Power down the PC after all SMM modules have been loaded, (2) Freeze the memory modules using cold spray, (3) Write a DXE module that reads the residual data and stores it in the SPI flash (configure it to run during the initial stages of the DXE phase, e.g., using apriori files), and (4) Boot up the PC again. An essential consideration is that the module responsible for reading the residual data must run before the BDS phase. This is because the residual SMRAM data are overwritten with new data during the DXE phase. As a premise of SmmPack, as stated in Sect. 3, SEC and PEI phases are protected at the hardware level to prevent tampering. Thus, attackers must read the residual data through a DXE module. Furthermore, if the DXE module is added, the PCR value changes, halting the boot process when SmmPackSmm is executed. Therefore, the DXE module must be executed during the initial stages of the DXE phase before SmmPackSmm is executed.

To insert a DXE module that reads the residual data, attackers must flash the BIOS of the system. Typically, flashing a BIOS requires several minutes to complete. However, modern DDR3 and DDR4 memory modules struggle to retain data for more than a minute, even when adequately cooled [5,34]. Therefore, it is considered impractical to obtain residual SMRAM data through a cold boot attack on the same system.

Furthermore, most systems employ the Memory Overwrite Request (MOR) mechanism [30], which zero-clears the memory when a clean shutdown is not performed, as in the case of a cold reset. For these two reasons, a cold boot attack on the same system cannot retrieve the decrypted SMM modules.

Cold Boot Attack (Memory Transplant). In this scenario, an attacker can freeze the memory, transplant it to another PC containing the DXE module that reads the residual SMRAM data, and then boot. MOR bit is not effective, because performing a clean shutdown and setting the bit to zero on the target PC beforehand would prevent memory zero-clearing during POST.

5.4 High Cost of DMA and Cold Boot Attacks

Although SmmPack may not prevent DMA and cold boot attacks involving memory transplant, these methods may not be viable depending on the presence of specific security mechanisms. Furthermore, even without those, attempting these acquisition methods against SmmPack remains a highly challenging task.

DMA Difficulties. There are two non-negligible reasons why the retrieval of decrypted SMM modules or keys using DMA is an extremely challenging task. First, the window of time between the initialization and the locking of SMRAM is very brief. Consequently, it is difficult to time the DMA operation. Attempts to insert a deadlock loop would alter the PCR values, preventing the loading of keys and halting the execution in SmmPackSmm. This limits the duration for which DMA can be performed, thereby increasing the cost for attackers.

Second, modern CPUs include various security mechanisms for DMA protection, including the DMA Protected Range (DPR) [17]. The memory region specified by DPR is inaccessible to DMA, and typically, DPR covers the memory areas used by SMRAM. If this security mechanism is active, attackers cannot use DMA to retrieve decrypted SMM modules or keys. Even if an attacker attempts to modify CPU registers to alter DPR settings, executing firmware to achieve this change requires changing the BIOS firmware, which alters PCR values. Consequently, if this security mechanism is enabled, DMA is unavailable.

Cold Boot Attack Difficulties. Performing cold boot attacks requires consideration of various security mechanisms present in memory modules and system-on-chips (SoCs). DDR3 and later DRAM modules have a memory-scrambling feature that allows only scrambled data in memory, which is descrambled only within the SoC. Although descrambling of DDR3 and DDR4 DRAM modules are possible [5,34], this will significantly increase the attacker's cost.

Moreover, certain SoCs employ technologies such as Intel's Total Memory Encryption (TME) [16], Total Memory Encryption Multi Key (TME-MK) [20], and AMD's Secure Memory Encryption (SME) [18]. With these features enabled, data are always stored in an encrypted state in the memory and decrypted within the SoC only. Cold boot attacks are thwarted when these features are active. Although these mechanisms can often be disabled by the user from the BIOS setup screen, if these features are designed such that they cannot be disabled by design, attackers will be unable to perform cold boot attacks.

6 Performance Evaluation

In this section, we evaluate the impact of SmmPack on system performance and demonstrate that the introduction of SmmPack is realistic.

6.1 Evaluation Method

Both the speed and size of the overhead depend on the number of SMM modules to be packed. The number of SMM modules varies depending on the BIOS. In the case of the BIOS used in this experiment, there were 39 SMM modules. Measurements of clock cycles and size in bytes were calculated using this number of modules; however, the overhead was represented as a percentage.

The boot-time delay caused by SmmPack is the sum of the execution times of SmmPackSmm and the decryption stubs of multiple packed modules. The execution time was measured by obtaining the number of clock cycles using rdtsc instruction. Also, the execution time of decryption stubs is proportional to the size of the text section to be decrypted. The SMM module included in the BIOS used in our experiment had a text section size of approximately 3000 to 50000 bytes, so we measured the number of clock cycles required to decrypt 3000, 20000, and 50000 bytes. All clock cycle counts were calculated by repeating the same work five times and taking the average value. The clock cycles measured are shown in seconds when executed on a 2.0 GHz CPU.

The increase in size was calculated by the sum of the size of SmmPackSmm itself and the sizes of the decryption stubs added to the packed SMM modules. These sizes are fixed, but the total size of the decryption stubs scales linearly with the number of SMM modules being packed.

6.2 Speed Overhead

The required clock cycle counts and the corresponding execution times on a 2.0GHz CPU for using SmmPack are shown in Table 4. The number of clock cycles required for SmmPackSmm execution was 460702429.0. This is about 0.23 s in a 2.0GHz CPU.

In terms of the decryption time, assuming 39 SMM modules, each with a text section size of 20000 bytes, the overhead was approximately 0.13 s. When combined with the execution time of SmmPackSmm, the total overhead in boot time was approximately 0.36 s. This is approximately 2.77% of the entire boot time in our experimental environment and can be considered a realistic value.

6.3 Size Overhead

The size of SmmPackSmm was 11648 bytes, and the size of the decryption stub was 188 bytes. If 39 SMM modules are packed, the total size of the BIOS will increase by 18980 bytes. This was approximately equal to the size of the average UEFI module. This was only 0.1% of the total data stored in the UP Squared Board Pro's SPI flash chip. Therefore, the increase in size owing to the introduction of SmmPack is negligible and can be considered a realistic value.

Table 4. Average clock cycles and time required in SmmPack.

Measurement target	Clock cycles	Time (s)
SmmPackSmm (whole)	460702429.0	0.23035
SmmPackSmm (key retrieval)	460638762.5	0.23032
Unpack 3000 bytes	1019979.0	0.00051
Unpack 20000 bytes	6766215.4	0.00338
Unpack 50000 bytes	16914954.6	0.00846

7 Discussion

In this section, we discussed the value of SmmPack as well as the management and adoption, and BIOS updates regarding SmmPack.

7.1 Value of SmmPack

There may be various opinions on whether the obfuscation provided by SmmPack justifies its cost. Furthermore, considering the practical implementation of SmmPack, there would likely be a need for discussions that extend beyond the page count of this paper. However, the true value of this study lies in the fact that it is the first to apply obfuscation to system firmware. We believe that the value of this study lies in shedding light on the effects, costs, the setup of threat models, and identifying critical points of focus when evaluating obfuscation systems towards system firmware.

7.2 Management and Adoption of SmmPack

To introduce SmmPack, OEM needs to perform certain tasks before and after shipping the BIOS. The tasks performed before shipping the BIOS are as follows: (1) determine the key for the BIOS, (2) pack each SMM module, (3) obtain the PCR value when SmmPackSmm is executed, and (4) seal the key into the TPM of the PC that uses the BIOS.

There are multiple methods to perform (4). One approach is to boot the OS with the unpacked SMM module included in the BIOS and use tools such as tpm2-tools on top of the OS to extract the key. After sealing, power off the system, replace the BIOS with the BIOS containing packed SMM module, and verify if the system boots properly. Alternatively, the code that seals the key to the TPM can be executed as a UEFI module without booting the OS. One important consideration is to restrict access to the key solely to the platform firmware entity. Permitting access to the key by the owner entity can lead to the erasure of the key, because of the TPM initialization process executed by the operating system. The task that the OEM should perform after the BIOS is shipped is the BIOS update, as described in the following subsection.

Regarding scalability, SmmPack can pack a single SMM module by running a single-packer program. This task can be performed simultaneously with the SMM module build process by incorporating it at the end of the build tool, such as EDK2. Additionally, because SmmPack uses the same key and PCR values for PCs of the same model, it does not require complex key management or provisioning tasks. Moreover, it can be integrated into the existing BIOS update process as explained in the following subsection. Therefore, it can be said that scalability is not compromised.

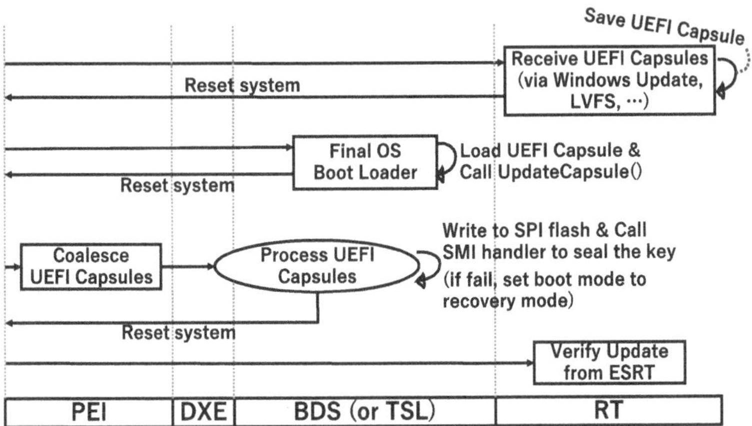

Fig. 6. UEFI capsule update with SmmPack (Figure adapted from [35] for this paper.).

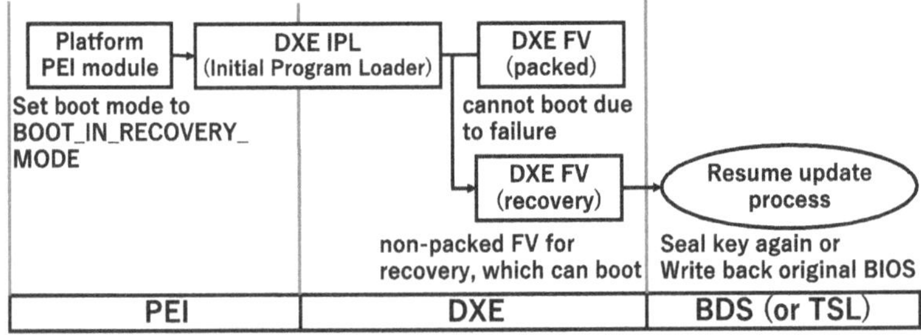

Fig. 7. Recovery from update failure.

7.3 BIOS Update

To prevent the analysis of the modules obtained from BIOS update file, they must be distributed in a packed state. Moreover, the key and the new PCR value must be stored in the receiver's TPM without being read by the receiver. The key received from the BIOS update server must be sealed in SMM because placing the key in a normal memory region during the update process can be easily read. Therefore, it is desirable to receive the key in an encrypted state and decrypt it inside the SmmPackSmm's SMI handler. The difficulty in extracting data from SMRAM is demonstrated in Sect. 5.

UEFI capsule update [33] can still be performed, even with the adoption of SmmPack. Figure 6 illustrates the flow of the update process. The capsule creator first constructs a capsule by including the packed SMM modules, a new encrypted key, and a new PCR value. When the PC receiving the capsule reaches the stage where the coalesce process is done and finishes writing the capsule contents to the SPI flash, it passes the encrypted key and new PCR value to the SMI handler mentioned above to seal the key.

If the key sealing process fails, recovery can still be performed. Figure 7 illustrates the flow of the recovery process. SmmPack targets SMM modules in the DXE phase, leaving the PEI modules untouched. This allows the system to boot, with or without SmmPack, up to the PEI phase. During the normal recovery process (without SmmPack), when the BIOS update process fails, a specific PEI module retrieves a recovery DXE FV from an external storage device and uses it to boot and perform the remaining update operations [33]. With SmmPack, after booting from the recovery image, the sealing process should be performed again at this point. It should be noted that the DXE FV for recovery needs to include an SMI handler that seals the key as described above. If the sealing process still fails, it is possible to write the original BIOS back. In the case of a failure occurring after the original key has been undefined but before sealing the new key, it is recommended to save the original key value in another NV space of the TPM before performing the undefine operation. Once the new key has been confirmed to be successfully sealed, the original key can be deleted. Therefore, the recovery process can also be performed with SmmPack adopted.

7.4 Shared Code Problem

In reality, a non-negligible amount of shared code is found among different vendors [9,14]. Therefore, even if a company adopts SmmPack, an attacker can analyze the vulnerabilities of the BIOS code of other companies that have not adopted SmmPack and apply the exploit to the former company, if the code they use is shared and this fact is known. However, the SMM modules that are most security-conscious are those that will be introduced in the future. When these new SMM modules are introduced in the future, SmmPack obfuscation can be applied to make it significantly more difficult for attackers to analyze their vulnerabilities. In addition, even if a packed SMM module contains shared code, it is not possible to determine whether the shared code is actually present in that module until its contents are analyzed.

8 Conclusion

We developed the first obfuscation mechanism in platform firmware, SmmPack. SmmPack is a packing method that uses a key sealed inside the TPM as a security mechanism to increase the cost for attackers who aim to acquire and analyze vulnerabilities in SMM modules. Experimental results using a prototype implementation showed that SmmPack had no adverse effects on the original module, and the increase in boot time and BIOS size was realistic. Furthermore, we clarified the management and adoption methods for SmmPack as well as the procedure for applying BIOS updates.

Acknowledgements. We would like to express our sincere gratitude to Xeno Kovha for his invaluable expertise and guidance on UEFI security.

References

1. RWEverything read & Write everything. http://rweverything.com/ (2017)
2. Unified extensible firmware interface forum. https://uefi.org/ (2021)
3. Halderman, J.A., et al.: Lest we remember: cold boot attacks on encryption keys. In: Proceedings of 17th USENIX Security Symp, pp. 91–98 (2008)
4. Ant: deitybounce ant product data. https://www.eff.org/files/2014/01/06/20131230-appelbaum-nsa_ant_catalog.pdf (2013)
5. Bauer, J., Gruhn, M., Freiling, F.C.: Lest we forget: cold-boot attacks on scrambled ddr3 memory. In: Proceedings of the Third Annual DFRWS Europe, pp. s65–s74 (2016)
6. Binarly: [BRLY-2021-004] SMM callout vulnerability in SMM driver on multiple HP devices (SMM arbitrary code execution). https://www.binarly.io/advisories/BRLY-2021-004/index.html (2021)
7. Binarly: [BRLY-2021-032] The heap buffer overflow vulnerability in child SW SMI handler on multiple HP devices. https://www.binarly.io/advisories/BRLY-2021-032/index.html (2021)
8. Binarly: [BRLY-2021-040] SMM Callout Vulnerability In SMM Driver On Multiple HP Devices. https://www.binarly.io/advisories/BRLY-2021-040/index.html (2021)
9. Binarly: firmware supply chain is hard(coded). https://www.binarly.io/posts/Firmware_Supply_Chain_is_Hard(coded)/index.html (2021)
10. Binarly: [BRLY-2022-016] Stack overflow vulnerability In SMI handler. https://www.binarly.io/advisories/BRLY-2022-016/index.html (2022)
11. Bulygin, Y., Gorobets, M., Furtak, A., Bazhaniuk, A.: Fractured backbone: breaking modern OS defenses with firmware attacks. https://www.blackhat.com/docs/us-17/wednesday/us-17-Bulygin-Fractured-Backbone-Breaking-Modern-OS-Defenses-With-Firmware-Attacks.pdf (2017)
12. chipsec: chipsec. https://github.com/chipsec/chipsec (2023)
13. Collberg, C., Davidson, J., Giacobazzi, R., Gu, Y.X., Herzberg, A., Wang, F.Y.: Toward digital asset protection. In: Proceedings of the 2011 IEEE Intelligent Systems, pp. 8–13 (2021)
14. Hudson, T., Kovah, X., Kallenberg, C.: Thunderstrike 2: Sith strike a macbook firmware worm. https://legbacore.com/Research_files/ts2-blackhat.pdf#page=22 (2015)
15. Intel Corporation: Intel® hardware shield - below-the-OS security. https://www.intel.com/content/dam/www/central-libraries/us/en/documents/below-the-os-security-white-paper.pdf, May 2021
16. Intel corporation: intel® hardware shield - intel® total memory encryption. https://www.intel.com/content/dam/www/central-libraries/us/en/documents/white-paper-intel-tme.pdf (2021)
17. Intel corporation: intel® trusted execution technology (intel® txt). https://cdrdv2-public.intel.com/315168/315168_TXT_MLE_DG_rev_017_4.pdf (2023)
18. Kaplan, D., Powell, J., Woller, T.: AMD memory encryption. https://www.amd.com/system/files/TechDocs/memory-encryption-white-paper.pdf (2021)
19. kokke: tiny-aes-c. https://github.com/kokke/tiny-AES-c, December 2021
20. Lin, J.: Multi-key total memory encryption on windows 11 22h2. https://techcommunity.microsoft.com/t5/windows-os-platform-blog/multi-key-total-memory-encryption-on-windows-11-22h2/ba-p/3683043 (2022)
21. MachineHunter: Smmpack. https://github.com/MachineHunter/SmmPack (2023)

22. Malhotra, A.: Amd ryzen™ pro 5000 series mobile processors making defenses count: Designing for substantial depth. https://www.amd.com/system/files/documents/amd-security-white-paper.pdf (2021)
23. shop, U.: Up squared pro atom quad core 04/64. https://up-shop.org/up-squared-pro-atom-quad-core-0464.html (2023)
24. tpm2 software: tpm2-tools. https://github.com/tpm2-software/tpm2-tools (2023)
25. The MITRE corporation: CVE search results. https://cve.mitre.org/cgi-bin/cvekey.cgi?keyword=smm (2023)
26. TheSecMaster: be aware about these six unpatched SMM vulnerabilities in HP enterprise devices. https://thesecmaster.com/be-aware-about-these-six-unpatched-smm-vulnerabilities-in-hp-enterprise-devices/, September 2022
27. TianoCore community: 38. SW SMI confused deputy SmramSaveState.c. https://edk2-docs.gitbook.io/security-advisory/sw-smi-confused-deputy-smramsavestate_c (2021)
28. TianoCore community: Tcg trusted boot chain in edk ii. https://tianocore-docs.github.io/edk2-TrustedBootChain/release-1.00/3_TCG_Trusted_Boot_Chain_in_EDKII.html, March 2021
29. TianoCore community: edk2. https://github.com/tianocore/edk2 (2023)
30. Trusted computing group: TCG platform reset attack mitigation specification. https://www.trustedcomputinggroup.org/wp-content/uploads/Platform-Reset-Attack-Mitigation-Specification.pdf (2008)
31. Trusted computing group: https://trustedcomputinggroup.org/ (2023)
32. Wojtczuk, R.: Analysis of the attack surface of windows 10 virtualization-based security. https://www.blackhat.com/docs/us-16/materials/us-16-Wojtczuk-Analysis-Of-The-Attack-Surface-Of-Windows-10-Virtualization-Based-Security.pdf (2016)
33. Yao, J., Zimmer, V.J.: A tour beyond bios capsule update and recovery in edk ii. https://github.com/tianocore-docs/Docs/raw/master/White_Papers/A_Tour_Beyond_BIOS_Capsule_Update_and_Recovery_in_EDK_II.pdf (2016)
34. Yitbarek, S.F., Aga, M.T., Das, R., Austin, T.: Cold boot attacks are still hot: security analysis of memory scramblers in modern processors. In: Proceedings of the 2017 IEEE HPCA Symp, pp. s65–s74 (2016)
35. Zimmer, V., Kinney, M., Hughes, R.: Capsule update LVFS: improving system firmware updates. https://archive.fosdem.org/2020/schedule/event/firmware_culisfu/attachments/slides/3709/export/events/attachments/firmware_culisfu/slides/3709/FOSDEM_2020_Intel_Capsule_Update.pdf (2020)

Presshammer: Rowhammer and Rowpress Without Physical Address Information

Jonas Juffinger[1(✉)], Sudheendra Raghav Neela[1], Martin Heckel[2], Lukas Schwarz[1], Florian Adamsky[2], and Daniel Gruss[1]

[1] Graz University of Technology, Graz, Austria
jonas.juffinger@iaik.tugraz.at
[2] Hof University of Applied Sciences, Hof, Germany

Abstract. Modern DRAM is susceptible to fault attacks that undermine the entire system's security. The most well-studied disturbance effect is Rowhammer, where an attacker repeatedly opens and closes (*i.e.*, hammers) different rows, which can lead to bitflips in adjacent rows. Different hammering strategies include double-sided, hammering two rows sandwiching a victim row, and one-location, hammering a single row. One-location Rowhammer requires no physical address information, as any location in memory is mapped to a DRAM row, and no relation between rows is required for hammering. The recently discovered Rowpress differs from Rowhammer by not hammering rows but keeping them open longer, evident by a disjoint set of affected memory locations.

In this paper, we examine the differences between four attack variants: one-location Rowhammer, a one-location Rowpress variant we developed, double-sided Rowhammer, and double-sided Rowpress on a set of 12 DDR4 modules. Our methodology is to hammer and press the exact same set of physical memory locations in all attack variants. Surprisingly, our results show that on 4 out of 12 DDR4 modules, we were only able to reproduce double-sided Rowhammer but none of the other attack variants. On 2 DDR4 modules, we were able to reproduce all attack variants. We find that the number of unique bitflip locations ranges from 161 to 15 612, when hammering the exact same set of physical memory locations. Our one-location Rowhammer attack induces roughly the same amount of bitflips as double-sided Rowhammer, however, only 61.8 % of bitflip locations overlap. We explain this by one-location Rowhammer inducing bitflips due to the Rowhammer as well as the Rowpress effect, making the differentiation of both methods difficult, therefore, calling it Presshammer. Based on our observed bitflips, we develop the first end-to-end one-location Rowpress attack. One-location Rowpress requires only minimal physical address information that an attacker can acquire through a same-row same-bank side-channel attack. Our end-to-end attack escalates to kernel privileges within less than 10 min.

F. Maggi et al. (Eds.): DIMVA 2024, LNCS 14828, pp. 460–479, 2024.
https://doi.org/10.1007/978-3-031-64171-8_24

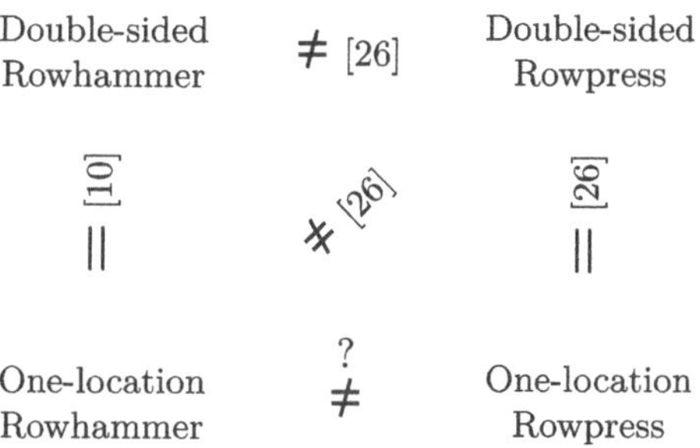

Fig. 1. Claims and observations from different papers. The equal sign indicates that these variants exploit the same root cause, and the unequal sign indicates that these variants have a different root cause. The reference indicates which work this claim is based on.

1 Introduction

A fundamental assumption for secure software systems is the correctness of the underlying hardware. A long line of work has shown that physical glitching could violate this assumption. However, for a long time [2], it was not clear whether it could also be violated in scenarios without a physical attacker. The discovery of Rowhammer as a security issue in 2014 [21] showed that this assumption can be fatally flawed. Rowhammer is a disturbance effect in the DRAM, leading to data corruption that can undermine the entire system's security. To trigger the Rowhammer effect, an attacker repeatedly opens and closes (*i.e.*, hammers) different rows, leading to row activation operations in the DRAM [21]. Different Rowhammer strategies have been discovered, such as single-sided hammering [21], double-sided decoys [9,17] which all open and close different rows in the same bank, triggering row activations. One-location Rowhammer [10] works slightly differently, as it does not activate multiple rows but instead performs frequent accesses to one single row. This approach has the advantage of not requiring physical address information, as any location in memory maps to a DRAM row, and no relation between rows is required for hammering.

Recently Luo et al. [26] discovered an effect in DDR4, called Rowpress. Rowpress, instead of accessing a single location in a row and then another one in a different row, accesses a larger number of offsets in one row before switching to another row. The goal of this access pattern is to keep the row open as long as possible, as this can induce bitflips due to RAS clobber, *i.e.*, the passing gate effect [12,15,52]. Luo et al. [26] report that Rowpress and Rowhammer bitflips belong to largely disjoint sets of memory locations. Rowpress uses physical address information to identify the rows and the offsets within the rows to target.

Comparing the open-source implementations of Rowpress [26] and one-location Rowhammer [10], it is interesting to see that both implementations `main-algo1` and `main-algo2` by Luo et al. indeed hammer two rows[1], albeit multiple offsets within a row, in a Flush+Reload [21] loop, whereas Gruss et al. only run a Flush+Reload loop on a single address. Hence, following the definition of Rowpress, the attack loop by Gruss et al. should, similarly, actually press a row and not hammer it. However, it is not clear whether the one-location hammering and Rowpress implementations on CPUs, exploit the same or different physical root causes: Gruss et al. [10] reported a difference in bitflip rates between single-sided, double-sided and one-location hammering but no significant difference in bitflip locations, indicating that they all exploit the same root cause. Combining these reports leads to a contradiction, as illustrated in Fig. 1. Thus, the relation between the different Rowhammer and Rowpress variants remains unclear.

In this paper, we examine the different Rowhammer and Rowpress variants and find that these effects are not easy to separate. We run Rowhammer and Rowpress tests on 12 DDR4 modules to assess the prevalence of Rowhammer and Rowpress effects on current systems. We find that double-sided Rowhammer works on only 6 out of 12 DDR4 modules. We implement a one-location Rowpress variant, which strictly presses only a single row in contrast to published proof of concepts. However, we find that Rowhammer and Rowpress variants other than double-sided Rowhammer only work on two DDR4 modules in our setup, both with the published proof of concept (PoC) as well as our own implementations. Both results pose a significant difference from the numbers reported by Jattke et al. [17] and Luo et al. [26].

We compare the four attack variants on all DDR4 modules with bitflips in any of the attacks. We hammer and press the exact physical memory locations on the same system in all attack variants by using a setup with fixed memory mappings so that physical addresses remain identical across different Rowhammer and Rowpress tools and runs. The number of unique bitflip locations ranges from 161 to 15 612 on the affected modules when hammering the exact same set of physical memory locations. Our one-location Rowhammer attack induces roughly the same amount of bitflips (14 264) as double-sided Rowhammer (15 612), however, only 61.8 % of bitflip locations overlap. This could be explained by the one-location Rowhammer pattern inducing bitflips due to the Rowhammer as well as the Rowpress effect, making it actually a *Presshammer* attack. We did observe significantly less bitflips with the original Rowpress pattern by Luo et al. [26] that accesses multiple offsets per row than with one-location Rowhammer, even though both keep a row open for a prolonged time.

Based on our observed Presshammer bitflips, we develop a novel one-location Presshammer approach, which requires almost no physical address information and show that an attacker can mount an end-to-end attack by using the same-bank same-row side channel. We demonstrate that our one-location Presshammer attack is practical by obtaining kernel privileges through a page-table corruption attack [44]. The attack runtime with one-location Presshammer is less than 10 min on average.

[1] https://github.com/CMU-SAFARI/RowPress/blob/main/demonstration/main-algo1.cpp#L143.

In summary, we make the following contributions in this work:

1. We systematically compare one-location and double-sided Rowhammer and Rowpress.
2. We study the prevalence of both effects across 12 DDR4 modules and find that 6 of these modules are affected by double-sided Rowhammer, but only 2 by all attack variants, with double-sided Rowhammer flipping the most locations and one-location Rowpress the fewest.
3. We show that our one-location Rowhammer pattern flip bits at locations only partly overlapping with the locations flipped by double-sided Rowhammer. 61.8 % of bitflip locations flipped by one-location Rowhammer also flipped with double-sided Rowhammer, while flipping roughly the same number of bits overall, 14 264 and 15 612 respectively.
4. We present the first one-location Rowpress end-to-end attack, reliably escalating to kernel privileges within less than 10 min of attack time.

Outline. Sect. 2 provides background on DRAM, Rowhammer, and Rowpress. Section 3 presents a systematic comparison between Rowhammer and Rowpress, including our prevalence study. Section 4 shows how one-location Rowpress can be mounted without physical address information. Section 5 presents a PoC attack on page tables, allowing an attacker to escalate to kernel privileges. We discuss mitigations in Sect. 6 before concluding in Sect. 7.

2 Background

This section provides background on DRAM, Rowhammer, and Rowpress.

2.1 DRAM

DRAM is the main memory used in modern computers. DRAM capacities have increased substantially with shrinking process sizes from a few kilobytes in the 1990 s to multiple gigabytes in personal computers and even terabytes in larger server systems today. The DRAM has a very high latency compared to the processor, and a single DRAM chip also has a limited bandwidth. Consequently, modern DRAM has a significant amount of parallelism to improve the bandwidth, positively influencing the average latency. DDR4 has 16 DRAM banks that operate in parallel, combined in a rank. There are single- and dual-rank DDR4 modules. Finally, the DRAM modules are connected to the processor via so-called channels. Today, dual-channel is the most widely used configuration in personal computers. Server CPUs typically have four or eight channels. Banks are divided into *rows* of *cells*, *i.e.*, the transistors and capacitors storing the data. As DRAM cell charges deplete over time, frequent refreshes are required to prevent data loss [18]. Typically, each cell is refreshed every 64 ms. Rowhammer mitigations like TRR can trigger additional row refreshes if they detect Rowhammer attacks. To access data, the memory controller has to activate the row, which moves the data from the row into the row buffer from where it can be read and written.

2.2 Rowhammer

In 2014, Kim et al. [21] discovered the *Rowhammer effect*. An attacker can induce bitflips in the DRAM from software through disturbance errors triggered by frequent row activations. Rowhammer attacks have been demonstrated in sandboxed environments [34,44], in native environments [10,44,45], in virtual machines [16,37,53], in JavaScript [3,11,38], on mobile devices [8,23,48], over the network [25,46], to influence or steal neural networks [36,47,55], and to steal cryptographic keys [7,24,28]. Rowhammer is getting worse with every new DRAM generation and increasing density. LPDDR4 requires less than a 10th of the activations than DDR3 memory to hammer [20,30].

The Rowhammer effect induces a bitflip by discharging the capacitor in a neighboring cell. Once a cell has lost sufficient charge, the value read during the next load will no longer be the original value [21,24,50]. A a correlation exists between the data in the aggressor rows and the resultant bitflips; typically, the victim row adopts the values of the aggressor row.

There are different variants of Rowhammer: double-sided Rowhammer [44], single-sided Rowhammer [21], one-location Rowhammer [10] and half-double Rowhammer [23]. Double-sided Rowhammer uses two rows called aggressors in the same DRAM bank, which sandwich another row in between, inducing more bitflips than single-sided Rowhammer. Half-double Rowhammer hammers one row further apart than double-sided Rowhammer and is dependent on the additional refreshes performed by the TRR mitigation to induce bitflips.

Gruss et al. [10] introduced one-location Rowhammer intending to defeat the mitigations published at the time. The significant difference from the other Rowhammer variants is that one-location hammering only accesses a single location in the DRAM, which should not cause row conflicts. Instead, the aggressor is kept open through repeated accesses to the exact location until it gets closed by the memory controller. When the row is closed, one-location Rowhammer reopens the row immediately. Gruss et al. [10] hypothesized that this reopening might be the underlying trigger to the disturbance effect they discovered (Fig. 2).

Rowpress. Rowpress was discovered by Luo et al. [26]. While it is similar to Rowhammer, it induces bitflips in rows through continuous and sustained activation of the same rows. Luo et al. [26] verified that bitflips are induced by holding the rows open for a long time using a specialized FPGA platform. As the memory controller of a CPU would normally instead close the row soon after the memory access, Luo et al. [26] use a unique access pattern with different offsets in the same row. This causes the memory controller to delay precharge commands up to a maximum of $9 \times t_{REFI}$. In a Rowpress attack, the attacker chooses offsets and repetitions such that the $9 \times t_{REFI}$ delay is approached. Luo et al. [26] find that the Rowpress read disturbance phenomenon is widespread across numerous DRAM modules and manufacturers. Their paper also analyzed and compared the flipped bits caused by Rowpress and the ones caused by Rowhammer. Their data shows that the flipped bits differ for both attacks; therefore, Luo et al.

[26] conclude that Rowpress is caused by another effect than Rowhammer. This effect is either called *RAS clobber* [15,52] or the *passing gate effect* [12].

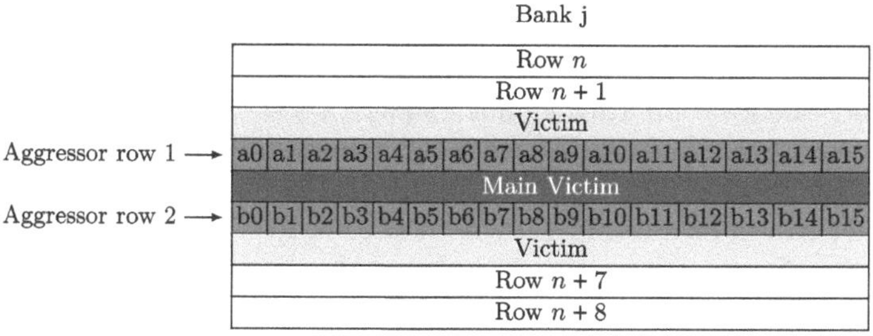

Bank j

Row n
Row $n+1$
Victim
Aggressor row 1 → a0 a1 a2 a3 a4 a5 a6 a7 a8 a9 a10 a11 a12 a13 a14 a15
Main Victim
Aggressor row 2 → b0 b1 b2 b3 b4 b5 b6 b7 b8 b9 b10 b11 b12 b13 b14 b15
Victim
Row $n+7$
Row $n+8$

Fig. 2. Rowpress activates the 16 addresses in the first aggressor row (a0–a15) one after another. Then, all 16 addresses in the second aggressor row (b0–b15) are accessed. This keeps the rows open for a long period of time, causing bitflips due to the passing gate effect.

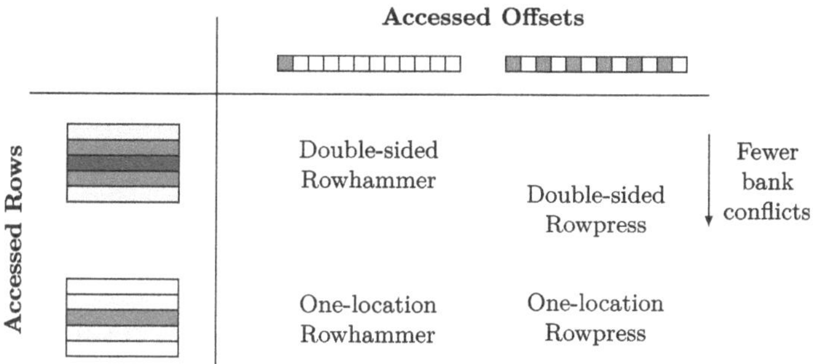

Fig. 3. Overview of the tested variants. Rowhammer attacks access only a single offset per DRAM row. Typically, this is offset 0. Rowpress accesses multiple offsets per DRAM row to keep the row open longer. The double-sided variants access two rows with the victim sandwiched between, however, with a lot lower frequency than double-sided Rowhammer. The one-location variants access only a single row, not explicitly causing bank conflicts.

3 Systematic Comparison of Rowhammer and Rowpress

The key difference separating one-location hammering from other Rowhammer methods is that it does not provoke bank conflicts with other rows in the same DRAM bank, *i.e.*, it is hammering only one location [10]. Similarly, Rowpress also works by keeping a row open as long as possible. With double-sided Rowpress, the open row is switched after some time, causing row conflicts but less than double-sided Rowhammer. One-location Rowpress keeps, like one-location Rowhammer, one row open. Figure 3 visualizes these similarities and differences.

Luo et al. [26] reported that different cells are affected by Rowpress than Rowhammer, hinting at another underlying cause. Rowpress exploits the passing gate effect [12] not the Rowhammer effect. Comparing the cells flipped by Rowpress, double- and one-location Rowhammer shows a less clear picture. While one-location Rowhammer and double-sided Rowhammer flip roughly the same number of bits, the flip locations only overlap by 61.8 %. This hints that one-location Rowhammer can induce bitflip with both effects, Rowhammer and Rowpress.

While the Rowpress paper suggested that one-location Rowpress is possible, it is essential to note that the published Rowpress PoC only contains the double-sided Rowpress variant [26]. Therefore, we used the double-sided Rowpress PoC to develop a one-location Rowpress PoC which presses in only one single location.

3.1 Differences Between One-Location Rowhammer and Rowpress

There are multiple differences between the shown one-location Rowhammer by Gruss et al. [10] and Rowpress by Luo et al. [26]. Listings 1.1, 1.2, 1.3, and 1.4 show pseudo-codes of the different Rowhammer and Rowpress methods.

Number of Accessed Rows. One-location Rowhammer accesses only a single row in one DRAM bank. While Luo et al. [26] mention Rowpress with a single row "one-location Rowpress", their PoC code performs only double-sided Rowpress. We modify this double-sided Rowpress code to support one-location pressing.

Number of Cache Block Reads. One-location Rowhammer accesses only one cache block in a row repeatedly. Typically, it is the first offset in the row. Rowpress accesses multiple cache blocks in a row.

TRR Evasion. Gruss et al. [10] did not actively evade TRR in their one-location implementation. Luo et al. [26] employ a simple TRR evasion similar to TRRespass [9]. They synchronize their hammer loop to `REF` and access dummy rows on the same bank to decrease the chance that TRR tracks the correct aggressor rows. However, as we show in Sect. 3.6, we were not able to reproduce Rowpress with this simple TRR evasion method.

3.2 Comparison Methodology

The goal is to record the exact cells that flip when hammering or pressing the same physical range in memory. We run the following four Rowhammer and Rowpress methods to find out whether the same or different cells flip with these methods: one-location Rowhammer (OL RH), single-sided Rowpress from now on called one-location Rowpress (OL RP), double-sided Rowhammer (DS RH), double-sided Rowpress (DS RH).

To be able to compare the flipped cells we have to ensure that we always hammer the exact same physical memory range. We use a 1 GiB huge page configured with Linux kernel command-line parameters. From this huge page we use the first 100 MiB region for all our tests. We used multiple CPUs to hammer and press all our DIMMs: Intel Core i7-6700K, Intel Core i9-9900K, Intel Core i9-10900K. Our study comprises 12 different DRAM DIMMs without ECC. Like Luo et al. [26] we disabled the cache prefetcher for all experiments.

Our goal is not to find the best hammer method in terms of row accesses or time passed for the attack. We want to find all cells flippable with each method to be able to measure the overlap between methods as an indication of how separated the effects are that they trigger. Therefore, we hammer our 100 MiB region multiple times with enough accesses per hammer loop to ensure we found the bulk of all flippable cells for each method. For each bitflip, we store the physical address, exact bit location, and data before and after the flip.

To find out whether one-location Rowpress flips different bits than one-location Rowhammer, we run the latter with every page offset that we also use for one-location Rowpress. We verify that the hammered offset within a row accessed does not influence the flips in the neighboring victim rows.

3.3 One-Location Rowhammer

```
for (iter = 0 ; iter < NUM_ITER ; iter++):
  for (i = 0 ; i < NUM_AGGR_ACTS ; i++):
    // access a single cache block of the aggressor row
    // to keep the aggressor row open longer
    for (j = 0 ; j < NUM_READS ; j++):
      *AGGRESSOR[x];
      clflushopt (AGGRESSOR[x]);
```

Listing 1.1. One-location Rowhammer code.

One-location Rowhammer repeatedly accesses one offset x in a single row. Varying this offset does not influence the flipped bits. We were only able to reproduce one-location Rowhammer on two identical DIMMs. They are two DDR4 modules with an early TRR implementation.

3.4 Rowpress

We use the improved Rowpress algorithm published in the extended version [27] of Rowpress [26] in Appendix G. It interleaves the aggressor accesses and flushes. Luo et al. [27] measured four times as many induced bitflips than with the original version that first accessed all aggressors and then flushed all of them.

```
for (iter = 0 ; iter < NUM_ITER ; iter++):
  for (i = 0 ; i < NUM_AGGR_ACTS ; i++):
    // access multiple cache blocks of the aggressor row
    // to keep the aggressor row open longer
    for (j = 0 ; j < NUM_READS ; j++):
      *AGGRESSOR[j];
      clflushopt (AGGRESSOR[j]);
```

Listing 1.2. The one-location Rowpress code used for this evaluation

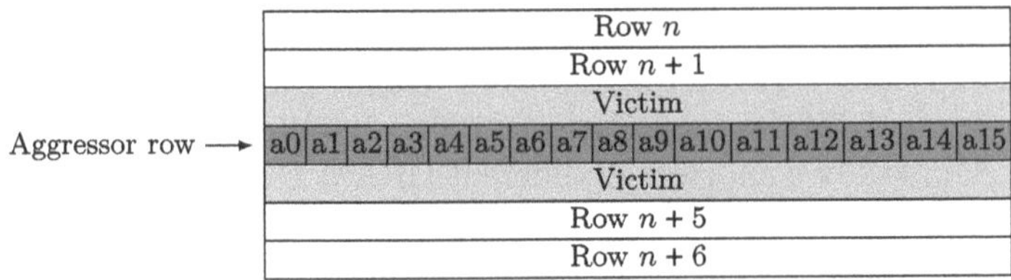

Fig. 4. Abstract representation of one-location Rowpress. We activate addresses a0 to a15 repeatedly to induce bitflips without provoking row conflicts.

One-location Rowpress accesses only offsets within a single row, as shown in Listing 1.2 and Fig. 4. We were only able to reproduce one-location Rowpress on two identical DIMMs, the same where we reproduced one-location Rowhammer.

```
for (iter = 0 ; iter < NUM_ITER ; iter++):
  sync_with_ref();

  for (i = 0 ; i < NUM_AGGR_ACTS ; i++):
    // access multiple cache blocks in each aggressor row
    // to keep the aggressor row open longer
    for (j = 0 ; j < NUM_READS ; j++):
      *AGGRESSOR1[j];
      clflushopt (AGGRESSOR1[j]);
    for (j = 0 ; j < NUM_READS ; j++):
      *AGGRESSOR2[j];
      clflushopt (AGGRESSOR2[j]);

  perform_dummy_row_accesses();
```

Listing 1.3. The double-sided Rowpress code used for this evaluation

The double-sided Rowpress code accesses offsets in two rows in the same bank. The code by Luo et al. [26] performs simple TRR evasion. They are syncing the row accesses to REF and access some dummy rows after the main row accesses, as shown in Listing 1.3. We found that this evasion is not enough for all but the same two DIMMs where we were able to reproduce the one-location methods.

3.5 Double-Sided Rowhammer

```
for (iter = 0 ; iter < NUM_ITER ; iter++):
  for (i = 0 ; i < NUM_AGGR_ACTS ; i++):
    // frequently open and close each aggressor row
    for (j = 0 ; j < NUM_READS ; j++):
      *AGGRESSOR1[0];
      *AGGRESSOR2[0];
      clflushopt (AGGRESSOR1[0]);
      clflushopt (AGGRESSOR2[0]);
```

Listing 1.4. The double-sided Rowhammer code used for this evaluation

To hammer DRAM with the double-sided pattern, we used Blacksmith by Jattke et al. [17]. Blacksmith is the most sophisticated TRR evasion method to date. With Blacksmith we successfully hammered 6 of our 12 tested DIMMs.

3.6 Results

Table 1 shows an overview of all tested DIMMs and the successful hammer and press methods. We were only able to induce bitflips with Rowpress patterns on two of our tested DIMMs, D1 and D2. On DIMMs D3, D4, D6, and D9, Blacksmith [17] found double-sided Rowhammer patterns that induced bitflips. We did not try the one-location variants of Rowhammer and Rowpress if we did not achieve bitflips with both double-sided variants.

Table 1. The table shows whether a Rowhammer or Rowpress method was able to cause bitflips in any of our tested DIMMs. We indicate success with (✓), not causing bitflips with(✗) or not tested with (~). We did not test the one-location (OL) variants if double-sided Rowpress (DL RP) was unsuccessful. * On D3 to D12, we used Blacksmith to hammer double-sided with TRR evasion.

	DIMM	Size	MT/s	Double Sided		One Location	
				RH*	RP	RH	RP
DDR4	D1	8 GB	2133	✓	✓	✓	✓
	D2	8 GB	2133	✓	✓	✓	✓
	D3	8 GB	2400	✓	✗	~	~
	D4	8 GB	2400	✓	✗	~	~
	D5	8 GB	2133	✗	✗	~	~
	D6	8 GB	2666	✓	✗	~	~
	D7	8 GB	2666	✗	✗	~	~
	D8	8 GB	2666	✗	✗	~	~
	D9	8 GB	2666	✓	✗	~	~
	D10	8 GB	2666	✗	✗	~	~
	D11	8 GB	3000	✗	✗	~	~
	D12	8 GB	2133	✗	✗	~	~

Our theory on why we only saw bitflips induced by the Rowpress pattern on 2 DIMMs is that TRR on modern DDR4 DIMMs has evolved to be more difficult to evade. Blacksmith can only do it by accessing a very large number of aggressor rows. For example, the best pattern on D3 repeatedly accesses 60 different rows during one refresh interval to evade TRR. However, as Rowpress tries to keep a row open for a long time, fewer row activations are possible per refresh interval. Luo et al. [26] caused Rowpress bitflips on a DIMM manufactured in 2018. Our DIMMs D1 & D2 are of similar age. Whether a blacksmith-like TRR evasion with Rowpress is possible on modern DDR4 memory, is up to future work.

Table 2 shows the detailed results of all DIMMs where we induced bitflips. On D1, we hammered the same physical address space with all methods listed. On D3, D4, D6, and D9, we used the best pattern found by Blacksmith.

Table 2. The number of bitflips per 100 MB of memory. RP-8 and RP-16 donate Rowpress patterns with 8 and 16 cache line accesses per row. * On D3 to D11, we used Blacksmith to hammer double-sided for TRR evasion.

DIMM	Double Sided			One Location		
	RH*	RP-8	RP-16	RH	RP-8	RP-16
D1 & D2	15 612	2174	4264	14 264	161	0
D3	13 656	0	0	0	0	0
D4	15 786	0	0	0	0	0
D6	157	0	0	0	0	0
D9	3040	0	0	0	0	0

		RH		RP	
		DS	OL	DS	OL
RH	DS	100	56.5	13.7	1.0
	OL	61.8	100	14.5	1.1
RP	DS	98.1	95.3	100	1.0
	OL	96.3	98.8	81.4	100

Fig. 5. Detailed results of DIMM D1 & D2. Each number represents the fraction in percent of exact bitflip locations of one method (y-axis) found in the other method (x-axis). For example, 61.8 % of bitflip locations flipped by one-location Rowhammer is also in the set of locations flipped by double-sided Rowhammer.

On DIMM D1 and D2, we were able to flip bits with all tested methods. Most importantly, we induced bitflips in the same physical address range with all methods. This allows us to analyze precisely whether the tested Rowhammer and Rowpress methods flipped bits in the same cells on this DIMM. We also verified that the offset we access within a row does not influence the bitflips we see when hammering. Figure 5 shows a confusion matrix of all bitflips caused by Rowhammer and Rowpress. Because Rowpress with 8 accesses per row also causes bitflips with 16 offsets, we use its numbers for this analysis.

Based on our results it remains inconclusive, whether one-location Rowhammer [10] is, in fact, Rowpress. We measure roughly the same amount of bitflips when hammering with a double-sided and one-location pattern. However, only 61.8 % of bitflip locations found by one-location Rowhammer flipped also with double-sided Rowhammer. The comparison between double-sided and one-location Rowpress also shows a reduction in the number of bitflips and a difference in bitflip locations. Comparing Rowpress and Rowhammer variants, we find that virtually all Rowpress bitflip locations are in the sets of both Rowhammer variants. However, this is likely also influenced by the smaller number of bitflips we found with all Rowpress variants. A possible explanation for our results could be that one-location Rowhammer causes both effects simultaneously very effectively, actually making it a *Presshammer* attack. It keeps a single row open, inducing bitflips due to RAS clobber in some cells, while the opening and closing of the row that is still happening is enough to induce bitflips in other cells.

4 Unprivileged One-Location Presshammer Attack

As our one-location Rowpress follows the same semantics as one-location Rowhammer, we can similarly reduce the requirements. Most Rowhammer attacks require at least partial knowledge of physical address bits above the page offset. Therefore, the OS hides physical address information from user programs to respond to these Rowhammer attacks [22]. While it is possible to reverse engineer the full DRAM addressing functions [1,33,51] and additionally uncover row addressing and physical address bits [23,24,43,45] using the bank conflict timing side-channel, these methods can be tedious and take time.

A	B	C	D	C	D	A	B	A	B	C	D	C	D	A	B	A	B	C	D	C	D	A	B	A	B	C	D	C	D	A	B

Fig. 6. The offsets within a DRAM row are parts of different memory pages `A` to `D`. Each part has the size of two cache lines, 128 bytes. This example is based on the DRAM addressing functions of a Skylake CPU [33].

For our one-location Presshammer attack, we only need the addresses of all offsets that map to the same row in the same bank. Due to the DRAM addressing functions, one page does not map to one row. On virtually all recent systems, different logical pages are interleaved on one row, as shown in Fig. 6, based on the DRAM addressing functions [33], to increase memory throughput. Thus, the missing parts of each page are simply mapped to other rows in other banks. To perform a Rowpress attack, we must access multiple offsets within one row, spread across various pages. Instead of using the physical address information to find these offsets in the correct locations, we use the bank-conflict timing side-channel in a reduced form.

Taking a random page `A`, we want to find the pages `B` to `D` that map to the same row. Because we work on virtual memory and do not have any knowledge about the underlying mapping to the physical memory, close and "far away" pages based on the virtual address are not necessarily reflected in the physical addresses. We increase this chance by allocating a large chunk of memory. After filling single-page holes in the physical memory, the buddy allocator in Linux typically returns chunks of physically contiguous memory [23]. This increases the performance of our approach but is not strictly necessary.

Two assumptions for our algorithm hold on all current x86 systems:

A1 A DRAM row is 8 KiB (2^{13} B) large.

A2 The physical address bits used for row addressing start high enough to have a single page always in the same row across banks, *i.e.*, a page is never in two rows in the same bank.

The algorithm also works with 4 KiB rows with tiny modifications.

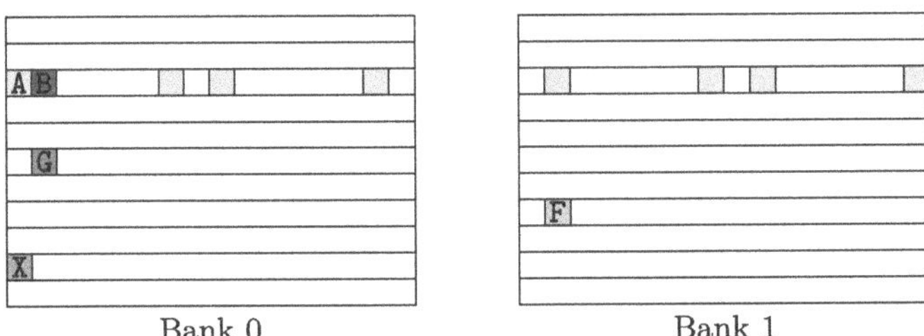

Fig. 7. Visualization of the two requirements to find page offsets in the same row. `F[1]` does not have a bank conflict with `X[0]`, violating **R1**. `G[1]` does have a bank conflict with `X[0]` but also with `A[0]`, violating **R2**. `B` fulfills both requirements.

First, we have to find a page `X` that is in the same bank as `A`. Whether page `A` and `X` are in the same bank is easily detectable with the bank-conflict timing side channel. To not interfere with the following search for pages `B` to `D`, we start the search for page `X` a few hundred virtual pages away from page `A`. We denote current bank-conflict candidate page as `S`. We repeatedly flush and reload `A[0]` and `S[0]` while measuring the timing to detect a bank conflict. With page `A` and `X`, finding pages `B` to `D` is based on two requirements, also shown in Fig. 7:

R1 Because `B` to `D` must be in the same row as page `A`, they must be in the same bank as `X`. We search for pages that have a bank conflict (same bank, but a different row) with page `X`.

R2 Because `B` to `D` are in the same row as page `A`, there must not be a bank conflict between `B` to `D` and `A`.

We iterate over chunks of 64 B to find `B` to `D`, starting with an offset of 64. We check **R1** by flushing and reloading `S[offset]` and `X[0]`. We try pages above and below `A` until we find the page conflicting with `X`. Then, we check **R2** by flushing and reloading `S[offset]` and `A[0]`. If there is no conflict, we found the page mapped to the first offset. We repeat these steps until all offsets are filled. If we found `B` to `D` once, checking the following offsets is quick because we no longer have to search for new pages. During the search, we also mark already "used" offsets to speed up the search the more rows we already found.

Because we exclusively work with virtual memory, there is a chance that any of the required pages to fill a row is not mapped in our address space. However, as shown by Luo et al. [26], Rowpress does not need all offsets to press efficiently, and the offsets we find are sufficient to mount Rowpress. However, the rate of bitflips will be reduced as we do not control the data for the entire row, which is known to influence the number of bitflips [5,24].

5 Proof-of-Concept Presshammer Attack on Page Tables

Luo et al. [26] did not present an end-to-end Rowpress exploit, as their focus was on the scientific evaluation of the effect, not the security impact. Consequently, they did not discuss a threat or attacker model and simply used 1 GiB huge pages to demonstrate the effect on real systems. We bridge this gap by showing that Rowpress can be used to attack a system without using 1 GiB or 2 MiB huge pages to get additional physical address information or any knowledge of the DRAM mapping functions.

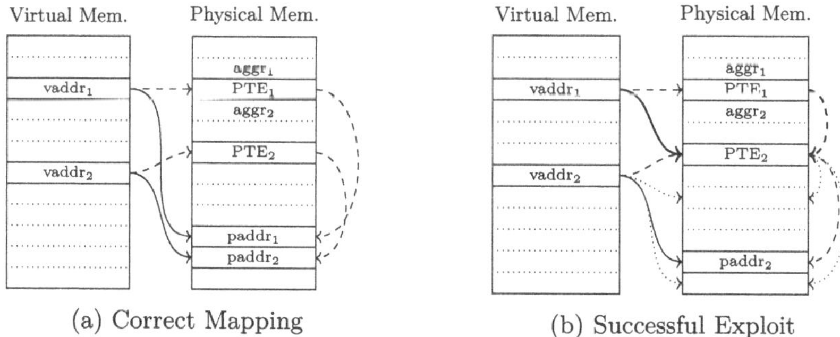

Fig. 8. A simplified example of page table flipping with only one level of page tables is shown. After successful exploitation (b) by hammering PTE_1, PTE_2 is writeable through $\texttt{vaddr}_1$. This makes the whole physical memory accessible by changing PTE_2 and then accessing the new page through $\texttt{vaddr}_2$.

We use Presshammer to attack operating system page tables based on the exploit first shown by Seaborn et al. [44] and many following [5,11,17,23,48,53,57]. The proof-of-concept exploit of Seaborn et al. [44] still relied on full access to the virtual to physical address translations, while we use the algorithm from Sect. 4 to get all the information we need. The attack aims to use a bitflip to change a page frame number (PFN) in a page table to point to another page table. This gives the attacker write access to one of its own page tables and, through it, access to the whole memory. Figure 8 shows the mappings before and after a successful exploitation. The exploitation technique for Presshammer does, at large, not change from the Rowhammer exploit. We ran the exploit on an Intel Core i7-6700K with Ubuntu 20.04.4 LTS using Linux version 5.4.210.

DRAM Row Finding. We use the algorithm described in Sect. 4 to find the aggressor rows. On our machine with 16 GiB of memory, 4 GiB were used by the operating system, and the exploit allocates the remaining 12 GiB of memory. When allocating enough memory, the buddy allocator of the Linux kernel starts to return chunks of physically contiguous memory [23,48].

A single measurement is insufficient to get a definitive result because of system noise when using the bank conflict side-channel. Software on other cores of

Table 3. Results of our row finding algorithm using a different number of measurements for the bank-conflict side channel. There are 128 offsets per row. We search for 10 000 rows and discard the first 100. The search radius is ±50 pages.

# Measurements	Duration per Row	Found Cache Lines	Correct Cache Lines
250	40 ms, σ = 17	124, σ = 4.9	119, σ = 14
500	45 ms, σ = 15	128, σ = 0.8	125, σ = 16
1000	116 ms, σ = 40	125, σ = 4.2	123, σ = 4.3
2500	260 ms, σ = 109	125, σ = 4.9	124, σ = 5.2

the CPU also accesses the memory. However, more measurements take longer, which is relevant if we want to check many rows for bitflips.

We ran the algorithm with 250 to 2500 measurements per address and analyzed the correctness of the result, which can be seen in Table 3. Taking 500 measurements gives the best results. We see that the standard deviation for the duration increases significantly for a higher number of measurements. This suggests more interrupts from the operating system that worsen the result.

Memory Templating. We verified that Rowpress bitflips are repeatable, like Rowhammer bitflips [21]. That means if an aggressor row flips a bit in a neighboring victim row, pressing this aggressor will flip the same bit again if the flip direction stays the same. This helps because we can template the memory in a first step, the exploit searches for aggressors that flip bits in exploitable locations. Bitflips are exploitable if they flip bits in the PFN range of a page table entry (PTE). We count bits 12 to 32 of a PTE (bits 0 to 20 of the PFN) as exploitable; this corresponds to a neighboring page up to a page 4 GiB away.

The exploit searches 10 aggressor victim pairs before continuing. This gives a higher exploit probability later because Rowpress can, like Rowhammer, only flip bits in one direction. Due to that, there is only a 50% chance that an aggressor-victim pair works when filled with an actual page table entry.

The cache prefetcher in the CPU [14] tries to predict which addresses will be loaded next and fetches them into the cache for increased performance. Luo et al. [26] turned off the prefetcher for their experiments. However, this is not possible with our threat model. We evade the prefetcher by shuffling the offsets we access. This decreases the chance that the prefetcher can detect a pattern in our accesses.

Page Table Spraying. Filling most of the memory with page tables is the easiest way to increase the chances of successful exploitation in two ways. First, it increases the chance of putting a page table in one of the previously found victim rows. Second, it increases the chance of a changed PFN pointing to another page table. Van der Veen et al. [48] used a more deterministic technique to place the page table called *Phys Feng Shui.* They place a page table in a victim page by precisely massaging the Linux page allocator. However, for this Presshammer PoC exploit, we chose the more straightforward spraying method.

Page table spraying is performed by mapping the same shared memory file repeatedly. This puts the shared memory file's content into the memory only once, while having to create new page tables for every mapping. After finding the aggressor and victim pairs, the exploit still has most of the physical memory reserved from the templating phase. Every time before mapping the shared memory file again, the exploit frees exactly as many random pages as required for the newly to-be-created page tables. After creating around 80% of all page tables, the exploit frees the victim rows one after another. When the page table spraying is finished, as much memory as possible is filled with page tables.

Page Table Flipping. After spraying the page tables, the exploit presses the aggressor rows one after another to induce bitflips in the victim PFNs. Because the victim rows are not mapped anymore, we cannot simply check for a bitflip. Instead, the exploit checks all mappings of the shared memory file if the content is the expected content. If not, a PFN was changed. If the content looks like a page table, the exploit changes one page-table entry of the page table and checks through all mappings again. When finding another page with changed content, the exploit found $\texttt{vaddr}_1$ and $\texttt{vaddr}_2$, as shown in Fig. 8b.

Exploitation. With all previous steps performed successfully, the exploit now has full read and write access to the whole physical address space. The original proof-of-concept exploit code from Seaborn et al. [44] used this access to patch a victim binary and change its behavior. For a real exploit, this could be any binary owned by root with the `setuid` sticky bit set to run code with superuser privileges. Without access to the virtual to physical address translations, the exploit has to scan the memory for the victim data it wants to attack. The exploit can also search for exploitable kernel structures to, e.g., give the process or user higher privileges or directly inject attacker-controlled code into the kernel. For our proof-of-concept exploit, we just dump the physical memory into a file.

While reading the whole physical memory, it is crucial to evict the TLB entry of the PTE modified by the exploit. If the exploit does not evict the TLB entry, all reads from $\texttt{vaddr}_2$ will return the same data. To evict it, we keep half of the mappings we used during the exploit and iterate over them. We free the other half to relieve the memory pressure on the system and prevent an out-of-memory situation. Dumping the memory to disk is limited by the disk write bandwidth.

We ran the exploit 5 times on our machine. Of these 5 runs, 3 were successful. It took, on average, 513 s (n=3, σ=125) to gain full access to the physical memory. One run was unsuccessful because the operating system did not put a page table in a victim row or no bit flipped in the PFN. The other run was unsuccessful because it did not point to another of our page tables after changing a PFN by row pressing it. If unsuccessful, the exploitation can just be tried again.

6 Discussion of Mitigations and Related Work

The large number of Rowhammer attacks published [3,5,7,8,10,11,16,23–25,28,34,36–38,44–48,53,55] motivated research on mitigations for DRAM disturbance errors. While many have been proposed in the academic community [4,13,19,29,31,32,39–42,49,54,56], only a few, like TRR [9], have found their way into practice. As Luo et al. [26] noted, Rowpress is also mitigated by TRR and other mitigations. However, they also note that similar to Rowhammer attacks bypassing TRR [9,17,23], Rowpress can also bypass it. And the same is true for many other proposed countermeasures. As CPUs cannot keep a row open longer than $9 \times t_{REFI}$, Rowpress always also causes a high number of row activations similar to Rowhammer. We show in this paper that this makes the two attacks hard to distinguish in practice, but this also means that we expect most countermeasures that are effective against current Rowhammer attacks to be also effective against Rowpress. In the worst case, with some additional tweaking to the detection threshold values. However, up to now, every deployed Rowhammer mitigation has been evaded eventually, and there is no guarantee that this will not be the case for Rowpress as well.

Our findings on the overlaps between Rowhammer and Rowpress effects indicate again [10] that mitigations must be principled not to be bypassable. That is, they cannot focus on the specific effect triggering the disturbance but must focus on preventing the effect of the disturbance. Mitigations that use this approach are, for instance, Error Correction Code (ECC) [5] and Chipkill [6]. ECC uses 8 redundancy bits per 64 bits (DDR4) or 32 bits (DDR5) of data. Chipkill can correct up to 8 bitflips, e.g., due to a chip failure [6]. However, ECC has already been bypassed with Rowhammer attacks, and Chipkill suffers from a significant overhead. As Qureshi [35] noted, it is essential to take an entirely new approach to mitigate Rowhammer and other upcoming DRAM disturbance effects. The earliest work in this direction is IVEC [13], which uses a hash instead of an error-correcting code and flips bits in corrupted memory until the hash matches again. With a similar approach, Saileshwar et al. [39] also propose a modification to ECC memory using a MAC instead of the error-correcting code to detect and correct a broader range of bitflips. Juffinger et al. [19] extended this approach into an error-correction system based on lightweight cryptography that effectively transforms DRAM disturbance errors from reliability issues into performance penalties, which, in the worst case, lead to a denial of service. Recently, Olgun et al. [29] introduced a Rowhammer mitigation based on a single per-row hardware counter to track row activations across all banks. This minimizes the hardware overhead while solving the issue of a small number of counters in TRR.

7 Conclusion

Many works analyzed DRAM disturbance effects and reported a wide range of prevalence estimations. In our set of 12 DDR4 modules, we can reproduce double-sided Rowhammer on 6 modules. All other Rowhammer and Rowpress variants only worked on 2 modules. For a fair comparison between the attack variants, we evaluated the number of unique bitflip locations for all variants, ranging from 161 to 15 612. However, 95.3% to 98.8% of Rowpress unique bitflip locations are also flipped by double-sided and one-location Rowhammer.

This surprising result raises questions about the difference between the effects these attack implementations trigger and indicates that they might partially exploit the same root cause. However, this result also allows us to craft the first unprivileged end-to-end one-location Rowpress attack, which uses the same-row same-bank side channel to escalate to kernel privileges within less than 10 min. Our work shows that further research is necessary to better understand the differences between Rowhammer and Rowpress implementations on real systems and their implications for security.

Acknowledgments. We thank Lukas Gerlach, Andreas Kogler, Stefan Gast, Patrick Rademacher, Sarah Rinderer and the reviewers for their feedback and help. This research was supported by the European Research Council (ERC project FSSec 101076409), the Austrian Science Fund (FWF SFB project SPyCoDe 10.55776/F85), the Deutsche Forschungsgemeinschaft (DFG) (503876675) and by the European Union (ROF-SG20-3066-3-2-2). Additional funding was provided by generous gifts from Red Hat and Google. Any opinions, findings, and conclusions or recommendations expressed in this paper are those of the authors and do not necessarily reflect the views of the funding parties.

References

1. Barenghi, A., Breveglieri, L., Izzo, N., Pelosi, G.: Software-only reverse engineering of physical DRAM mappings for rowhammer attacks. In: International Verification and Security Workshop (IVSW) (2018)
2. Biham, E., Shamir, A.: Differential Fault Analysis of Secret Key Cryptosystems. In: CRYPTO (1997)
3. Bosman, E., Razavi, K., Bos, H., Giuffrida, C.: Dedup est machina: memory deduplication as an advanced exploitation vector. In: S&P (2016)
4. Brasser, F., Davi, L., Gens, D., Liebchen, C., Sadeghi, A.R.: Can't touch this: software-only mitigation against rowhammer attacks targeting kernel memory. In: USENIX Security (2017)
5. Cojocar, L., Razavi, K., Giuffrida, C., Bos, H.: Exploiting correcting codes: on the effectiveness of ECC memory against rowhammer attacks. In: S&P (2019)
6. Dell, T.J.: A white paper on the benefits of chipkill-correct ECC for pc server main memory. Technical Report, IBM Microelectronics (1997)
7. Fahr Jr, M., et al.: When frodo flips: end-to-end key recovery on FrodoKEM via Rowhammer. In: CCS (2022)
8. Frigo, P., Giuffrida, C., Bos, H., Razavi, K.: Grand Pwning unit: accelerating microarchitectural attacks with the GPU. In: S&P (2018)
9. Frigo, P., et al.: TRRespass: exploiting the many sides of target row refresh. In: S&P (2020)
10. Gruss, D., et al.: Another flip in the wall of rowhammer defenses. In: S&P (2018)
11. Gruss, D., Maurice, C., Mangard, S.: Rowhammer.js: a remote software-induced fault attack in JavaScript. In: DIMVA (2016)
12. Hong, S., et al.: DSAC: low-cost rowhammer mitigation using in-DRAM stochastic and approximate counting algorithm. arXiv preprint (2023)
13. Huang, R., Suh, G.E.: IVEC: Off-chip memory integrity protection for both security and reliability. In: ISCA (2010)

14. Intel: Intel 64 and IA-32 architectures software developer's manual, Volume 1: basic architecture (2016)
15. Ito, Y., He, Y.: Apparatus and methods for refreshing memory (2019). U.S. Patent 11062754B2
16. Jang, Y., Lee, J., Lee, S., Kim, T.: SGX-bomb: locking down the processor via Rowhammer attack. In: SysTEX (2017)
17. Jattke, P., van der Veen, V., Frigo, P., Gunter, S., Razavi, K.: BLACKSMITH: Rowhammering in the frequency domain. In: S&P, November 2021
18. JEDEC solid state technology association: low power double data rate 4 (2017). http://www.jedec.org/standards-documents/docs/jesd209-4b
19. Juffinger, J., Lamster, L., Kogler, A., Eichlseder, M., Lipp, M., Gruss, D.: CSI: Rowhammer - cryptographic security and integrity against Rowhammer. In: S&P (2023)
20. Kim, J.S., Patel, M., Yağlıkçı, A.G., Hassan, H., Azizi, R., Orosa, L., Mutlu, O.: Revisiting RowHammer: an experimental analysis of modern DRAM devices and mitigation techniques. In: ISCA (2020)
21. Kim, Y., et al.: Flipping bits in memory without accessing them: an experimental study of DRAM disturbance errors. In: ISCA (2014)
22. Kirill A. Shutemov: Pagemap: do not leak physical addresses to non-privileged Userspace (2015). https://git.kernel.org/cgit/linux/kernel/git/torvalds/linux.git/commit/?id=ab676b7d6fbf4b294bf198fb27ade5b0e865c7ce
23. Kogler, A., et al.: Half-double: hammering from the next row over. In: USENIX Security (2022)
24. Kwong, A., Genkin, D., Gruss, D., Yarom, Y.: RAMBleed: reading bits in memory without accessing them. In: S&P (2020)
25. Lipp, M., et al.: Nethammer: inducing rowhammer faults through network requests. In: SILM Workshop (2020)
26. Luo, H., et al.: RowPress: amplifying read disturbance in modern DRAM Chips. In: ISCA (2023)
27. Luo, H., et al.: RowPress: amplifying read disturbance in modern DRAM chips. arXiv:2306.17061 (2024)
28. Mus, K., Doröz, Y., Tol, M.C., Rahman, K., Sunar, B.: Jolt: Recovering TLS signing keys via Rowhammer faults. In: S&P (2023)
29. Olgun, A., et al.: ABACuS: all-bank activation counters for scalable and low overhead RowHammer mitigation. In: USENIX Security (2024)
30. Orosa, L., et al.: A deeper look into RowHammer's sensitivities: experimental analysis of real DRAM chips and implications on future attacks and defenses. In: MICRO (2021)
31. Park, J.H., et al.: Row hammer reduction using a buried insulator in a buried channel array transistor. IEEE Trans. Electron Devices **69**(12), 6710–6716 (2022)
32. Park, Y., Kwon, W., Lee, E., Ham, T.J., Ahn, J.H., Lee, J.W.: Graphene: strong yet lightweight row hammer protection. In: MICRO (2020)
33. Pessl, P., Gruss, D., Maurice, C., Schwarz, M., Mangard, S.: DRAMA: exploiting DRAM addressing for cross-CPU attacks. In: USENIX Security (2016)
34. Qiao, R., Seaborn, M.: A new approach for Rowhammer attacks. In: HOST (2016)
35. Qureshi, M.: Rethinking ECC in the era of row-hammer. In: DRAMSec (2021)
36. Rakin, A.S., Chowdhuryy, M.H.I., Yao, F., Fan, D.: DeepSteal: advanced model extractions leveraging efficient weight stealing in memories. In: S&P (2022)
37. Razavi, K., Gras, B., Bosman, E., Preneel, B., Giuffrida, C., Bos, H.: Flip Feng Shui: hammering a needle in the software stack. In: USENIX Security (2016)

38. de Ridder, F., Frigo, P., Vannacci, E., Bos, H., Giuffrida, C., Razavi, K.: SMASH: synchronized many-sided Rowhammer attacks from JavaScript. In: USENIX Security (2021)
39. Saileshwar, G., Nair, P.J., Ramrakhyani, P., Elsasser, W., Qureshi, M.K.: Synergy: rethinking secure-memory design for error-correcting memories. In: HPCA (2018)
40. Saileshwar, G., Wang, B., Qureshi, M., Nair, P.J.: Randomized row-swap: mitigating row hammer by breaking spatial correlation between aggressor and victim rows. In: ASPLOS. pp. 1056–1069 (2022)
41. Saxena, A., Saileshwar, G., Juffinger, J., Kogler, A., Gruss, D., Qureshi, M.: PT-guard: integrity-protected page tables to defend against breakthrough Rowhammer attacks. In: DSN (2023)
42. Saxena, A., Saileshwar, G., Nair, P.J., Qureshi, M.: AQUA: scalable rowhammer mitigation by quarantining aggressor rows at runtime. In: MICRO) (2022)
43. Schwarz, M., Gruss, D., Weiser, S., Maurice, C., Mangard, S.: Malware guard extension: using SGX to conceal cache attacks. In: DIMVA (2017)
44. Seaborn, M., Dullien, T.: Exploiting the DRAM Rowhammer bug to gain kernel privileges. In: Black Hat USA (2015)
45. Tatar, A., Giuffrida, C., Bos, H., Razavi, K.: Defeating software mitigations against rowhammer: a surgical precision hammer. In: RAID (2018)
46. Tatar, A., Krishnan, R., Athanasopoulos, E., Giuffrida, C., Bos, H., Razavi, K.: Throwhammer: Rowhammer attacks over the network and defenses. In: USENIX ATC (2018)
47. Tol, M.C., Islam, S., Adiletta, A.J., Sunar, B., Zhang, Z.: Don't Knock! DSN, Rowhammer at the backdoor of DNN models (2023)
48. van der Veen, V., et al.: Drammer: deterministic Rowhammer attacks on mobile platforms. In: CCS (2016)
49. van der Veen, V., et al.: GuardION: practical mitigation of DMA-based Rowhammer attacks on ARM. In: DIMVA (2018)
50. Walker, A.J., Lee, S., Beery, D.: On DRAM Rowhammer and the physics of insecurity. IEEE Trans. Electron Devices **68**(4), 1400–1410 (2021)
51. Wang, M., Zhang, Z., Cheng, Y., Nepal, S.: Dramdig: a knowledge-assisted tool to UncoverDRAM address mapping. In: Design Automation Conference (DAC) (2020)
52. Wolff, G.D.: Word line cache mode (2019), u.S. Patent 10366733B1
53. Xiao, Y., Zhang, X., Zhang, Y., Teodorescu, R.: One bit flips, one cloud flops: cross-VM row hammer attacks and privilege escalation. In: USENIX Security (2016)
54. Yaglikci, A.G., et al.: BlockHammer: preventing RowHammer at low cost by black-listing rapidly-accessed DRAM rows. In: HPCA (2021)
55. Yao, F., Rakin, A.S., Fan, D.: DeepHammer: depleting the intelligence of deep neural networks through targeted chain of bit flips. In: USENIX Security (2020)
56. Yauglikcci, A.G., et al.: HiRA: hidden row activation for reducing refresh latency of off-the-Shelf DRAM chips. In: MICRO (2022)
57. Zhang, Z., Cheng, Y., Liu, D., Nepal, S., Wang, Z., Yarom, Y.: PThammer: cross-user-kernel-boundary Rowhammer through implicit accesses. In: MICRO (2020)

Cyber Physical Systems and IoT

SecMonS: A Security Monitoring Framework for IEC 61850 Substations Based on Configuration Files and Logs

Onur Duman[1(✉)], Mengyuan Zhang[2], Lingyu Wang[1], and Mourad Debbabi[1]

[1] Concordia University, Montreal, QC, Canada
Onur.Duman@Kristiania.no

[2] Vrije Universiteit Amsterdam, Amsterdam, The Netherlands

Abstract. Substations are critical components of the smart grid since compromising them can lead to significant consequences, such as blackouts. Threat modeling aims to model different ways critical networks, such as substations, can be attacked. Attack graphs are commonly used for modeling threats and there is a huge literature on attack graphs. However, attack graph generation is still an open problem, and attack graphs are usually generated based on static configurations. To overcome those challenges, this paper provides an attack graph-based threat modeling and Markov Decision Process (MDP)-based monitoring framework for substations, named SecMonS. Specifically, we first generate static attack graphs based on substation configuration language (SCL) descriptions of intelligent electronic devices (IED). Second, we generate automaton models for modeling the behaviors of IEDs directly from log files that contain Generic Object Oriented Substation Event (GOOSE) protocol parameter values. Third, we enhance static attack graphs with automaton models to ensure that those threat models contain updated information based on real-world attacks. Fourth, we tackle the state space explosion faced by MDP models which are utilized for identifying potential physical consequences of attacks to integrate physical aspects of attacks into the threat modeling. Lastly, we evaluate the practicality of SecMonS through simulations using data both from a public data set and randomly generated events for different types of attacks.

Keywords: smart grid · substations · SCL · attack graph · log

1 Introduction

Substations, which are used to change voltage levels while providing protection, monitoring, and control within the smart grid, are critical components. This is due to the fact that compromising substations may lead to severe physical consequences, including widespread blackout scenarios (e.g., the 2015 Ukraine attack [1]). In the 2015 Ukraine attack, cyber attackers targeted seven substations and

F. Maggi et al. (Eds.): DIMVA 2024, LNCS 14828, pp. 483–502, 2024.
https://doi.org/10.1007/978-3-031-64171-8_25

this attack resulted in a blackout affecting 225,000 customers [1]. This cyber attack with physical consequences demonstrates the feasibility of such attacks against substations and calls for solutions that safeguard them.

An effective method for protecting substations is to identify potential attacker strategies at their early stage and stop them before any consequences materialize. In the literature, researchers propose to utilize threat modeling to obtain an overview of potential attacks and security threats [21]. Attack graphs [6,30] are visual models used to model different ways critical systems can be attacked through attacker actions (e.g., vulnerability exploits). Those models are commonly used in the literature to develop quantitative security metrics or to identify ways to systematically harden [10] critical systems. A typical attack graph consists of two types of nodes which are exploits and conditions. Figure 4 shows an example attack graph. In this attack graph, there are two types of nodes, which are exploits (in ovals) and conditions. Exploits (e.g., $\langle V\_XCBR, IED100, IED100\rangle$ in Fig. 4) represent vulnerabilities that can be exploited by the attacker and conditions (e.g., $\langle IED100, Network\rangle$ in Fig. 4) represent conditions required for exploiting those vulnerabilities or conditions gained after. Most existing tools for attack graph generation in the literature, such as MulVAL [26] and CAULDRON [18], are designed to generate attack graphs based on static network configurations. More specifically, directly applying traditional attack graph generation to substations poses two challenges. First, the dynamic nature of the substations, which refers to changes in Generic Object Oriented Substation Event (GOOSE) protocol parameter values (i.e., STATUS and ALARM parameters) depending on states of IEDs, is missing in the static network configuration-based attack graph generation. Researchers have attempted to address this issue by incorporating alerts into the generation process [23]. However, the sheer volume of alerts received by security operation centers, which include both legitimate alerts and false alarms, poses a challenge to derive the real threats from noise [4]. Secondly, most attack graph modeling approaches do not consider the physical consequences of cyber-physical attacks. However, to effectively secure substations, it becomes critical to develop a solution that dynamically models substations while considering the physical impacts of attacks.

In this paper, we propose a novel framework for attack graph-based threat modeling and Markov Decision Process (MDP)-based monitoring in substations in which attack graphs are first generated from Substation Configuration Language (SCL) files and then refined using information from log files. In addition, we consider the potential physical consequences of attacks by utilizing an MDP in which the attacker is modeled as an agent aiming to maximize the amount of physical damage to the power system. In summary, our contributions are as follows:

- We provide a threat modeling and monitoring tool, namely SecMonS, which generates attack graphs showing potential attack paths, based on SCL files, visualizing parameters the attacker can tamper with in a way that those attack graphs can later be enhanced with real instances of attacks.

- We generate a visual summary of many log events as an automaton model in which the number of states depends on the number of GOOSE protocol parameter changes which is utilized to enhance the static attack graph with real instances of attacks so that we can utilize the enhanced attack graph for identifying how to harden substations.
- We tackle the state-space explosion problem faced by MDP models that are generated for identifying potential physical consequences of attacks.
- We demonstrate the effectiveness of SecMonS through simulations. Our results show that the runtime of our tool is practical for real-world implementation for attack graph-based threat modeling and MDP-based monitoring in power systems.

1.1 Motivating Example

A typical substation network consists of Intelligent Electronic Devices (IEDs). Configurations of IEDs that contain names of parameters sent by IEDs are defined using SCL files. Logs contain actual parameter values sent by IEDs. Let's assume that the power operator relies on attack graphs to model potential attack paths against the substation. Figure 1 illustrates an operator's challenges in modeling and monitoring the substations' security posture. Firstly, the dynamic aspect of substations (such as GOOSE [9] parameters that are currently being attacked, detailed in Sect. 2.2) is not integrated within the traditional attack graph framework. Secondly, many log events are generated in substations and it is not feasible to go through all those log events manually in order to identify attack paths utilized by attackers. Thirdly, attack graphs are based on known attacks and those models need to be updated manually as new attack paths are discovered. So, enhancing attack graphs with information from ongoing instances of attacks is challenging. Lastly, in order to analyze the potential physical impacts of attacks, MDP models and attack graph models can be utilized together in a way that attack graphs can feed probabilities and attacker states to the MDP. However, MDP models, which are commonly used for contingency analysis (analyzing what-if scenario in power systems [5,37]), suffer from state space explosion since they start from the state in which the attacker has not compromised any hosts.

In order to tackle those challenges, SecMonS first generates attack graphs visualizing parameters the attacker tampers with directly from SCL files. After that, it generates a visual summary of many log events by modeling the behavior of the system both when the system is and is not under attack. Consequently, it enhances the static attack graph generated in the previous step using information about the system behavior. Lastly, it generates MDP models from the starting point of the attacker in order to tackle the MDP state space explosion problem.

The rest of this paper is organized as follows. Section 2 provides background information. Section 3 presents the threat model considered in this paper. Section 4 presents the framework, SecMonS. Section 5 provides a case study. Section 6 provides simulation results. Section 7 provides a literature review. Finally, Sect. 8 concludes the paper.

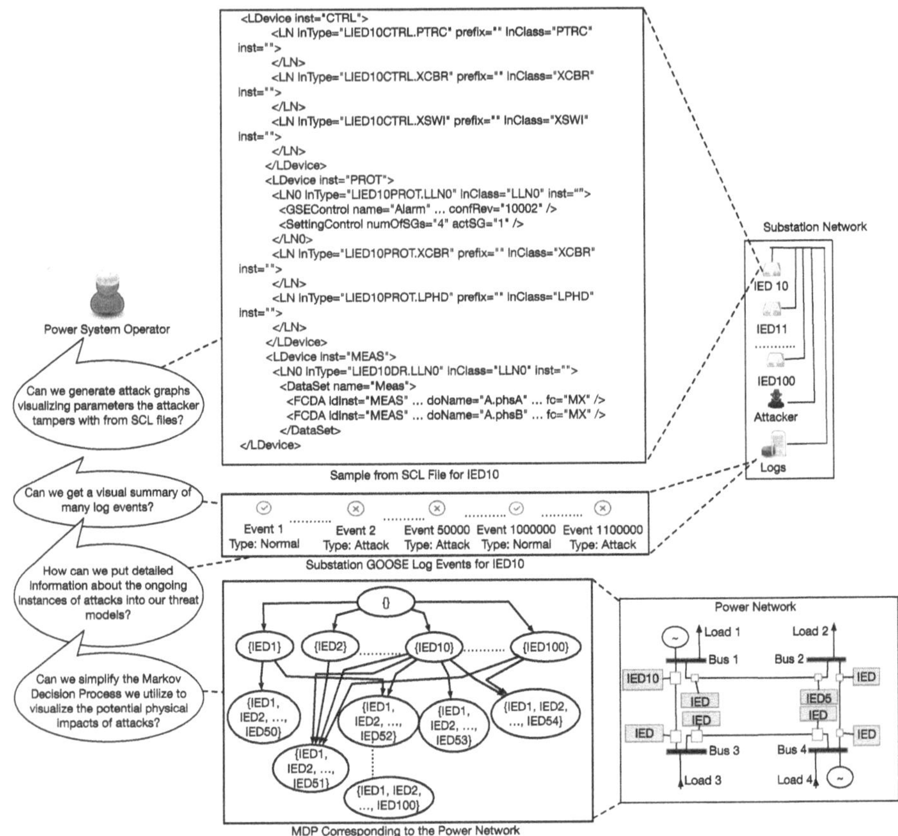

Fig. 1. Motivating Example

2 Background

In this section, we first explain SCL files and their usage. Second, we explain how the GOOSE protocol, in which parameters included in SCL files are sent periodically, works in substations. In addition, we explain the attack dataset [9] utilized in this paper along with the dataset generation tool [8] both of which contain sample SCL files and log entries including GOOSE parameter values. Finally, we explain an open-source tool, namely Flexfringe [35], for building automaton models which model parameter changes in the GOOSE protocol.

2.1 Substation Configuration Language (SCL)

SCL files are utilized for information exchange between IED vendors and power utilities [29]. An important part of IEC 61850 (which is an essential standard for enabling communication between IEDs from different manufacturers) is SCL [11]. There are four types of SCL files [11], which are: system specification description (SSD), substation configuration description (SCD), IED capability description

(ICD), and configured IED description (CID). SSD files describe the entire power system, including many substations. SCD files contain the description of one substation. ICD files contain descriptions of IEDs inside one substation. CID files contain a description of a single IED. Given the importance of SCL files, there is a rich literature on threat model generation based on SCL files (such as [13,14,22], a detailed review of the related work is given in Sect. 7). In this paper, we define SCL-based attack graphs demonstrating how the attacker can perform cyber attacks by tampering with communication parameters defined in SCL files.

2.2 Generic Object Oriented Substation Event (GOOSE) Protocol

In substations, GOOSE protocol is utilized for protection, control, and automation [31]. GOOSE messages are exchanged between IEDs in the multicast mode, a message sent by one IED is received by all other IEDs in the same network. Data values in a GOOSE message are contained in the Protocol Data Unit (PDU), which contains a sequence number, state number, timestamp, message type, and data values. Figure 2 shows example GOOSE PDUs for each types of GOOSE messages obtained from the dataset provided in [9].

Fig. 2. Sample GOOSE Protocol Data Units [9] for each message types: STATUS, ALARM, and MESAUREMENT

In order to attack the GOOSE protocol, the attacker needs to gain full access to IEDs and be able to tamper with messages sent in the substation network [9]. In [9], the authors mention different attack types against the GOOSE protocol which are denial of service, message suppression, data manipulation, and control manipulation. In the denial of service (DoS) attack, a targeted IED is prevented from receiving legitimate GOOSE packets by being flooded. In message suppression, the attacker modifies GOOSE headers, such as timestamps, to prevent IEDs from receiving legitimate GOOSE packets. In data manipulation, the attacker modifies data values such as measurements. In control manipulation, control messages are tampered with by the attacker. In addition, the authors in [9] generate a GOOSE attack dataset for different attack scenarios. Different attack scenarios are simulated using power system testbeds. Firstly, packet captures of the normal behavior with and without varying loads are provided along with SCL files. After that, packet captures for attack scenarios are provided. As a follow-up work for [9], the authors also develop a packet capture generation tool in which the user defines SCL parameters and attack steps and then generates traces containing GOOSE events [8]. In this paper, we leverage the GOOSE dataset provided in [9] for simulations.

2.3 Flexfringe

Flexfringe is an automaton generation tool developed to analyze software or protocol behavior [35]. Analyzing the behavior of a given software can aid in its maintenance and testing. Flexfringe is utilized to generate attack graphs from logs in [23]. In [23], the authors generate attack graphs from alert logs without requiring expert input or knowledge to analyze frequently used attacker strategies. In this paper, we utilize Flexfringe [35] to generate automaton models modeling the GOOSE protocol behavior under the normal scenario and attack scenarios. This allows the operators to visually see deviations from the normal behavior visually rather than going through many log entries. In addition, we utilize automaton models to enhance static attack graphs.

3 Threat Model

The following describes different aspects of our threat model.

- **Types of Attackers:** In this paper, we assume the attackers are already in the IED network. GOOSE protocol was chosen as a case study and SecMonS can be utilized for other protocols, such as Sampled Values (SV) [17]. Steps taken by the attacker in order to connect to the substation network are ignored in this paper. Any attacks that are based on tampering with messages sent between IEDs can be modeled using SecMonS.
- **Attackers' Capability:** The attackers are assumed to be capable of tampering with protocol parameters sent between IEDs [9]. Such parameter tampering is captured in log files since parameter values are periodically sent between IEDs. However, log files themselves are assumed to be protected using existing integrity protection techniques, such as those based on blockchains [16,27]. In particular, the authors in [27] propose a blockchain-based log tampering protection system called Logchain which avoids log tampering by sealing logs cryptographically and adding them to a hierarchical ledger and claim that their solution provides immutability of log files. In [16], the authors propose a log storage management protocol utilizing blockchain and signature chain in order to ensure data integrity and data tamper resistance for logs and claim that their protocol is applicable to IoT (Internet of Things) devices.

4 SecMonS Framework

In this paper, we provide an online attack graph-based threat modeling and MDP-based monitoring framework named "Security Monitoring for Substations" (SecMonS). SecMonS utilizes static and dynamic information. Static information includes SCL files for IEDs, and dynamic information includes log files and power system dynamics, such as loads on transmission lines. By taking those into account, SecMonS first generates SCL-based attack graphs (Sect. 4.1). After that, it generates automaton models from log files according to GOOSE parameters sent by IEDs for the normal and attack behaviors of the system. Those models are then used to identify GOOSE parameter changes without going through log files manually (Sect. 4.2). The SCL-based attack graph is further enhanced based on log based automaton models (Sect. 4.3). As the last step, SecMonS generates the MDP model [5,37] corresponding to the substation starting from compromised IEDs to analyze the physical impacts of cyber attacks. Inputs to the MDP generation include probabilities obtained from the enhanced attack graph, transmission line capacities, and loads on transmission lines after each IED compromise (Sect. 4.4). Figure 3 shows the overview of the framework. This section details each module of SecMonS.

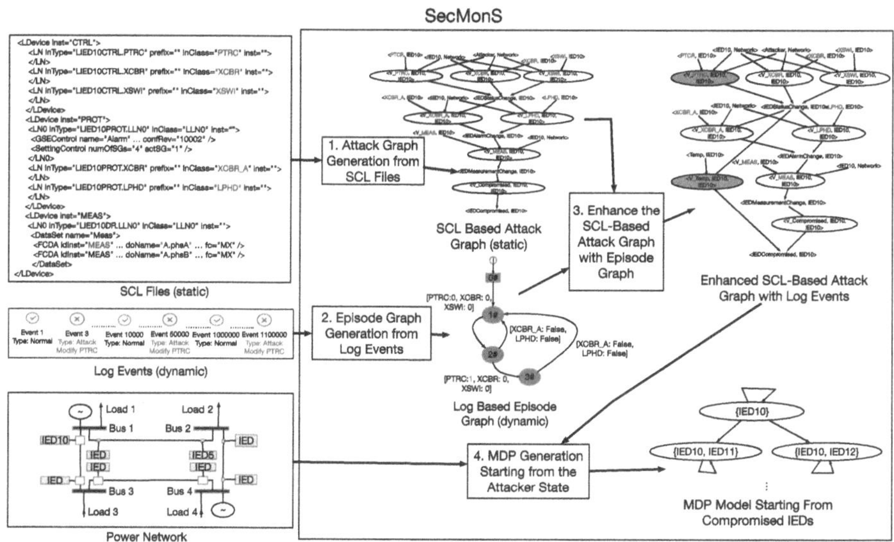

Fig. 3. Overview of SecMonS

4.1 SCL-Based Attack Graph Generation

The initial phase in the SecMonS involves generating attack graphs from SCL files. SCL files contain names of parameters exchanged between IEDs but those files do not contain any information about how those IEDs can be attacked. As an example, in a GOOSE data manipulation attack [9], the attacker modifies GOOSE parameters, such as current measurements on transmission lines, parameters showing the status of each IED, and any alarms generated by IEDs. In the SCL-based attack graph, each GOOSE parameter is modeled as an exploit, showing the attacker's action of modifying that GOOSE parameter.

The SCL file on the left of Fig. 4 contains all GOOSE parameters the attacker can tamper with in three elements which are "Status", "Alarm", and "Meas" as shown in the Extensible Markup Language (XML) elements shown on the left of Fig. 4. The corresponding static SCL-based attack graph on the right of Fig. 4 contains all parameters in those elements inside the SCL file. Based on this, the attack graph in Fig. 4 on the right contains all the possible parameters the attacker can tamper with according to the SCL file in Fig. 4 on the left. According to the dataset in [9], GOOSE messages are sent in the following order according to message types: STATUS, ALARM, and MEASUREMENT. In the SCL file we provide as an example in Fig. 4, STATUS messages include parameter names XCBR, XSWI, and PTRC. Moreover, ALARM messages include the parameter names PIOC, XCBR, LPHD, and PSCH. Lastly, MEASUREMENT messages include the parameter name MMXU. On the contrary, in attack graphs [6,30], we have attack paths containing exploits and conditions and the attacker takes attack paths by exploiting vulnerabilities in the order given by the attack graph in order to reach the goal condition (e.g., $\langle V\_Compromised, IED100, IED100\rangle$ in Fig. 4 meaning that the attacker has compromised IED100). Based on that, in the SCL-based attack graph, the first exploit is one of the STATUS parameters, after that the second exploit is one of the ALARM parameters and the last exploit is one of the MEASUREMENT parameters. In this case, attack paths contain one STATUS parameter, one ALARM parameter, and one MEASUREMENT parameter in order (e.g., V_XCBR -> V_PIOC -> V_MMXU is one of the attack paths in Fig. 4). After generating the SCL-based attack graph, which is the starting point for attack graph-based threat modeling, the next step is to update it with newly discovered attack paths utilizing automaton models.

4.2 Log-Based Automaton Model

Flexfringe [35] provides an automaton model based on a given list of events. In the case of substation logs, the list of events refers to GOOSE parameter values. We also consider that attackers have the capability to tamper with GOOSE parameters sent between IEDs. Generally, we take into account any parameters (STATUS, ALARM, MEASUREMENT) at any time. Therefore, it is not necessary for attackers to start with STATUS messages. We generate automaton models using Flexfringe for the normal behavior without varying load and for the attack behavior according to the dataset in [9].

As shown in Fig. 5 on the left, during the normal behavior without load variation, GOOSE Status values sent are $\{1, 1, 0, 1, False\}$, GOOSE Alarm values sent are $\{False, 1, False, False, False\}$. The automaton model generated by Flexfringe [35] also contains state numbers and number of times each state is visited, such as 1#600 shows that the state 1 has been visited 600 times. We ignore those during the generation and usage of log-based automaton models. The next step is to analyze the attack dataset to identify GOOSE data values sent by IEDs

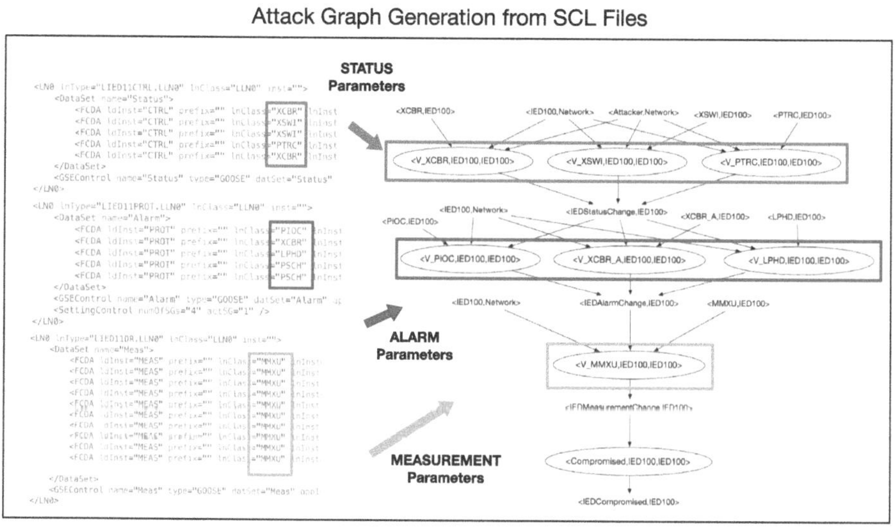

Attack Graph Generation from SCL Files

STATUS Parameters

ALARM Parameters

MEASUREMENT Parameters

Fig. 4. Static Attack Graph Generation From SCL File for IED11 Taken from [9]

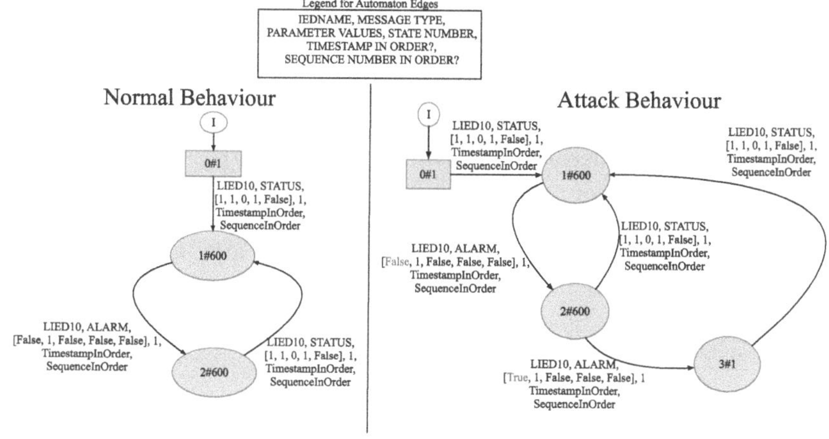

Fig. 5. Automaton Model for The Normal Behavior (Left) and Attack Behavior (Right)

during an attack. Figure 5 on the right shows the automaton model resulting from one change in the alarm parameters from $\{False, 1, False, False, False\}$ to $\{True, 1, False, False, False\}$. Another use case of the automaton models is to detect deviations from normal behavior in IEDs, as indicated by log file analysis. If an IED exhibits abnormal activity, it may be indicative of an ongoing attack. Consequently, operators can identify such IEDs as potentially compromised units. This information is required for the generation of the MDP model.

4.3 Enhancing the SCL-Based Attack Graph with Automaton

After building automaton models for the normal and attack behaviors, the next step is to enhance the SCL-based attack graph with parameters changed by the attacker. All of those exploits that refer to GOOSE parameters are assumed to be zero-day [36] with probability 0.08 (as detailed in [12]) in the SCL-based attack graph since the vulnerability the attacker utilizes to change those parameters is unknown. Log-based automaton model for the attack shows which parameters the attacker has changed during the attack. In that case, the exploit becomes known, and its probability becomes 0.59 since the average CVSS score of known vulnerabilities is 5.9 [2] at the time of writing this paper. Those exploits are shown in blue in Fig. 6.

As depicted in the static SCL-based attack graph, the attacker is required to first tamper with a STATUS parameter, followed by an ALARM parameter, and finally a MEASUREMENT parameter. This suggests that the attacker must compromise all three types of parameters. However, the attack dataset described in [9] suggests that attackers may only need to tamper with one or two parameters. To address this, we introduced intermediate exploits, enabling the representation of attack paths where the attacker targets only one or two of these parameters. This adaptation not only addresses the original graph's limitation, which mandated tampering with all three parameters in a set order but also upholds the fundamental semantics of the attack graph model [6,30]. Those nodes are exploit nodes with probability 1.0 since those are exploits that require zero effort from the attacker. As an example, in Fig. 6, we add two intermediate exploits shown in blue color ($\langle V\_Inter, IED11, IED11\rangle$), one for tampering with a STATUS parameter and another for tampering with a MEASUREMENT parameter to model the fact that the attacker can still reach the goal by tampering with one ALARM parameter ($\langle V\_XCBR, IED11, IED11\rangle$ in Fig. 4). By leveraging the merged attack graph model, we enhance the static SCL-based attack graph, which outlines potential attack paths, with actual paths undertaken by the attacker.

4.4 Generating the MDP Model

After generating the merged attack graph, the next step is to generate an MDP model in which the attacker is modeled as an agent aiming to maximize the physical damage. A typical MDP model [32] consists of the following entities: a finite set of states, a finite set of actions, a transition probability function which is the probability of moving to another state from one state using an action, and a reward function. In SecMonS, states correspond to the set of compromised IEDs. As an example, the state $\{IED2, IED7\}$ in Fig. 9 refers to the attacker state in which the attacker has compromised IED2 and IED7. Actions correspond to vulnerability exploits which lead the attacker to add one more host to the set of compromised hosts. As an example, in order to go from the state $\{IED2\}$ to $\{IED2, IED7\}$, the attacker needs to compromise IED7 given that the attacker has compromised IED2. In order to get the transition probability of this action,

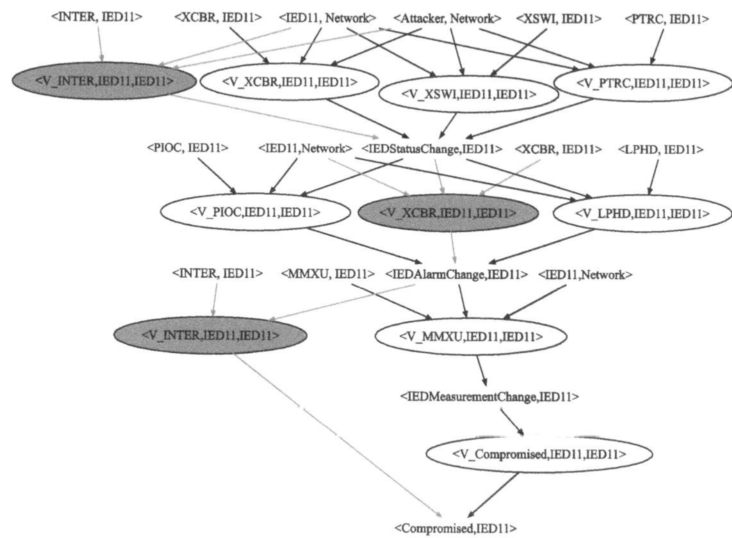

Fig. 6. Merged Attack Graph According to SCL File and Logs

we need to query the enhanced attack graph by asking the question: *What is the probability of compromising IED7 given that IED2 has been compromised?*. The reward function is the sum of transmission line overloads. MDP models have been utilized to identify the potential physical damages to transmission lines in [5,37]. Different from those, we generate the MDP model by starting from the state the attacker is in rather than the state in which none of the IEDs are compromised.

5 Case Study: IEEE 4-Bus

In this section, we demonstrate how SecMonS can be leveraged for attack graph-based threat modeling and MDP-based monitoring in the IEEE 4-Bus system obtained from pandapower as case4gs [3]. Each bus corresponds to a substation, and each transmission line contains circuit breakers at the beginning and end, controlled by IEDs. Figure 7 shows such a configuration.

Assuming that each IED has the same configuration as IED10 in the dataset [9], the first step is to generate SCL-based attack graphs for each bus by merging SCL-based attack graphs generated for IEDs connected to the bus system. The next step is to generate log-based automaton models for each IED to determine changed parameters. After determining the changed parameters, the next step is to generate the enhanced attack graph. Assuming that IED1 has the STATUS parameter modified, and IED6 has the STATUS and ALARM parameters modified, the enhanced attack graph for substation 1 is shown in Fig. 8.

The next step is to generate MDP models for each substation after generating the SCL-based attack graphs and the enhanced attack graphs. In this section, we explain the MDP generation for Bus 1. MDP models for other buses can be generated in a similar manner. For MDP generation, a potential starting point (which

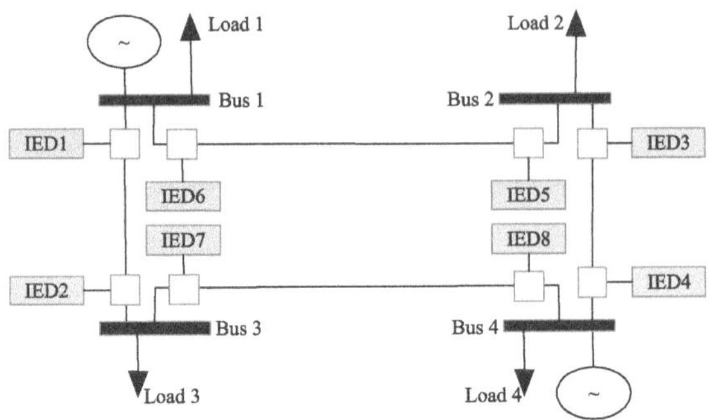

Fig. 7. IEEE 4Bus system with IEDs

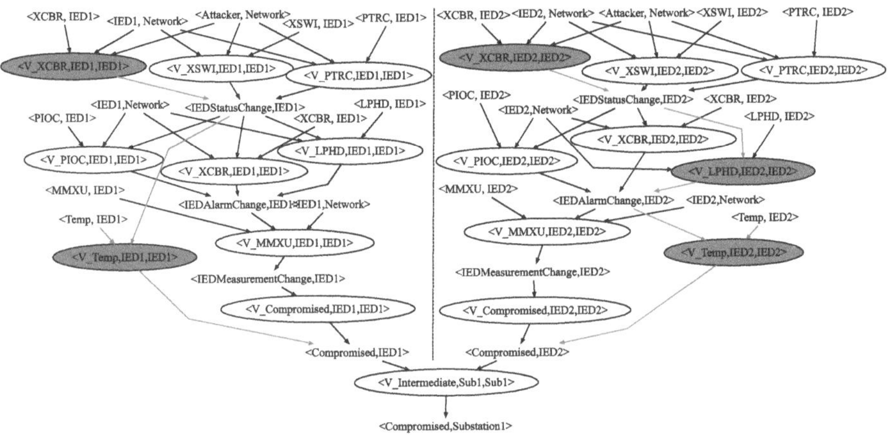

Fig. 8. Enhanced Attack Graph for Substation 1

is usually the set of compromised IEDs according to automaton models) is provided by the operator. Probabilities are calculated by converting the enhanced attack graph into a Bayesian network such that probabilities of all white exploits are 0.08 [12], assuming that those exploits are zero-day. The probabilities of blue exploits are 0.59 since those represent exploits in attack paths the attacker takes. Initial conditions (such as $\langle IED1, Network\rangle$ in Fig. 8), which are conditions without preconditions, are assumed to be always true. Exploit nodes are true when all of their preconditions are true (i.e., $\langle V\_MMXU, IED1, IED1\rangle$ is true when $\langle IED1, Network\rangle$, $\langle IEDAlarmChange, IED1\rangle$ and $\langle MMXU, IED1\rangle$ are all true in Fig. 8). After converting the attack graph in Fig. 8 into a Bayesian network, the following probabilities are obtained: $\Pr(\langle Comp., IED1\rangle \mid \{\}) = 0.64$, $\Pr(\langle Comp., IED6\rangle \mid \{\}) = 0.42$, $\Pr(\langle Comp., Substation1\rangle \mid \{\langle Comp., IED1\rangle\}) = 0.42$, $\Pr(\langle Comp., Substation1\rangle \mid \{\langle Comp., IED6\rangle\}) = 0.64$.

These probabilities are utilized for state transitions in the MDP model. After that, each state's rewards are calculated using the power system analysis tool named Pandapower [3]. In Pandapower, each transmission line can be disabled by setting the parameter "in_service" to False before calculating loads on each transmission line. Loads on transmission lines are calculated during the power flow and provided as the parameter "loading_percent" in Pandapower [3]. In order to calculate the reward function, we calculate the sum of loads that exceed the capacity on each transmission line. As an example, if IED1 in Fig. 7 trips, the following load percentages are obtained for each transmission line (TL): TL From Bus1 to Bus2: 57.61 %, TL From Bus1 to Bus3: 0.0 % since this transmission line is disabled, TL From Bus2 to Bus4: 41.80 %, TL From Bus3 to Bus4: 104.55 %. Based on these calculations, the amount of reward when IED1 is maliciously tripped is 4.55 since it causes 4.55 % overload on transmission lines. Rewards for each state are calculated as follows: R({}) = 0.0, R($\langle Comp., IED1\rangle$) = 4.55, R($\langle Comp., IED6\rangle$) = 0.0, R($\langle Comp., Substation1\rangle$) = 1000 which is the maximum reward since this causes a blackout. By taking the probabilities and rewards, Fig. 9 provides the MDP model starting from the state in which none of the IEDs are in the power system are compromised on the left, and the MDP model starting from the state in which IED1, IED7, and IED8 are compromised on the right.

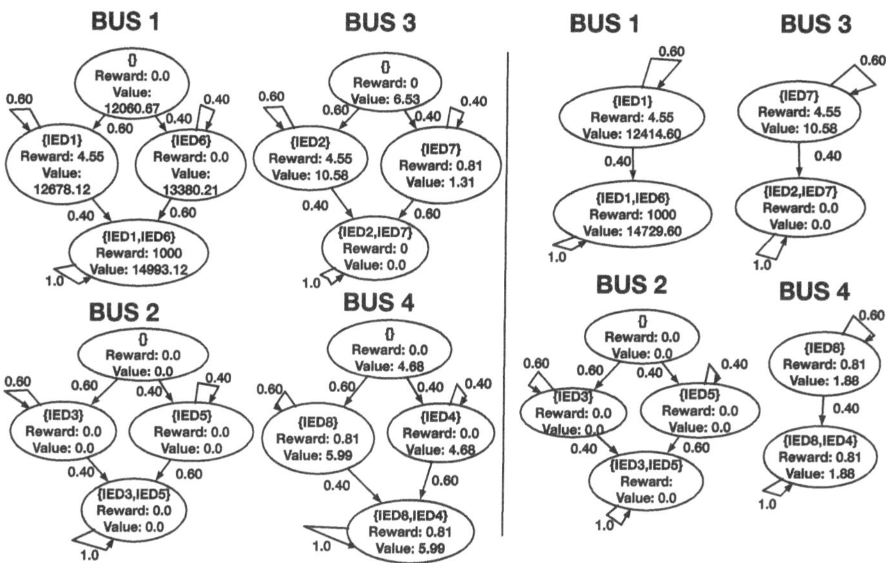

Fig. 9. MDP models for each bus starting from the beginning (left) and MDP models considering IED1, IED7, and IED8 are compromised (right)

In order to solve the MDP model, the value iteration function is utilized with a discount factor of 0.95 (the authors in [5] experimented with varying discount factors, in this paper, we assume that the difference between obtaining the reward at present or in the future does not have a huge difference from the attacker's perspective). State values calculated by the value iteration function are also shown in Fig. 9. As it can be seen, even though attacking IED1 gives a non-zero immediate reward to the attacker, attacking IED6 has more value in terms of the future reward (value function 13380.21 vs. value function 12678.12). Even though we chose the IEEE 4-Bus system as a case study, we experiment with larger bus systems with simulations in Sect. 6 to demonstrate that SecMonS can help with the state space explosion problem faced by MDP models.

6 Simulations

In order to evaluate the practicality of SecMonS, our simulations aim to estimate the execution time for SCL-based attack graph generation, log-based automaton model generation, merged graph generation, and MDP generation for different bus systems and different percentages of compromised IEDs. Our attack graphs follow the standard definitions [6,30] whose correctness is well understood in the literature [12,36]. Therefore, our static SCL-based attack graphs will correctly capture the threats in a substation as long as the given SCL file is accurate. Also, our enhancement of this static attack graph with log events will not alter its structure or correctness, and the correctness of the details added to the attack graph depends on the integrity and accuracy of the logs. Finally, the MDP model also follows standard definitions [5,37] to contain the set of compromised IEDs as the states. Given that the attacker is connected to the IED network, the attacker can attack any combination of IEDs, so the attacker can be in any of those MDP states. In the simulations, We aim to conduct a systematic evaluation of the usability of SecMonS. Utilizing enhanced attack graphs and MDP models generated by SecMonS for systematically hardening substations against attacks on GOOSE protocol and simulating attackers before and after hardening to analyze the usefulness of SecMonS are left as future work. Simulations are conducted on MacOS Ventura 13.4 with an Apple M1 Max processor and 64 GB of memory. Figure 10 provides simulation results related the generation of attack graphs and automaton models.

Our first simulation aims to determine if SecMonS can be leveraged for large substations with many IEDs. For that purpose, we simulate the SCL-based graph generation according to the number of IEDs by using SCL files in the dataset provided by the authors in [9]. The dataset [9] contains 18 SCL files. In this simulation, we vary the number of IEDs from 1 to 18. For each number of IEDs, we choose a random set of IEDs according to the number of IEDs in that experiment and perform 100 experiments for each number of IEDs and calculate the average execution time and the average attack graph size, measured as the number of nodes. Results of this simulation are provided in Fig. 10a.

Results and Implications: As it can be seen from Fig. 10a, both the attack graph size and the execution time increase linearly as the number of IEDs increases. Even for 18 IEDs, the execution time is around 40 milliseconds. This result implies that SecMonS can be utilized for substations with many IEDs.

Our second simulation aims to identify the relationship between the number of log entries and the automaton model generation time. According to a survey

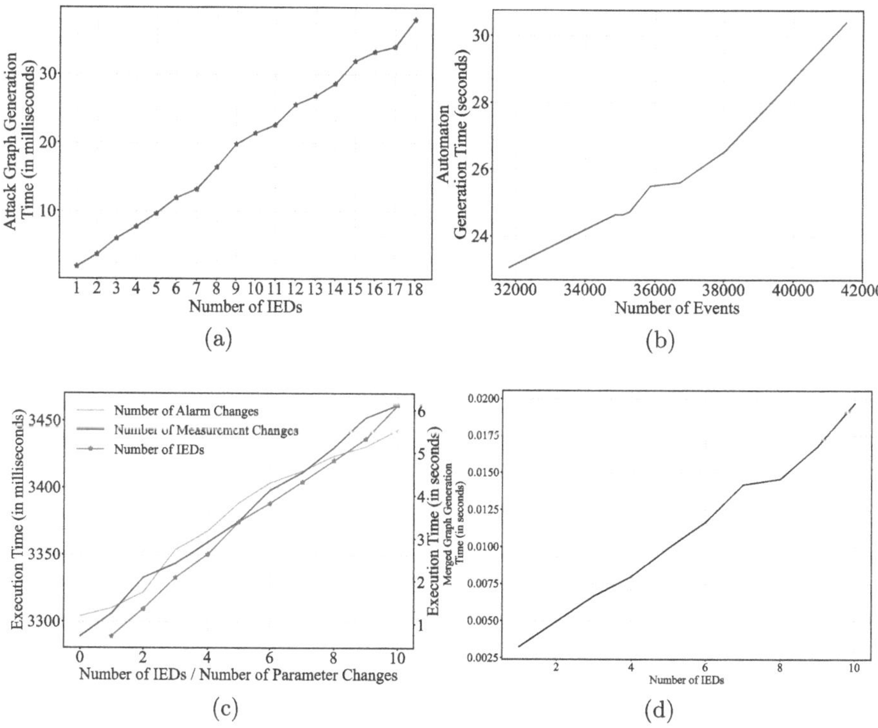

Fig. 10. Simulation Results for Attack Graph and Automaton Generation (a) SCL-based Attack Graph Generation Time and Number of Nodes Using SCL Files Provided in [9] (b) Number of Log Entries vs Automaton Generation Time for the Dataset [9] (c) Number of IEDs and Parameter Changes vs Automaton Generation Time (Simulated Data) (d) Number of IEDs ve Merged Graph Generation Time (Simulated Data)

[4], 27 percent of information technology professionals receive more than 1 million security alerts daily. Given that the number of security alerts can grow, it is essential to have a reasonable automaton model generation time. We simulate log-based automaton generation time using captures provided in [9]. Results of this simulation are provided in Fig. 10b.

Results and Implications: As seen in Fig. 10b, automaton model generation time grows linearly with the number of events and it is around 30 seconds for 42000 events. There are small variations around 36,000 events since logs containing attacks require different processing times based on the number of parameters changed during the attack.

Our third simulation aims to identify the relationship between the number of parameter changes and execution time for simulated data. The goal is to identify how the generation time for the log-based automaton model changes according to the number of changed parameters to identify the scalability of SecMonS for coordinated attacks in which many parameters may be changed concurrently. Results of this simulation are provided in Fig. 10c.

Results and Implications: As seen in Fig. 10c, as the number of IEDs increases, automaton model generation time increases linearly. Even for a substation with 10 IEDs, the automaton model can be generated in 6 seconds. As the number of tampered parameters increases, automaton model generation time increases linearly. This result shows that SecMonS can be utilized for coordinated attacks in which many GOOSE parameters in many IEDs can be tampered with by attackers simultaneously.

Our fourth simulation aims to determine the merged graph generation time according to the number of IEDs and changed parameters. Merged graph generation should be done more frequently compared to SCL-based graph generation since merged graphs utilize the automaton model which can change more frequently. Based on this, merged graph generation should be done in a reasonable time. Results of this simulation are provided in Fig. 10d.

Results and Implications: As seen in Fig. 10d, merged graph generation time and size grow almost linearly according to the number of IEDs. This result shows that merged graphs can be generated in a reasonable time (such as 20 seconds for 10 IEDs) for modeling the security posture of IEC 61850 substations.

Our last simulation aims to determine MDP generation time and MDP number of states for different percentages of compromised IEDs for IEEE 9-Bus, 33-Bus, and 39-Bus from pandapower [3]. Figure 11 shows simulation results.

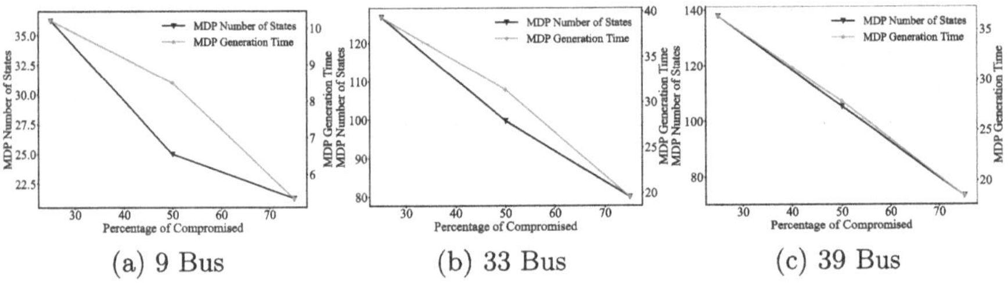

(a) 9 Bus (b) 33 Bus (c) 39 Bus

Fig. 11. Simulation Results for MDP Generation, Percentage of Compromised IEDs vs MDP Number of States and MDP Generation Times for Different Bus Systems

Results and Implications: As seen in Fig. 11a, Fig. 11b, and Fig. 11c, MDP generation time decreases exponentially with respect to the percentage of compromised IEDs. These results imply that starting the MDP model from IEDs, which the attacker has already compromised, can reduce the execution time no matter how big the power system is. In addition, the decrease in MDP number of states and MDP generation time becomes almost linear as the power system size increases. This shows the importance of generating MDP models from the starting state of the attacker for large power systems.

7 Related Work

Alert correlation refers to the process of finding relationships between many alerts [19]. Each alert can refer to a step taken by the attacker. There is a rich literature on alert correlation. In [7], the authors present a system to correlate alerts and generate attack graphs by finding compromised entities and attacker steps. In [20], the authors utilize system logs to detect threats, predict attacker actions, and then generate attack graphs based on predictions. In [24], the authors build attack graphs using intrusion alerts directly without relying on expert knowledge for forensic analysis of prior attacks. In [25], the authors develop a framework that generates attack graphs based on log events and without prior expert knowledge, use it for a case study including alerts from a security competition and conclude that the attackers often follow shorter attack paths even after they discover long ones. In [28], the authors propose a cyber attack situational awareness tool for substations using deep learning. Different from those, in this paper, we generate attack graphs based on both static (configuration files) and dynamic information (event logs) and then merge them to create an enhanced attack graph which is based on both possible attack steps and attack steps that have been taken by attackers.

Threat modeling based on SCL files has also received attention. In [22], the authors create a language for creating threat models and use those models to generate attack graphs. In [14], the authors utilize hybrid attack graphs to model an attack in which the attacker causes the transformer to overheat. In [13], the authors utilize a modeling tool called PowerLang to model IT and OT infrastructures in the power domain and conclude that their tool can weak points in the power domain. Our work differs from [13,14,22] in a way that our attack graph specifically focuses on attacks against GOOSE protocol rather than the whole substation. In addition, the authors in [14,22], and [13] generate attack graphs for different attack scenarios. In our case, a static attack graph is generated first and then it is enhanced based on real instances of attacks.

Anomaly detection in IEC 61850 substations has been studied in the literature. In [15], the authors perform anomaly detection for the GOOSE protocol by analyzing one packet at a time or multiple packets. In [33], the authors detect anomalies in GOOSE and SV protocols by identifying violated security rules. In [4], the authors provide detection for replay attacks and demonstrate the validity of their solution. In [34], the authors utilize machine learning to distinguish normal packets from abnormal ones and demonstrate the validity of their approach. Different from those, in this paper, we first build static SCL-based attack graphs and then enhance those from real instances of attacks to provide comprehensive and realistic threat models for systematic hardening.

MDP models have been utilized in the context of securing the smart grid [5,37]. In [37], the authors extend power system contingency analysis to consider cyber attacks in addition to natural failures. They utilize the MDP model in which states are the set of compromised hosts, and the reward is the total amount of transmission line overloads. In [5], the authors extended [37] to consider concurrent contingencies and experimented with different reward functions.

In this paper, we build MDP models from attackers' starting points for each Bus to reduce the number of states. In addition, rather than assigning the probability of exploits based on their severity in CVSS (like in [5,37]), we retrieve MDP probabilities from the generated merged graph in order to take relationships between vulnerabilities into account.

8 Conclusion

In this paper, we have developed an attack graph-based threat modeling and MDP-based monitoring framework, SecMonS, that utilizes static and dynamic information. This framework monitors the security of power systems concerning cyber attacks with physical consequences. The goal of SecMonS is to enhance static threat models with information in an automated manner to make the gap between threat models and real attacks as small as possible. Our simulation results demonstrate that SecMonS can be deployed to big power networks and it can be utilized for attack graph-based threat modeling and MDP-based monitoring. Our future work includes integrating SecMonS with power system simulators, utilizing enhanced attack graphs and MDP models generated by SecMonS for systematically hardening substations against attacks on the GOOSE protocol, and simulating attackers before and after hardening to analyze the usefulness of SecMonS, tackling the state space explosion problem in automaton models in addition to MDP models, applying SecMonS to other attack scenarios in substations such as tripping circuit breakers, integrating SecMonS with anomaly detection systems for substations so that threat models are based on as many possible attacks against substations as possible including newly discovered attacks, enhancing SecMonS to recommend possible actions for preventing attacks, and applying SecMonS to a smart grid testbed.

Acknowledgement. The authors thank the anonymous reviewers for their valuable comments. The research reported in this paper is funded by the NSERC/Hydro-Québec Thales Research Chair in Smart Grid Security and the NSERC Discovery Grant N01035.

References

1. Analysis of the cyber attack on the Ukrainian power grid. https://www.sans.org/webcasts/analyzing-ukrainian-power-grid-cyber-attacks-102007, Accessed 29 May 2023
2. Current CVSS score distribution for all vulnerabilities. https://www.cvedetails.com/cvss-score-distribution.php, Accessed 29 May 2023
3. Power system test cases. https://pandapower.readthedocs.io/en/v2.4.0/networks/power_system_test_cases.html, Accessed 2 July 2023
4. Survey: 27 percent of IT professionals receive more than 1 million security alerts daily. https://www.imperva.com/blog/27-percent-of-it-professionals-receive-more-than-1-million-security-alerts-daily, Accessed 3 July 2023

5. Akaber, P., et al.: Cases: concurrent contingency analysis-based security metric deployment for the smart grid. IEEE Trans. Smart Grid **11**(3), 2676–2687 (2019)
6. Ammann, P., Wijesekera, D., Kaushik, S.: Scalable, graph-based network vulnerability analysis. In: Proceedings of the 9th ACM Conference on Computer and Communications Security, pp. 217–224 (2002)
7. Bhattarai, B., Huang, H.: Steinerlog: prize collecting the audit logs for threat hunting on enterprise network. In: Proceedings of the 2022 ACM on Asia Conference on Computer and Communications Security, pp. 97–108 (2022)
8. Biswas, P.P., Li, Y., Tan, H.C., Mashima, D., Chen, B.: An attack-trace generating toolchain for cybersecurity study of iec61850 based substations. In: 2020 IEEE International Conference on Communications, Control, and Computing Technologies for Smart Grids (SmartGridComm), pp. 1–7. IEEE (2020)
9. Biswas, P.P., Tan, H.C., Zhu, Q., Li, Y., Mashima, D., Chen, B.: A synthesized dataset for cybersecurity study of IEC 61850 based substation. In: 2019 IEEE International Conference on Communications, Control, and Computing Technologies for Smart Grids (SmartGridComm), pp. 1–7. IEEE (2019)
10. Borbor, D., Wang, L., Jajodia, S., Singhal, A.: Optimizing the network diversity to improve the resilience of networks against unknown attacks. Comput. Commun. **145**, 96–112 (2019)
11. Code, P.: Communication networks and systems for power utility automation–part 6: Configuration description language for communication in electrical substations related to IEDS (2010)
12. Duman, O., Zhang, M., Wang, L., Debbabi, M., Atallah, R.F., Lebel, B.: Factor of security (FOS): quantifying the security effectiveness of redundant smart grid subsystems. IEEE Trans. Dependable Secure Comput. **19**(2), 1018–1035 (2020)
13. Hacks, S., Katsikeas, S., Ling, E., Lagerström, R., Ekstedt, M.: powerlang: a probabilistic attack simulation language for the power domain. Energy Inf. **3**, 1–17 (2020)
14. Hawrylak, P.J., Haney, M., Papa, M., Hale, J.: Using hybrid attack graphs to model cyber-physical attacks in the smart grid. In: 2012 5th International Symposium on Resilient Control Systems, pp. 161–164. IEEE (2012)
15. Hong, J., Liu, C.C., Govindarasu, M.: Detection of cyber intrusions using network-based multicast messages for substation automation. In: ISGT 2014, pp. 1–5. IEEE (2014)
16. Hsu, C.L., Chen, W.X., Le, T.V.: An autonomous log storage management protocol with blockchain mechanism and access control for the internet of things. Sensors **20**(22), 6471 (2020)
17. Ingram, D.M., Schaub, P., Taylor, R.R., Campbell, D.A.: Performance analysis of IEC 61850 sampled value process bus networks. IEEE Trans. Ind. Inform. **9**(3), 1445–1454 (2012)
18. Jajodia, S., Noel, S., Kalapa, P., Albanese, M., Williams, J.: Cauldron mission-centric cyber situational awareness with defense in depth. In: 2011-MILCOM 2011 Military Communications Conference, pp. 1339–1344. IEEE (2011)
19. Levshun, D., Kotenko, I.: A survey on artificial intelligence techniques for security event correlation: models, challenges, and opportunities. Artif. Intell. Rev. **56**, 8547–8590 (2023)
20. Li, T., Jiang, Y., Lin, C., Obaidat, M.S., Shen, Y., Ma, J.: Deepag: attack graph construction and threats prediction with bi-directional deep learning. IEEE Trans. Dependable Secure Comput. **20**(1), 740–757 (2022)

21. Ling, E., Lagerström, R., Ekstedt, M.: A systematic literature review of information sources for threat modeling in the power systems domain. In: Rashid, A., Popov, P. (eds.) CRITIS 2020. LNCS, vol. 12332, pp. 47–58. Springer, Cham (2020). https://doi.org/10.1007/978-3-030-58295-1_4
22. Ling, E.R., Ekstedt, M.: A threat modeling language for generating attack graphs of substation automation systems. Int. J. Crit. Infrastruct. Prot. **41**, 100601 (2023)
23. Nadeem, A., Verwer, S., Moskal, S., Yang, S.J.: Alert-driven attack graph generation using S-PDFA. IEEE Trans. Dependable Secure Comput. **19**(2), 731–746 (2021)
24. Nadeem, A., Verwer, S., Moskal, S., Yang, S.J.: Enabling visual analytics via alert-driven attack graphs. In: Proceedings of the 2021 ACM SIGSAC Conference on Computer and Communications Security, pp. 2420–2422 (2021)
25. Nadeem, A., Verwer, S., Yang, S.J.: Sage: intrusion alert-driven attack graph extractor. In: 2021 IEEE Symposium on Visualization for Cyber Security (VizSec), pp. 36–41. IEEE (2021)
26. Ou, X., Govindavajhala, S., Appel, A.W., et al.: Mulval: a logic-based network security analyzer. In: USENIX Security Symposium, vol. 8, pp. 113–128. Baltimore, MD (2005)
27. Pourmajidi, W., Miranskyy, A.: Logchain: Blockchain-assisted log storage. In: 2018 IEEE 11th International Conference on Cloud Computing (CLOUD), pp. 978–982. IEEE (2018)
28. Presekal, A., Ştefanov, A., Rajkumar, V.S., Palensky, P.: Attack graph model for cyber-physical power systems using hybrid deep learning. IEEE Trans. Smart Grid **19**(2), 731–746 (2023)
29. Rencelj Ling, E., Ekstedt, M.: Generating threat models and attack graphs based on the IEC 61850 system configuration description language. In: Proceedings of the 2021 ACM Workshop on Secure and Trustworthy Cyber-Physical Systems, pp. 98–103 (2021)
30. Sheyner, O., Haines, J., Jha, S., Lippmann, R., Wing, J.M.: Automated generation and analysis of attack graphs. In: Proceedings 2002 IEEE Symposium on Security and Privacy, pp. 273–284. IEEE (2002)
31. Silveira, P., Silva, E.F., Galletta, A., Lopes, Y.: Security analysis of digitized substations: a systematic review of goose messages. Internet Things **22**, 100760 (2023)
32. Sutton, R.S., Barto, A.G.: Reinforcement Learning: An Introduction. MIT Press, Cambridge (2018)
33. Ustun, T.S., Farooq, S.M., Hussain, S.S.: A novel approach for mitigation of replay and masquerade attacks in smartgrids using IEC 61850 standard. IEEE Access **7**, 156044–156053 (2019)
34. Ustun, T.S., Hussain, S.S., Ulutas, A., Onen, A., Roomi, M.M., Mashima, D.: Machine learning-based intrusion detection for achieving cybersecurity in smart grids using IEC 61850 goose messages. Symmetry **13**(5), 826 (2021)
35. Verwer, S., Hammerschmidt, C.A.: Flexfringe: a passive automaton learning package. In: 2017 IEEE International Conference on Software Maintenance and Evolution (ICSME), pp. 638–642. IEEE (2017)
36. Wang, L., Jajodia, S., Singhal, A., Cheng, P., Noel, S.: k-zero day safety: a network security metric for measuring the risk of unknown vulnerabilities. IEEE Trans. Dependable Secure Comput. **11**(1), 30–44 (2013)
37. Zonouz, S., Davis, C.M., Davis, K.R., Berthier, R., Bobba, R.B., Sanders, W.H.: Socca: a security-oriented cyber-physical contingency analysis in power infrastructures. IEEE Trans. Smart Grid **5**(1), 3–13 (2013)

FaultGuard: A Generative Approach to Resilient Fault Prediction in Smart Electrical Grids

Emad Efatinasab[1(✉)], Francesco Marchiori[2], Alessandro Brighente[2], Mirco Rampazzo[1], and Mauro Conti[2,3]

[1] Department of Information Engineering, University of Padova, Padua, Italy
emad.efatinasab@phd.unipd.it, mirco.rampazzo@unipd.it

[2] Department of Mathematics, University of Padova, Padua, Italy
francesco.marchiori.4@phd.unipd.it, {alessandro.brighente,mauro.conti}@unipd.it

[3] Faculty of Electrical Engineering, Mathematics and Computer Science, Delft University of Technology, Delft, The Netherlands

Abstract. Predicting and classifying faults in electricity networks is crucial for uninterrupted provision and keeping maintenance costs at a minimum. Thanks to the advancements in the field provided by the smart grid, several data-driven approaches have been proposed in the literature to tackle fault prediction tasks. Implementing these systems brought several improvements, such as optimal energy consumption and quick restoration. Thus, they have become an essential component of the smart grid. However, the robustness and security of these systems against adversarial attacks have not yet been extensively investigated. These attacks can impair the whole grid and cause additional damage to the infrastructure, deceiving fault detection systems and disrupting restoration.

In this paper, we present **FaultGuard**, the first framework for fault type and zone classification resilient to adversarial attacks. To ensure the security of our system, we employ an Anomaly Detection System (ADS) leveraging a novel Generative Adversarial Network training layer to identify attacks. Furthermore, we propose a low-complexity fault prediction model and an online adversarial training technique to enhance robustness. We comprehensively evaluate the framework's performance against various adversarial attacks using the IEEE13-AdvAttack dataset, which constitutes the state-of-the-art for resilient fault prediction benchmarking. Our model outclasses the state-of-the-art even without considering adversaries, with an accuracy of up to 0.958. Furthermore, our ADS shows attack detection capabilities with an accuracy of up to 1.000. Finally, we demonstrate how our novel training layers drastically increase performances across the whole framework, with a mean increase of 154% in ADS accuracy and 118% in model accuracy.

F. Maggi et al. (Eds.): DIMVA 2024, LNCS 14828, pp. 503–524, 2024.
https://doi.org/10.1007/978-3-031-64171-8_26

1 Introduction

Smart grids represent a transformative paradigm in the realm of energy distribution [7,32]. Through advanced technologies, they aim to enhance electrical grids' efficiency, reliability, and sustainability. Unlike traditional power distribution systems, smart grids leverage real-time data, communication networks, and intelligence control mechanisms to optimize electricity generation, distribution, and consumption. As such, they enable a bidirectional flow of information between utilities and customers, forming a responsive energy ecosystem [32]. The significance of smart grids lies in their ability to address the challenges posed by the evolving energy landscape. Indeed, they facilitate the integration of renewable resources such as solar and wind, mitigating the impact of their variability and contributing to the overall sustainability of the energy sector. Given their importance in the current energy landscape, ensuring the security of smart grids is imperative. Indeed, their interconnection and dependence on digital communication expose them to potential cyber threats and vulnerabilities [33]. As smart grids increasingly rely on data-driven technologies, robust security measures are indispensable to safeguard confidentiality, integrity, and availability across the energy infrastructure. Despite the many papers in the literature proposing new models and methodologies for various aspects of smart grids [3,11,14,22,29,31,35,41,49], a notable gap exists in addressing their security considerations. As the proposed models increasingly rely on Artificial Intelligence (AI) and Machine Learning (ML), it is imperative to address the inherent vulnerabilities that these methodologies suffer from, such as adversarial attacks.

Contribution. To reduce this gap in the literature, we propose **FaultGuard**, a framework for fault type and fault zone prediction in smart grids resilient to adversarial attacks. Unlike many studies focusing on enhancing predictive capabilities, we emphasize resiliency and incorporate robust security layers. In particular, compared to the state-of-the-art [5], we *(i)* add an Anomaly Detection System (ADS) and *(ii)* employ adversarial training in different parts of the system. The ADS detects adversarial attacks toward the fault prediction system, showing an accuracy of up to 1.000 when paired with an adversarial learning training technique. Our new learning technique shows an average improvement of the ADS of 154% compared to its counterpart that has not been trained with adversarial learning. We then develop a low-complexity fault prediction system outperforming the state-of-the-art [5]. To increase the resilience of our fault prediction system against adversarial attacks, we propose and employ *online adversarial training* during its training phase. This procedure shows a mean increase in the model's accuracy of up to 118% when the system is under attack. We evaluate our model on the IEEE13-AdvAttacks dataset [5], a simulated dataset based on the IEEE-13 test node feeder. Our results show that our model outclasses the state-of-the-art, reaching an accuracy of up to 0.958. Our contributions can be summarized as follows.

- We propose **FaultGuard**, a resilient framework for predicting fault types and zones in smart grids capable of withstanding adversarial attacks.

- We propose a single-layer Gated Recurrent Unit (GRU) architecture that outclasses the state-of-the-art in fault type and fault zone prediction.
- We propose an ADS capable of detecting complex adversarial attacks generated with different amounts of adversarial noise.
- We propose an online adversarial training technique, showing how including a subset of adversarial samples in the training process drastically increases the accuracy of the models under attack.
- We evaluate our models, attacks, and defenses on a publicly available dataset, showing the efficacy of the attacks in unrestricted scenarios and the capabilities of our defenses.
- We make the code of our systems, attacks, and the dataset available at: https://github.com/emadef1/FaultGuard/.

Organization. The paper is organized as follows. In Sect. 2, we mention the challenges and limitations of the related works in the literature. Our system and threat models are proposed in Sect. 3. The methodology for our attacks is discussed in Sect. 4, while the details of our model implementation are discussed in Sect. 5. In Sect. 6, we evaluate our model, attacks, and defenses. We report the takeaways of this study in Sect. 7, and Sect. 8 concludes this work.

2 Related Works

While the literature has extensively discussed and implemented fault prediction models on smart grids, their security and robustness have not been thoroughly studied. Indeed, these models have been shown to be vulnerable to adversarial attacks. For instance, Ardito et al. [5] investigated the robustness of fault type and zone classification systems against adversarial attacks. They conducted evaluations through dataset releases, benchmarking, and assessments of smart grid failure prediction systems under adversarial assaults.

Numerous papers have delved into fault detection and classification methodologies within Smart Grids [3,11,14,22,29,31,35,41,49]. As outlined by Saha et al. [36], the categorization of fault location methodologies in power systems includes traditional, observant, and intelligent approaches. This paper specifically focuses on intelligent approaches for fault detection, utilizing smart sensors or expert systems. These intelligent methods involve various techniques, such as expert systems, ML, and Deep Learning (DL), all aimed at identifying faults within the system. Indeed, Artificial Neural Networkss (ANNs) have been extensively explored in the literature for identifying and predicting faults [1,6,13,15,17,24,40,45]. Shadi et al. [38] leveraged Recurrent Neural Network (RNN) and Long Short-Term Memory (LSTM) models within a real-time hierarchical architecture to accurately pinpoint and localize faults. Bhattacharya et al. [8] developed a framework for intelligent fault analysis, leveraging Support Vector Machines (SVMs) and LSTMs. Zhang et al. [48] introduced a method leveraging the attention mechanism, Bidirectional GRU, and a dual structure network to analyze data from diverse perspectives. Thukaram et al. [43] proposed a hybrid approach combining SVM and ANN architectures. In their method, the SVM streamlines the relationship between measurements and fault distance. Tree-based methods like Random Forests (RFs) have emerged as highly favored techniques for fault location due to their versatility and low variance [16,34]. Also, Sapountzoglou et al. [37] proposed a gradient-boosting tree model to detect, identify, and localize faults within low-voltage smart distribution grids. Wilches-Bernal et al. [47] introduced an innovative fault location and classification algorithm, leveraging mathematical morphology in conjunction with RFs. Majidi et al. [30] introduced a fuzzy-c clustering approach to identify potential fault points. Ghaemi et al. [18] introduced an ensemble approach to enhance the precision of fault node localization. Their method is designed to leverage the strengths of SVMs, k-Nearest Neighborss (kNNs), and RFs.

3 System and Threat Model

We now delve into the system and threat model of our study. In the former, we disclose the standard functionality of the system in adversary-free environments. In the latter, we discuss the possible attacker's capabilities and the assumption of their system knowledge.

System Model. In an unthreatened scenario (i.e., without attackers aiming to disrupt the system), the model inputs the data from the smart grid infrastructure. We assume having two fault prediction models, one for each task: fault type prediction and fault zone prediction. In the former, the model objective is to determine the type of voltage sags faults, which can be asymmetric phase-to-phase (LL), single-phase-to-ground (LG), two-phase-to-ground (LLG), or symmetric three-phase-to-ground (LLLG or LLL). In the latter, the model objective is determining the geographical zone where the fault occurred. We assume the models have been trained on an uncorrupted dataset and are finally deployed into the system.

Threat Model. As we aim to provide efficient defenses against adversaries targeting ML models in the smart grid, we delineate our threat model encompassing the most favorable scenarios for the attacker. Thus, while aiming to compromise the fault prediction model, we assume the adversary can successfully infiltrate the system and inject data into the grid. There are various ways in which an attacker can achieve this, as exploiting known or new vulnerabilities was demonstrated to be an effective way to gain remote access [12,42]. Once an adversary has gained access to the infrastructure, they aim to compromise fault type or fault zone prediction using adversarial examples. In the former scenario, the attacker intends to cause the misclassification of potential faults, potentially prompting inappropriate recovery actions by grid operators, leading to catastrophic consequences. In the latter scenario, the adversary manipulates fault prediction models, targeting fault zone prediction in smart grids. This results in recovery teams being dispatched erroneously to the wrong zone, amplifying the impact on operational efficiency and necessitating robust security measures to safeguard smart grid applications.

We can define two scenarios based on the attacker's knowledge of the exchanged data and the smart grid models.

- *White-box Scenario*: the attacker has access to the data used for testing the model and the model's architecture and parameters. This is the most favorable scenario for the attacker, who can leverage this intelligence to craft powerful adversarial samples. Furthermore, having access to the model weights allows the adversary to tune the attack parameters offline.
- *Gray-box Scenario*: the attacker has access to the data used for testing the model, but not the model's architecture or parameters. This scenario is more challenging for an adversary aiming to use adversarial samples, as it would require them to be transferable among different model architectures. However, several studies have demonstrated the difficulties of this task and showed the inefficiency of using surrogate models for generating adversarial noise to test samples [2,4].

It is worth noting that while the white-box scenario is the most favorable from the attacker's perspective, the gray-box scenario is more achievable in real-world implementations. Indeed, an adversary can obtain the model architecture from either *(i)* the known implementation disclosed by the manufacturer, or *(ii)* having direct access to the system input/output and use model extraction techniques [19,23]. As such, model parameters can be protected by *(i)* not publicly disclosing the model architecture and training dataset used and *(ii)* using model obfuscation techniques. Instead, gaining data is a more accessible technique for the attacker, as many entry points are present across the infrastructure. Indeed, many IoT devices and networks compose the smart grid. With the integration of data coming from sustainable energy producers, adversaries can collect data in various parts of the system.

4 Attacks

We now discuss the attacks that we employ against fault type and zone prediction systems in smart grids. Our attacks are different depending on the assumptions of the attacker's knowledge, namely, white-box scenario (Sect. 4.1) and gray-box scenario (Sect. 4.2).

4.1 White-Box Scenario

In our white-box threat model, the adversary possesses full knowledge of the data and the trained model. Thus, we analyze prominent adversarial attacks to reveal vulnerabilities in ML models. We focus on specific attacks highlighted in the literature for their significance and capacity to uncover weaknesses.

- *Fast Gradient Sign Method (FGSM):* swiftly crafts adversarial examples by leveraging the sign of the gradient of the loss function. Recognized for computational efficiency, it is a foundational benchmark for evaluating model robustness [20].
- *Basic Iterative Method (BIM):* extends FGSM through an iterative application, introducing small perturbations at each step to enhance attack potency. Provides insights into cumulative perturbation effects for nuanced robustness evaluation [27].
- *Carlini & Wagner (CW):* a sophisticated attack that formulates adversarial example crafting as an optimization problem, seeking minimal perturbations for misclassification with minimal perceptibility. Challenges models with minimal perturbations, assessing resistance against imperceptible adversarial examples [9].
- *Randomized Fast Gradient Sign Method (RFGSM):* introduces randomness into FGSM iterations by incorporating random noise, enhancing attack diversity. Explores the impact of variability in adversarial perturbations, providing insights into model robustness against unpredictable attacks [44].
- *Projected Gradient Descent (PGD):* employing an iterative optimization approach akin to BIM, PGD includes a projection step to confine perturbations within a defined constraint set. It stands out for crafting potent adversarial examples, allowing rigorous examination of model robustness under stringent conditions [28].

4.2 Gray-Box Scenario

In the gray-box scenario, the adversary has gained access solely to the data and cannot get access to the prediction models. Exploiting this limited access, the adversary employs a Generative Adversarial Network (GAN) to synthesize malicious data that closely mirrors authentic instances. GANs are a class of artificial intelligence algorithms that consist of two neural networks, a generator and a discriminator, trained simultaneously to generate realistic data. In this way, the attacker can lead the fault prediction model towards the detection of a specific fault type or zone. This evasion strategy involves training a GAN model on real data, enabling synthetic data generation that resembles legitimate smart grid data. By successfully training the GAN, the adversary acquires a powerful tool (i.e., the trained generator), which can then be employed to inject the generated data into the smart grid system. Introducing maliciously generated data designed to mimic real data poses a nuanced challenge, showcasing the adversarial capabilities of GANs in evading detection.

5 FaultGuard

In this section, we present our proposed FaultGuard framework, which is graphically shown in Fig. 1. We consider two primary data sources for prediction: the legitimate sensor data gathered in the smart grid and the malicious data injected by possible adversaries. We employ an ADS for detecting adversarial samples in the input data. If samples are detected as malicious, they are discarded. Instead, if the data appears legitimate, they are fed as input to the fault prediction model. Furthermore, we employ online adversarial training to boost our system resilience towards possible attacks. Finally, our model generates a prediction for each task it is trained on: fault type and zone prediction.

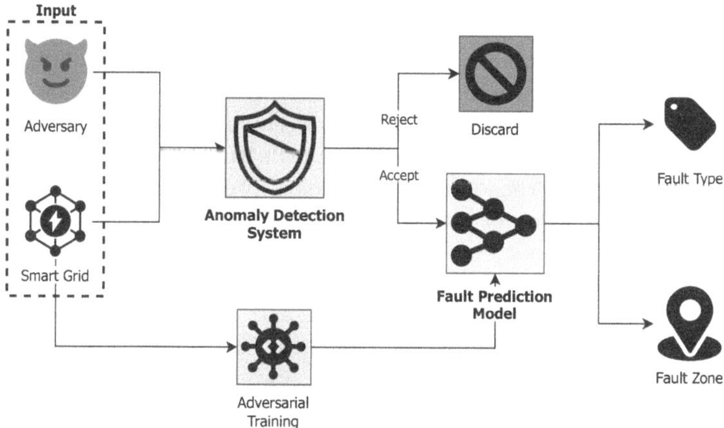

Fig. 1. FaultGuard framework.

We present our GAN-based ADS in Sect. 5.1, and the fault prediction system in Sect. 5.2. While we list the components of our pipeline in order or appearance, it is worth noting that, in real-world scenarios, the first step would be training the fault prediction system. Indeed, our ADS uses the trained prediction model in its implementation. After training both the model and the ADS, components can be organized as detailed in Fig. 1.

5.1 Anomaly Detection System

As shown in Fig. 1, we employ an ADS before feeding the input to our fault prediction system. The aim of the ADS is to detect and reject adversarial attacks while allowing legitimate samples. We use a GAN to achieve this. Our GAN model is characterized by neural networks featuring linear input and output layers. In particular, we leverage the discriminator for anomaly detection after training.

Architecture. The generator model creates synthetic data that mimic legitimate data patterns. It achieves this through four fully connected layers, with neurons varying from 51 (i.e., the number of features) to 128. The discriminator model serves a dual purpose as an ADS and an authenticity evaluator. It is tasked with evaluating the genuineness of incoming data by discerning whether it is authentic grid data (real) or artificially generated by the generator model (fake). Comprising five fully connected layers, the discriminator's neuron count ranges from 51 to 512. More details on our implementation and hyper-parameters are publicly available in our GitHub repository.

Training. Recognizing the imperative to fortify the discriminator's capabilities, we introduce a novel layer of training in the traditional GAN training process. An overview of this process is shown in Fig. 2. The first training steps adhere to the standard procedure and, as such, start with training the discriminator on real data (step ①). Once the discriminator's loss is backpropagated, we generate fake data with the generator. We do this by starting with a random tensor of latent inputs, which the generator model consequently processes to create the fake inputs (step ②). We evaluate these samples with the discriminator and subsequently backpropagate the loss (step ③). Before proceeding with the training of the generator, we first add our novel layer of training. To increase the adversarial detection capabilities of our discriminator, we feed adversarial samples with the FGSM and BIM attack (step ④). These samples are derived from the real data the discriminator was previously trained on and generated through the fault prediction model. In this way, the discriminator can increase its capabilities in detecting FGSM and BIM samples (step ⑤); however, the transferability property of these attacks allows the ADS also to detect other types of attacks [2]. We call this layer of training *adversarial learning*, whose contributions are evaluated in Sect. 6.4. Finally, we train the generator by generating another batch of fake

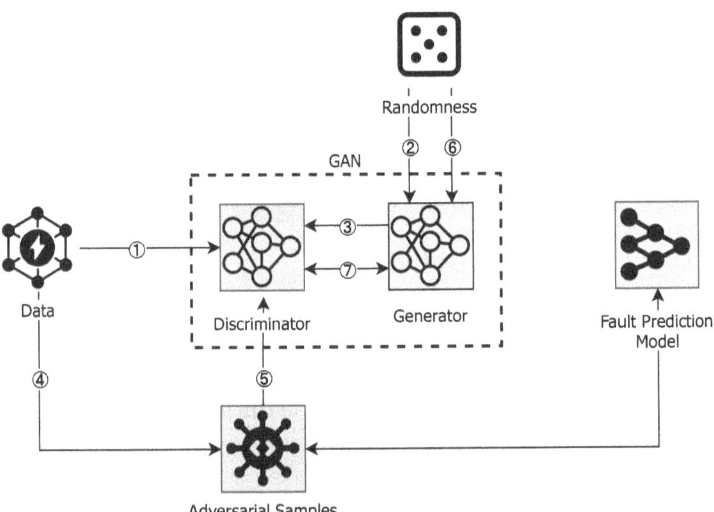

Fig. 2. Schema of our GAN-based ADS training process for each epoch.

data (step ⑥), computing the loss on the discriminator, and backpropagating it to the generator model (step ⑦). This process is repeated for 100 epochs, with a learning rate of 2×10^{-4} for each model.

5.2 Fault Prediction System

In this section, we comprehensively explore the novel model we introduced for fault type and zone prediction tasks. The fault prediction system constitutes the ultimate stage in our framework, working with data that has undergone thorough legitimacy verification by the ADS. While the preceding work, delineated in the dataset introduction paper, included a benchmarking model utilizing a Multi-Layer Perceptron (MLP), its performance was not the primary focus of the study [5]. Indeed, it is worth noting that the causality between consecutive data samples is lost by using a MLP. The oversight in considering the temporal relationships among data points could contribute to the sub-optimal performance observed in benchmarking the model.

Architecture. Recognizing the importance of enhanced performance in real-world scenarios, we propose a novel model tailored for superior efficacy in fault type and fault zone classification. Focusing on practicality and efficiency, we design our fault prediction model to maintain simplicity. The core architecture of our model centers on a one-layer bidirectional GRU comprising 220 neurons. This design enables the model to adeptly capture dependencies in both forward and backward directions within the temporal sequence. Following the GRU layer, we introduced a dropout layer with a rate of 0.5. This layer mitigates overfitting and enhances the model's resilience to the subtle changes that characterize adversarial samples. Post-dropout, the output of the GRU layer transforms a linear layer housing 440 neurons. Subsequently, an element-wise sigmoid activation function is applied, yielding an output vector with dimensions equal to the number of considered classes.

Training. We use cross-entropy as a loss function for our model training. Our optimization process employs the Adam optimizer with a learning rate of 1×10^{-3}, a value empirically selected for efficient convergence. In our case, the training regimen spans a fixed number of epochs set to 80. While employing the ADS contributes a significant layer of security, we have developed an additional strategy to fortify our system. In fortifying the resilience of our fault prediction model, we have devised a strategy involving adversarial training, specifically incorporating both BIM and FGSM attacks. Traditionally, adversarial training involves generating attacks on a pre-trained model using diverse algorithms and subsequently augmenting the training dataset. In contrast, our approach integrates the attack generators directly into the model training process. We call this approach *online adversarial training*. In this scenario of online adversarial training, the adversary adapts over time, becoming more sophisticated as the model improves. This adaptive training strategy challenges the model with increasingly difficult adversarial examples, forcing it to improve its robustness continually. Our adversarial training methodology entails a single, comprehensive training loop. The model is sequentially trained within this loop on real and adversarial data generated by FGSM and BIM attacks. We dynamically create adversarial inputs for each batch of inputs and corresponding labels using FGSM and BIM attacks with a specific epsilon value of 0.2. This value controls the amount of noise added to the samples and, in our case, is large enough to cause misclassification but small enough to maintain some level of perceptibility in the perturbed examples. The model's robustness is tested against realistic perturbations within a reasonable range of what an attacker might apply in real-world scenarios by choosing a moderate epsilon value (i.e., from 0.05 to 0.5). Subsequently, the model undergoes forward passes using these adversarial inputs, and losses for both attacks are computed against the original labels. The backward pass is executed following the forward passes, and the model's gradients are updated using the optimizer. We aggregate the total training loss for the epoch by summing the adversarial losses obtained from FGSM and BIM attacks, scaling the cumulative loss based on the batch size processed during each iteration. This step ensures an appropriate scaling of the total loss relative to the batch size. The accumulated loss is the foundation for computing the average training loss after each epoch. By integrating this new layer of training into the model, we can enhance its performance on real-world data, akin to a form of data augmentation.

6 Evaluation

We now delve into the evaluation of the attacks and FaultGuard system. Our evaluation comprehends all scenarios detailed in the previous sections. We first give details on the dataset used for our evaluation in Sect. 6.1. To provide a baseline evaluation to discuss the success of our attacks and defenses, we evaluate our models on different tasks in Sect. 6.2. We then evaluate our attacks in Sect. 6.3, and finally study the capabilities of FaultGuard in Sect. 6.4.

6.1 Dataset

In the electrical industry, a variety of simulation programs are being used to address fault prediction challenges, including PSCAD [10,25], MATLAB Simulink [46], RSCAD [39], and MATPOWER [21]. Despite the extensive use of these simulation tools in smart grid failure prediction systems, there is a lack of publicly available datasets generated from these tools. So we turn to the dataset introduced by Ardito et al. [5], the only publicly available dataset including substantial simulated fault data rooted in the IEEE-13 test node feeder. The IEEE-13 node test feeder includes a 4.16 kV voltage generator, 13 buses for fault simulation, and three-phase signal measurement. The distribution system is divided into four zones, which are used to identify the location of a fault that has occurred. This dataset comprises 51 features and two target classes: fault label and fault zone, incorporating both traditional and renewable energy sources. It has been carefully curated to serve as a benchmark for assessing the effectiveness of adversarial attacks against fault prediction systems in smart electrical grids. Moreover, we introduce a robust windowing technique to handle our data effectively. This involves partitioning the dataset into segments, each of a predefined size. These windows are created by iteratively traversing the data with a step size equal to half of the window size. Specifically, we choose a window size of 16 s for our dataset. To enhance the dataset's quality and optimize it for our prediction models, we conduct essential preprocessing steps, chief among them being normalization. This critical process ensures consistent data quality and mitigates potential biases that may arise from variations in feature magnitudes. We divide our dataset into three subsets: training (85% of the dataset), validation (5% of the dataset), and test (10% of the dataset).

6.2 Baseline Evaluation

In this phase, we look at the evaluation of our fault prediction system. Initially, we gauge the baseline performance of our system without incorporating countermeasures or exposure to adversarial attacks. The training phase involves utilizing the training data, and subsequently, we evaluate the effectiveness of our GRU-based fault prediction system on the test set. The results are notable, with our model achieving a mean accuracy of 0.604 ± 0.01 for fault type prediction and an accuracy of 0.958 ± 0.01 for fault zone prediction. This performance marks a substantial improvement (a mean 33.11% increase) compared to the state-of-the-art [5]. When replicating this model from the literature, we implemented our preprocessing methods and adjusted the seed and computing environment to match those of our proposed model. These modifications may have influenced the discrepancies observed between the reported results in the paper and our findings. Also, we have chosen to integrate classical ML algorithms such as XGBoost, Random Forest, and Decision Tree into our analysis for two primary reasons. Firstly, they serve as a baseline for comparison against our proposed models, enabling us to gauge the performance and efficacy of our approaches. Secondly, their inclusion underscores the significance of considering causality between data points in this task, highlighting the importance of leveraging advanced techniques to capture temporal dependencies within the data. The detailed results are presented comprehensively in Table 1.

Table 1. Comparison of the model's accuracy.

Model	Accuracy	
	Fault Type	Fault Zone
MLP reproduced from [5]	0.407	0.800
MLP claimed by [5]	0.460	0.710
Decision Tree	0.522	0.818
Random Forest	0.543	0.831
XGBoost	0.560	0.841
GRU	**0.604**	**0.958**

Combinatorial Accuracy. While our fault prediction system demonstrates superior evaluation accuracy compared to existing literature, the practical implementation necessitates a nuanced approach to address misclassification events. Issuing notifications to grid operators for each detected fault could potentially result in multiple false alarms, leading to operational challenges. To address this concern, we adopt a strategy where we wait for the identification of multiple consecutive fault data batches before triggering a notification. This approach introduces the concept of *combinatorial accuracy*, which considers the number of consecutive faulty batches required to initiate an alert. This concept is crucial for balancing the trade-off between efficient fault detection and minimizing false alarms, ensuring system robustness in real-world scenarios. The formula for combinatorial accuracy reflects a geometric distribution and is expressed as follows:

$$combinatorial\_accuracy = \left(1 - (1 - accuracy)^{batches}\right). \tag{1}$$

This formulation ensures that a higher accuracy value is associated with each notification, providing a more reliable indication of actual fault occurrences. The relationship between this accuracy value and the number of consecutive faulty batches is thoroughly analyzed and illustrated in Fig. 3, offering valuable insights into the system's performance under this combinatorial accuracy framework. As evident, the simple strategy of awaiting confirmation from another faulty batch of data significantly enhances the model's accuracy for fault type prediction, improving from 0.604 to a score of 0.843. Likewise, for fault zone classification, accuracy rises from 0.958 to 0.998. This approach drastically reduces the probability of issuing a false alarm notification to grid operators. Specifically, the probability is minimized to 13.7% for fault type prediction and 0.2% for fault zone prediction. This underscores our methodology's effectiveness in elevating accuracy and mitigating the risk of generating false alarms, contributing to a more reliable fault prediction system. Therefore, we opt for a notification delay parameter of two batches of unauthorized data, as it strikes a balanced trade-off between minimizing the false alarm rate and the timely data collection.

6.3 Attacks Evaluation

We now evaluate our attacks against the fault prediction models. We divide the evaluation into the scenarios discussed in the threat model in Sect. 3, i.e., white-box attacks and gray-box attacks.

White-box Evaluation. In this evaluation, we comprehensively assess the efficacy of the white-box attacks, as discussed in Sect. 4. To implement these attacks, we leverage the TorchAttacks library [26], probing the baseline system to evaluate the susceptibility of our model without incorporating any countermeasures or defenses. The attacks are executed with varying epsilon values, signifying the strength of each attack and the degree of perturbation introduced. Specifically, we explore epsilon values ranging from 0.05 to 0.50. The outcomes of these

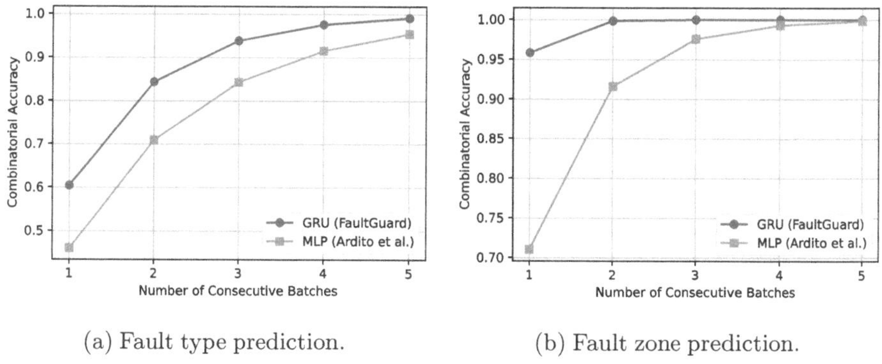

(a) Fault type prediction.

(b) Fault zone prediction.

Fig. 3. Combinatorial accuracy of fault prediction tasks when delaying notification by the number of consecutive faulty batches.

attacks across different tasks are visually presented in Fig. 4. Notably, even with a minimal epsilon of 0.05, a significant decline in the performance of all models is evident. In this case, the accuracy of the prediction model drops to (an average of) 0.155 for fault type and 0.467 for fault zone. For the reproduced MLP model from [5], the accuracy drops to (an average of) 0.070 for fault type and 0.178 for fault zone under the FGSM attack. This evaluation underscores the baseline system's vulnerability to white-box attacks, shedding light on the need for robust defenses and countermeasures to fortify our fault prediction model against adversarial threats.

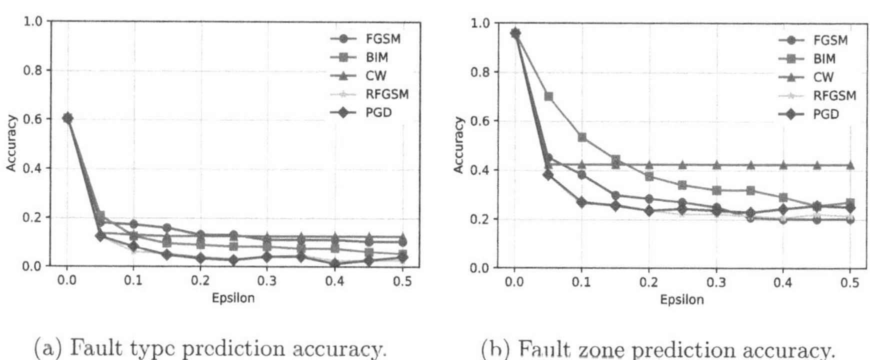

(a) Fault type prediction accuracy.

(b) Fault zone prediction accuracy.

Fig. 4. Model's accuracy at varying epsilon values on the white-box attacks.

Gray-box Evaluation. In this section, we thoroughly examine the effectiveness of our gray-box attack, as detailed in Sect. 4.2. This attack is executed using the generator component within our GAN model, originally designed for anomaly detection. The key aspect of our approach is the use of binary cross-entropy loss, a fundamental element in the GAN framework. The successful training of our GAN relies on specific hyper-parameters, including 150 training epochs and a learning rate of 2×10^{-4}, which governs the optimization process. After completing the training of our GAN, we generate a considerable volume of synthetic malicious data, totaling around 1500 batches. These synthetic data samples, crafted from random noise by our trained generator, are fed into our fully trained fault prediction system for classification. The results show that the generator produces data that can be classified as belonging to 9 out of 11 classes in the fault type prediction task and 3 out of 4 classes in the fault zone prediction task. Effectively, this translates to an Attack Success Rate (ASR) of 0.818 for fault type prediction and 0.750 for fault zone prediction. This underscores that even without prior knowledge of the fault prediction models, an adversary with access to data can create deceptive, malicious data that can elude classic ADS employed in the smart grid.

6.4 FaultGuard Evaluation

In this section, we evaluate the performance of FaultGuard under the attacks described in the previous section. First, we evaluate the effectiveness of the ADS module against white-box attacks. Then, we evaluate our ADS against white-box and gray-box attacks.

ADS Evaluation. In this section, we conduct an experimental evaluation of our ADS, a critical component of our defense strategy against GAN-based attacks and various white-box attacks, as detailed in Sect. 5. We initiate the process by utilizing the training dataset to train our GAN model. We retain the trained discriminator, a crucial element for our subsequent evaluation. The evaluation involves merging generated malicious data with authentic test data from our dataset, simulating malicious attempts alongside real data. These merged datasets are then subjected to the discriminator for analysis, utilizing a predefined threshold (0.5) for discerning determinations between legitimate and anomalous data. The results underscore the discriminator's effectiveness, achieving a mean accuracy rate of 0.991 ± 0.005 standard deviation when classifying real data from malicious data generated by our GAN in the fault type prediction task and a mean accuracy rate of 0.972 ± 0.005 standard deviation in the fault zone prediction task. We subject our model to different white-box adversarial attacks in the subsequent phase. We generate adversarial attacks targeting the fault type and fault zone prediction models, individually applying these attacks to the models. The results are detailed in Table 2, highlighting the ADS's resilience and the contributions of our adversarial learning training layer. The ADS achieves a mean accuracy rate of 0.979 ± 0.050 standard deviation when classifying real data from malicious data generated by our white-box attacks in the fault type prediction task and a mean accuracy rate of 0.821 ± 0.050 standard deviation in the fault zone prediction task.

Table 2. ADS accuracy for each attack and considering the Adversarial Learning (AL) layer. Results are averaged for each ϵ value.

Task	AL	Accuracy				
		FGSM	BIM	CW	RFGSM	PGD
Fault Type	✗	0.674 ± 0.032	0.261 ± 0.067	0.386 ± 0.006	0.703 ± 0.009	0.700 ± 0.007
	✓	$\mathbf{1.000 \pm 0.000}$	$\mathbf{1.000 \pm 0.000}$	$\mathbf{0.897 \pm 0.000}$	$\mathbf{1.000 \pm 0.000}$	$\mathbf{1.000 \pm 0.000}$
Fault Zone	✗	0.274 ± 0.032	0.241 ± 0.011	0.232 ± 0.000	0.300 ± 0.004	0.293 ± 0.008
	✓	$\mathbf{1.000 \pm 0.000}$	$\mathbf{0.999 \pm 0.001}$	0.108 ± 0.002	$\mathbf{1.000 \pm 0.000}$	$\mathbf{1.000 \pm 0.000}$

Despite the ADS's robust performance against various white-box adversarial attacks, it exhibits vulnerability to the CW attack, a sophisticated adversarial technique known for its intricacy. While the ADS may occasionally miss some batches of adversarial attacks, particularly those crafted using the intricate CW technique, the subsequent layer of defense comes into play. The adversarial learning layer ensures the fault prediction models' resilience against these advanced attacks. In only one case (i.e., fault zone + AL against CW attacks), we notice a drop in performance with the inclusion of our novel layer. This standalone case is caused by the missing inclusion of the CW attack in the layer, and more details are given in Sect. 7. As elucidated in upcoming sections, our fault prediction models showcase remarkable resilience even when confronted with the CW attack, successfully averting a significant decline in system performance. The multi-layered defense strategy underscores a comprehensive approach aimed at enhancing the overall robustness and effectiveness of the fault prediction system in smart electrical grids.

Fault Prediction Model Evaluation. This section comprehensively evaluates our fault prediction systems, augmented with online adversarial training as a defense mechanism. As observed in Sect. 6.3, our model, while surpassing the state-of-the-art, remains vulnerable to white-box adversarial attacks. We introduce a novel training layer to fortify our models against such attacks, extensively discussed in Sect. 5.2. After integrating this new training layer, we assess our models' performance in fault type and zone prediction critical tasks. The outcomes of this evaluation are presented in Fig. 5. Notably, the resilience of our models against adversarial attacks, particularly in comparison to models without the defense mechanism, exhibits a substantial improvement. This enhancement is particularly pronounced when facing sophisticated attacks like CW, which have demonstrated the ability to bypass our ADS system. The results underscore the efficacy of the introduced adversarial training layer in bolstering the model's robustness, significantly mitigating the impact of adversarial attacks on fault prediction performance. This defense mechanism is a crucial safeguard, ensuring our fault prediction systems' continued reliability and effectiveness, even in the face of sophisticated and intricate adversarial challenges. A summary of the results of our models and their improvement when paired with the defense mechanism is shown in Table 3.

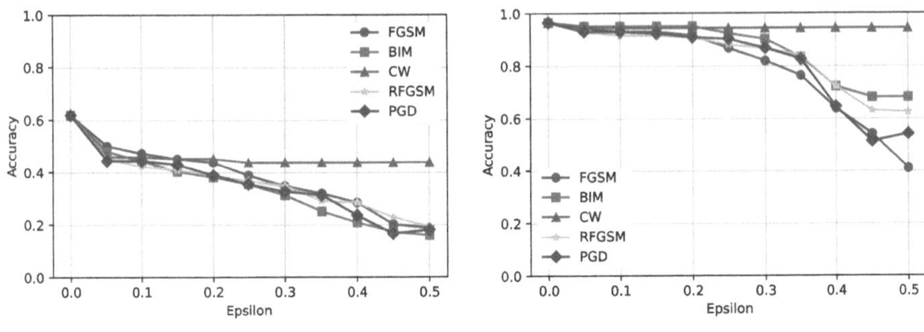

(a) Fault type prediction accuracy. (b) Fault zone prediction accuracy.

Fig. 5. Model's accuracy at varying epsilon values on the white-box attacks when equipped with the defense mechanism.

Table 3. Comparison of model performance under adversarial attacks ($\epsilon = 0.05$) before and after employing our Online Adversarial Training (OAT).

Task	OAT	Accuracy					
		Baseline	FGSM	BIM	CW	RFGSM	PGD
Fault Type	✗	0.604	0.180	0.208	0.138	0.125	0.125
	✓	**0.618**	**0.500**	**0.479**	**0.458**	**0.451**	**0.444**
Fault Zone	✗	0.958	0.451	0.701	0.423	0.381	0.381
	✓	**0.965**	**0.944**	**0.951**	**0.951**	**0.930**	**0.930**

Computational Cost. All experiments in this paper have been conducted on Kaggle as a cloud resource with the following configurations: Intel Xeon (2.20 GHz), NVIDIA Tesla P100 (3584 Cuda cores, 16 GB), 30 GB of RAM, Linux Debian with Python 3.10.12[1]. The defense strategies have shown commendable performance through lightweight models and two white-box adversarial attacks. However, a notable trade-off is the introduction of additional computational load. The ADS and fault prediction systems have been augmented with two adversarial data generators and incorporated two new loss calculations for the prediction models to enhance robustness. As a result, there is a significant increase in computational demands, leading to longer training times for the models. The changes significantly affect training times, as depicted in Table 4, comparing standard models to those fortified with defense mechanisms. Despite the increased computational cost, it is crucial to consider the enhanced robustness and resilience brought by these defenses. Furthermore, this training procedure is needed only when initially deploying the model in the system and does not require maintenance. This computational trade-off underscores the ongoing challenge of bal-

[1] Additional details on packages' versions are available at: https://github.com/emadef1/FaultGuard/blob/main/requirements.txt.

Table 4. Model training times with and without Online Adversarial Training (OAT).

Task	OAT	Training Time
Fault Type	✗	1.18 min
	✓	78.90 min
Fault Zone	✗	1.29 min
	✓	95.00 min

ancing model efficiency with the imperative to defend against adversarial threats in the smart grid domain.

7 Takeaways

Fault prediction systems are a well-researched topic in the literature. However, researchers often neglect the security aspect of these systems, making their implementation problematic due to the high chances of errors when dealing with adversaries. Therefore, we present a summary of the key takeaway messages, making it easier for practical implementation in real-world scenarios. By doing so, we assist practitioners in effectively applying these systems and offer researchers recommended best practices for their studies.

Takeaway 1 – *Fault prediction systems are vulnerable to adversarial attacks, regardless of the scenario's assumptions on the attacker's knowledge.*

As discussed in Sect. 6.3, adversarial attacks are particularly effective against the models tested in this study. Indeed, model accuracy dropped by up to 74.34% with an epsilon value of just 0.05. While these values are valid only for white-box scenarios, it is worth noting that even in gray-box scenarios (i.e., having access to the data), the attacker has great leverage on the system, reaching ASRs up to 0.818. As such, adversarial attacks require particular consideration when designing a ML-based fault prediction system.

Takeaway 2 – *Complex attacks, such as CW, are more difficult to be detected by the ADS.*

As shown in Table 2, our ADS can detect most attacks with perfect accuracy, except for the CW attack. While in the fault type prediction task we obtained an accuracy of 0.897 with the addition of the adversarial learning layer, in the fault zone task, including this layer effectively worsened its performance (from 0.232 to 0.108). It is noteworthy that adversarial learning is performed only with the FGSM and BIM attacks. However, even when including CW in the training process, there were no improvements w.r.t. the standard ADS model against the same attack. For these reasons, complex attacks require special attention, as improperly implementing the ADS might not be enough to detect them.

Takeaway 3 – *Complex attacks, such as CW, are more effective against good-performing models.*

Another correlation regarding the CW attack that can be extracted from our evaluation is that attacks are more effective against good-performing models. Indeed, the fault type prediction model (baseline accuracy of 0.604) was less affected by the attacks w.r.t. the fault zone prediction model (baseline accuracy of 0.958). As such, while researchers and practitioners strive to reach the highest possible accuracy score, the threat of adversarial attacks becomes more pronounced. This highlights the paradoxical relationship between model performance and susceptibility to attacks. Consequently, the pursuit of high accuracy should be accompanied by a heightened awareness of potential vulnerabilities and the implementation of robust defense mechanisms.

Takeaway 4 – *Different forms of adversarial training greatly improve the models' resistance against adversaries.*

One of the significant contributions of this paper is the proposal of novel adversarial training techniques. Indeed, including the adversarial learning layer on the ADS significantly improved its performance, and the online adversarial training performed on the fault prediction model made it more resilient against attacks. These findings underscore the effectiveness and versatility of diverse adversarial training methodologies in improving model robustness and defense capabilities, contributing valuable insights to advancing secure and resilient machine learning models.

8 Conclusions

In the smart grid, fault prediction systems are promising tools that can ensure energy delivery and reliability. However, despite the growing interest in the literature, the security aspect of these systems is often neglected. This oversight can lead to safety issues and delays, making the implementation of those systems counterproductive.

Contribution. In this paper, we introduced **FaultGuard**, a resilient framework designed for fault type and zone classification tasks. To ensure the security of our system, we incorporated an ADS with a unique GAN training layer to detect attacks. Additionally, our approach involved a low-complexity fault prediction model and employed an online adversarial training technique to bolster robustness. We thoroughly evaluated the framework's performance, assessing its resilience against diverse adversarial attacks using the publicly available IEEE13-AdvAttack dataset, a simulated dataset derived from the IEEE-13 test node feeder. FaultGuard outclassed the state-of-the-art, reaching accuracy values up to 0.958 and being natively resilient against adversarial attacks. Furthermore, our ADS detected attack attempts with accuracies up to 1.000.

Future Work. While still outclassing the state-of-the-art and other ML models in the same task, improving the accuracy for fault type prediction is needed to further strengthen its contribution. Indeed, the accuracy value achieved in this study is highly dependent on the dataset employed. As such, creating a richer dataset is the most significant way to improve these results. Also, since this paper focused on evasion attacks towards fault prediction systems, studying poisoning attacks might provide powerful insights into the resilience of these systems. By delineating a new system and threat model accounting for this threat, it might be possible to study the vulnerabilities of the models and thus improve their security. Finally, by combining these results with those obtained in this paper, we would be able to provide a complete overview of the resilience of ML-based fault prediction models, aiding practitioners in safely deploying these systems.

References

1. Al-Shaher, M.A., Sabry, M.M., Saleh, A.S.: Fault location in multi-ring distribution network using artificial neural network. Electric Power Syst. Res. **64**(2), 87–92 (2003)
2. Alecci, M., Conti, M., Marchiori, F., Martinelli, L., Pajola, L.: Your attack is too dumb: formalizing attacker scenarios for adversarial transferability. In: Proceedings of the 26th International Symposium on Research in Attacks, Intrusions and Defenses, pp. 315–329 (2023)
3. Andresen, C., Torsæter, B.N., Haugdal, H., Uhlen, K.: Fault detection and prediction in smart grids, pp. 1–6 (2018). https://doi.org/10.1109/AMPS.2018.8494849
4. Apruzzese, G., Anderson, H.S., Dambra, S., Freeman, D., Pierazzi, F., Roundy, K.: Real attackers don't compute gradients: bridging the gap between adversarial ml research and practice. In: 2023 IEEE Conference on Secure and Trustworthy Machine Learning (SaTML), pp. 339–364 (2023)
5. Ardito, C., Deldjoo, Y., Di Noia, T., Di Sciascio, E., Nazary, F.: Ieee13-advattack a novel dataset for benchmarking the power of adversarial attacks against fault prediction systems in smart electrical grid. In: Proceedings of the 31st ACM International Conference on Information & Knowledge Management, CIKM 2022, pp. 3817-3821. ACM (2022). https://doi.org/10.1145/3511808.3557612
6. Aslan, Y.: An alternative approach to fault location on power distribution feeders with embedded remote-end power generation using artificial neural networks. Electr. Eng. **94**, 125–134 (2012)
7. Bayindir, R., Colak, I., Fulli, G., Demirtas, K.: Smart grid technologies and applications. Renew. Sustain. Energy Rev. **66**, 499–516 (2016)
8. Bhattacharya, B., Sinha, A.: Intelligent fault analysis in electrical power grids. In: 2017 IEEE 29th International Conference on Tools with Artificial Intelligence (ICTAI), pp. 985–990 (2017). https://doi.org/10.1109/ICTAI.2017.00151
9. Carlini, N., Wagner, D.: Towards evaluating the robustness of neural networks. In: 2017 IEEE Symposium on Security and Privacy (SP), pp. 39–57. IEEE Computer Society, May 2017. https://doi.org/10.1109/SP.2017.49
10. Chakraborty, S., Das, S.: Application of smart meters in high impedance fault detection on distribution systems. IEEE Trans. Smart Grid **10**(3), 3465–3473 (2019). https://doi.org/10.1109/TSG.2018.2828414

11. Chen, K., Huang, C., He, J.: Fault detection, classification and location for transmission lines and distribution systems: a review on the methods. High Voltage **1**(1), 25–33 (2016). https://doi.org/10.1049/hve.2016.0005
12. Chen, T.M., Abu-Nimeh, S.: Lessons from stuxnet. Computer **44**(4), 91–93 (2011)
13. Coser, J., do Vale, D.T., Rolim, J.G.: Design and training of artificial neural networks for locating low current faults in distribution systems. In: 2007 International Conference on Intelligent Systems Applications to Power Systems, pp. 1–6 (2007). https://doi.org/10.1109/ISAP.2007.4441599
14. De La Cruz, J., Gómez-Luna, E., Ali, M., Vasquez, J.C., Guerrero, J.M.: Fault location for distribution smart grids: literature overview, challenges, solutions, and future trends. Energies **16**(5) (2023). https://doi.org/10.3390/en16052280
15. Dehghani, F.: CIRED -open access. Proc. J. **2017**(3), 1134–1137 (2017)
16. El Mrabet, Z., Sugunaraj, N., Ranganathan, P., Abhyankar, S.: Random forest regressor-based approach for detecting fault location and duration in power systems. Sensors **22** (2022). https://doi.org/10.3390/s22020458
17. Farias, P.E., de Morais, A.P., Rossini, J.P., Cardoso, G.: Non-linear high impedance fault distance estimation in power distribution systems: a continually online-trained neural network approach. Electric Power Syst. Res. **157**, 20–28 (2018). https://doi.org/10.1016/j.epsr.2017.11.018
18. Ghaemi, A., Safari, A., Afsharirad, H., Shayeghi, H.: Accuracy enhance of fault classification and location in a smart distribution network based on stacked ensemble learning. Electric Power Syst. Res. **205**, 107766 (2022). https://doi.org/10.1016/j.epsr.2021.107766
19. Gong, X., Wang, Q., Chen, Y., Yang, W., Jiang, X.: Model extraction attacks and defenses on cloud-based machine learning models. IEEE Commun. Mag. **58**(12), 83–89 (2020)
20. Goodfellow, I.J., Shlens, J., Szegedy, C.: Explaining and harnessing adversarial examples (2015)
21. He, M., Zhang, J.: A dependency graph approach for fault detection and localization towards secure smart grid. IEEE Trans. Smart Grid **2**(2), 342–351 (2011). https://doi.org/10.1109/TSG.2011.2129544
22. Hussain, N., Nasir, M., Vasquez, J.C., Guerrero, J.M.: Recent developments and challenges on ac microgrids fault detection and protection systems-a review. Energies **13** (2020). https://doi.org/10.3390/en13092149
23. Jagielski, M., Carlini, N., Berthelot, D., Kurakin, A., Papernot, N.: High accuracy and high fidelity extraction of neural networks. In: 29th USENIX security symposium (USENIX Security 20), pp. 1345–1362 (2020)
24. Javadian, S., Nasrabadi, A., Haghifam, M.R., Rezvantalab, J.: Determining fault's type and accurate location in distribution systems with dg using MLP neural networks. In: 2009 International Conference on Clean Electrical Power, pp. 284–289 (2009). https://doi.org/10.1109/ICCEP.2009.5212044
25. Jiang, H., Zhang, J.J., Gao, W., Wu, Z.: Fault detection, identification, and location in smart grid based on data-driven computational methods. IEEE Trans. Smart Grid **5**(6), 2947–2956 (2014). https://doi.org/10.1109/TSG.2014.2330624
26. Kim, H.: Torchattacks: a pytorch repository for adversarial attacks. arXiv preprint arXiv:2010.01950 (2020)
27. Kurakin, A., Goodfellow, I., Bengio, S.: Adversarial examples in the physical world (2017)
28. Madry, A., Makelov, A., Schmidt, L., Tsipras, D., Vladu, A.: Towards deep learning models resistant to adversarial attacks (2019)

29. Mahfouz, M.M., El-Sayed, M.A.: Smart grid fault detection and classification with multi-distributed generation based on current signals approach. IET Gener. Transm. Distrib. **10**(16), 4040–4047 (2016). https://doi.org/10.1049/iet-gtd.2016.0364
30. Majidi, M., Arabali, A., Etezadi-Amoli, M.: Fault location in distribution networks by compressive sensing. IEEE Trans. Power Delivery **30**(4), 1761–1769 (2015). https://doi.org/10.1109/TPWRD.2014.2357780
31. Mousa, M., Abdelwahed, S., Klüss, J.: Review of diverse types of fault, their impacts, and their solutions in smart grid, pp. 1–7, April 2019. https://doi.org/10.1109/SoutheastCon42311.2019.9020355
32. Muqeet, H.A., Liaqat, R., Jamil, M., Khan, A.A.: A state-of-the-art review of smart energy systems and their management in a smart grid environment. Energies **16**(1), 472 (2023)
33. Nafees, M.N., Saxena, N., Cardenas, A., Grijalva, S., Burnap, P.: Smart grid cyber-physical situational awareness of complex operational technology attacks: a review. ACM Comput. Surv. **55**(10), 1–36 (2023)
34. Okumus, H., Nuroglu, F.M.: A random forest-based approach for fault location detection in distribution systems. Electr. Eng. **103**(1), 257–264 (2021)
35. Rahman Fahim, S., K. Sarker, S., Muyeen, S.M., Sheikh, M.R.I., Das, S.K.: Microgrid fault detection and classification: machine learning based approach, comparison, and reviews. Energies **13** (2020). https://doi.org/10.3390/en13133460
36. Saha, M.M., Izykowski, J.J., Rosolowski, E.: Fault location on power networks. Springer Science & Business Media (2009)
37. Sapountzoglou, N., Lago, J., Raison, B.: Fault diagnosis in low voltage smart distribution grids using gradient boosting trees. Electric Power Syst. Res. **182**, 106254 (2020). https://doi.org/10.1016/j.epsr.2020.106254
38. Shadi, M.R., Ameli, M.T., Azad, S.: A real-time hierarchical framework for fault detection, classification, and location in power systems using PMUS data and deep learning. Int. J. Electr. Power Energy Syst. **134**, 107399 (2022). https://doi.org/10.1016/j.ijepes.2021.107399
39. Shafiullah, M., Abido, M.A.: S-transform based FFNN approach for distribution grids fault detection and classification. IEEE Access **6**, 8080–8088 (2018). https://doi.org/10.1109/ACCESS.2018.2809045
40. Souza, J., Rodrigues, M., Schilling, M., Do Coutto Filho, M.: Fault location in electrical power systems using intelligent systems techniques. IEEE Trans. Power Delivery **16**, 59–67 (2001). https://doi.org/10.1109/61.905590
41. Stefanidou-Voziki, P., Sapountzoglou, N., Raison, B., Dominguez-Garcia, J.: A review of fault location and classification methods in distribution grids. Electric Power Syst. Res. **209**, 108031 (2022). https://doi.org/10.1016/j.epsr.2022.108031
42. Sullivan, J.E., Kamensky, D.: How cyber-attacks in ukraine show the vulnerability of the us power grid. Electr. J. **30**(3), 30–35 (2017)
43. Thukaram, D., Khincha, H., Vijaynarasimha, H.: Artificial neural network and support vector machine approach for locating faults in radial distribution systems. IEEE Trans. Power Delivery **20**(2), 710–721 (2005). https://doi.org/10.1109/TPWRD.2005.844307
44. Tramèr, F., Kurakin, A., Papernot, N., Goodfellow, I., Boneh, D., McDaniel, P.: Ensemble adversarial training: attacks and defenses (2020)
45. Usman, M.U., Ospina, J., Faruque, M.O.: Fault classification and location identification in a smart distribution network using ann. In: 2018 IEEE Power & Energy Society General Meeting (PESGM), pp. 1–6 (2018). https://doi.org/10.1109/PESGM.2018.8586471

46. Wang, X., et al.: High impedance fault detection method based on variational mode decomposition and teager-kaiser energy operators for distribution network. IEEE Trans. Smart Grid **10**(6), 6041–6054 (2019). https://doi.org/10.1109/TSG.2019.2895634
47. Wilches-Bernal, F., Jiménez-Aparicio, M., Reno, M.J.: An algorithm for fast fault location and classification based on mathematical morphology and machine learning. In: 2022 IEEE Power & Energy Society Innovative Smart Grid Technologies Conference (ISGT), pp. 1–5 (2022). https://doi.org/10.1109/ISGT50606.2022.9817473
48. Zhang, F., Liu, Q., Liu, Y., Tong, N., Chen, S., Zhang, C.: Novel fault location method for power systems based on attention mechanism and double structure GRU neural network. IEEE Access **8**, 75237–75248 (2020). https://doi.org/10.1109/ACCESS.2020.2988909
49. Zidan, A., Khairalla, M., Abdrabou, A.M., Khalifa, T., Shaban, K., Abdrabou, A., El Shatshat, R., Gaouda, A.M.: Fault detection, isolation, and service restoration in distribution systems: State-of-the-art and future trends. IEEE Trans. Smart Grid **8**(5), 2170–2185 (2017). https://doi.org/10.1109/TSG.2016.2517620

Wireless Modulation Identification: Filling the Gap in IoT Networks Security Audit

Florent Galtier[1(✉)], Guillaume Auriol[1,2], Vincent Nicomette[1,2], Paul L. R. Olivier[1], Romain Cayre[3], and Mohamed Kaâniche[1]

[1] CNRS, LAAS, 7 Avenue du Colonel Roche, 31400 Toulouse, France
{florent.galtier,guillaume.auriol,vincent.nicomette,paul.olivier, mohamed.kaaniche}@laas.fr

[2] Université de Toulouse, INSA, LAAS, 31400 Toulouse, France

[3] Eurecom, 450 Route des Chappes, 06410 Biot, France
romain.cayre@eurecom.fr

Abstract. The massive deployment of IoT devices comes with the creation of many wireless communication protocols designed to support various applications. However, while some of these protocols are public and widely used, such as Bluetooth Low Energy or Enhanced ShockBurst, the specifications of other protocols are only partially, if at all, publicly available, making it difficult for security researchers to study them, especially for auditing purposes. In this paper, we address this issue by presenting an innovative, easy-to-use and protocol-agnostic toolkit to analyze unknown network communications. The toolkit is designed to automatically infer their physical layer characteristics and to extract the binary content of their frames. We conducted experiments to identify the modulation parameters for various wireless communications used by off-the-shelf devices as well as for randomly generated protocols, considering both FSK and ASK modulations. In addition, our results show that the proposed toolkit is capable of successfully detecting covert channels in wireless environments. We also conducted a case study on an undocumented proprietary wireless mouse protocol.

Keywords: IoT · Intrusion Detection · Radio Frequency Identification · Secure Communication · Cyber-Physical Systems · Security and Privacy

1 Introduction

Industrial and home networks are increasingly relying on wireless networks, either to ease communication, save space or reduce costs. However, such networks are more exposed to security attacks than traditional wired networks, due to the use of unconfined and open radio transmissions.

In such systems, attackers could easily eavesdrop on communications or inject malicious transmissions among legitimate ones. In contrast, such attacks on

F. Maggi et al. (Eds.): DIMVA 2024, LNCS 14828, pp. 525–544, 2024.
https://doi.org/10.1007/978-3-031-64171-8_27

wired networks are mitigated by using robust communication systems and by monitoring anomalies in transmission frames. However, auditing and monitoring wireless networks is hindered by the large number and heterogeneity of protocols. They can have radically different physical layers and use specific channels to communicate, sometimes adding channel hopping. Moreover, an attacker willing to stealthily exfiltrate data could communicate through a custom crafted protocol designed for escaping the scrutiny of potential monitoring measures.

Extracting binary streams from known protocols is possible with available tools [1,2], but such methods are lacking for more generic use cases. Auditing unfamiliar wireless protocols involves laborious tasks to identify their modulations and parameters, then to recover the binary streams and higher-layer protocols.

With the diverse protocols in IoT and numerous proprietary ones, auditing wireless communications becomes increasingly difficult. Many IoT devices have flaws that make them vulnerable to attacks. Furthermore, proprietary protocols often rely on security through obscurity, further complicating audits. Securing wireless networks is therefore a pressing challenge. Developing new tools to analyze protocols in a broader way is crucial for improving wireless security audits.

In this paper, we propose to address this challenge by designing and implementing a novel modulation identification approach and analysis module suitable for different types of modulations, and easily extensible. The proposed approach also allows to retrieve the binary payload of known protocols, and identify fields of interest in unknown ones. More precisely, this approach with the designed toolkit is able to detect transmissions and infer their parameters and content, without making any assumption. It combines reception over a wide frequency band (for example higher than 10MHz) and different modulation detection algorithms to estimate the parameters used in the transmission. This tool can then be enriched with grammar estimation modules to become an assistant for wireless protocol reverse-engineering, or a covert channel attack detection system to identify exfiltrated data. It only requires for the operator manual parameterization and estimation of the noise level beforehand, and is entirely automated afterwards. Furthermore, the source code of the tool and the experiments will be made public to allow reproducibility.

Overall, the following contributions can be highlighted in this paper:

- A novel approach to identify and analyze modulations in signals, infer their parameters and retrieve basic information about the higher-layer protocols.
- An open-source tool implementing this approach.
- The validation of the approach and test of the tool on well-known as well as on randomly generated protocols, by demonstrating its capability to detect covert channels, and describing a full case study on a wireless mouse.

The remainder of this paper is structured as follows. Section 2 discusses the state-of the art about wireless protocol identification and transmission detection in wireless networks. Then, Sect. 3, briefly describes the theoretical background behind our tool. The implementation is presented in detail in Sect. 4, and its validation on different application use cases is addressed in Sect. 5. It also presents a case study on a wireless mouse using an undocumented and proprietary protocol, and shows how our tool can help a reverse-engineer. The limitations of and possible improvements are discussed in Sect. 6. Finally, Sect. 7 concludes and presents future directions for our work.

2 Related Works

Automated detection of transmissions has always been a major concern in security because of its importance in the design of security monitoring systems.

Several steps are needed to retrieve the information of interest from the signal: first, detecting when and at which frequency frames are transmitted; then, identifying the modulation type, identifying its parameters; and finally, demodulating and analyzing the binary stream to extract information on the protocol's frame format, and on the higher-layer content of the communications. The first step (detection of frame transmission) is well-documented in the state of the art, and one of the most common methods, which we also use, is based on an amplitude threshold. Regarding the second step (identification of the modulation type), several modulation detection techniques exist in the state of the art. Some approaches such as [3] or [4] focus on specific modulations and their unique characteristics to detect them. The main issue with these approaches is their high false positive rates, as they ignore the other modulations. Also, as with other protocol detection methods, their focus on specific modulations makes them unusable in a multiprotocol monitoring context, unless a number of them are combined. Some other works are based on the comparison of the received signals with models of the different modulations. For example in [5] and [6], probabilistic models of the modulations are built, and then compared to signals, using goodness-of-fit tests such as Kolmogorov-Smirnov. The main limitation of this type of approach is the potential difficulty of building a model that fits a specific modulation. In this paper, we describe an approach based on the detection of symbols in the signals by computing the autocorrelation of different representations fitting different modulation types, allowing the addition of extensions for new modulation types if needed. This method is simpler than building models for goodness-of-fit tests, but it strongly depends on the hypothesis that a stable periodicity can be highlighted by the autocorrelation of the signal with the correct modulation representation. However, this hypothesis is, in general, met with protocols using pulse-shaping filters to increase the spectral efficiency, which are predominant.

Existing solutions take full advantage of partial to complete knowledge of a set of protocols they want to monitor in order to ease their detection and collect additional metadata about the transmissions. Some approaches focus on a single protocol, and their goal is rather to detect attacks on upper layers [7]. Others take a broader perspective, considering different possibilities and performing a case disjunction based on protocol-dependent heuristics to identify the type of the emissions [1,2]. However, they do not allow detecting unknown protocols, especially outside pre-defined channels. This exposes them to covert channel attacks that exploit the knowledge of the monitored channels.

Other works focus on specification-agnostic anomaly detection. For instance, in RIDS [8], the authors trained an auto-encoder in an attack-free environment to detect unusual activities in the monitored frequencies. However, these approaches suffer from a trade-off between detection accuracy and the amount of processed data. This does not allow performing a deeper analysis of the anomaly beyond the identification of its occurrence time and associated frequencies. The method and tool presented in this paper aim at addressing the limitations of both of these approaches in the context of a more generalized covert channel detection.

Finally, a high and growing number of works focus on the use of Neural Networks-based classification systems to identify the modulation type, for example [9–13]. However, such methods require huge resources for their training, and suffer from a lack of explainability because of the black-box aspect of neural networks. Moreover, none of those methods allows reversing the modulation parameters and recovering the binary stream, unlike the method we propose.

Our approach can be applied to a wide range of classic digital modulations, such as standard Frequency Shift Keying, Phase Shift Keying or Amplitude Shift Keying. Compared to the state-of-the art, our detection approach makes minimal assumptions about the signal, relying on state-of-the-art frame and channel detection and a dictionary containing various modulation types and protocols to identify the physical layer used and extract the raw bitstreams. Furthermore, its genericity allows for the easy addition of new modulations and protocols, and the extracted metadata can be used as a set of features for auditing unknown signals or detecting covert channel emissions.

Once the modulation parameters have been identified and the frame demodulated, another important aspect of the analysis of the emissions is the inference of the packet format. As stated by J. Duchêne in [14], a large majority of the state of the art field-identification approaches are based on the PI Project [15], an approach from 2004 based on the use of bioinformatics algorithms, Needleman-Wunsch [16] and UPGMA [17,18]. In the case of unknown protocols, we also use them in our approach. These methods often use specific field delimiters that are known in advance [19,20], or make some assumptions on the protocol to simplify the analysis [21]. However, the goal of our approach is primarily to extract a maximum number of information from the protocol without intervention and without making assumptions on the packet format. To our knowledge, one of the most advanced approaches is Netzob [22]. It is also based on bioinformatics algorithms to align and delimit fields, and then on applying the L* language inference algorithm [23] to estimate the protocol's grammar. However, this tool is relatively complex, and lacks recent updates, pushing us towards implementing a simpler and lighter approach. Finally, we can also cite some approaches using the Voting Experts algorithm [24] to segment packets into fields. For instance, in IPART [25], the authors introduce a specialized version of the algorithm that allows them to do a basic reverse-engineering of industrial protocols. However, the assumption behind this algorithm is that all packets should come from the same protocol, which makes it more complex to implement in our case.

3 Analysis Workflow

Figure 1 illustrates the three main parts of our approach: 1) detecting where the packets are in the raw signal received from an SDR (Software Defined Radio), 2) analyzing the physical layer to demodulate, and 3) finally extracting information from the demodulated stream.

3.1 Emission Detection

The first step, represented by the **Emission detection** block, is to find where in time and frequency the signals are emitted. We first use a simple amplitude-based detection, called a *squelch*. This isolates parts of the signal with an amplitude above a certain threshold. The squelch threshold is based on the SDR's noise profile, and must be evaluated beforehand on a signal containing only noise. Then, we compute the Fourier Transform of each signal isolated this way, to detect the presence of amplitude peaks that highlight the frequencies in use. For each estimated central frequency, we then keep a "baseband" copy of the frame, i.e., a version that we have re-centered on the identified central frequency.

To demodulate and isolate the different channels, we need to estimate the bandwidth for each detected frame. We first make a coarse estimate using the previous Fourier transform, by identifying the width of the corresponding peak. This estimate is then used to pass the signal through a state-of-the-art Butterworth low-pass filter to avoid interferences from neighbouring channels. Since our estimate is not accurate enough at this point, the considered cut-off frequency used for the low-pass filter is twice the distance to the edge of the peak in the Fourier transform. This results in including more noise, but prevents accidentally removing interesting parts of the signal. Note that if this cut-off frequency is too large compared to the sample rate, and in this case the Shannon criteria would not be met, no filter is applied.

3.2 Physical Layer Analysis

In this section, we explain the different steps of the Physical Layer Analysis, as shown on Fig. 1. This process is separated in two distinct workflows, one in the general case, and another on the On-Off-Keying (OOK) modulation, which has the particularity of using non-emission periods to code information.

General Case - Modulation Identification. In the general case, we test the isolated signals against different types of modulation, as represented in the **Modulation analysis** block.

To precisely estimate the bandwidth, we first identify the symbol period. For different modulation types, we implement a first function that extracts the physical quantity corresponding to it. For example with an amplitude modulation, the amplitude is extracted from the signal. Then, we implement a second function that projects all symbols on a single one. For instance, we take the absolute value if the symbols are opposites.

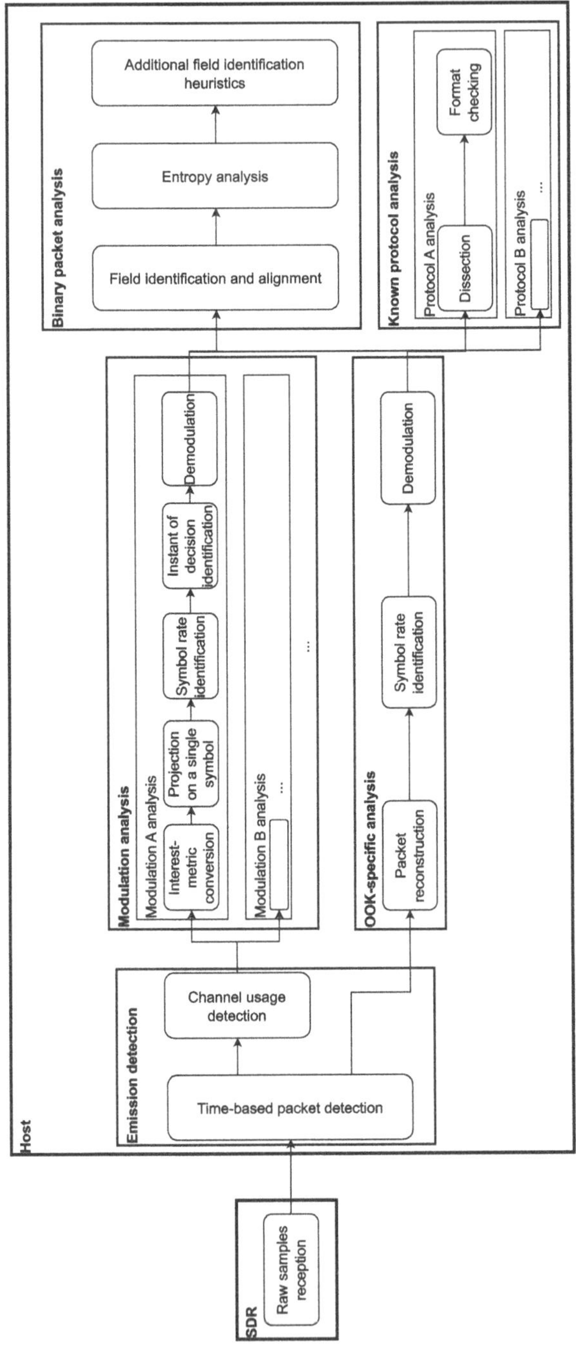

Fig. 1. Approach architecture overview

Second, we compute the autocorrelation on this projection to highlight the presence of periodicity, which would come from the presence of regular symbols in the stream. Indeed, the autocorrelation is the correlation between the signal and shifted versions of itself. Therefore, if the signal contains recurring patterns at regular intervals, it correlates better with itself shifted by multiples of this interval, resulting in regular peaks in the autocorrelation. When such periodicity is found, we deduce the number of samples per symbol and bandwidth.

General Case - Binary Stream Extraction. After identifying a modulation and filtering the signal, it is possible to automatically recover the binary stream.

All signals are made of series of symbols shaped with a pulse-shaping filter, that transforms pulses corresponding to binary data into a modulated signal. Accordingly, we build a bench of different usual filters, and correlate each of them with the received signal. If the chosen filter is similar to the original pulse-shaping filter that was used, the noise correlates significantly less than the original signal, partially reducing the noise's impact. The quality of the correlation is evaluated thanks to eye diagrams : all pairs of two successive symbols are overlapped in order to see when the symbols are the farthest from each other. Eye diagram examples in a practical case can be found in Fig. 2. The correct moment of decision for demodulation is selected where the eye-shape is the most open, thus maximizing the distance between symbols. The correct filter is chosen where the ratio between the height of the shape, linked to the signal power, to the width of its lines, linked to the noise impact on the signal, is kept minimal, as it maximizes the signal-to-noise ratio. Additionally, the eye diagrams are used to estimate the moment of decision to demodulate the signal, which is the instant where the signal-to-noise ratio is maximized.

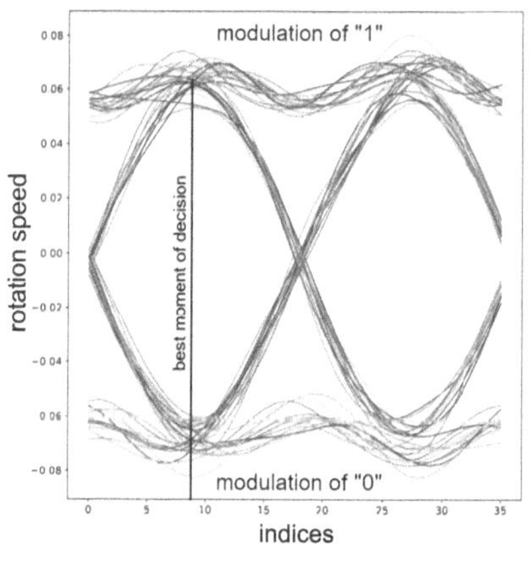

(a) eye diagram - correct filter

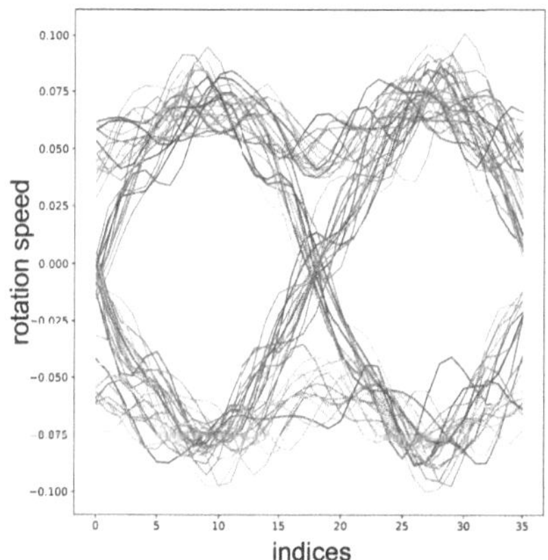

(b) eye diagram - less adapted filter

Fig. 2. Eye diagrams for GFSK, with both the right filter and a less adapted one

The Figs. 2a and 2b represent eye diagrams for a Gaussian Frequency Shift Keying modulation, applying a matching filter and a less adapted one. The eye diagram then represents the different possible transitions in rotation speed between symbols representing binary 0 or 1. The moment of decision is then chosen as the ninth sample of each period, because, as seen in Fig. 2a, the symbols for 0 and 1 are concentrated each in a single spot at this moment, meaning it minimizes the risks of demodulation errors. Additionally, as seen in Fig. 2b, a less adapted filter keeps a significant amount of noise in the signal, reducing the margin between the two symbols.

OOK-Specific Analysis. The amplitude-detection method has the drawback of not detecting correctly packets sent with an OOK, since this modulation uses sequences of periods with and without emissions to code zeros or ones. More precisely, a period without emission and the "emission" of a sequence of zeros are similar, making it difficult to detect when a packet begins or ends with amplitude only. Thus, we added an OOK-specific workflow, based on a "fusion" algorithm, that builds aggregates of packets previously detected based on their amplitudes that are near enough. More precisely, series of packets that are separated by periods of time of the same order of magnitude as the length of individual packets are "fused" together as a new packet. Note that the pre-fusion packets are still present, the new "fused" packet being treated as a new one altogether. This step is represented in Fig. 1 by the **Packet reconstruction** sub-block in the OOK-specific processing.

To estimate the symbol rate in this case, for a given packet, we estimate the minimal length of an emission or non-emission period. This allows us to demodulate the packet by estimating, for each of those periods, the number of times they contain the shortest one. However, this method relies on the assumption that the packet contains at least an isolated "1" or "0" bit, meaning the shortest period divides all others. To be able to demodulate packets that do not meet this condition, we also test integer divisors of the minimal size. Then, using a score on symbol length stability (meaning how much the lengths deviate from multiples of the estimated symbol length), we choose the most probable option.

3.3 Link Layer Analysis

The Link Layer analysis was also split in two parts : one for analyzing unknown protocols to extract information from the packet format, the other to dissect packets according to known protocols that correspond to the received bitstream.

To facilitate the addition of new protocols, the known protocols are regrouped in a dictionary. Each element representing a protocol is a set of functions allowing to identify whether a binary stream corresponds or not to this protocol.

First, the received bitstream is tested against a dictionary of known protocols, by trying to dissect the data according to the protocol's specification, and then evaluating if the packet is valid. Note that, by default, the protocols are chosen depending on the modulation they should use. However, it is also possible to

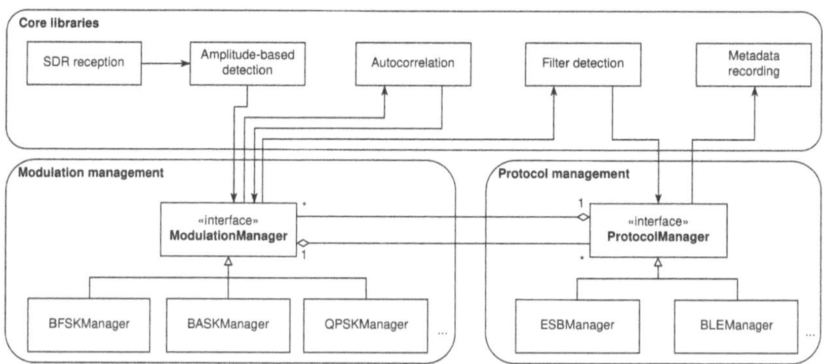

Fig. 3. Tool architecture overview

test any protocol for any modulation, in case an attacker would use a protocol with an unintended modulation.

Then, the bitstream is analysed to try to find fields of interest. To do so, we first need to identify fields that are at different positions due to variable-length fields. This was done using the Needleman-Wunsch alignment algorithm, that aligns similar fields, as used in some works of the state of the art such as Netzob [22], a tool to assist in simple protocol reverse engineering.

Once fields are aligned, we compute their entropy to put them in three classes:

- **fixed-value fields**, that could correspond to addresses
- **low-entropy variable fields**, for instance length fields or unencrypted data
- **high-entropy variable fields**, or **pseudo-random fields**, that could correspond to encrypted data

Finally, variable length fields with low entropy are tested against different heuristics to identify more precisely their types. In the current state of our works, two heuristics are tested : correlation between fields and packet length to identify length fields, and linear evolution of a given field for sequence numbers.

4 Implementation

We implemented the approach into a new tool, mainly developed in C and Python, and based on Scapy[1] and the SoapySDR[2] library. It consists of three main components: a set of core libraries, a modulation dictionary and a protocol dictionary for each modulation[3]. This architecture is represented in Fig. 3. Please note that this illustration doesn't cover the particular ASK-OOK processing, which is implemented via a set of independent methods in the tool.

[1] https://scapy.net/.

[2] https://github.com/pothosware/SoapySDR.

[3] BASK: Binary Amplitude Shift Keying, BFSK: Binary Frequency Shift Keying, QPSK: Quadrature Phase Shift Keying (4 equidistant phase values).

4.1 Core Libraries

The core libraries of our implementation are split into five blocks. The first block (*SDR Reception*) is a reception module implemented in C, using the SoapySDR library for SDR control. This allows our code to be easily used with any SDR compatible with this library. The module receives sample buffers before sending them to the next block.

The following block (*Amplitude-based detection*) isolates by amplitude the received frames. The IQ samples corresponding to the frames are analysed to isolate the frequency channels in use, before being saved in a file for further offline processing. The signal is then analysed by different Modulation Managers, presented in more detail in 4.2.

Each Modulation Manager then passes the representation of the signal to the *Autocorrelation* block, that aims at identifying the symbol period of a possible frame that it would contain. If the estimated symbol period is stable enough, the Modulation Manager then sends the signal representation along with the symbol period to the *Filter detection* block.

The *Filter detection* block passes the signal representation through the filter bench described in Sect. 3 to identify the filter shape minimizing the noise and get the moment of decision to demodulate the signal. The demodulated stream that results from this block is then sent to the Protocol Managers (as described in 4.3) that are linked with the current Modulation Manager, along with the analysis for eventual unknown protocols.

Once the Protocol Managers have analyzed the binary stream, all the extracted metadata is then saved by the *Metadata recording* block in a structure for further use by different applications, such as the one presented in 5.3.

4.2 Modulation Managers

To be able to transparently manage different modulations, and to allow for an easy enrichment of our tool with new modulations, we defined an interface called "ModulationManager", that all Modulation Manager classes must implement. To implement this interface, and be able to interact with the rest of the tool, those classes need to implement four functions:

- `get_label`: a function without arguments that returns a string, which must uniquely identify the modulation
- `get_mod`: a function that takes as input raw IQ samples, and returns its representation depending on the corresponding modulation
- `get_modulation_index`: a function that takes as input the representation from `get_mod` and returns a measure of the distance between the symbols for the modulation, for further use as an identifying feature
- `eye_diagram_precomputing`: a function that projects the result from `get_mod` on two opposite symbols to discern the moment of decision in eye-diagrams

All Modulation Managers must then be added to the modulation dictionary of the tool to be used by the core libraries. As of now, only binary Frequency Shift

Keying and Amplitude Shift Keying are implemented in the tool. However, it is possible to easily add new modulations by implementing those four functions.

4.3 Protocol Managers

To validate the reception of frames from known protocols, we needed to be able to analyse the binary streams respectively to different possible protocols, for each modulation. We then created, similarly to the "ModulationManager" interface, a "ProtocolManager" interface that all Protocol Manager classes must implement. Two methods must be implemented:

- `get_label`: a function similar to the one for Modulation Managers, that also needs to return a uniquely identifying string for the protocol
- `check_protocol`: a function checking whether a valid frame for the protocol is present in the binary stream

More precisely, the `check_protocol` function takes as inputs the raw binary stream extracted from the signal, the symbol rate that was identified and the frequency on which the signal was detected. If the binary stream contains a valid frame for the protocol, the function must then return a boolean with the value True, a Scapy class that corresponds to the frame type to compute later the corresponding scapy packet, and the truncated binary stream. If not, it returns a boolean with the value False and two empty values (*None* in Python).

For each Modulation Manager, we associate corresponding protocols[4] in individual protocol dictionaries. Once a modulation is identified and a binary stream extracted, the stream is tested against all available protocols for the modulation. If the binary stream matches a given protocol, its metadata is then augmented with metadata returned by `check_protocol`. Let us note that, if a protocol that has not been implemented in scapy is added, it is possible to add user-defined scapy dissectors to the tool in a dedicated directory.

As for the Modulation Managers, this dictionnary is user defined and must be completed beforehand by implementing the needed ProtocolManagers.

4.4 Link Layer Analysis

In addition to the previously mentioned Protocol Managers, all frames identified for any given modulation are analyzed following the workflow described in 3.3. The resulting assumptions on the different fields are then also saved as metadata linked to the corresponding signal.

---

[4] We have chosen to link modulations with protocols that use them, but our approach allows linking any protocol manager to any modulation manager.

5 Experiments

This section describes the experiments conducted to validate our signal analysis approach, and to implement basic defensive algorithms based on the results. The source of our tool, along with the code for the experiments, is openly available on https://gitlab.laas.fr/fgaltier/wireless-analysis-toolset.

We first checked that the approach works as intended on well known wireless protocols. We then evaluated the accuracy of the parameter estimation on randomly generated wireless protocols. Finally, we demonstrated the usability of our approach in defensive applications, by building on top of it a whitelist and blacklist-based covert-channel detection.

All our experiments were conducted using an Ettus USRP B200mini as receiver, and when needed, we used a LimeSDR USB to emit the signals.

5.1 Validation on Known Protocols

We first tested our approach against well known wireless protocols: 1) Bluetooth Low Energy[5], or BLE, often used by sensors or for short wireless commands, and integrated in most modern Bluetooth chips, and 2) Enhanced ShockBurst[6], or ESB, used by several wireless keyboards. We also tested it on IEEE 802.15.4 communications in 2.4GHz O-QPSK, using a GFSK demodulator to simplify the approach. Indeed, as it was shown in various works [26,27], there is a functional equivalence between this version of O-QPSK and a GFSK, requiring minor changes on the demodulation process. These protocols were chosen because of the availability of their modulation parameters, both using GFSK in the 2.4GHz ISM band. In fact, this combination of band and modulation is one of the most frequent in the IoT. The goal of this experiment is to validate whether our approach is able to detect correctly emissions from well-known and widespread protocols, without significant performance degradation compared to a traditional receiver. As such, we compare the number of frames received correctly by our approach and by a traditional receiver, and evaluate the errors in the estimation of the different parameters of the emission.

For this experiment, we sent frames with known parameters with both protocols in a realistic environment with other emission sources, including BLE and Wi-Fi communications. The signal was received by an USRP B200mini listening at 2405MHz with a sample rate of 10Msps, during 10 s. Note that, even if the duration of the capture is quite short, the trace collected actually contains several hundreds of frames and is sufficiently rich to assess the relevance of our tool. We first computed which frames a standard receiver could demodulate with a correct CRC as a reference. We then compared the results of our approach with these, by evaluating the difference between the number of frames that we correctly received with the number of reference frames. For BLE experiments, we collected several hundreds of BLE signals from the everyday environment

[5] https://www.bluetooth.com/specifications/specs/core-specification-5-4/.
[6] As described in nRF24L01+ Product Specification v1.0.

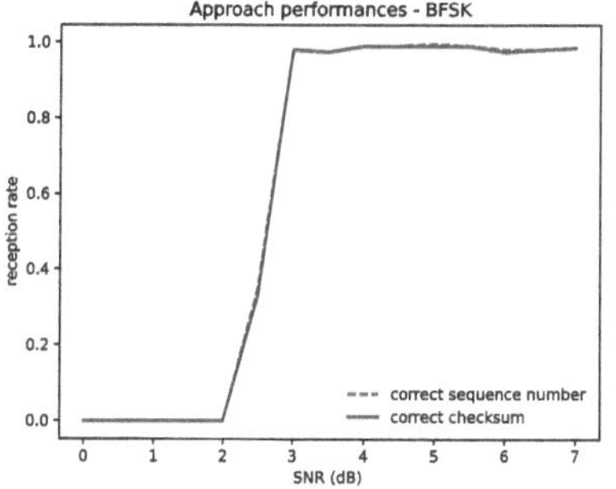

(a) Performances of our approach on a 2-FSK

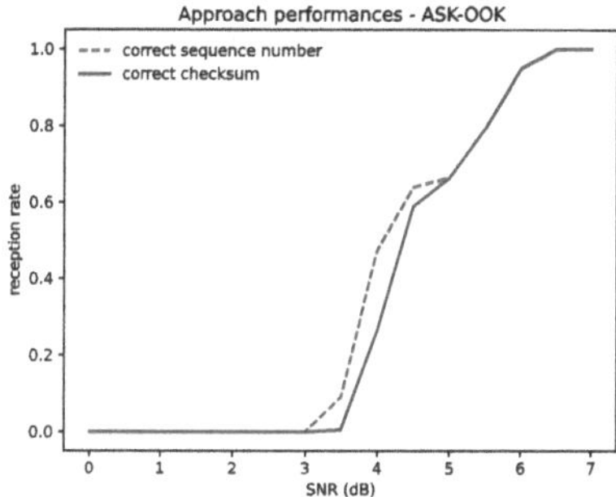

(b) Performances of our approach on a 2-ASK (OOK)

Fig. 4. Performances versus SNR in 2-FSK and 2-ASK

of our laboratory, including phones, computers, and some smaller devices such as smartwatches. The ESB experiments were conducted only with a dongle and a corresponding mouse, the devices using this protocol being limited to mice, keyboards and dongles, and also included several hundred received frames. The IEEE 802.15.4 communications were generated using ZigBee devices, as ZigBee uses this protocol as MAC layer. We used a XBee device, a Philips Hue Bridge and a Philips Hue lightbulb. The ratio of frames correctly decoded by our approach over the frames correctly decoded by the hardcoded receiver is 99.7% (955 correctly received with our approach versus 958 with the standard demodulator) for BLE and 98.8% for ESB (89 correctly received with our approach versus 90 with the standard demodulator). For ZigBee, we compared the results of our approach with a GFSK demodulator with hard-coded carrier frequency and binary rate. In this case, we received 85 out of 85 frames with our approach. These results show the relevance of our approach: the demodulation performances on known protocols are very close to those with hardcoded values.

5.2 Validation on Random Protocols

We then tested our approach on unknown protocols, generated with random modulation parameters. To test the efficiency of our approach on different modulations, we chose the BFSK and ASK-OOK modulations. The goal of this experiment is to evaluate, on emissions with non-standard parameters, whether our approach would still be able to detect and analyze the received frames.

This was done in two steps :

1. testing the approach on simulated emissions with controlled noise, to estimate our sensitivity to the Signal to Noise Ratio (SNR)
2. testing the approach on real signals with randomly chosen parameters, to estimate the reception quality in more realistic conditions

Table 1. Results of our approach for randomly generated modulation parameters

	FSK		ASK	
	Average	Standard deviation	Average	Standard deviation
reception accuracy	94.67%	/	96.14%	/
frequency error	14.1 kHz	286 Hz	9.93 kHz	116 Hz
binary rate error	0.227 bits/s	1.04 bits/s	919 bits/s	808 bits/s

In the case of the simulated emissions, we ran our algorithm on 10 samples of 100 packets for each modulation, each time with SNR values between 0 and 7 dB. For both modulations, the packets used central frequencies randomly chosen between 433 and 435 MHz, and the simulated reception was made at 434 MHz with a sample rate of 4 Msps. The BFSK packets were using between 4 and 16 samples per symbol, and had a modulation index between 0.4 and 0.6. The ASK-OOK packets were using between 33 and 64 samples per symbol. The results are shown on Figs. 4a and 4b.

In the case of real-life emissions, we sent packets from a YARD Stick One (radio dongle with pre-implemented BFSK and ASK-OOK modulations) and received them with an USRP B200 mini. The performances of our approach are shown in Table 1. We included a checksum in the payload of the protocol, and used it to calculate the reception accuracy by comparing the number of packets with a correct checksum with the number of packets actually received (estimated from sequence numbers in the first and last received packets). In the case of ASK-OOK, we received 281 packets, among which 274 with a valid checksum. In the case of BFSK, we received 242 packets, among which 231 with a valid checksum.

The metrics used to assess the quality of the reception are defined as follows:

- **reception accuracy :** proportion of frames with valid checksum
- **frequency error :** error on the carrier estimation
- **bit rate error :** error on the bit-rate estimation

Overall, the results show a very good reception accuracy of our approach for both frequency and amplitude modulations. The frequency error and bit rate error are also very good since their value is at the maximum several kHz for the frequency and a few bits per second for the bit rate, whereas the bandwidth and bitrate are in the MHz range. As a consequence, the desynchronization due to the bitrate error only appears in frames long enough for the successive shifts to become significant, and the frequency error being negligible compared to the bandwidth, it is not enough to hinder the demodulation process. This is further demonstrated by the good reception accuracy, which shows that these errors do not generate significant bitflips.

#status	type	label	freq	bitrate
AUTHORIZED	Modulation	2ASK	*	*
AUTHORIZED	Protocol	BLE	[2402e6,2404e6,2406e6,2408e6,2410e6]	1e6
AUTHORIZED	Protocol	BLE	[2402e6,2404e6,2406e6,2408e6,2410e6]	2e6
AUTHORIZED	Modulation	2FSK	[2402e6]	1e6
AUTHORIZED	Modulation	2FSK	[2402e6]	2e6
ALERT	Protocol	ESB	*	*

Fig. 5. Allowed and forbidden communication flows

5.3 Covert Channel Detection

We implemented a covert channel detection mechanism on top of our implementation as a real-life example of a defense mechanism. Covert channel attacks are critical as they may take several forms, while being quite easy to deploy and difficult to detect in absence of a monitoring system listening on a large frequency band, such as the one proposed in this paper. For example, an attacker could visit a company and try to exfiltrate confidential information using a device communicating with a protocol not used by the company legitimate devices and thus likely not monitored. This scenario can even be deployed by compromising for instance an employee's smartwatch to make it exfiltrate sensitive data.

This experiment was carried out in a realistic office environment, with numerous legitimate communications in BLE and WiFi. We did not take any particular measure to reduce the noise implied by those communications, so that the experiment took place in real-life conditions.

For this experiment, we added a module that takes as input the metadata from our approach, and a configuration file to specify what should raise an alert. The configuration file uses "AUTHORIZED" and "ALERT" rules for a set of a modulation or protocol, a set of frequencies and a bit-rate, as shown in Fig. 5. The rules are evaluated sequentially on each frame until the conditions for one of them are met, or if none of them matches the frame, a default rule is applied, either raising an alert for all unknown packets or for none of them.

We used this configuration file for an experiment in which we emitted ESB frames with a commercial emitter (supposed to be illegitimate), along with BLE frames (supposed to be legitimate) in several sequences of 10 s in the first 10MHz of the 2.4–2.5 GHz band.

The results of this experiment confirm the relevance of our tool: out of 768 ESB frames that were sent, 659 frames raised alerts, thus with a detection rate of 85.8%. No false positives were detected during this experiment. Even if some ESB frames were not correctly identified, this detection rate is sufficient to actually identify the covert channel. Note that as our approach allows providing the demodulated binary streams corresponding to the frames transmitted, it can also be used for simultaneous detection of malicious payloads in various protocols.

5.4 Case Study - Wireless 2.4 GHz Mouse

To illustrate the full workflow of our tool, this section presents the analysis of a wireless mouse using a proprietary protocol. We used our tool to 1) identify the modulation of the physical layer, and 2) reverse the resulting binary streams. This experiment was carried out by making the mouse communicate with a computer while performing different actions. We then captured the mouse's emissions using an USRP B200, and transmitted them to our tool.

First, we looked at several frequencies of the 2.4 GHz band to detect emissions from the mouse, in a realistic environment. Doing so, we identified several frequencies on which the dongle associated with the mouse was emitting, notably 2400 MHz and 2480 MHz. The mouse chooses a channel among those each time it is turned on. Our tool quickly identified that the modulation in use was a GFSK, at 1 Mbps, allowing it to extract the corresponding binary streams.

The entropy analysis performed using our tool allowed us to identify that all packets followed a same format : first, a series of `0xAA` bytes, followed by four fixed bytes, a variable quantity of variable fields, and finally a postamble of `0xFF`. We hypothesized that the fixed field at the beginning corresponds to an address. Then, by observing that the two last bytes before the postamble had an entropy reflecting the entropy of the previous variable fields, we assumed it to be a CRC. Our tool also identified that the packet did not contain a size field.

Finally, some of the longest packets contained numerous `0x5A` bytes, or near values. This could highlight the use of whitening, a common practice in wireless communications consisting in masking the data with a xor to even out the distribution of 1 and 0 bits in the packet. However, we could not immediately check the integrity of our reception, to help validate our hypotheses on valid packets, as we could not find the exact parameters of the CRC algorithm.

In parallel, we found that the dongle was exposed as a `MosArt Semiconductor Corp. 2.4 G INPUT DEVICE`, indicating that the device might use the "MosArt" protocol identified by M. Newlin in MOUSEJACK [28]. Using the reversed parameters, we could validate the use of whitening with a repeated `0x5A` byte, along with the use of a `XMODEM-16` CRC. We also validated that the format did not contain any size field, as our tool indicated. This also allowed us to find out some differences between our mouse's protocol and the one reverse-engineered.

Thanks to our entropy analysis, we identified the frame sequences corresponding to mouse movement and scrolling. Moreover, it allowed us to detect, depending on the input, which field corresponded to which action. Interestingly, this showed us that the payload of movement frames was different from MOUSEJACK version. Indeed, the movement coordinates were alternating between X-coordinates and Y-coordinates bytes instead of being two contiguous two-byte values. We also identified a seemingly empty field after the frames corresponding to scrolling, that was not described by M. Newlin.

Overall, this case study shows the capacity of our tool to extract information from unknown signals, to help reverse-engineering unknown protocols. It validates the results from the previous experiments, showing the capability of our tool to help analyzing devices. However, it also shows the need for further work in the higher-layer analysis, notably on the detection and reverse of CRC. To do so, we plan to integrate into our tool an error-tolerant variant of CRC

RevEng [29], a CRC algorithm finder. This could allow to automatically, with an heuristic to find CRC fields, or semi-automatically, by specifying the field manually, identify the CRC parameters.

6 Limitations and Discussion

Some modulation schemes use closely overlapping channels, like OFDM. This situation introduces new technical difficulties in the demodulation process, that need to be addressed by specific mechanisms integrated in the specifications. Thus, our approach, without prior knowledge of the communication scheme that was used and the associated mechanisms, or their parameters, is not suitable for such schemes. Several approaches have been studied to detect OFDM communications and estimate their parameters, as presented in [30], which separates them in four main categories: identification based on the inter-frame spacing, with the help of the cyclic prefix [31] or not [32], techniques modifying the emitters to add detectable information [33], and techniques based on the pilot symbols used by the scheme [34]. However, since our objective is to analyze the modulation without making any assumptions on the signal or modifying the communicating objects, we cannot use any of those methods in our protocol-agnostic approach. Moreover, all these approaches do not allow recovering the bitstream from an OFDM signal in realistic conditions with unknown parameters. Indeed, it would require to correct the channel selectivity of the wireless medium, which is done by knowing the values sent as pilot or reference symbols, and thus an assumption on the transmitted data. To our knowledge, none of the existing approaches allow estimating the original versions of those pilot or reference symbols. However, OFDM signals still carry specific characteristics, such as the fixed-lengths blocks separated by cyclic prefixes or guard intervals, or reference symbols. Such characteristics could be detected by the use of methods relying on cyclostationarity or autocorrelation to detect periodic repetitions, knowing usual sample rates and OFDM block sizes. Therefore, an OFDM-specific detection module could be built and integrated in our approach, even without going as far as to demodulate the original data, for example for covert channel detection purposes.

Some other schemes, such as Chirp Shift Keying (CSK), used in LoRaWAN, do not directly transmit symbols as values of specific parameters of the signal. For example, this modulation uses "chirps", symbols where the signal transitions from a frequency to another, and use temporally shifted versions of those chirps to modulate data (the shift corresponding to the value to modulate). In the case of those modulations, it is then needed to add a translation layer to the ModulationManager. In the example of CSK, it would amount to translating the signal into an array containing at each instant the corresponding CSK symbol if the chirp began at this instant. We are still integrating such modulations, and building the corresponding translation layers, in particular for CSK.

Let us also note that frequency hopping schemes are currently supported, but only as independent frames. Indeed, we did not integrate any means of identifying whole communications, so each one is analyzed on its own. We plan to add such functionality in the future, but it is not currently under development.

Finally, it is to note that our approach focuses on modulations carrying digital data, as its goal is to translate a signal into corresponding binary streams. It is therefore unsuitable for detecting transmissions using analog data.

We also identified several other areas for future work. During our experiments, we were limited in the wideness of the band we listened to by our processing speed. However, as explained in Sect. 3, it would have been possible to maximize the efficiency of the hardware used by implementing the first steps isolating the frames and channels in the SDR's FPGA, which would then send to the computer host the already isolated frames, reducing the risks of overflows in the communication buffers between the two. Note also that this approach is not designed to run in real-time, since we need to receive the entire frames to analyse them, especially for the steps using the autocorrelation of the frame's signals. The performance of our approach can be improved by implementing the frame and channel isolation into the SDR's FPGA, which would reduce the amount of data to send to the computer to allow for a larger band simultaneous reception.

Our reverse engineering module is, for now, limited to a few specific kinds of fields, but could be augmented with more heuristics to detect other fields of interest, in particular frequent header fields or CRC fields.

7 Conclusion

In this paper, we presented a tool and experiments to fill the gap in the current state of the art of wireless protocol auditing and wireless covert channel attack detection, by implementing a method for detecting and analysing the physical layer of such wireless emissions. The different experiments we carried out show that this tool is able to successfully retrieve data from known protocols, and to efficiently detect the parameters of randomly generated ones. Furthermore, this tool also allows to reverse engineer the extracted data.

This approach constitutes an important step towards the design of a unified means for reverse-engineering the various and heterogeneous protocols that can be used in the domain of IoT. Indeed, as of now, we lack such a tool that would allow saving substantial amounts of time in the audit of unknown wireless devices, and analysis of unknown emissions in a monitored environment.

For future work, we plan to investigate several topics. Regarding the physical analysis, we plan to add a PSK analysis to complement the ASK et FSK. We also plan to work on the extension of the tool, by enriching the reverse engineering capabilities from the protocol analysis point of view.

Acknowledgements. This work has been partially supported by the French National Research Agency under the France 2030 label (Superviz ANR-22-PECY-0008). The views reflected herein do not necessarily reflect the opinion of the French government.

References

1. Li, A., et al.: Demodulation-free protocol identification in heterogeneous wireless networks. Comput. Commun. **55**, 102–111 (2015). https://linkinghub.elsevier.com/retrieve/pii/S0140366414003041

2. Lakshminarayanan, K., Sapra, S., Seshan, S., Steenkiste, P.: Rfdump: an architecture for monitoring the wireless ether. In: Proceedings of the 5th International Conference on Emerging Networking Experiments and Technologies, pp. 253–264 (2009)
3. Chen, Q., Wang, Y., Bostian, C.W.: Universal classifier synchronizer demodulator. In: 2008 IEEE International Performance Computing and Communications Conference (IPCCC). Los Alamitos, CA, USA, IEEE Computer Society, December 2008. https://doi.ieeecomputersociety.org/10.1109/PCCC.2008.4745076
4. Mototolea, D., Youssef, R., Radoi, E., Nicolaescu, I.: Non-cooperative low-complexity detection approach for FHSS-GFSK drone control signals. IEEE Open J. Commun. Soc. **1**, 401–412 (2020)
5. Azim, A.W., Khalid, S.S., Abrar, S.: Modulation classification based on modified Kolmogorov-Smirnov test. In: 2013 IEEE 9th International Conference on Emerging Technologies (ICET), pp. 1–6. IEEE (2013)
6. Hao, S., Wang, N., Sun, B., Liu, M., Dong, M., Wang, H.: Modulation classification using a goodness of fit test. In: Journal of Physics: Conference Series, vol. 1169, no. 1. IOP Publishing, p. 012068 (2019)
7. Helluy-Lafont, É., Boé, A., Grimaud, G., Hauspie, M.: Bluetooth devices fingerprinting using low cost SDR. In: 2020 Fifth International Conference on Fog and Mobile Edge Computing (FMEC), pp. 289–294. IEEE (2020)
8. Gimenez, P.-F., Roux, J., Alata, E., Auriol, G., Kaâniche, M., Nicomette, V.: Rids: Radio intrusion detection and diagnosis system for wireless communications in smart environment. ACM Trans. Cyber-Phys. Syst. **5**(3), 1–1 (2021)
9. Liu, X., Yang, D., El Gamal, A.: Deep neural network architectures for modulation classification. In: 2017 51st Asilomar Conference on Signals, Systems, and Computers, pp. 915–919. IEEE (2017)
10. Zhang, Z., Wang, C., Gan, C., Sun, S., Wang, M.: Automatic modulation classification using convolutional neural network with features fusion of SPWVD and BJD. IEEE Trans. Sig. Inf. Process. Netw. **5**(3), 469–478 (2019)
11. Peng, S., et al.: Modulation classification based on signal constellation diagrams and deep learning. IEEE Trans. Neural Netw. Learn. Syst. **30**, 718–727 (2019)
12. West, N.E., O'Shea, T.: Deep architectures for modulation recognition. In: 2017 IEEE International Symposium on Dynamic Spectrum Access Networks (DySPAN), pp. 1–6 (2017)
13. O'Shea, T., Roy, T., Clancy, T.C.: Over-the-air deep learning based radio signal classification. IEEE J. Sel. Top. Sig. Process. **12**, 168–179 (2017)
14. Duchêne, J., Le Guernic, C., Alata, E., Nicomette, V., Kaâniche, M.: State of the art of network protocol reverse engineering tools. J. Comput. Virol. Hacking Tech. **14**(1), 53–68 (2018)
15. Beddoe, M.A.: Network protocol analysis using bioinformatics algorithms. Toorcon **26**(6), 1095–1098 (2004)
16. S. B. Needleman and C. D. Wunsch, "A general method applicable to the search for similarities in the amino acid sequence of two proteins," *Journal of Molecular Biology*, vol. 48, no. 3, pp. 443–453, 1970. [Online]. Available: https://www.sciencedirect.com/science/article/pii/0022283670900574
17. R. R. Sokal and C. D. Michener, "A statistical method for evaluating systematic relationships." *Univ. Kansas, Sci. Bull.*, vol. 38, pp. 1409–1438, 1958
18. Nei, M., Tajima, F., Tateno, Y.: Accuracy of estimated phylogenetic trees from molecular data. J. Mol. Evol. **19**(2), 153–170 (1983)
19. W. Cui, J. Kannan, and H. J. Wang, "Discoverer: Automatic protocol reverse engineering from network traces." in *USENIX Security Symposium*, 2007, pp. 1–14

20. J. Antunes, N. Neves, and P. Verissimo, "Reverx: Reverse engineering of protocols," 2011
21. W. Cui, V. Paxson, N. Weaver, and R. H. Katz, "Protocol-independent adaptive replay of application dialog." in *NDSS*, 2006
22. G. Bossert, F. Guihéry, G. Hiet *et al.*, "Netzob github repository," 2012. [Online]. Available: https://github.com/netzob/netzob/tree/master/netzob
23. Angluin, D.: Learning regular sets from queries and counterexamples. Inf. Comput. **75**(2), 87–106 (1987)
24. Cohen, P., Adams, N., Heeringa, B.: Voting experts: An unsupervised algorithm for segmenting sequences. Intelligent Data Analysis **11**(6), 607–625 (2007)
25. Wang, X., Lv, K., Li, B.: Ipart: an automatic protocol reverse engineering tool based on global voting expert for industrial protocols. Int. J. Parallel Emergent Distrib. Syst. **35**(3), 376–395 (2020)
26. R. Cayre, F. Galtier, G. Auriol, V. Nicomette, M. Kaâniche, and G. Marconato, "Wazabee: attacking zigbee networks by diverting bluetooth low energy chips," in *2021 51st Annual IEEE/IFIP International Conference on Dependable Systems and Networks (DSN)*. IEEE, 2021, pp. 376–387
27. Pasupathy, S.: Minimum shift keying: A spectrally efficient modulation. IEEE Commun. Mag. **17**(4), 14–22 (1979)
28. M. Newlin, "Mousejack, keysniffer and beyond: Keystroke sniffing and injection vulnerabilities in 2.4 ghz wireless mice and keyboards," *DEFCON*, 2016
29. G. Cook, "Crc reveng website." [Online]. Available: https://reveng.sourceforge.io/
30. M.-R. Oularbi, S. Gazor, S. Houcke, and A. Aissa-El-Bey, "Ofdm system identification using pilot tone signature," in *International Workshop on Systems, Signal Processing and their Applications, WOSSPA*. IEEE, 2011, pp. 95–98
31. Oner, M., Jondral, F.: On the extraction of the channel allocation information in spectrum pooling systems. IEEE J. Sel. Areas Commun. **25**(3), 558–565 (2007)
32. Bouzegzi, A., Ciblat, P., Jallon, P.: New algorithms for blind recognition of ofdm based systems. Signal Process. **90**(3), 900–913 (2010)
33. Sutton, P.D., Nolan, K.E., Doyle, L.E.: Cyclostationary signatures in practical cognitive radio applications. IEEE J. Sel. Areas Commun. **26**(1), 13–24 (2008)
34. Socheleau, F.-X., Houcke, S., Ciblat, P., Aïssa-El-Bey, A.: Cognitive ofdm system detection using pilot tones second and third-order cyclostationarity. Signal Process. **91**(2), 252–268 (2011)

Extended Abstract: Assessing GNSS Vulnerabilities in Smart Grids

Sine Canbolat(✉), Clemens Fruböse(✉), Ghada Elbez, and Veit Hagenmeyer

Karlsruhe Institute of Technology, KIT, Karlsruhe, Germany
{sine.canbolat,clemens.fruboese,ghada.elbez,veit.hagenmeyer}@kit.edu

Abstract. For the future power grid, the need for time-synchronized measurement data of high resolution also comes with the drawback of an increasing reliance on Global Navigation Satellite Signals (GNSSs) for time synchronization. Although the manipulation of GNSS signals has been extensively investigated, the question of risk magnitude of such attacks on the power grid is still open: Apart from proof-of-principle examples that require the attacker to have profound knowledge of the system, the risk to real world systems remains largely undetermined. In this paper, we present a vulnerability analysis of a time server used in International Electrotechnical Commission (IEC) 61850-based substations to five GNSS spoofing attacks. We demonstrate that using a GNSS simulator, both quick-conspicuous and slow-stealthy attacks are successful. We assess the difficulty in the scope of a qualitative risk assessment. Our findings can support quantitative risk analysis, and hence decision-makers in ensuring grid stability.

Keywords: Smart Grid (SG) · Global Navigation Satellite System (GNSS) · Time Synchronization Attack (TSA) · Precision Time Protocol (PTP) · Vulnerability Analysis (VA)

1 Introduction

The transition to renewable energy sources requires a transformation of the electric power infrastructure to adapt to the fluctuating nature of renewable energy, which in turn requires greater grid flexibility. Smart Grids (SGs) represent essential components of the energy infrastructure, employing sophisticated communication technologies to facilitate real-time monitoring and management of the electrical grid [5]. More specifically, numerous SG applications, including power-oscillation monitoring, phase-angle monitoring, protection and fault localization, require data with precise and accurate timestamps. For instance, the IEEE standard for phasor measurements requires a time accuracy of 1 microsecond [7]. In certain situations, an incorrect time stamp can result in a misjudgment of which line tripped, potentially causing false automated actions to be triggered [16]. To keep the measurement devices synchronized, Global Navigation Satellite Systems (GNSSs) are nowadays by far the most-used means of providing accurate time

F. Maggi et al. (Eds.): DIMVA 2024, LNCS 14828, pp. 545–555, 2024.
https://doi.org/10.1007/978-3-031-64171-8_28

services due to the accuracy up to few nanoseconds and high availability [8]. Vulnerability awareness is an important aspect when considering countermeasures to enhance the SG's security posture. Previous studies have shown that satellite-navigation-based systems are vulnerable to spoofing attacks resulting in increased attention to signal manipulation in recent years [14,15,18,20]. Understanding Advanced Persistent Threats (APTs) targeting critical infrastructure [1] can be further supported by vulnerability awareness. Our approach relies on qualitatively assessing the attack difficulty to increase awareness of GNSS vulnerabilities in the energy domain. For this purpose, five experiments are carried out as an example in our International Electrotechnical Commission (IEC) 61850-based substation test-bed.

Related Work. It has been shown already in 2013 that the measurements of Phasor Measurement Units (PMUs) are very sensitive to time spoofing attacks [21]. Numerous contributions about the impact of such spoofing have been published [4,10,11,16]. However, the considerations of the impact were mostly exploratory and/or theoretical, often requiring deep knowledge of the network by the attacker. The impacts, most often, comprise state estimation errors and miscalculation of faults. The scenario of time-critical measurement values for protection purposes has also been discussed recently [2]. Additionally, some authors investigated the most vulnerable PMUs for Time Synchronization Attacks (TSAs) [4]. Several approaches to detect spoofing have been proposed. These range from special antenna setups and signal analysis techniques to machine learning bad-data-detection algorithms which raise alarms when time-sensitive measurement values are inconsistent with other measurement techniques [9,12,13,18].

Additionally, the Precision Time Protocol (PTP) is built on a master-slave principle, and hence naturally susceptible to adequate attacks [1]. We also investigate the internal judgement of the time server itself regarding its calculated offset of the time signal from the internal clock. This could serve as basis for an alarm framework with minimal required changes to existing systems in the field. This approach is often overlooked when proposing TSA countermeasures. Both single- and multi-constellation attacks were conducted. A single-constellation attack only simulates signals from a specific satellite constellation, exemplified in our research by GPS. Conversely, a multi-constellation attack targets signals from multiple satellite constellations, including GPS and Galileo in our analysis.

Contributions. We describe an attacker setup used to manipulate the time of a commercial time server, which could be deployed in a digital substation. Within this paper, we present the following contributions:

1. We successfully conducted five experiments, ranging from simple asynchronous to synchronous multi-constellation attacks, using GNSS signal generator software. However, quick time shifts are in general not stealthy.
2. We assess the challenges presented by these attack scenarios within the scope of IEC 61850-based substation vulnerability analysis. Furthermore, these findings can also be applicable to other domains that require precise timing.

3. We provide a detailed description of our next steps regarding an impact analysis of TSAs in IEC 61850-based substation as extended work.

The remainder of the paper is organized as follows: First, we present an experimental setup and the results of GNSS spoofing attacks in Sect. 2. We continue with a qualitative risk assessment in Sect. 3. Thereafter, a discussion regarding results and possible extended work is addressed in Sect. 4. Finally, the conclusion is presented in Sect. 5.

2 Experimental Setup and Results

GNSS signals. GNSS navigation is based on calculating the distance to several satellites with the propagation time of their light-speed signals. In order for the receiver to obtain the satellite's current position and broadcast time, the so-called navigation message is continuously sent to earth by each satellite. Using GPS L1 C/A as example, we briefly review the working principle:
The information bits are modulated on the carrier frequency (e.g. 1575.42 MHz for GPS L1) using Binary Phase-Shift Keying (BPSK). Additionally, using Direct Sequence Spread Spectrum (DSSS) technique, a pseudo-random chipping code is modulated on the signal, where the increased bandwidth improves the robustness to interference.
The signal $r_i(t)$ at the receiver, sent by satellite i, can be modelled as

$$r_i(t) = A_i N_i(t) C_i(t) \cos(2\pi f_i t + \phi_i),$$

with $A_i, N_i, C_i, f_i, \phi_i$ referring to signal amplitude, navigation message, chipping code, Doppler-adjusted carrier frequency and phase offset, respectively. Note that navigation message and chipping code for unauthenticated signals are publicly available, whereas the other quantities can be calculated by the simulator software. For successful demodulation, the receiver aligns its internally generated chipping code in terms of time and Doppler shift to the received signal. Accordingly, a spoofing signal has to fulfill precise requirements in terms of time synchronization and Doppler shift for the receiver to seamlessly lock onto it. Due to the attenuation of the GNSS signal because of the distance, counterfeit signals can obtain comparable signal strengths from several kilometers [15,20].

Description of IEC 61850-Based Substation Setup. The IEC 61850 standard serves as a communication protocol that outlines how devices within a substation interact with each other and interface with more advanced systems like energy management systems. Using the standard enables inter alia interoperability, analyzing the cyber-physical security, monitoring and controlling since it includes comprehensive guidelines for designing automation systems for electrical substations [6]. Our test-bed at KASTEL Security Lab Energy hence implements this standard in the digital substation part.

GNSS Spoofing Setup. Our experimental setup comprises three main components: system under test, spoofing hardware and validation hardware, as illustrated in Fig. 1. The system under test is a commercial time server for time synchronization tasks in a digital substation. It distributes PTP messages within the

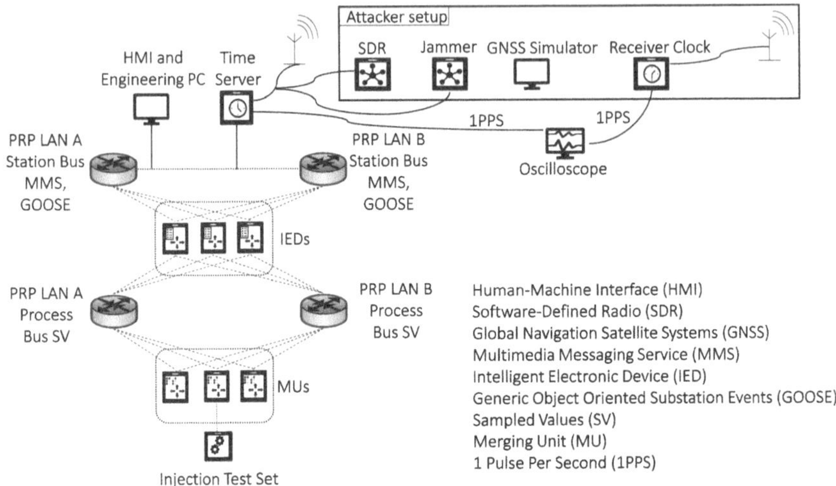

Fig. 1. Experimental setup at KASTEL Security Lab Energy. For converting analog into digital data Merging Units (MUs) are used whose time stamps are susceptible to time manipulation attacks in the same way as PMUs.

network to provide accurate time stamps to measurement and control devices. The time server is connected to both a GNSS antenna and the spoofer's signal via cable using a signal combiner. Due to regulations, it is not permitted to broadcast in restricted frequency bands over-the-air. The spoofer's setup consists of a computer with powerful GPU to generate the signals, low-cost GNSS receivers to enable generation of GNSS signals time-synchronized to live sky and a Software-Defined Radio (SDR) to convert synthetized GNSS signals into analog signals.

For validation, we use an oscilloscope to track the 1 Pulse-Per-Second (1PPS) signals and a network readout to assess the PTP messages concerning the clock quality (PTP clock class, clock accuracy) announced by the time server.

There are several possible approaches to manipulate a GNSS receiver. They differ in effectiveness and obfuscation level which come at the cost of different requirements for hardware, software and planning. An asynchronous attack broadcasts synthesized or pre-recorded GNSS signals. Introducing a jamming period to block reception of the genuine signal can cause the receiver to lose track of the previously found correlation. The receiver can lock onto a counterfeit signal that is subsequently broadcast. A synchronous spoofing attack firstly aligns the genuine signal to the real signal to a time accuracy of few nanoseconds and matching Doppler shifts of few Hertz. After the receiver is locked to the higher intensity counterfeit signal, the signal properties are smoothly changed such that the receiver follows that correlation.

In addition to that, the method employing multiple synchronized simulators, an array of transmitting antennas or even drones to replicate the angles of arrival of the signals is a highly-sophisticated spoofing. To our knowledge no such powerful attacks have been reported yet, although researchers discuss that possibility

[12]. Notice however, that also this attack leaves traces and could be detected if the signal is extensively analyzed. The main concern is, that most receivers for civil usage in the energy grid do not have even the basic capabilities to withstand a simple attack [7]. To introduce a time manipulation, we modify the broadcast time using small time shifts in the GNSS simulation. Hence, we can manipulate the time server while the signal content is self-consistent. This results in a time server manipulation without the receiver losing lock to the GNSS signal. Our attacks thus cover both asynchronous and synchronous levels.

Experimental Results. We experimented on a commercial time server, using GPS signals for single-constellation and adding Galileo for multi-constellation spoofing attacks. The results of the conducted experiments are summarized in Table 1.

We have shown that a commercial time server is vulnerable to all conducted GNSS spoofing attacks: The simple asynchronous attack (Attack No. 1) could manipulate the time server's output by few minutes with an attack duration of less than 5 min. Since the time accuracy requirements in the energy grid are in the order of microseconds, the success of such a basic attack should raise concern. Note however that such an attack can be detected by a low-cost receiver using basic signal analyzing. Additionally, the PTP clock properties (namely the clock class indicating the synchronization level and the estimated accuracy) which are announced by the time server itself change, such that measurement devices could equip measured data with appropriate uncertainty values or alarms. The synchronous multi-constellation attack (Attack No. 4) could successfully steer the time server from the true time at a rate of about $50ns/s$ which lead to an offset of $10\mu s$ in less than 5 min. Again, the PTP clock properties deteriorated such that measurement values could be accompanied with an alarm. A stealthy synchronous multi-constellation attack (Attack No. 5) was capable to slowly spoof the time without deterioration of the announced clock properties. However, the attainable time shift is only about $1\mu s/hour$, such that unusual system states arising from these measurements, can be investigated early. The maximal attainable time drift in this way is confined by the maximal steering speed of the local oscillator (Fig. 2b).

To further quantify the consequences, it is important to take into account that quick time shifts are non-stealthy, as these made the time server announce a deteriorated clock class. The required attack duration of a stealthy attack is dependent on the maximal steering speed of the internal oscillator, such that only less inert time servers are vulnerable to quick-stealthy attacks.

3 Qualitative Risk Assessment

Due to the possible serious consequences of time spoofing and sophisticated attacks on electrical substations, an adequate risk assessment is needed. It is however worth noticing that the last two steps, risk evaluation and risk treatment, of a risk assessment process [17] are excluded from the present paper. Context establishment of our study is defined as an IEC 61850-based substation

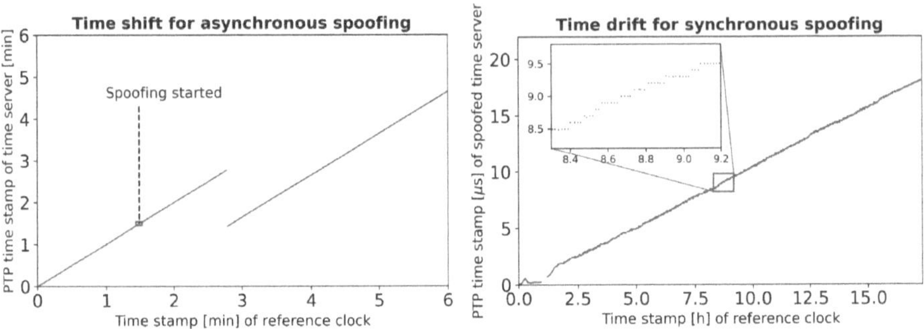

Fig. 2. (a) Announced PTP time stamps of synchronization messages with a time spoofing starting at 93 s, (Attack No. 1) (b) Time drift of time server due to a stealthy synchronous time spoofing, attack starting at 0 s, (Attack No. 5). Switching to the fake signal induces jitters, the step-wise manipulation can be seen in close-up.

Table 1. Overview of Experimental Results

Attack No.	Experiment Description	Time Manipulation	PTP Message Changes	Web Interface Warning	Detection of Low-Cost Receiver
1	Simple asynchronous single-constellation (only GPS)	✓	✓	✓	✓
2	Jamming + asynchronous single-constellation (only GPS)	✓	✓	✓	✓
3	Jamming + asynchronous multi-constellation (GPS and Galileo)	✓	✓	✓	✓
4	Synchronous multi-constellation (GPS and Galileo)	✓	✓	X	X
5	Slow synchronous multi-constellation (GPS and Galileo)	✓	X	X	X

(Fig. 1) as target and GNSS interference as attack means. After assessing the initial outcomes within the present paper, there is the opportunity to broaden the context in future research to attacks on the time distribution protocol [1]. An Attack Execution Graph (AEG) is used to help identifying potential risks. Figure 3 illustrates possible threat scenarios by outlining the sequence of actions an attacker might take to compromise a time synchronization in an IEC 61850-based substation. With the help of the present AEG, it can be possible to identify vulnerabilities in GNSS that might be exploited by attackers. The skills of "Spoofing" and "Jamming", depicted in thick black outline, represent the two main steps in the present experiment. The presented AEG will be extended and then the rest of the actions will be executed in further study.

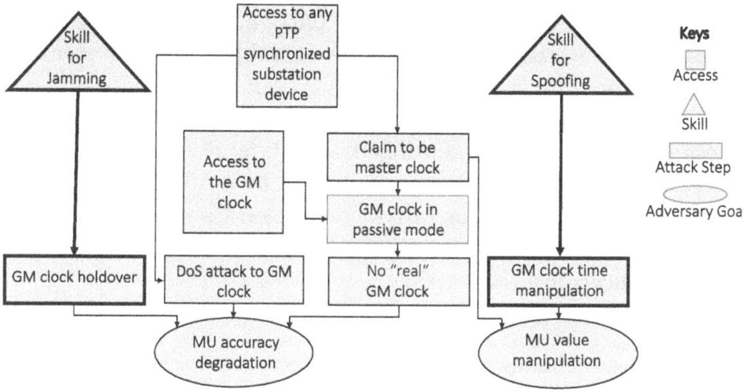

Fig. 3. Representation of Attack Execution Graph (AEG) for time synchronization in an IEC 61850-based substation

Due to the lack of historical data regarding GNSS attacks on time servers and to take into account the trade-off view of an attacker, attack difficulty is taken into consideration as a relevant qualitative measure. Difficulty refers to the budget, required technical skill of an attacker and legal obstacles (such as restricted GNSS software) in the presented experiment. We have conducted a difficulty assessment based on 'Required Skill Level' and 'Legal Obstacles'. 'Required Skill Level' considers the ease of the attack, ranging from low-medium to very high, while the scale for 'Legal Obstacles' (such as restricted software) spans from none to high. The results are depicted in Table 2.

Table 2. Assessment of difficulty for conducted GNSS time manipulation attacks (Attack No. 1–5). Attack No. 6 for comparison.

Attack No.	Experiment Description	Approximate Hardware Cost (EUR)	Required Skill Level	Legal Obstacles
1	Simple asynchronous single-constellation	250-1000	low-medium	none
2	Jamming + asynchronous single-constellation			
3	Jamming + asynchronous multi-constellation	15,000-20,000	medium	medium
4	Synchronous multi-constellation	20,000-150,000	medium-high	medium
5	Slow synchronous multi-constellation			
6	Synchronous multi-antenna multi-constellation	40,000-300,000	very high	medium

To assess the impact, often-mentioned consequences are gathered in Table 3. One might be tempted to use both tables to assign a price tag to certain impacts. While the approximate difficulty of a successful TSA can indeed be read off from the table, we argue that the connection between attack and impact on the grid is not covered in a satisfactory manner. In particular, impacts relying on time shifts in the order of milliseconds were not attainable in a quick-stealthy way in our experiment. Since the energy grid is a complex system, choosing worst-case scenarios as illustrative examples is prevalent. We introduce —without claiming completeness— the commonly mentioned threats and consequences associated with TSAs [7–9,21], which we plan to provide a more detailed assessment of in future research. Despite the inconclusive connection between the attack and the resulting consequences, the relative simplicity of GNSS spoofing attacks makes the overall risk concerning.

Table 3. Commonly mentioned possible consequences of TSAs.

Description	Threat	Asset(s)	Consequence
Distribution grid PMU data manipulation	Logging data errors	Logging files	Wrong fault location/ reconstruction
	State-estimation errors	Analyzing algorithms	Line overload/ underload (grid instability)
Transmission grid PMU phase angle manipulation	Incorrect wide-area-monitoring measures	Monitoring algorithms	Line overload/ underload (grid instability)
	Power line disconnection	Substation circuit breakers	Blackout in substantial areas of the grid

4 Results and Discussion

We have demonstrated that time spoofing on a commercial, high-grade time server can achieve time jumps even with simple means. Both jumps in the range of milliseconds —which are dangerous in the energy grid timescale— and traceless time drifts are possible. We assessed the difficulty of these GNSS attacks in terms of capital expenditures and skill requirements. Our study raises concerns about the time server's behavior to obey the observed GNSS time and quickly adjust the distributed time signals, if the observed GNSS time changes rapidly. As the internal oscillator is of high grade, switching to holdover in such cases would impede attackers attempting to manipulate time rapidly. We acknowledge that recently more manufacturers have offered resilient GNSS receiver chips. We however assume that most already deployed time servers are not equipped with these and it is questionable if they will be replaced.

To mitigate GNSS spoofing, we recommend using resilient receivers and/or authenticated GNSS signals such as the Galileo OSNMA, Galileo PRS and GPS CHIMERA schemes which offer a high spoofing resistance [19]. For any TSA, it makes sense to flag measurement values with the estimated accuracy according to the internal oscillators. This can then be taken into account by bad-data protection schemes prior to actions based on those measurement values. Although the investigated GNSS receiver was very susceptible to spoofing attacks, its decent internal oscillator limits quick-stealthy time shifts which can take days. Consequences which rely on quick time shifts of milliseconds as studied in [10,11,21] did not show to be stealthy in our experiment. We thus stress taking the duration of the attack into consideration when investigating the risk of stealthy spoofing attacks.

Extended Work. The initial findings from the experimental study indicate that we can proceed with our work in a more comprehensive manner. Our initial focus for quantifying the impact of GNSS spoofing attacks will involve considering the distribution grid using our previously developed methods [3]. At first, the distribution grid will be most strongly affected by the change to a more flexible grid which includes the frequent automatic reaction to measurement data. Additionally, GNSS receivers of stand-alone micro-PMUs in the distribution grid may be more vulnerable than proper time servers and hence easier to spoof. Using numerical simulations of several small electrical grid setups with appliances relying on time stamped measurement we plan to demonstrate the level of impact of TSAs. This will be done in terms of the amount of required successful time manipulations and worst-case scenarios—fitting to an attacker's mindset. In addition to GNSS spoofing, it's crucial to assess the risk of TSAs more broadly. This includes attacks on time beyond GNSS spoofing, as illustrated in the AEG (see Fig. 3). We hence plan to conduct Best-Master-Clock-Algorithm attacks interfering with the basic principles of the time synchronization protocol. For those attacks, we assess the vulnerability in terms of difficulty for an attacker. This two-pronged approach can explore how to adapt risk evaluation and treatment procedures tailored to attacker-driven framework for IEC 61850-based substations and the SG in general.

5 Conclusion

The potential risks of time manipulation attacks on the energy grid have yet to be systematically explored, despite the significant value such an investigation would offer to grid operators. We experimentally assessed the vulnerability of a commercial time server to GNSS attacks, uncovering instances of unnoticed but slow attacks that did not alter the announced PTP-clock quality. Combining these findings with a future systematic study of impacts enables a quantitative risk assessment of TSAs for the grid of the future.

Acknowledgement. We thank John Seron Seiquera for experimental assistance, Dr. Mohammed Ramadan for constructive feedback, and the reviewers and our shepherd for their helpful comments and suggestions. This work was supported by funding from the topic Engineering Secure Systems of the Helmholtz Association (HGF), by KASTEL Security Research Labs (structure 46.23.02) and by the German Research Foundation (DFG) Research Training Group GRK 2153: Energy Status Data - Informatics Methods for its Collection, Analysis and Exploitation.

References

1. Alghamdi, W., Schukat, M.: Practical implementation of APTs on PTP time synchronisation networks. In: 2020 31st Irish Signals and Systems Conference (ISSC), IEEE (2020)
2. Biswal, C., Sahu, B.K., Mishra, M., Rout, P.K.: Real-time grid monitoring and protection: a comprehensive survey on the advantages of phasor measurement units. Energies **16**(10), 4054 (2023)
3. Canbolat, S., Elbez, G., Hagenmeyer, V.: A new hybrid risk assessment process for cyber security design of smart grids using fuzzy analytic hierarchy processes. at-Automatisierungstechnik **71**(9), 779–788 (2023)
4. Delcourt, M., Shereen, E., Dän, G., Le Boudec, J.Y., Paolone, M.: Time-synchronization attack detection in unbalanced three-phase systems. IEEE Trans. Smart Grid **12**(5), 4460–4470 (2021)
5. Dileep, G.: A survey on smart grid technologies and applications. Renew. Energy **146**, 2589–2625 (2020)
6. Elbez, G., Keller, H.B., Hagenmeyer, V.: A cost-efficient software testbed for cyber-physical security in IEC 61850-based substations. In: 2018 IEEE International Conference on Communications, Control, and Computing Technologies for Smart Grids (SmartGridComm), IEEE (2018)
7. European union agency for cybersecurity, Stergiopoulos, G., Moulinos, K.: Power sector dependency - Attacks against time sensitive services. European Network and Information Security Agency (2020). https://doi.org/10.2824/496449
8. Falletti, E., Margaria, D., Marucco, G., Motella, B., Nicola, M., Pini, M.: Synchronization of critical infrastructures dependent upon GNSS: current vulnerabilities and protection provided by new signals. IEEE Syst. J. **13**(3), 2118–2129 (2018)
9. Gao, W., Li, H., Zhong, M., Lu, M.: An underestimated cybersecurity problem: quick-impact time synchronization attacks and a fast-triggered detection method. IEEE Trans. Smart Grid **14**, 4784–4798 (2023)
10. Geetha, S.J., Meghwani, A., Chakrabarti, S., Rajawat, K., Terzija, V.: Spoofing attack on synchrophasor GPS clock: impact and detection in power system state estimation. Int. J. Electr. Power Energy Syst. **134**, 107396 (2022)
11. Konstantinou, C., Sazos, M., Musleh, A.S., Keliris, A., Al-Durra, A., Maniatakos, M.: GPS spoofing effect on phase angle monitoring and control in a real-time digital simulator-based hardware-in-the-loop environment. IET Cyber-Phys. Syst. Theor. Appl. **2**(4), 180–187 (2017)
12. Laverty, D.M., Kelsey, C., O'Raw, J.B.: GNSS time signal spoofing detector for electrical substations. IEEE Trans. Smart Grid **13**(2), 1468–1477 (2021)
13. Lu, Q., Feng, X., Zhou, C.: A detection and weakening method for GNSS time-synchronization attacks. IEEE Sens. J. **21**(17), 19069–19077 (2021)
14. Meng, Q., Hsu, L.T., Xu, B., Luo, X., El-Mowafy, A.: A GPS spoofing generator using an open sourced vector tracking-based receiver. Sensors **19**(18), 3993 (2019)
15. Miljanovic, S., et al.: Experimental testing and impact analysis of jamming and spoofing attacks on professional GNSS receivers. In: Proceedings of International Conference on Localization and GNSS (ICL-GNSS) (2022)
16. Pignati, M., Zanni, L., Romano, P., Cherkaoui, R., Paolone, M.: Fault detection and faulted line identification in active distribution networks using synchrophasors-based real-time state estimation. IEEE Trans. Power Delivery **32**(1), 381–392 (2016)

17. Refsdal, A., Solhaug, B., Stølen, K.: Cyber-risk management. Springer, Cham (2015)
18. Tippenhauer, N.O., Pöpper, C., Rasmussen, K.B., Capkun, S.: On the requirements for successful GPS spoofing attacks. In: Proceedings of the 18th ACM conference on Computer and Communications Security (2011)
19. Wesson, K., Rothlisberger, M., Humphreys, T.: Practical cryptographic civil GPS signal authentication. NAVIGATION: J. Inst. Navig. **59**(3), 177–193 (2012)
20. Wu, Z., Zhang, Y., Yang, Y., Liang, C., Liu, R.: Spoofing and anti-spoofing technologies of global navigation satellite system: a survey. IEEE Access **8**, 165444–165496 (2020)
21. Zhang, Z., Gong, S., Dimitrovski, A.D., Li, H.: Time synchronization attack in smart grid: impact and analysis. IEEE Trans. Smart Grid **4**(1), 87–98 (2013)

Author Index

F. Maggi et al. (Eds.): DIMVA 2024, LNCS 14828, pp. 557–558, 2024.
https://doi.org/10.1007/978-3-031-64171-8

Zeitfracht Medien GmbH
Ferdinand-Jühlke-Straße 7
99095 Erfurt, Deutschland
produktsicherheit@kolibri360.de